Number Theory for the Millennium II

Number Theory for the Millennium II

Edited by

M. A. Bennett, B. C. Berndt,
N. Boston, H. G. Diamond,
A. J. Hildebrand, and W. Philipp

A K Peters
Natick, Massachusetts

Editorial, Sales, and Customer Service Office

A K Peters, Ltd.
63 South Avenue
Natick, MA 01760
www.akpeters.com

Library of Congress Cataloging-in-Publication Data

Millennial Conference on Number theory (2000 : University of Illinios at Urbana-Champaign)
    Number theory for the millennium / edited by M. A. Bennett ...[et al.].
      p. cm.
    Includes bibliographical references.
    ISBN 1-56881-126-8 (v. 1) – ISBN 1-56881-146-2 (v. 2) – ISBN 1-56881-152-7 (v. 3)
     1. Number theory–Congresses. I. Bennett, (Michael A.), 1966- II. Title.

  QA241 .M533 2000
  512'.7–dc21

                       2002029343

Printed in Canada
06 05 04 03 02   10 9 8 7 6 5 4 3 2 1

Contents of Volume 1

Contents of Volume 2

Contents of Volume 3

Preface

The Millennial Conference on Number Theory was held May 21 - 26, 2000 on the campus of the University of Illinois at Urbana-Champaign. The meeting was organized by M. A. Bennett, B. C. Berndt, N. Boston, H. G. Diamond, and A. J. Hildebrand of the UIUC Mathematics Department, and W. Philipp, of the UIUC Statistics Department. A total of 276 mathematicians from 30 countries were present at the meeting.

The conference featured 157 talks, of which 19 were one-hour plenary talks, 73 were half-hour invited talks given in four parallel sessions, and 65 were contributed talks given in five parallel sessions. In addition, there were three 40 minute talks on history and reminiscences, and several after dinner speeches honoring Professor Emeritus Paul Bateman on his 80th birthday. Other evening activities included a reception, a banquet, and musical and theatrical performances.

The conference was underwritten by financial grants from the Number Theory Foundation, the National Science Foundation, the National Security Agency, the College of Liberal Arts and Sciences of the University of Illinois, and the Institute for Mathematics and Applications. Also, the University of Illinois Mathematics Department provided financial and logistical support for the conference along with its facilities in Altgeld Hall. We thank the staff and graduate students of the Mathematics Department for the many hours of work they contributed to the conference. Finally, we recognize the efforts of Betsy Gillies, Conference Coordinator of the Mathematics Department, who worked tirelessly to ensure that the conference would run smoothly. We are grateful to all these organizations and people for their assistance and support.

Urbana, Illinois
7 December 2001

M. A. Bennett
B. C. Berndt
N. Boston
H. G. Diamond
A. J. Hildebrand
W. Philipp

List of Contributors

R. C. Baker, Department of Mathematics, Brigham Young University, Provo, UT 84602, USA, baker@math.byu.edu

Michel Balazard, C.N.R.S., Laboratoire d'algorithmique arithmétique, Mathématiques, Université de Bordeaux 1, 351, Cours de la Libération, 33405 Talence Cedex, France, balazard@math.u-bordeaux.fr

Daniel J. Bernstein, Department of Mathematics, University of Illinois at Chicago, Chicago, IL 60607-7045, USA, djb@math.uic.edu

F. Beukers, Department of Mathematics, Rijksuniversiteit te Utrecht, 3508 TA Utrecht, The Netherlands, beukers@math.uu.nl

Jean Bourgain, School of Mathematics, Institute for Advanced Study, Princeton, NJ 08540, USA, bourgain@ias.edu

John Boxall, Département de Mathématiques et de Mécanique, CNRS FRE 2271, Université de Caen, Boulevard Maréchal Juin, B.P. 5186, 14032 Caen Cedex, France, boxall@math.unicaen.fr

David W. Boyd, Department of Mathematics, University of British Columbia, Vancouver, BC V6T 1Z2, Canada, boyd@math.ubc.ca

Yann Bugeaud, Université Louis Pasteur, U.F.R. de Mathématiques, 7, rue René Descartes, 67084 Strasbourg, France, bugeaud@math.u-strasbg.fr

David A. Cardon, Department of Mathematics, Brigham Young University, Provo, UT 84602, USA, cardon@math.byu.edu

Heng Huat Chan, Department of Mathematics, National University of Singapore, Kent Ridge, Singapore 119260, Republic of Singapore, matchh@nus.edu.sg

Robin Chapman, School of Mathematical Sciences, University of Exeter, Exeter EX4 4QE, United Kingdom, rjc@maths.ex.ac.uk

Ted Chinburg, Department of Mathematics, University of Pennsylvania, Philadelphia, PA 19104, USA, ted@math.upenn.edu

Fan Chung, Department of Mathematics, University of California at San Diego, San Diego, CA 92110, USA, fan@math.ucsd.edu

Todd Cochrane, Department of Mathematics, Kansas State University, Manhattan, KS 66506, USA, cochrane@math.ksu.edu

J. B. Conrey, American Institute of Mathematics, 360 Portage Ave., Palo Alto, CA 94306, USA, conrey@aimath.org

Morley Davidson, Department of Mathematics, Kent State University, Kent, OH 44240, USA, davidson@math.kent.edu

Harold G. Diamond, Department of Mathematics, University of Illinois, 1409 West Green Street, Urbana, IL 61801, USA, diamond@math.uiuc.edu

Karl Dilcher, Department of Mathematics and Statistics, Dalhousie University, Halifax, Nova Scotia B3H 3J5, Canada, dilcher@mathstat.dal.ca

Darrin Doud, Department of Mathematics, Brigham Young University, Provo, UT 84602, USA, doud@math.byu.edu

Minkin Eie, Department of Mathematics, National Chung Cheng University, Ming-Hsiung Chia-Yi 621, Taiwan, mkeie@math.ccu.edu.tw

P.D.T.A. Elliott, Department of Mathematics, University of Colorado, Boulder, CO 80309, USA, pdtae@euclid.colorado.edu

Christian Elsholtz, Institut für Mathematik, Technische Universität Clausthal, Erzstrasse 1, D-38678 Clausthal-Zellerfeld, Germany, elsholtz@math.tu-clausthal.de

Paul Erdős[†], Hungarian Academy of Sciences, Budapest, H-1053, Hungary

Ronald Evans, Department of Mathematics, University of California at San Diego, La Jolla, CA 92093-0112, USA, revans@ucsd.edu

Jan-Hendrik Evertse, Universiteit Leiden, Mathematisch Instituut, Postbus 9512, 2300 RA Leiden, The Netherlands, evertse@math.leidenuniv.nl

Michael Filaseta, Mathematics Department, University of South Carolina, Columbia, SC 29208, USA, filaseta@math.sc.edu

Kevin Ford, Department of Mathematics, University of Illinois, Urbana, IL 61801, USA, ford@math.uiuc.edu

D. Eric Freeman, Department of Mathematics, University of Colorado, Boulder, CO 80309, USA, freem@euclid.colorado.edu

Emiliano Gómez, Department of Mathematics, University of California at Berkeley, Berkeley, CA 94720, USA, emgomez@math.berkeley.edu

F. G. Garvan, Department of Mathematics, University of Florida, Gainesville, FL 32611, USA, frank@math.ufl.edu

[†]Paul Erdős passed away on September 23, 1996.

Ronald Graham, Department of Computer Science and Engineering, University of California at San Diego, La Jolla, CA 92093-0114, graham@ucsd.edu

David Grant, Department of Mathematics, University of Colorado, Boulder, CO 80309, USA, grant@euclid.colorado.edu

George Greaves, School of Mathematics, Cardiff University, 23 Senghenydd Rd., Cardiff CF2 4YH, United Kingdom, greaves@cf.ac.uk

James Lee Hafner, IBM Research, Almaden Research Center, K53-B2, 650 Harry Road, San Jose, CA 95120, USA, hafner@almaden.ibm.com

H. Halberstam, Department of Mathematics, University of Illinois, 1409 West Green Street, Urbana, IL 61801, USA, heini@math.uiuc.edu

Glyn Harman, Department of Mathematics, Royal Holloway, University of London, Egham Hill, Egham, Surrey TW20 0EX, United Kingdom, G.Harman@rhul.ac.uk

Charles Helou, Penn State Univ., 25 Yearsley Mill Rd, Media, PA 19063, USA, cxh22@psu.edu

Doug Hensley, Department of Mathematics, Texas A&M University, College Station, TX 77843, USA, doug.hensley@math.tamu.edu

Noriko Hirata-Kohno, Nihon University, College of Science and Technology, Department of Mathematics, Surugudai, Tokyo 101-8308, Japan, hirata@math.cst.nihon-u.ac.jp

C. Hooley, School of Mathematics, Cardiff University, 23 Senghenydd Rd., Cardiff CF2 4YH, United Kingdom

James G. Huard, Department of Mathematics and Statistics, Canisius College, Buffalo, NY 14208, USA, huard@canisius.edu

M. N. Huxley, School of Mathematics, University of Cardiff, 23, Senghenydd Road, Cardiff CF24 4YH Wales, United Kingdom, huxley@cf.ac.uk

Georg Illies, Fachbereich Mathematik, Universität-Gesamthochschule Siegen, 57068 Siegen, Germany, illies@math.uni-siegen.de

J. P. Keating, School of Mathematics, University of Bristol, Bristol BS8 1TW, England, j.p.keating@bristol.ac.uk

A. Kumchev, Department of Mathematics, University of Toronto, Toronto, ON M5S 3G3, Canada, kumchev@math.toronto.edu

M. B. S. Laporta, Dipartimento di Matematica e Appl. "R. Caccioppoli", Complesso Universitario di Monte S. Angelo, Via Cinthia, 80126 Napoli, Italy, laporta@matna2.dma.unina.it

Xian-Jin Li, Department of Mathematics, Brigham Young University, Provo, UT 84602, USA, xianjin@math.byu.edu

Wen-Ching Winnie Li, Department of Mathematics, Pennsylvania State University, University Park, PA 16802, USA, wli@math.psu.edu

Lisa Lorentzen, Department of Mathematical Sciences, Norwegian University of Science and Technology, N-7491 Trondheim, Norway, lisa@math.ntnu.no

Lutz G. Lucht, Institut für Mathematik, Technische Universität Clausthal, Erzstraße 1, 38678 Clausthal-Zellerfeld, Germany, lucht@math.tu-clausthal.de

Eugenijus Manstavičius, Vilnius University, Department of Mathematics and Informatics, Naugarduko Str. 24, LT 2600 Vilnius, Lithuania, eugenijus.manstavicius@maf.vu.lt

Greg Martin, Department of Mathematics, University of Toronto, Toronto, ON M5S 3G3, Canada, gerg@math.toronto.edu

Kohji Matsumoto, Graduate School of Mathematics, Nagoya University, Chikusa-ku, Nagoya 464-8602, Japan, kohjimat@math.nagoya-u.ac.jp

Jeffrey L. Meyer, Department of Mathematics, Syracuse University, 215 Carnegie Hall, Syracuse, NY 13244-1150, USA, jlmeye01@mailbox.syr.edu

Maurice Mignotte, Université Louis Pasteur, U.F.R. de Mathématiques, 7, rue René Descartes, 67084 Strasbourg, France, mignotte@math.u-strasbg.fr

Siguna Müller, University of Klagenfurt, Department of Mathematics, A-9020 Klagenfurt, Austria, siguna.mueller@uni-klu.ac.at

V. Kumar Murty, Department of Mathematics, University of Toronto, Toronto, ON M5S 3G3, Canada, murty@math.toronto.edu

Ram Murty, Department of Mathematics, Queen's University, Kingston, ON K7L 3N6, Canada, murty@mast.queensu.ca

J.-L. Nicolas, Institut Girard Desargues, UMR 5028, Université Claude Bernard (Lyon 1), F-69622 Villeurbanne Cedex, France, jlnicola@in2p3.fr

Pace P. Nielsen, Department of Mathematics, Brigham Young University, Provo, UT 84602, USA

Cormac O'Sullivan, Department of Mathematics and Computer Science, Bronx Community College, City University of New York, Bronx, NY 10453, cormac@math.umd.edu

Yao Lin Ong, Department of Mathematics, National Chung Cheng University, Ming-Hsiung Chia-Yi 621, Taiwan

Ken Ono, Department of Mathematics, University of Wisconsin, Madison, WI 53706, USA, ono@math.wisc.edu

Zhiming M. Ou, Department of Basic Science, Beijing University of Posts and Telecommunications, Beijing 100876, People's Republic of China

Matthew A. Papanikolas, Department of Mathematics, Brown University, Providence, RI 02912-1917, USA, map@math.brown.edu

Georgios Pappas, Department of Mathematics, Michigan State University, East Lansing, MI 48824, USA, pappas@math.msu.edu

Scott T. Parsell, Department of Mathematics, Penn State University, University Park, PA 16802, USA, parsell@alum.mit.edu

Bernadette Perrin-Riou, Mathématiques, Bat. 425, Université Paris-Sud, F-91 405 Orsay, France, bpr@math.u-psud.fr

Bjorn Poonen, Department of Mathematics, University of California, Berkeley, CA 94720-3840, USA, poonen@math.berkeley.edu

Alfred J. van der Poorten, Centre for Number Theory Research, Macquarie University, Sydney, Australia 2109, alf@mpce.mq.edu.au

Igor E. Pritsker, Department of Mathematics, 401 Mathematical Sciences, Oklahoma State University, Stillwater, OK 74078-1058, USA, igor@math.okstate.edu

Robert A. Rankin[†], Department of Mathematics, University of Glasgow, Glasgow G12 8QT, Scotland, UK

David P. Roberts, Division of Science and Mathematics, University of Minnesota-Morris, Morris, MN 56267, USA, roberts@mrs.umn.edu

Fernando Rodriguez-Villegas, Department of Mathematics, University of Texas at Austin, Austin, TX 78712, USA, villegas@math.utexas.edu

Michael Rubinstein, Department of Mathematics, University of Texas at Austin, Austin, TX 78712, USA, miker@fireant.ma.utexas.edu

[†]Robert Rankin passed away on January 27, 2001.

A. Sárközy, Department of Algebra and Number Theory, Eötvös Loránd University, Pázmány Péter sétány 1/C, H-1117 Budapest, Hungary, sarkozy@cs.elte.hu

C. J. Smyth, Department of Mathematics and Statistics, University of Edinburgh, JCMB, King's Buildings, Mayfield Road, Edinburgh EH9 3JZ, United Kingdom, chris@maths.ed.ac.uk

N. C. Snaith, School of Mathematics, University of Bristol, Bristol BS8 1TW, England, n.c.snaith@bristol.ac.uk

Blair K. Spearman, Department of Mathematics and Statistics, Okanagan University, College, Kelowna, BC V1V 1V7, Canada, bspearman@okanagan.bc.ca

Anupam Srivastav, Department of Mathematics and Statistics, State University of New York, Albany, NY 12222, USA, anupam@math.albany.edu

Kenneth B. Stolarsky, Department of Mathematics, University of Illinois, 1409 West Green Street, Urbana, IL 61801, USA, stolarsk@math.uiuc.edu

Martin J. Taylor, UMIST, Manchester, M60 1QD, United Kingdom, Martin.Taylor@umist.ac.uk

Jeffrey Lin Thunder, Department of Mathematics, Northern Illinois University, DeKalb, IL 60115, USA, jthunder@math.niu.edu

R. Tijdeman, Mathematical Institute, Leiden University, Postbus 9512, 2300 RA Leiden, The Netherlands, tijdeman@math.leidenuniv.nl

Alain Togbé, Department of Mathematics and Comp.   Sci., Greenville College, 315 E. College Ave., Greenville, IL 62246, USA, atogbe@greenville.edu

R. C. Vaughan, Department of Mathematics, McAllister Building, Pennsylvania State University, University Park, PA 16802, USA, rvaughan@math.psu.edu

V. H. Vu, Department of Mathematics, University of California at San Diego, La Jolla, CA 92093-0112, USA, vanvu@ucsd.edu

Samuel S. Wagstaff, Jr., Center for Education and Research in Information Assurance and Security, and Department of Computer Sciences, Purdue University, West Lafayette, IN 47907-1398, USA, ssw@cerias.purdue.edu

Lynne H. Walling, Department of Mathematics, University of Colorado, Boulder, CO 80309, USA, `walling@euclid.colorado.edu`

P. G. Walsh, Department of Mathematics, University of Ottawa, 585 King Edward, Ottawa, ON K1S 5B6, Canada, `gwalsh@mathstat.uottawa.ca`

Kenneth S. Williams, Centre for Research in Algebra and Number Theory, School of Mathematics and Statistics, Carleton University, Ottawa, ON K1S 5B6, Canada, `williams@math.carleton.ca`

H. C. Williams, Department of Mathematics and Statistics, University of Calgary, Calgary, AB T2N 1N4, Canada, `williams@math.ucalgary.ca`

Siman Wong, Department of Mathematics, University of Massachusetts, Amherst, MA 01003, USA, `siman@math.umass.edu`

T. D. Wooley, Department of Mathematics, University of Michigan, East Hall, 525 E. University Avenue, Ann Arbor, MI 48109-1109, USA, `wooley@math.lsa.umich.edu`

Wen-Bin Zhang, Department of Mathematics, University of the West Indies, Mona, Kingston 7, Jamaica, `wbzhang@uwimona.edu.jm`

Zhiyong Zheng, Department of Mathematics, Tsinghua University, Beijing 100084, People's Republic of China, `zzheng@math.tsinghua.edu.cn`

Coverings of the Integers Associated with an Irreducibility Theorem of A. Schinzel

Michael Filaseta[1]

1 Introduction

The purpose of this paper is to give a partially expository account of results related to coverings of the integers (defined below) while at the same time making some new observations concerning a related polynomial problem. The polynomial problem we will consider is to determine whether for a given positive integer d there exists an $f(x) \in \mathbb{Z}^+[x]$ such that $f(x)x^n + d$ is reducible over the rationals for every non-negative integer n. We begin with some background material.

A *covering of the integers* is a system of congruences $x \equiv a_j \pmod{m_j}$, with a_j and m_j integral and $m_j \geq 1$, such that every integer satisfies at least one of the congruences. Four examples are as follows:

$x \equiv 0 \pmod 2$	$x \equiv 0 \pmod 2$	$x \equiv 0 \pmod 2$	$x \equiv 0 \pmod 2$
$x \equiv 1 \pmod 2$	$x \equiv 1 \pmod 4$	$x \equiv 2 \pmod 3$	$x \equiv 0 \pmod 3$
	$x \equiv 3 \pmod 8$	$x \equiv 1 \pmod 4$	$x \equiv 1 \pmod 4$
	$x \equiv 7 \pmod{16}$	$x \equiv 1 \pmod 6$	$x \equiv 3 \pmod 8$
	$\vdots$	$x \equiv 3 \pmod{12}$	$x \equiv 7 \pmod{12}$
			$x \equiv 23 \pmod{24}$

Two open problems concerning coverings are

Open Problem 1. For every $c > 0$, does there exist a finite covering with distinct moduli and with the minimum modulus $\geq c$?

Open Problem 2. Does there exist a finite covering consisting of distinct odd moduli > 1?

We shall call a covering as in the second problem an "odd covering". According to Richard Guy [3], Paul Erdős has offered $500 for a proof or

[1]The author gratefully acknowledges support from the National Security Agency.

1

disproof that a c exists as in the first problem and has offered \$25 for a proof that there is no odd covering. John Selfridge has offered \$900 for an explicit example of an odd covering. In private communication, Selfridge has indicated to the author that he will now pay \$2000 for an explicit odd covering. Observe that in the odd covering problem no direct financial gain is made for a non-constructive proof that an odd covering exists. In [5], R. Morikawa announced that a covering exists as in the first problem with $c = 24$.

We stress the importance of the word "finite" in the above problems with a simple example of an infinite covering with relatively prime odd moduli that are arbitrarily large. Fix $c > 0$, and let $M = \{m_1, m_2, \dots\}$ be an arbitrary infinite set of relatively prime integers $> c$ (for example, M could be the set of primes $> c$). Let $a_1, a_2, \dots$ be some ordering of the integers. Then the infinite system $x \equiv a_j \pmod{m_j}$ clearly covers the integers.

One of the now classical examples of the use of coverings is in a disproof that Erdős gave of the following conjecture.

Polignac's Conjecture.   For every sufficiently large odd integer $k > 1$, there is a prime p and an integer n such that $k = 2^n + p$.

The Prime Number Theorem would suggest that this is a reasonable conjecture, but small $k > 1$ not the sum of a prime and a power of two are easy to find. The smallest such k is 127 and the smallest composite k is 905. Erdős' argument is based on the last example given of a covering in the first display above. A variation on Erdős' argument is as follows. One considers any positive integer k satisfying the congruence $k \equiv 1 \pmod 2$ so that k is odd and the congruences $k \equiv 1 \pmod 3$, $k \equiv 1 \pmod 7$, $k \equiv 2 \pmod 5$, $k \equiv 8 \pmod{17}$, $k \equiv 11 \pmod{13}$, and $k \equiv 121 \pmod{241}$ (such k exist by the Chinese Remainder Theorem). The key idea then is to consider what happens if n satisfies one of the congruences in the last covering displayed above. For example, if $n \equiv 7 \pmod{12}$, then $2^n \equiv 2^7 \equiv 11 \pmod{13}$ so that $k - 2^n$ is divisible by 13. Since every integer n satisfies one of the congruences listed above, one can deduce that $k - 2^n$ must be divisible by at least one element of $S = \{3, 7, 5, 13, 17, 241\}$. From an analytic point of view, we are through. The Chinese Remainder Theorem gives that for some $\delta > 0$ and for x sufficiently large, there are $> \delta x$ different $k \le x$ as above. On the other hand, for any such k, the only possible prime values of $k - 2^n$ are elements of S so that k will be of the form $2^n + s$ where $s \in S$. Since $|S| = 6$, one easily deduces that there are $\ll (\log x)^6$ such $k \le x$. It follows that a positive proportion of positive integers k cannot be written in the form $2^n + p$.

Another classical application of the use of coverings was given by Wa-claw Sierpiński.

Sierpiński's Theorem. A positive proportion of odd positive integers ℓ satisfy $\ell \times 2^n + 1$ is composite for all non-negative integers n.

It is unknown what the smallest such ℓ is. It is probably $\ell = 78557$ (attributed to Selfridge). Extensive computations are being done to check that for each odd $\ell < 78557$ there is a prime of the form $\ell \times 2^n + 1$, but it may be some time before they are completed. There are apparently 19 values of ℓ left to eliminate at the time of writing this paper; see

$$\texttt{http://vamri.xray.ufl.edu/proths/sierp.html}.$$

The question of whether there are infinitely many prime Fermat numbers is related to the existence of even ℓ as above. For example, if $F_n = 2^{2^n} + 1$ is composite for $n \geq 5$, then for $\ell = 2^{17} = 131072$ one has $\ell \times 2^n + 1$ is composite for all non-negative integers n.

Andrzej Schinzel noted Sierpiński's Theorem follows from the above solution to Polignac's Conjecture. Take $\ell = -k$. For each $n \geq 0$, consider $p \in \{3, 5, 7, 13, 17, 241\}$ such that $\ell + 2^{239n} \equiv 0 \pmod{p}$. Then $\ell + 2^{239n} \equiv \ell + 2^{-n} \pmod{p}$ so that $\ell \times 2^n + 1 \equiv 0 \pmod{p}$. By considering $k < 0$ satisfying the congruences imposed on k above, one obtains the result of Sierpiński.

In this paper, we consider a polynomial variant of Sierpiński's result. In its most simplest form, the problem we consider is as follows.

The Analogous Polynomial Problem.   Find $f(x) \in \mathbb{Z}[x]$ with $f(1) \neq -1$ such that $f(x)x^n + 1$ is reducible over the rationals for all $n \geq 0$.

The condition $f(1) \neq -1$ makes the problem non-trivial. Otherwise, one could simply consider any $f(x)$ of degree > 1 satisfying $f(1) = -1$ as then $f(x)x^n + 1$ will always have the factor $x - 1$. Schinzel [7] first considered this problem in a slightly different form with $f(x)x^n + 1$ replaced by $x^n + f(x)$ (and with the added condition $f(0) \neq 0$). His version was chosen as an approach to understanding a conjecture of Turán that every reducible polynomial is in some sense near an irreducible polynomial. Modifying an example of Schinzel's, we note that

$$f(x) = 5x^9 + 6x^8 + 3x^6 + 8x^5 + 9x^3 + 6x^2 + 8x + 3$$

has the property that $f(x)x^n + 12$ is reducible for all $n \geq 0$. The argument for this is based on the third covering example displayed in the second opening paragraph. Indeed, that $f(x)x^n + 12$ is reducible for all $n \geq 0$ can be obtained by noting that every non-negative integer satisfies at least one of the congruences given there and that the following implications all hold:

$$n \equiv 0 \pmod{2} \implies f(x)x^n + 12 \equiv 0 \pmod{x+1}$$
$$n \equiv 2 \pmod{3} \implies f(x)x^n + 12 \equiv 0 \pmod{x^2 + x + 1}$$
$$n \equiv 1 \pmod{4} \implies f(x)x^n + 12 \equiv 0 \pmod{x^2 + 1}$$
$$n \equiv 1 \pmod{6} \implies f(x)x^n + 12 \equiv 0 \pmod{x^2 - x + 1}$$
$$n \equiv 3 \pmod{12} \implies f(x)x^n + 12 \equiv 0 \pmod{x^4 - x^2 + 1}.$$

Observe that the moduli on the right are the cyclotomic polynomials $\Phi_2(x)$, $\Phi_3(x)$, $\Phi_4(x)$, $\Phi_6(x)$, and $\Phi_{12}(x)$, respectively. The implications are easily justified. For example, if $n \equiv 1 \pmod{6}$, then $x^n \equiv x \pmod{\Phi_6(x)}$ so that by a direct computation

$$f(x)x^n + 12 \equiv f(x)x + 12 \equiv 0 \pmod{\Phi_6(x)}.$$

The dual role of 12 here (as the least common multiple of the moduli in the covering and as the constant term in $f(x)x^n + 12$ for $n > 0$) is misleading. The main new result in this paper is in fact a demonstration that a suitable covering (considerably more complicated than those given in the examples above) can give rise to an $f(x) \in \mathbb{Z}^+[x]$ with $f(x)x^n + 4$ reducible for all $n \geq 0$. More generally, we obtain

Theorem 1. *Let d be a positive integer divisible by 4. There is an $f(x) \in \mathbb{Z}^+[x]$ with $f(x)x^n + d$ reducible for all $n \geq 0$.*

Observe that the condition $f(1) \neq -d$, that would prevent trivial examples of polynomials $f(x)$ with $f(x)x^n + d$ having $x - 1$ as a factor for all $n \geq 0$, is replaced by the condition that $f(x)$ have positive coefficients. Our condition seemingly makes a stronger result, but we really only choose this formulation to simplify the statement of the result.

We are left with the question

Open Problem 3. For what positive integers d does there exist an $f(x) \in \mathbb{Z}[x]$ with $f(1) \neq -d$ such that $f(x)x^n + d$ is reducible over the rationals for all $n \geq 0$?

The author is unable to establish more d with this property than what the theorem states above. Schinzel [7] already established a result that at least suggests that the existence of such $f(x)$ when $d = 1$ would be difficult to establish. We modify his ideas to show

Theorem 2. *Let d be an odd positive integer. If there is an $f(x) \in \mathbb{Z}[x]$ with $f(1) \neq -d$ and $f(x)x^n + d$ reducible for all $n \geq 0$, then there is an odd covering of the integers.*

We emphasize that this can be obtained by a simple variation on Schinzel's work in [7] and that, in fact, Schinzel's work gives a necessary and sufficient condition (somewhat more complicated than the existence of an odd covering) for such an $f(x)$ to exist in the case that $d = 1$. Nevertheless, we will give a proof of the above result, a proof that is similar in flavor but still somewhat different from Schinzel's original work on the subject.

2   A Result Concerning Cyclotomic Polynomials

We make use of the notation

$$\Phi_n(x) = \prod_{d \mid n} (x^d - 1)^{\mu(n/d)} \tag{2.1}$$

so that $\Phi_n(x)$ denotes the nth cyclotomic polynomial. It is well-known and easy to establish from (2.1) that

$$\Phi_{pn}(x) = \begin{cases} \Phi_n(x^p) & \text{if } p \mid n \\ \Phi_n(x^p)/\Phi_n(x) & \text{if } p \nmid n. \end{cases} \tag{2.2}$$

We also set $\zeta_n = e^{2\pi i/n}$. Throughout this paper, d will denote a positive integer.

The example given by Schinzel of a polynomial $f(x) \in \mathbb{Z}[x]$ with $f(1) \neq -12$ and $f(x)x^n + 12$ reducible for all $n \geq 0$ has a clear connection to cyclotomic polynomials. Indeed, we saw in the introduction that there is a finite list of polynomials P, all cyclotomic, such that each $f(x)x^n + 12$ is divisible by some element of P. To demonstrate the important role of cyclotomic polynomials in our investigations here, suppose that $f(x)x^n + d$ is divisible by $g(x)$ for some irreducible $g(x) \in \mathbb{Z}[x]$ and for at least two different non-negative integers n, say $n = u$ and $n = v$ with $u > v$. Then $g(x)$ also divides

$$\big(f(x)x^u + d\big) - \big(f(x)x^v + d\big) = f(x)x^v\big(x^{u-v} - 1\big).$$

Since $d \neq 0$, we deduce that $g(x)$ divides $x^{u-v} - 1$ and, therefore, is cyclotomic. It follows that if we want to establish that $f(x)x^n + d$ is reducible for all non-negative integers n by finding a finite list of polynomials P such that each $f(x)x^n + d$ is divisible by an element of P, then P must contain cyclotomic polynomials. Another simple result in this direction is the following:

Lemma 1. *Suppose $f(x)x^a + d$ is divisible by $\Phi_m(x)$ for some positive integer m. Then $f(x)x^n + d$ is divisible by $\Phi_m(x)$ if and only if $n \equiv a$ (mod m).*

Proof. Let $F(x) = f(x)x^n + d$. If $n \equiv a$ (mod m), then clearly $F(\zeta_m) = f(\zeta_m)\zeta_m^a + d = 0$ so that $F(x)$ is divisible by $\Phi_m(x)$. If $F(x)$ is divisible by $\Phi_m(x)$, the equality

$$0 = \zeta_m^{n-a}\big(f(\zeta_m)\zeta_m^a + d\big) - F(\zeta_m) = d\big(\zeta_m^{n-a} - 1\big)$$

implies $n \equiv a$ (mod m). $\qquad\qquad\square$

The lemma alluded to in the title of this section is the following.

Lemma 2. *Let n and m be positive integers with $n > m$. If n/m is not a power of a prime, then for every integer a, there exist $u(x)$ and $v(x)$ in $\mathbb{Z}[x]$ satisfying*

$$\Phi_n(x)u(x) + \Phi_m(x)v(x) = a. \qquad (2.3)$$

If for some prime p and some positive integer t we have $n/m = p^t$, then there exist $u(x)$ and $v(x)$ in $\mathbb{Z}[x]$ satisfying (2.3) if and only if $p \mid a$.

Momentarily, we turn to some preliminary lemmas (some quite well-known) that will not only give us what we need for the above result but also keep our arguments self-contained. Schinzel has pointed out, however, that Lemmas 3-6 and the proof of Lemma 2 can be replaced by applications of a result of Tom Apostol [1] on the resultant of two cyclotomic polynomials. Therefore, we present here Schinzel's alternative approach as well as the author's more self-contained approach. We begin with Schinzel's argument for Lemma 2. The remaining use of Apostol's work, as suggested by Schinzel, is presented in Concluding Remarks at the end of this paper.

First Proof of Lemma 2. Let $R(n, m)$ denote the resultant of $\Phi_n(x)$ and $\Phi_m(x)$. Then it is well-known that there exist polynomials $u_1(x)$ and $v_1(x)$ in $\mathbb{Z}[x]$ such that

$$\Phi_n(x)u_1(x) + \Phi_m(x)v_1(x) = R(n, m).$$

In the case that n/m is not a power of a prime, Apostol's work [1] implies that $R(n, m) = \pm 1$. We deduce immediately that, for every integer a, there are polynomials $u(x)$ and $v(x)$ in $\mathbb{Z}[x]$ satisfying (2.3).

In the case that $n/m = p^t$ as in the statement of the lemma, Apostol establishes that $R(n, m) = \pm p^{\phi(m)}$. Note that $n = p^t m$ implies $\phi(m) \mid \phi(n)$. We consider the polynomial

$$h(x) = \frac{1}{p}\left(\Phi_n(x) - \Phi_m(x)^{\phi(n)/\phi(m)}\right).$$

From (2.2), it follows that $\Phi_n(x) \equiv \Phi_m(x)^{\phi(n)/\phi(m)} \pmod{p}$ so that $h(x) \in \mathbb{Z}[x]$. Observe that the resultant of $\Phi_m(x)$ and $h(x)$ can be expressed as

$$\prod_{\substack{1 \le k \le m \\ \gcd(k,m)=1}} h(\zeta_m^k) = p^{-\phi(m)} \prod_{\substack{1 \le k \le m \\ \gcd(k,m)=1}} \Phi_n(\zeta_m^k) = p^{-\phi(m)} R(n,m) = \pm 1.$$

Hence, there exist polynomials $u_2(x)$ and $v_2(x)$ in $\mathbb{Z}[x]$ such that

$$h(x)u_2(x) + \Phi_m(x)v_2(x) = 1.$$

Given the definition of $h(x)$, we deduce

$$\Phi_n(x)u_2(x) + \Phi_m(x)\left(pv_2(x) - \Phi_m(x)^{\phi(n)/\phi(m)-1}u_2(x)\right) = p.$$

We deduce in this case that, for every integer a divisible by p, there are polynomials $u(x)$ and $v(x)$ in $\mathbb{Z}[x]$ satisfying (2.3). Furthermore, if polynomials $u(x)$ and $v(x)$ in $\mathbb{Z}[x]$ exist satisfying (2.3), then we must have $a \equiv 0 \pmod{p}$ since otherwise $\Phi_n(x)$ and $\Phi_m(x)$ would be relatively prime in the finite field with p elements contradicting that $R(n, m)$ is 0 in this field. $\square$

Lemma 3. *Suppose n and m are integers with $n/m = p^r$ for some prime p and some positive integer r. Then $\Phi_n(\zeta_m) = pw$ for some unit $w \in \mathbb{Z}[\zeta_m]$.*

Proof. We consider three cases: (i) $n = pm$ and $p \nmid m$, (ii) $n = p^r m$ with $r > 1$ and $p \nmid m$, and (iii) $n = p^u t$ and $m = p^v t$ with $u > v > 0$. Let ξ denote an arbitrary primitive mth root of 1 (so $\xi \in \mathbb{Z}[\zeta_m]$). For (i), observe that (2.2) implies $\Phi_n(x) = \Phi_m(x^p)/\Phi_m(x)$. Hence,

$$\Phi_n(\xi) = \lim_{x \to \xi} \frac{\Phi_m(x^p)}{\Phi_m(x)} = \frac{p\xi^{p-1}\Phi_m'(\xi^p)}{\Phi_m'(\xi)}.$$

Since ξ^{pj} and ξ^j range over the same set of values for $1 \le j \le m$ and $\gcd(j, m) = 1$, we deduce that $\Phi_m'(\xi^p)$ and $\Phi_m'(\xi)$ have the same norm in the field $\mathbb{Q}(\zeta_m)$ over $\mathbb{Q}$. Also, the norm of ξ is ± 1. We deduce that the norm of $\Phi_n(\xi)$ is $\pm p^{\phi(m)}$. On the other hand,

$$\Phi_n(x) = \frac{\Phi_m(x^p)}{\Phi_m(x)} = \prod_{\substack{1 \le k \le m \\ \gcd(k,m)=1}} \left(\frac{x^p - \xi^{kp}}{x - \xi^k}\right)$$

$$= \prod_{\substack{1 \le k \le m \\ \gcd(k,m)=1}} \left(x^{p-1} + \xi^k x^{p-2} + \xi^{2k} x^{p-3} + \cdots + \xi^{(p-1)k}\right).$$

By considering the factor corresponding to $k = 1$, we deduce that $\Phi_n(\xi) = pu$ for some $u \in \mathbb{Z}[\zeta_m]$. It follows that there are $u_k \in \mathbb{Z}[\zeta_m]$ such that $\Phi_n(\zeta_m^k) = pu_k$ for each positive integer $k \le m$ with $\gcd(k,m) = 1$. Therefore,

$$p^{\phi(m)} \prod_{\substack{1 \le k \le m \\ \gcd(k,m)=1}} u_k = \prod_{\substack{1 \le k \le m \\ \gcd(k,m)=1}} \Phi_n(\zeta_m^k) = \pm p^{\phi(m)}.$$

We deduce that each u_k is a unit in $Z[\zeta_m]$. Hence, (i) follows.

For (ii), use (2.2) again to obtain $\Phi_n(\zeta_m) = \Phi_{pm}\left(\zeta_m^{p^{r-1}}\right)$ and apply the argument for (i) with $\xi = \zeta_m^{p^{r-1}}$. For (iii), use (2.2) as before to obtain $\Phi_n(\zeta_m) = \Phi_{p^u - v_t}\left(\zeta_m^{p^v}\right) = \Phi_{p^u - v_t}(\zeta_t)$. Now, cases (i) and (ii) imply $\Phi_n(\zeta_m) = pw$ for some unit $w \in \mathbb{Z}[\zeta_t] \subseteq \mathbb{Z}[\zeta_m]$ (since $\zeta_t = \zeta_m^{p^v}$). □

Lemma 4. *Let m be an integer > 1. Then*

$$\Phi_m(1) = \begin{cases} p & \text{if } m = p^r \text{ for some } r \in \mathbb{Z}^+ \\ 1 & \text{otherwise.} \end{cases}$$

Proof. Clearly, $\Phi_p(1) = p$. If $m = p^r k$ with k and r positive integers such that $p \nmid k$, then (2.2) implies $\Phi_m(1) = \Phi_{pk}(1^{p^{r-1}}) = \Phi_{pk}(1)$. The lemma follows if $k = 1$. If $k > 1$, then applying (2.2) again we obtain $\Phi_m(1) = \Phi_{pk}(1) = \Phi_k(1^p)/\Phi_k(1) = 1$. □

Lemma 5. *Let m and ℓ be integers with $m \ge 1$ and $\ell \ge 0$. For $\alpha \in \mathbb{Q}(\zeta_m)$, let $N(\alpha) = N_{\mathbb{Q}(\zeta_m)/\mathbb{Q}}(\alpha)$ denote the norm of α. Then $N(\zeta_m^\ell - 1)$ is divisible by a prime p if and only if $m/\gcd(\ell,m)$ is a power of p.*

Proof. The idea is to apply Lemma 4 and use that

$$N(\zeta_m^\ell - 1) = \pm\Phi_{m/\gcd(\ell,m)}(1)^{\phi(m)/\phi(m/\gcd(\ell,m))}.$$

This last equation can be seen as follows. Note that $N(\zeta_m^\ell - 1) = \pm N(1 - \zeta_m^\ell)$. The value of $N(1 - \zeta_m^\ell)$ is the product of its $\phi(m)$ field conjugates $1 - \zeta_m^{k\ell}$ where $1 \leq k \leq m$ and $\gcd(k, m) = 1$. Observe that $1 - \zeta_m^\ell = 1 - \zeta_{m/d}^{\ell/d}$ where $d = \gcd(m, \ell)$, and $\zeta_{m/d}^{\ell/d}$ is a primitive (m/d)th root of unity. The field conjugates associated with $1 - \zeta_m^\ell$ are thus the same as the numbers $1 - \zeta_{m/d}^t$ where $1 \leq t \leq m/d$ and $\gcd(t, m/d) = 1$ with each $1 - \zeta_{m/d}^t$ appearing among the $1 - \zeta_m^{k\ell}$ precisely $\phi(m)/\phi(m/d)$ times (see Theorem 2-5 of William LeVeque's book [4]). Finally, observe that

$$\prod_{\substack{1 \leq t \leq m/d \\ \gcd(t, m/d) = 1}} \left(1 - \zeta_{m/d}^t\right) = \Phi_{m/d}(1).$$

The result now follows.   □

Observe that Lemma 5, using the notation there, implies that $\zeta_m^\ell - 1$ is a unit in $\mathbb{Z}[\zeta_m]$ if and only if $m/\gcd(\ell, m)$ is not a power of a prime.

Lemma 6. *Let $n > 1$ and a be positive integers with $\gcd(a, n) = 1$. Then the quotient $(\zeta_n^a - 1)/(\zeta_n - 1)$ is a unit in $\mathbb{Z}[\zeta_n]$.*

Proof. Let $\beta = (\zeta_n^a - 1)/(\zeta_n - 1)$. Observe that

$$\beta = 1 + \zeta_n + \zeta_n^2 + \cdots + \zeta_n^{a-1}$$

so that $\beta \in \mathbb{Z}[\zeta_n]$. The condition $\gcd(a, n) = 1$ implies that the set of values of ζ_n^{ja} and the set of values of ζ_n^j are the same as j varies over the positive integers $\leq n$ that are relatively prime to n. It follows that the norm of β in the field $\mathbb{Q}(\zeta_n)$ over $\mathbb{Q}$ is 1. Hence, β is a unit in $\mathbb{Z}[\zeta_n]$.   □

Second Proof of Lemma 2. Since $n > m$, there is a prime p and non-negative integers r and s with $r > s$ satisfying $n = p^r n'$ and $m = p^s m'$ for some integers n' and m' each not divisible by p. The condition $n/m = p^t$ in the lemma is satisfied precisely when $n' = m'$ (with $t = r - s$). We use that

$$\Phi_m(\zeta_n) = \prod_{d|m} \left(\zeta_n^d - 1\right)^{\mu(m/d)}.$$

Given that $d \mid m$, observe $n/\gcd(d, n)$ can be a power of a prime only if $n' \mid d$. It follows from Lemma 5 (see the comment after its proof) that if $n' \nmid d$, then $\zeta_n^d - 1$ is a unit in $\mathbb{Z}[\zeta_n]$. Thus, there is a unit $w \in \mathbb{Z}[\zeta_n]$ such that

$$\Phi_m(\zeta_n) = w \prod_{\substack{d|m \\ n'|d}} \left(\zeta_n^d - 1\right)^{\mu(m/d)}.$$

In particular, if $n' \nmid m'$, then $\Phi_m(\zeta_n)$ is a unit in $\mathbb{Z}[\zeta_n]$. If $n' \mid m'$, then we set $k = m'/n'$, take $d = p^j n' d'$ where $d' \mid k$, and rewrite the above to obtain

$$\Phi_m(\zeta_n) = w \prod_{j=0}^{s} \prod_{d' \mid k} \left(\zeta_{p^{r-j}}^{d'} - 1 \right)^{\mu(p^{s-j}k/d')}.$$

If $k > 1$ and $j \in \{s-1, s\}$, we use that

$$\sum_{d' \mid k} \mu(p^{s-j}k/d') = \pm \sum_{d' \mid k} \mu(k/d') = 0$$

to deduce that

$$\Phi_m(\zeta_n) = w \prod_{j=s-1}^{s} \prod_{d' \mid k} \left(\frac{\zeta_{p^{r-j}}^{d'} - 1}{\zeta_{p^{r-j}} - 1} \right)^{\mu(p^{s-j}k/d')}.$$

Note that for each $j \in \{0, 1, \ldots, s\}$, we have $\mathbb{Z}[\zeta_{p^{r-j}}] \subseteq \mathbb{Z}[\zeta_n]$. We deduce from Lemma 6 that if $n' \neq m'$, then $\Phi_m(\zeta_n)$ is a unit in $\mathbb{Z}[\zeta_n]$. In this case, for some $v_0(x) \in \mathbb{Z}[x]$, we obtain $\Phi_m(\zeta_n)v_0(\zeta_n) = 1$. In other words, $\Phi_m(x)v_0(x) - 1$ has ζ_n as a root. It follows that $\Phi_m(x)v_0(x) - 1 = \Phi_n(x)u_0(x)$ for some $u_0(x) \in \mathbb{Z}[x]$ so that (2.3) holds by multiplying through by a (i.e., taking $u(x) = -au_0(x)$ and $v(x) = av_0(x)$). If $n' = m'$, then we apply Lemma 3 to obtain that $\Phi_n(\zeta_m) = pw_0$ for some unit w_0 in $\mathbb{Z}[\zeta_m]$. Therefore, taking $u_0(\zeta_m) \in \mathbb{Z}[\zeta_m]$ to be the inverse of w_0, we obtain $\Phi_n(x)u_0(x) - p$ is divisible by $\Phi_m(x)$. We deduce in this case that if $p \mid a$, then (2.3) has a solution in polynomials $u(x)$ and $v(x)$ in $\mathbb{Z}[x]$. Lemma 3 also implies the necessity of having $p \mid a$ when $n' = m'$ as follows. Take $x = \zeta_m$ in (2.3) to obtain $\Phi_n(\zeta_m)u(\zeta_m) = a$. By Lemma 3, the norm of $\Phi_n(\zeta_m)u(\zeta_m)$ (in $\mathbb{Q}(\zeta_m)/\mathbb{Q}$) is divisible by p so that the norm of a must also be divisible by p. Thus, $p \mid a$, concluding the proof. □

3   How Coverings Produce Reducible Polynomials

Let d be a positive integer. Suppose we wish to find an $f(x)$ with positive integer coefficients such that $f(x)x^n + d$ is reducible for every non-negative integer n. The purpose of this section is to show how certain coverings of the integers can be used to obtain such $f(x)$. This is achieved through the following result.

Theorem 3. *Let d be a positive integer. Suppose that S is a system of congruences*

$$x \equiv 2^{j-1} \pmod{2^j} \qquad for\ j \in \{1, 2, \ldots, k\} \tag{3.1}$$

for some positive integer k together with

$$x \equiv a_j \pmod{m_j} \qquad \text{for } j \in \{1, 2, \ldots, r\} \qquad (3.2)$$

for some positive integer r satisfying:

(i) *The system S is a covering of the integers.*

(ii) *The moduli in (3.1) and (3.2) are all distinct and > 1.*

(iii) *For each $j \in \{1, 2, \ldots, r\}$,*

$$\left(\prod_{\substack{1 \le i \le r \\ i \ne j}} a(i, j) \right) \left(\prod_{i=1}^{k} b(i, j) \right) \quad \text{divides } d$$

where

$$a(i, j) = \begin{cases} p & \text{if } m_i/m_j = p^t \text{ for some prime } p \text{ and some integer } t \\ 1 & \text{otherwise} \end{cases}$$

and

$$b(i, j) = \begin{cases} p & \text{if } m_j/2^i = p^t \text{ for some prime } p \text{ and some integer } t \\ 1 & \text{otherwise.} \end{cases}$$

(iv) *The double product $\prod_{i=1}^{k} \prod_{j=1}^{r} b(i, j)$ divides d.*

Then there exists $f(x) \in \mathbb{Z}[x]$ with positive coefficients such that $f(x)x^n + d$ is reducible over the rationals for all non-negative integers n.

Proof. Fix d now as in the statement of the theorem. We consider as we may that $0 \le a_j < m_j$ in (3.2). Suppose for the moment that we have an $f(x)$ satisfying the system of congruences consisting of

$$f(x) \equiv d \pmod{w(x)}, \qquad (3.3)$$

where $w(x) = \prod_{j=1}^{k} \Phi_{2^j}(x)$, together with

$$f(x) \equiv -dx^{m_j - a_j} \pmod{\Phi_{m_j}(x)} \qquad \text{for } j \in \{1, 2, \ldots, r\}. \qquad (3.4)$$

Let n be a non-negative integer. By (i), n must satisfy at least one of the congruences in (3.1) and (3.2). If there is a $j \in \{1, 2, \ldots, k\}$ such that

$n \equiv 2^{j-1} \pmod{2^j}$, then for some $\ell \in \mathbb{Z}$ we have $n = 2^{j-1} + 2^j\ell$. Since $\Phi_{2^j}(x) = x^{2^{j-1}} + 1$, we obtain from (3.3) that

$$f(x)x^n + d \equiv d\big(x^{(2\ell+1)2^{j-1}} + 1\big) \equiv 0 \pmod{x^{2^{j-1}} + 1}.$$

We deduce that $f(x)x^n + d$ is divisible by $\Phi_{2^j}(x)$. On the other hand, if there is a $j \in \{1, 2, \ldots, r\}$ such that $n \equiv a_j \pmod{m_j}$, then $n = a_j + m_j\ell$ for some integer ℓ. Since $\Phi_{m_j}(x)$ divides $x^{m_j} - 1$, we obtain from (3.4) that

$$f(x)x^n + d \equiv -d\big(x^{(\ell+1)m_j} - 1\big) \equiv 0 \pmod{\Phi_{m_j}(x)}.$$

Thus, $f(x)x^n + d$ is divisible by $\Phi_{m_j}(x)$.

To finish the proof, it suffices to show that we can find an $f(x)$ with positive integral coefficients satisfying the congruences in (3.3) and (3.4) and such that, for every non-negative integer n, $f(x)x^n + d$ is not a constant times $\Phi_{2^j}(x)$ for $j \in \{1, 2, \ldots, k\}$ and not a constant times $\Phi_{m_j}(x)$ for $j \in \{1, 2, \ldots, r\}$. To do this, we show that there is an $f(x)$ with positive integral coefficients satisfying the congruences in (3.3) and (3.4) and such that $\deg f$ is greater than both 2^k and $\max_{1 \le j \le k}\{m_j\}$.

We apply Lemma 2 to deduce that, for i and j in $\{1, 2, \ldots, r\}$ with $i \neq j$, there are polynomials $u_{i,j}(x)$ and $v_{i,j}(x)$ in $\mathbb{Z}[x]$ such that

$$\Phi_{m_i}(x)u_{i,j}(x) + \Phi_{m_j}(x)v_{i,j}(x) = a(i,j). \tag{3.5}$$

Also, by that lemma, for $i \in \{1, 2, \ldots, k\}$ and $j \in \{1, 2, \ldots, r\}$, there are polynomials $u'_{i,j}(x)$ and $v'_{i,j}(x)$ in $\mathbb{Z}[x]$ such that

$$\Phi_{2^i}(x)u'_{i,j}(x) + \Phi_{m_j}(x)v'_{i,j}(x) = b(i,j). \tag{3.6}$$

We fix $j \in \{1, 2, \ldots, r\}$ and expand

$$c \prod_{\substack{1 \le i \le r \\ i \neq j}} \big(\Phi_{m_i}(x)u_{i,j}(x) + \Phi_{m_j}(x)v_{i,j}(x)\big)$$

$$\times \prod_{1 \le i \le k} \big(\Phi_{2^i}(x)u'_{i,j}(x) + \Phi_{m_j}(x)v'_{i,j}(x)\big)$$

where

$$c = \frac{d}{\left(\displaystyle\prod_{\substack{1 \le i \le r \\ i \neq j}} a(i,j)\right)\left(\displaystyle\prod_{i=1}^{k} b(i,j)\right)}.$$

From (iii), (3.5), and (3.6), we deduce that

$$\left(\prod_{\substack{1 \leq i \leq r \\ i \neq j}} \Phi_{m_i}(x) \right) w(x) u_j(x) + \Phi_{m_j}(x) v_j(x) = d$$

for some polynomials $u_j(x)$ and $v_j(x)$ in $\mathbb{Z}[x]$. Similarly, combining (iv) and (3.6), one obtains $u(x)$ and $v(x)$ in $\mathbb{Z}[x]$ satisfying

$$\left(\prod_{1 \leq j \leq r} \Phi_{m_j}(x) \right) u(x) + w(x) v(x) = d.$$

Let $M = 2^k m_1 m_2 \cdots m_r$, and let $k(x) \in \mathbb{Z}[x]$. It follows that

$$f(x) = \sum_{j=1}^{r} \left(\prod_{\substack{1 \leq i \leq r \\ i \neq j}} \Phi_{m_i}(x) \right) w(x) u_j(x) \left(-x^{m_j - a_j} \right)$$

$$+ \left(\prod_{1 \leq j \leq r} \Phi_{m_j}(x) \right) u(x) + k(x) \left(\frac{x^M - 1}{x - 1} \right)$$

satisfies the congruences in (3.3) and (3.4). Furthermore, we may take $k(x)$ appropriately (for example, $k(x) = b(x^s - 1)/(x-1)$ with b and s sufficiently large positive integers) so that $f(x)$ has positive integral coefficients and $\deg f$ is greater than both 2^k and $\max_{1 \leq j \leq k} \{m_j\}$. This completes the proof. $\qquad \square$

4 A Preliminary Covering and Theorem 1

In this section, we establish

Theorem 4. *There is a covering of the integers consisting of moduli $m_1, m_2, \ldots, m_r$ satisfying:*

(i) *For each positive integer m, there exist at most three $\ell \in \{1, 2, \ldots, r\}$ for which $m_\ell = m$.*

(ii) *Each m_ℓ is odd and > 1.*

(iii) *Each m_ℓ has at least two distinct prime factors.*

Observe that Theorem 4 implies that an odd covering exists if we allow up to three congruences for each odd modulus. The existence of an $f(x) \in \mathbb{Z}^+[x]$ such that $f(x)x^n + 2$ is reducible for all integers $n \geq 0$ would follow if one can establish the existence of a covering as in Theorem 4 but with "three" replaced by "two" in (i) above.

Before turning to the proof of Theorem 4, we explain how it is used with Theorem 3 to deduce Theorem 1. Let $x \equiv a_j \pmod{m_j}$ for $j \in \{1, 2, \ldots, r\}$ denote the r congruences given by Theorem 4. We suppose as we may (by (i) in Theorem 4) that if $m_i = m_j$ for some integers i and j with $1 \leq j < i \leq r$, then $i = j + 1$ or $i = j + 2$. In particular, for j fixed, the conditions $i \neq j$ and $m_j = m_i$ imply there are at most two possibilities for i. Define

$$m_j' = 2^j m_j \qquad \text{for } j \in \{1, 2, \ldots, r\}.$$

By the Chinese Remainder Theorem, for each $j \in \{1, 2, \ldots, r\}$, there is an integer b_j satisfying

$$b_j \equiv a_j \pmod{m_j} \qquad \text{and} \qquad b_j \equiv 0 \pmod{2^j}.$$

We consider the congruences

$$x \equiv 2^{j-1} \pmod{2^j} \qquad \text{for } j \in \{1, 2, \ldots, r\} \tag{4.1}$$

together with

$$x \equiv b_j \pmod{m_j'} \qquad \text{for } j \in \{1, 2, \ldots, r\}. \tag{4.2}$$

We show that these congruences form a system S of congruences satisfying the conditions of Theorem 3 (so $k = r$ in Theorem 3 and the m_j have been replaced by m_j' there) provided d is divisible by 4.

Let n be an arbitrary integer that does not satisfy one of the congruences in (4.1). Then $n \equiv 0 \pmod{2^r}$ (otherwise, n would satisfy the congruence in (4.1) corresponding to the largest positive integer j, necessarily $\leq r$, for which 2^{j-1} divides n). Also, since the congruences $x \equiv a_j \pmod{m_j}$ for $j \in \{1, 2, \ldots, r\}$ form a covering of the integers, $n \equiv a_j \pmod{m_j}$ for some $j \in \{1, 2, \ldots, r\}$. By the definition of b_j, we have for that choice of j that $x \equiv b_j \pmod{m_j'}$. Hence, n satisfies one of the congruences in (4.2). Thus, S satisfies the condition (i) of Theorem 3. Condition (ii) of Theorem 3 is easily checked for the congruences in (4.1) and (4.2). To verify conditions (iii) and (iv) of Theorem 3 for the congruences in (4.1) and (4.2), we alter the definitions of $a(i,j)$ and $b(i,j)$ accordingly so that m_i and m_j are replaced by m_i' and m_j'. Since m_j' is 2^j times the odd number m_j, if the ratio $m_j'/m_i' = p^t$ for some prime p and some integer t, then $p = 2$ and, consequently, $m_j = m_i$. Recall that for j fixed, the conditions $i \neq j$ and $m_j = m_i$ imply there are at most two possibilities for i. We deduce that for each $j \in \{1, 2, \ldots, r\}$,

$$\prod_{\substack{1 \leq i \leq r \\ i \neq j}} a(i,j) \quad \text{divides} \quad 4.$$

By conditions (ii) and (iii) in Theorem 4, each m_j and, hence, each m'_j has at least two odd prime divisors. It follows that $b(i,j) = 1$ for every choice of i and j in $\{1, 2, \ldots, r\}$. Conditions (iii) and (iv) of Theorem 3 now easily follow since d is divisible by 4.

We turn now to establishing Theorem 4.

Lemma 7. *Let p be a prime, and let E be a positive integer. Suppose that n is an integer which is not congruent to -1 modulo p^E. Then n satisfies at least one of the congruences*

$$x \equiv p^{e-1}(j+1) - 1 \pmod{p^e} \qquad \text{where } 0 \leq j \leq p-2 \text{ and } 1 \leq e \leq E.$$

Proof. Consider the positive integer e satisfying $p^{e-1} \| (n+1)$. Then $1 \leq e \leq E$, and for some integer $n' \not\equiv 0 \pmod{p}$ we have $n + 1 = p^{e-1}n'$. Let $j \in \{0, 1, \ldots, p-2\}$ be such that $n' \equiv j+1 \pmod{p}$. Then $n' = j+1+pn''$ for some integer n''. Thus, $n + 1 = p^{e-1}(j+1) + p^e n''$ so that n satisfies the congruence $x \equiv p^{e-1}(j+1) - 1 \pmod{p^e}$, as required. $\square$

Lemma 8. *Let p be a prime, and let E be a positive integer not divisible by p. Suppose that n is an integer $\equiv -1 \pmod{p^E}$. Then n satisfies at least one of the congruences*

$$x \equiv b_j \pmod{p^j E} \qquad \text{with } 1 \leq j \leq E,$$

where

$$b_j \equiv -1 \pmod{p^j} \qquad \text{and} \qquad b_j \equiv j \pmod{E}.$$

Proof. Consider $j \in \{1, 2, \ldots, E\}$ such that $n \equiv j \pmod{E}$. $\square$

Proof of Theorem 4. As we shall demonstrate, the specific covering is given by Tables 1 and 2 below. We begin by explaining the entries in the tables. Each modulus is of the form

$$m = q_1^{e(1)} q_2^{e(2)} \cdots q_{12}^{e(12)},$$

where q_k denotes the kth odd prime (so $q_1 = 3, q_2 = 5, \ldots, q_{12} = 41$) and the $e(k)$ denote non-negative integers. Each tuple under the heading "Congruences" in Table 1 indicates a list of congruences $x \equiv a \pmod{m}$ as follows. Fix a value of j from the range indicated in the third column of the table (each j will produce one or more congruences). We consider m of the form indicated above. Let T denote the set of positive integers k for which the kth component in the tuple is a "$*$". If $k \in T$, set $e(k) = 0$. If the kth component in the tuple is of the form $b : (t)$, then we define $b_k = b$ and set $e(k) = t$. If $k \notin T$ and the kth component in the tuple is not of the form $b : (t)$, then we define b_k as the value of that component and $e(k)$

is as indicated in Table 2, with each choice of $e(k)$ in Table 2 determining a different modulus m. For each such m, we determine a by the Chinese Remainder Theorem from the congruences

$$a \equiv b_k \pmod{q_k^{e(k)}}$$

where k ranges over those positive integers ≤ 12 that are not in T. As an example, we note that Row 1 in Table 1 corresponds to $41 \times 31 \times 3 = 3813$ congruences; further, if $e(1) = 1$, $e(2) = 3$, and $j = 1$, then the congruence determined by this row is $x \equiv 174 \pmod{375}$. Note that each of the three values of j in Row 1 gives a congruence modulo m where $m = 3^{e(1)}5^{e(2)}$ so that moduli can occur three times as indicated in (i) of Theorem 4. A quick look through Table 1 shows that the two sets of primes q_k with $e(k) > 0$ associated with congruences from any two distinct rows of Table 1 are distinct sets. One easily sees then that condition (i) of the theorem is satisfied by the complete collection of congruences given in the table. Similarly, it is easy to check that conditions (ii) and (iii) are satisfied by this collection of congruences. What remains to be established is that this collection of congruences, say $\mathcal{C}$, is in fact a covering of the integers.

We begin by considering the congruences $\mathcal{C}_1$ in Rows 2–5 of Table 1 with $e(1)$, $e(2)$, and $e(4)$ fixed but with $e(5)$ varying over its values in Table 2. By Lemma 7, if n is an integer satisfying

$$x \equiv 3^{e(1)-1} - 1 \pmod{3^{e(1)}}, \qquad x \equiv 5^{e(2)-1}4 - 1 \pmod{5^{e(2)}},$$
$$x \equiv 11^{e(4)-1} - 1 \pmod{11^{e(4)}} \tag{4.3}$$

and $x \not\equiv -1 \pmod{13^{23}}$, then n is *covered by* at least one of the congruences in $\mathcal{C}_1$ (i.e., n satisfies one of the congruences in $\mathcal{C}_1$). By Lemma 8, the congruences in Row 6 of Table 1 cover the integers $\equiv -1 \pmod{13^{23}}$. Thus, the congruences indicated in Table 1 cover every integer satisfying all three congruences in (4.3). Note that by the Chinese Remainder Theorem the congruences in (4.3) correspond to a single congruence modulo $3^{e(1)}5^{e(2)}11^{e(4)}$.

Observe that the last congruence in (4.3) is the same as

$$x \equiv 11^{e(4)-1}(j+1) - 1 \pmod{11^{e(4)}} \qquad \text{where } j = 0.$$

Consider now the congruences $\mathcal{C}_2$ in Rows 7–9 of Table 1 together with the single congruence corresponding to (4.3) with $e(1)$ and $e(2)$ fixed but with $e(4)$ varying over its values in Table 2. Appealing to Lemma 7 again, we deduce that if n is an integer satisfying

$$x \equiv 3^{e(1)-1} - 1 \pmod{3^{e(1)}}, \qquad x \equiv 5^{e(2)-1}4 - 1 \pmod{5^{e(2)}}, \tag{4.4}$$

	Congruences	j Range
1	$\left(3^{e(1)-1}-1,5^{e(2)-1}(j+1)-1,*,\ldots\right)$	$0 \leq j \leq 2$
2	$\left(3^{e(1)-1}-1,*,*,11^{e(4)-1}-1,13^{e(5)-1}(j+1)-1,*,\ldots\right)$	$0 \leq j \leq 2$
3	$\left(*,5^{e(2)-1}4-1,*,11^{e(4)-1}-1,13^{e(5)-1}(j+1)-1,*,\ldots\right)$	$3 \leq j \leq 5$
4	$\left(3^{e(1)-1}-1,5^{e(2)-1}4-1,*,11^{e(4)-1}-1,13^{e(5)-1}(j+1)-1,*,\ldots\right)$	$6 \leq j \leq 8$
5	$\left(*,*,*,11^{e(4)-1}-1,13^{e(5)-1}(j+1)-1,*,\ldots\right)$	$9 \leq j \leq 11$
6	$\left(*,*,*,*,-1:(j),*,*,j:(1),*,\ldots\right)$	$1 \leq j \leq 23$
7	$\left(3^{e(1)-1}-1,*,*,11^{e(4)-1}(j+1)-1,*,\ldots\right)$	$1 \leq j \leq 3$
8	$\left(*,5^{e(2)-1}4-1,*,11^{e(4)-1}(j+1)-1,*,\ldots\right)$	$4 \leq j \leq 6$
9	$\left(3^{e(1)-1}-1,5^{e(2)-1}4-1,*,11^{e(4)-1}(j+1)-1,*,\ldots\right)$	$7 \leq j \leq 9$
10	$\left(*,*,*,-1:(j),*,*,*,*,j:(1),*,\ldots\right)$	$1 \leq j \leq 29$
11	$\left(*,-1:(j),*,*,*,*,*,*,j:(1),*,*\right)$	$1 \leq j \leq 31$
12	$\left(3^{e(1)-1}2-1,*,7^{e(3)-1}(j+1)-1,*,\ldots\right)$	$0 \leq j \leq 2$
13	$\left(3^{e(1)-1}2-1,5^{e(2)-1}-1,7^{e(3)-1}(j+1)-1,*,\ldots\right)$	$3 \leq j \leq 5$
14	$\left(*,*,-1:(j),*,*,*,*,*,*,j:(1),*\right)$	$1 \leq j \leq 37$
15	$\left(*,5^{e(2)-1}2-1,7^{e(3)-1}(j+1)-1,*,\ldots\right)$	$3 \leq j \leq 5$
16	$\left(3^{e(1)-1}2-1,*,*,*,13^{e(5)-1}(j+1)-1,*,\ldots\right)$	$0 \leq j \leq 2$
17	$\left(3^{e(1)-1}2-1,5^{e(2)-1}3-1,*,*,13^{e(5)-1}(j+1)-1,*,\ldots\right)$	$3 \leq j \leq 5$
18	$\left(*,5^{e(2)-1}3-1,*,*,13^{e(5)-1}(j+1)-1,*,\ldots\right)$	$6 \leq j \leq 8$
19	$\left(3^{e(1)-1}2-1,*,7^{e(3)-1}4-1,*,13^{e(5)-1}(j+1)-1,*,\ldots\right)$	$9 \leq j \leq 11$
20	$\left(3^{e(1)-1}2-1,5^{e(2)-1}3-1,7^{e(3)-1}5-1,*,13^{e(5)-1}(j+1)-1,*,\ldots\right)$	$9 \leq j \leq 11$
21	$\left(*,5^{e(2)-1}3-1,7^{e(3)-1}6-1,*,13^{e(5)-1}(j+1)-1,*,\ldots\right)$	$9 \leq j \leq 11$
22	$\left(3^{e(1)-1}2-1,5^{e(2)-1}4-1,7^{e(3)-1}4-1,j:(1),*,\ldots\right)$	$0 \leq j \leq 2$
23	$\left(3^{e(1)-1}2-1,*,7^{e(3)-1}4-1,j:(1),*,\ldots\right)$	$3 \leq j \leq 5$
24	$\left(*,5^{e(2)-1}4-1,7^{e(3)-1}4-1,j:(1),*,\ldots\right)$	$6 \leq j \leq 8$
25	$\left(*,*,7^{e(3)-1}4-1,j:(1),*,\ldots\right)$	$9 \leq j \leq 10$
26	$\left(3^{e(1)-1}2-1,5^{e(2)-1}4-1,7^{e(3)-1}5-1,*,*,j:(1),*,\ldots\right)$	$0 \leq j \leq 2$
27	$\left(3^{e(1)-1}2-1,*,*,*,*,j:(1),*,*,*\right)$	$3 \leq j \leq 5$
28	$\left(*,5^{e(2)-1}4-1,*,*,*,j:(1),*,*,*\right)$	$6 \leq j \leq 8$
29	$\left(*,*,7^{e(3)-1}5-1,*,*,j:(1),*,*,*\right)$	$9 \leq j \leq 11$
30	$\left(3^{e(1)-1}2-1,5^{e(2)-1}4-1,*,*,*,j:(1),*,\ldots\right)$	$12 \leq j \leq 14$
31	$\left(3^{e(1)-1}2-1,*,7^{e(3)-1}5-1,*,*,j:(1),*,\ldots\right)$	$15 \leq j \leq 16$
32	$\left(3^{e(1)-1}2-1,5^{e(2)-1}4-1,7^{e(3)-1}6-1,*,*,*,j:(1),*,\ldots\right)$	$0 \leq j \leq 2$
33	$\left(3^{e(1)-1}2-1,*,*,*,*,*,j:(1),*,\ldots\right)$	$3 \leq j \leq 5$
34	$\left(*,5^{e(2)-1}4-1,*,*,*,*,j:(1),*,\ldots\right)$	$6 \leq j \leq 8$
35	$\left(*,*,7^{e(3)-1}6-1,*,*,*,j:(1),*,\ldots\right)$	$9 \leq j \leq 11$
36	$\left(3^{e(1)-1}2-1,5^{e(2)-1}4-1,*,*,*,*,j:(1),*,\ldots\right)$	$12 \leq j \leq 14$
37	$\left(3^{e(1)-1}2-1,*,7^{e(3)-1}6-1,*,*,*,j:(1),*,\ldots\right)$	$15 \leq j \leq 17$
38	$\left(*,5^{e(2)-1}4-1,7^{e(3)-1}6-1,*,*,*,j:(1),*,\ldots\right)$	$j = 18$
39	$\left(-1:(j),*,*,*,*,*,*,*,*,*,*,j:(1)\right)$	$1 \leq j \leq 41$

Table 1. The Covering for Theorem 4

$1 \le e(1) \le 41$	$1 \le e(2) \le 31$	$1 \le e(3) \le 37$	$1 \le e(4) \le 29$	$1 \le e(5) \le 23$

Table 2. The Exponents Ranges

and $x \not\equiv -1 \pmod{11^{29}}$, then n is covered by at least one of the congruences in $\mathcal{C}_2$ (and, hence, one of the congruences determined by Rows 2–9 of Table 1). By Lemma 8, the congruences in Row 10 of Table 1 cover the integers $\equiv -1 \pmod{11^{29}}$. Thus, the congruences indicated in Table 1 cover every integer satisfying both congruences in (4.4) (which, by the Chinese Remainder Theorem, corresponds to a single congruence modulo $3^{e(1)}5^{e(2)}$).

Observe that the last congruence in (4.4) is the same as

$$x \equiv 5^{e(2)-1}(j+1) - 1 \pmod{5^{e(2)}} \qquad \text{where } j = 3.$$

Consider now the congruences $\mathcal{C}_3$ in Row 1 of Table 1 together with the single congruence corresponding to (4.4) with $e(1)$ fixed but with $e(2)$ varying over its values in Table 2. By Lemma 7, if n is an integer satisfying

$$x \equiv 3^{e(1)-1} - 1 \pmod{3^{e(1)}}, \tag{4.5}$$

and $x \not\equiv -1 \pmod{5^{31}}$, then n is covered by at least one of the congruences in $\mathcal{C}_3$. By Lemma 8, the congruences in Row 11 of Table 1 cover the integers $\equiv -1 \pmod{5^{31}}$. Thus, the congruences indicated in Table 1 cover every integer satisfying (4.5).

The congruence in (4.5) is the same as $x \equiv 3^{e(1)-1}(j+1)-1 \pmod{3^{e(1)}}$ with $j = 0$. The idea now is to appeal to Lemmas 7 and 8 after showing that the congruences in Table 1 cover the integers satisfying $x \equiv 3^{e(1)-1}(j+1)-1 \pmod{3^{e(1)}}$ with $j = 1$. In other words, we will show that the congruences in Table 1 cover the integers satisfying $x \equiv 3^{e(1)-1}2-1 \pmod{3^{e(1)}}$. Then, by letting $e(1)$ vary over the values indicated in Table 2, Lemma 7 will imply that every integer $n \not\equiv -1 \pmod{3^{41}}$ is covered by a congruence from $\mathcal{C}$. Using the congruences corresponding to Row 39 of Table 1 and appealing to Lemma 8, we can then deduce that every integer is covered by some congruence from $\mathcal{C}$. Hence, the theorem will follow.

We cover (by congruences from $\mathcal{C}$) the integers satisfying simultaneously both

$$x \equiv 3^{e(1)-1}2 - 1 \pmod{3^{e(1)}} \quad \text{and} \quad x \equiv 5^{e(2)-1} - 1 \pmod{5^{e(2)}} \tag{4.6}$$

in a manner identical to the approach above. We apply Lemma 7 with the congruences in Rows 12 and 13; and then we apply Lemma 8 with the congruences in Row 14. We can likewise cover integers satisfying simultaneously

$$x \equiv 3^{e(1)-1}2 - 1 \pmod{3^{e(1)}} \quad \text{and} \quad x \equiv 5^{e(2)-1}2 - 1 \pmod{5^{e(2)}}$$

$$(4.7)$$

by applying Lemma 7 with the congruences in Rows 12 and 15 and Lemma 8 with the congruences in Row 14.

Lemma 7 with the congruences in Rows 16–19 and Lemma 8 with the congruences in Row 6 imply that the integers satisfying simultaneously the congruences

$$x \equiv 3^{e(1)-1}2 - 1 \pmod{3^{e(1)}}, \quad x \equiv 5^{e(2)-1}3 - 1 \pmod{5^{e(2)}},$$
$$\text{and} \quad x \equiv 7^{e(3)-1}4 - 1 \pmod{7^{e(3)}}$$

$$(4.8)$$

are covered by congruences from $\mathcal{C}$. Lemma 7 with the congruences in Rows 16, 17, 18, and 20 and Lemma 8 with the congruences in Row 6 imply that the integers satisfying simultaneously the congruences

$$x \equiv 3^{e(1)-1}2 - 1 \pmod{3^{e(1)}}, \quad x \equiv 5^{e(2)-1}3 - 1 \pmod{5^{e(2)}},$$
$$\text{and} \quad x \equiv 7^{e(3)-1}5 - 1 \pmod{7^{e(3)}}$$

$$(4.9)$$

are also covered. Lemma 7 with the congruences in Rows 16, 17, 18, and 21 and Lemma 8 with the congruences in Row 6 imply that the integers satisfying simultaneously the congruences

$$x \equiv 3^{e(1)-1}2 - 1 \pmod{3^{e(1)}}, \quad x \equiv 5^{e(2)-1}3 - 1 \pmod{5^{e(2)}},$$
$$\text{and} \quad x \equiv 7^{e(3)-1}6 - 1 \pmod{7^{e(3)}}$$

$$(4.10)$$

are covered. We apply now Lemma 7 with the congruences in Row 12 together with those given by (4.8), (4.9), and (4.10); and then we appeal to Lemma 8 with the congruences in Row 14. We deduce that all integers satisfying simultaneously the congruences

$$x \equiv 3^{e(1)-1}2 - 1 \pmod{3^{e(1)}} \quad \text{and} \quad x \equiv 5^{e(2)-1}3 - 1 \pmod{5^{e(2)}}$$

$$(4.11)$$

are covered by congruences from $\mathcal{C}$.

Since every integer is congruent to one of $0, 1, \ldots, 10$ modulo 11, the congruences given in Rows 22–25 cover all integers satisfying simultaneously the congruences

$$x \equiv 3^{e(1)-1}2 - 1 \pmod{3^{e(1)}}, \quad x \equiv 5^{e(2)-1}4 - 1 \pmod{5^{e(2)}},$$
$$\text{and} \quad x \equiv 7^{e(3)-1}4 - 1 \pmod{7^{e(3)}}.$$

$$(4.12)$$

Similarly, since every integer is congruent to one of $0, 1, \ldots, 16$ modulo 17, the congruences given in Rows 26–31 cover all integers satisfying simultaneously the congruences

$$x \equiv 3^{e(1)-1}2 - 1 \pmod{3^{e(1)}}, \quad x \equiv 5^{e(2)-1}4 - 1 \pmod{5^{e(2)}},$$
$$\text{and} \quad x \equiv 7^{e(3)-1}5 - 1 \pmod{7^{e(3)}}; \tag{4.13}$$

and since every integer is congruent to one of $0, 1, \ldots, 18$ modulo 19, the congruences given in Rows 31–38 cover all integers satisfying simultaneously the congruences

$$x \equiv 3^{e(1)-1}2 - 1 \pmod{3^{e(1)}}, \quad x \equiv 5^{e(2)-1}4 - 1 \pmod{5^{e(2)}},$$
$$\text{and} \quad x \equiv 7^{e(3)-1}6 - 1 \pmod{7^{e(3)}}. \tag{4.14}$$

We use Lemma 7 with the congruences given in Row 12 together with (4.12), (4.13), and (4.14) and we use Lemma 8 with the congruences in Row 14 to deduce that all integers satisfying simultaneously the congruences

$$x \equiv 3^{e(1)-1}2 - 1 \pmod{3^{e(1)}} \quad \text{and} \quad x \equiv 5^{e(2)-1}4 - 1 \pmod{5^{e(2)}} \tag{4.15}$$

are covered.

Finally, we appeal to the congruences in (4.6), (4.7), (4.11), and (4.15). We apply Lemma 7 with these and apply Lemma 8 with the congruences in Row 11 to obtain that every integer satisfying

$$x \equiv 3^{e(1)-1}2 - 1 \pmod{3^{e(1)}}$$

is covered by a congruence from $\mathcal{C}$. As discussed earlier in this proof, the theorem now follows. $\square$

No real attempt was made to keep the number of congruences in our proof for Theorem 4 at a minimum; the author feels that regardless any such covering for Theorem 4 must in some sense be complicated. We note that the number of congruences used in our proof is 6928899.

5   The Connection with the Odd Covering Problem

In this final section, we give a proof of Theorem 2. For this purpose, we define a non-zero polynomial $f(x) \in \mathbb{Q}[x]$ as being reciprocal if $f(x) =$

$\pm x^{\deg f} f(1/x)$. The non-reciprocal part of $f(x)$ is $f(x)$ removed of its irreducible reciprocal factors. For example, the non-reciprocal part of

$$2x^5 - 5x^4 + 9x^3 - 9x^2 + 5x - 2 = (x-1)(x^2 - x + 2)(2x^2 - x + 1)$$

is $(x^2 - x + 2)(2x^2 - x + 1) = 2x^4 - 3x^3 + 6x^2 - 3x + 2$. As this example illustrates, the non-reciprocal part of a polynomial may in fact be reciprocal (as only the *irreducible* reciprocal factors are removed).

We make use of the following result:

Lemma 9. *Let d be a positive integer, and let $f(x)$ be in $\mathbb{Z}[x]$. Suppose that n is sufficiently large (depending on f). Then the non-reciprocal part of $f(x)x^n + d$ is irreducible over $\mathbb{Q}$ or identically ± 1 unless one of the following holds:*

(i) $-f(x)/d$ *is a pth power in $\mathbb{Q}[x]$ for some prime p dividing n.*

(ii) $f(x)/d$ *is 4 times a 4th power in $\mathbb{Q}[x]$ and n is divisible by 4.*

The above lemma is key to the ideas in this section. Schinzel's argument in [7] also made use of this result. A proof of the lemma, which we do not include here, can be found in Schinzel [6]. An alternative proof has recently been given by Ford, Konyagin, and the author [2].

In addition, we make use of the following results.

Lemma 10. *Suppose that n_0 is a real number such that every integer $n \geq n_0$ satisfies at least one of the congruences*

$$x \equiv a_1 \pmod{m_1}, \ x \equiv a_2 \pmod{m_2}, \ldots, \ x \equiv a_r \pmod{m_r}$$

where the a_j's and m_j's are arbitrary integers with each $m_j > 0$. Then this system of congruences forms a covering of the integers (i.e., every integer $n < n_0$ also satisfies at least one of the congruences).

Proof. Let $M = \mathrm{lcm}(m_1, m_2, \ldots, m_r)$. Let $n \in \mathbb{Z}$. Consider a positive integer k such that $n + kM \geq n_0$. Then $n \equiv n + kM \equiv a_j \pmod{m_j}$ for some integer $j \in \{1, 2, \ldots, r\}$, establishing the lemma. $\square$

Lemma 11. *Let p be a prime, and let m be a positive integer such that p divides m. Then $x^p = \zeta_m$ has no solutions $x \in \mathbb{Q}(\zeta_m)$.*

Proof. Let $\zeta = \zeta_m$. The roots of $x^p - \zeta = 0$ are $\zeta_{pm}\zeta_p^k$ where $0 \leq k \leq p - 1$. Note that $\zeta_p = \zeta_m^{m/p} \in \mathbb{Q}(\zeta)$. Thus, $x^p = \zeta$ and $x \in \mathbb{Q}(\zeta)$ imply $\zeta_{pm} \in \mathbb{Q}(\zeta)$, a contradiction (for example, since ζ_{pm} is a root of an irreducible polynomial of degree $\phi(pm) = p\phi(m)$ which exceeds the degree of the extension $\mathbb{Q}(\zeta)$ over $\mathbb{Q}$). $\square$

Lemma 12. *Let d be a positive integer. Suppose that $-f(x)/d = g(x)^p$ for some prime p and $f(x)x^n + d$ is divisible by $\Phi_m(x)$ where $p \mid m$. Then $n \equiv 0 \pmod{p}$.*

Proof. We set $\zeta = \zeta_m$, and assume $p \nmid n$. Then there are integers u and v such that $-nu + pv = 1$. Since also $f(\zeta)\zeta^n + d = 0$, we deduce that $-f(\zeta)/d = \zeta^{-n}$. Hence,

$$\left(g(\zeta)^u \zeta^v\right)^p = \zeta^{-nu+pv} = \zeta.$$

Thus, $x^p = \zeta$ has a solution $x \in \mathbb{Q}(\zeta)$, contradicting Lemma 11. $\qquad\square$

Proof of Theorem 2. We suppose (as we may) that $f(0) \neq 0$. Since $x^{2^t} + 1 = \Phi_{2^{t+1}}(x)$ is irreducible for every $t \in \mathbb{Z}^+$, we deduce $f(x) \not\equiv 1$. Let $\tilde{f}(x) = x^{\deg f} f(1/x)$. Then each reciprocal factor $g(x)$ of $F(x) = f(x)x^n + d$ divides

$$f(x)\widetilde{F}(x) - dx^{\deg f}F(x) = f(x)\left(dx^{n+\deg f} + \tilde{f}(x)\right) - dx^{\deg f}\left(f(x)x^n + d\right)$$
$$= f(x)\tilde{f}(x) - d^2 x^{\deg f}.$$

In particular, there is a finite list of irreducible reciprocal factors that can divide $f(x)x^n + d$ as n varies. Each reciprocal non-cyclotomic irreducible factor divides at most one polynomial of the form $f(x)x^n + d$ (see the comment before Lemma 1). By Lemma 9, we deduce that there are $\Phi_{m_1}(x), \ldots, \Phi_{m_r}(x)$ such that if n is sufficiently large and both (i) and (ii) of Lemma 9 do not hold, then $\Phi_{m_j}(x) \mid \left(f(x)x^n + 1\right)$ for some j. Note that (ii) does not hold since otherwise $f(x)x^n + d$ could not be divisible by a cyclotomic polynomial (if $\Phi_m(x)$ were a factor, then $f(\zeta_m)\zeta_m^n = -d$, contradicting that the left side has even norm and the right side has odd norm) so that $f(x)x^n + d$ is irreducible by Lemma 9 whenever n is a sufficiently large prime. We may suppose that for each $j \in \{1, 2, \ldots, r\}$ there is an a_j such that $\Phi_{m_j}(x) \mid \left(f(x)x^{a_j} + 1\right)$. Let $\mathcal{P}$ denote the set of primes p for which $f(x)$ is minus a pth power. We remove from consideration any m_j divisible by a $p \in \mathcal{P}$ (but abusing notation we keep the range of subscripts). Then Lemmas 1, 10 and 12 imply that the congruences

$$x \equiv 0 \pmod{p} \text{ for } p \in \mathcal{P} \quad \text{and} \quad x \equiv a_j \pmod{m_j} \text{ for } j \in \{1, 2, \ldots, r\}$$

cover the integers.

Claim. Suppose $m_j = p^t m_0$ and $m_i = p^s m_0$, where p is a prime not dividing d, m_0 is an integer > 1 such that $p \nmid m_0$, and t and s are integers with $t > s \geq 0$. Then $a_j \equiv a_i \pmod{m_0}$.

For the moment, suppose the claim holds. Take $p = 2$ in the claim. Since d is odd, clearly p does not divide d. We replace $x \equiv a_j \pmod{m_j}$ and $x \equiv a_i \pmod{m_i}$ with $x \equiv a_j \pmod{m_0}$. If for some j there is no i as above, we still replace $x \equiv a_j \pmod{m_j}$ with $x \equiv a_j \pmod{m_0}$. Then we are left with a covering with moduli that are distinct odd numbers together with possibly powers of 2. Observe that $\sum_{j=1}^{\infty} 1/2^j = 1$ implies that there is an $a \in \mathbb{Z}$ and a $k \in \mathbb{Z}^+$ such that no integer satisfying $x \equiv a \pmod{2^k}$ satisfies one of the congruences in our covering with moduli a power of 2. Denote by $x \equiv a'_j \pmod{m'_j}$ the congruences with m'_j odd. Let u and v be integers such that

$$2^k u + v \left(\prod m'_j \right) = 1.$$

For any $n \in \mathbb{Z}$, consider the number $m = a + 2^k u(n - a)$. Then $m \equiv n \pmod{m'_j}$ for every m'_j and $m \equiv a \pmod{2^k}$. It follows that $n \equiv m \equiv a'_j \pmod{m'_j}$ for some m'_j. Therefore, every $n \in \mathbb{Z}$ satisfies one of the congruences $x \equiv a'_j \pmod{m'_j}$. So these congruences form an odd covering of the integers, and we are left with establishing the claim.

Let $k \in \mathbb{Z}^+ \cup \{0\}$ such that

$$a_i + (k - 1)m_i < a_j \le a_i + km_i.$$

Let $\ell = a_i + km_i - a_j$. Then $\ell \in [0, m_i)$. Since $\Phi_{m_i}(x)$ divides $f(x)x^{a_i + km_i} + d$ by Lemma 1 and $\Phi_{m_j}(x)$ divides $f(x)x^{a_j} + d$, we deduce that there are $u(x)$ and $v(x)$ in $\mathbb{Z}[x]$ such that

$$f(x)x^{a_i + km_i} + d = -\Phi_{m_i}(x)u(x)$$

and

$$f(x)x^{a_i + km_i} = f(x)x^{\ell + a_j} = -dx^{\ell} + \Phi_{m_j}(x)v(x).$$

Hence,

$$\Phi_{m_i}(x)u(x) + \Phi_{m_j}(x)v(x) = d(x^{\ell} - 1).$$

Letting $x = \zeta_{m_i}$ above and applying Lemma 3, we obtain $pw = d(\zeta_{m_i}^{\ell} - 1)$ for some $w \in \mathbb{Z}[\zeta_{m_i}]$. Applying Lemma 5 and using that $p \nmid d$, we deduce that m_0 divides ℓ. The definition of ℓ and the fact that m_0 divides both ℓ and m_i imply the claim. $\qquad\square$

Concluding Remarks. In the closing arguments above, we used Lemmas 3 and 5 to justify that m_0 divides ℓ. Schinzel has pointed out that instead one can apply the work of Apostol [1] on the resultants of two cyclotomic polynomials. By (2.2), $\Phi_{m_0}(x)$ divides both $\Phi_{m_i}(x)$ and $\Phi_{m_j}(x)$ modulo p. Since $p \nmid d$, the last equation displayed above implies $\Phi_{m_0}(x)$

divides $x^\ell - 1$ modulo p. Hence, $\Phi_{m_0}(x)$ and some divisor $\Phi_{\ell'}(x)$ of $x^\ell - 1$ in $\mathbb{Z}[x]$ have a factor in common modulo p. In other words, there is a positive integer ℓ' dividing ℓ such that the resultant of $\Phi_{m_0}(x)$ and $\Phi_{\ell'}(x)$ is divisible by p. Recall from above that $p \nmid m_0$. Apostol's work implies that ℓ'/m_0 is a power of p. It follows that m_0 divides ℓ' and, hence, ℓ.

The author expresses his gratitude to Andrzej Schinzel for taking an interest in this work and for supplying the author with alternative approaches to some of the arguments.

References

[1] T. M. Apostol, *Resultants of cyclotomic polynomials*, Proc. Amer. Math. Soc. **24** (1970), 457–462.

[2] M. Filaseta, K. Ford, and S. Konyagin, *On an irreducibility theorem of A. Schinzel associated with coverings of the integers*, Illinois J. Math. **44** (2000), 633–643.

[3] R. K. Guy, *Unsolved problems in number theory*, second ed., Springer-Verlag, New York, 1994.

[4] W. J. LeVeque, *Topics in number theory*, vol. II, Addison-Wesley, Reading, 1956.

[5] R. Morikawa, *Some examples of covering sets*, Bull. Fac. Liberal Arts Nagasaki Univ. (1981), 1–4.

[6] A. Schinzel, *On the reducibility of polynomials and in particular of trinomials*, Acta Arith. **11** (1965), 1–34, Errata, ibid., vol. 12.

[7] ———, *Reducibility of polynomials and covering systems of congruences*, Acta Arith. **13** (1967), 91–101.

Zero-free Regions for the Riemann Zeta Function

Kevin Ford[1]

1 Introduction

The methods of Korobov [12] and Vinogradov [26] produce a zero-free region for the Riemann zeta function $\zeta(s)$ of the following strength: for some constant $c > 0$, there are no zeros of $\zeta(s)$ for $s = \beta + it$ with $|t|$ large and

$$1 - \beta \leq \frac{c}{(\log|t|)^{2/3}(\log\log|t|)^{1/3}}. \tag{1.1}$$

The principal tool is an upper bound for $|\zeta(s)|$ near the line $\sigma = 1$. One form of this upper bound was given by Richert [18] as

$$|\zeta(\sigma + it)| \leq A|t|^{B(1-\sigma)^{3/2}}\log^{2/3}|t| \qquad (|t| \geq 3, \frac{1}{2} \leq \sigma \leq 1) \tag{1.2}$$

with $B = 100$ and A and unspecified absolute constant. Subsequently, (1.2) was proved with smaller values of B, the best published value being 18.497 [13] (the author has a new result [8] that (1.2) holds with $B = 4.45$, $A = 76.2$).

Table 1 shows the historical progression of zero-free regions for $\zeta(s)$ prior to the work of Vinogradov and Korobov.

Recently, versions of (1.1) with explicit constants c have been given, valid for $|t|$ sufficiently large. Popov [16] showed that (1.1) holds with $c = 0.00006888$. Heath-Brown [10] proved (1.1) with $c \approx 0.0269B^{-2/3}$, and he noted (but did not give details) that the methods of [9] could be used to improve 0.0269 to about 0.0467. The main object of this note is to improve the constant c as a function of B.

Theorem 1. *If* (1.2) *holds with a certain constant B, then for large $|t|$,* $\zeta(\beta + it) \neq 0$ *for*

$$1 - \beta \leq \frac{0.05507B^{-2/3}}{(\log|t|)^{2/3}(\log\log|t|)^{1/3}}.$$

[1]Research supported in part by National Science Foundation grant DMS-0070618.

Zero-free region	Reference
$1 - \beta \leq \dfrac{c}{\log|t|}$	de la Vallée Poussin [7], 1899
$1 - \beta \leq \dfrac{c \log\log|t|}{\log|t|}$	Littlewood [14], 1922
$1 - \beta \leq \dfrac{c}{(\log|t|)^{3/4+\varepsilon}}$	Chudakov [5], 1938

Table 1.

Pintz [15, Theorem 8] established a precise relationship between the location of the zeros of $\zeta(s)$ and the error term in the prime number theorem, which together with Theorem 1 implies that

$$|\pi(x) - \mathrm{li}(x)| \ll x \exp\{-d(\log x)^{3/5}(\log\log x)^{-1/5}\}, \quad d = 0.38124 B^{-2/5}. \tag{1.3}$$

Taking $B = 4.45$ (from [8]) in Theorem 1 gives the zero-free region (1.1) with $c = \frac{1}{49.13}$ and $d = 0.2098$ in (1.3). In addition, we prove a totally explicit zero-free region of type (1.1), with an explicit c and valid for all $|t| \geq 3$. This depends on both A and B in (1.2)), and may be used to give completely explicit bounds for prime counting functions (see, e.g., [20], [21], [17]). Cheng [1] proved (1.2) with $A = 175$ and $B = 46$ and used this to deduce that (1.1) holds for all $|t| \geq 3$ with the constant $c = 1/990$. In turn, this result was used to show [2] that for all $x > 10$,

$$|\pi(x) - \mathrm{li}(x)| \leq 11.88 x (\log x)^{3/5} \exp\{-\frac{1}{57}(\log x)^{3/5}(\log\log x)^{-1/5}\},$$

and that for $x \geq e^{e^{44.06}}$, there is a prime between x^3 and $(x+1)^3$ [3].

Theorem 2. *Suppose* (1.2) *holds,* $T_0 \geq e^{30000}$ *and* $\frac{\log T_0}{\log\log T_0} \geq \frac{1740}{B}$. *Suppose the zeros* $\beta + it$ *of* $\zeta(s)$ *with* $T_0 - 1 \leq t \leq T_0$ *all satisfy*

$$1 - \beta \geq \frac{M_1 B^{-2/3}}{(\log t)^{2/3}(\log\log t)^{1/3}}, \tag{1.4}$$

where

$$M_1 = \min\left(0.05507, \frac{0.1652}{2.9997 + \max_{t \geq T_0} X(t)/\log\log t}\right),$$

and

$$X(t) = 1.1585 \log A + 0.859 + 0.2327 \log \left(\frac{B}{\log \log t} \right)$$

$$+ \left(\frac{1.313}{B^{4/3}} - \frac{2.188}{B^{1/3}} \right) \left(\frac{\log \log t}{\log t} \right)^{\frac{1}{3}}.$$

Then (1.4) *is satisfied for all zeros with* $t \geq T_0$.

Since $\frac{0.1652}{2.9997} > 0.05507$, $M_1 = 0.05507$ when T_0 is sufficiently large. By classical zero density bounds (see, e.g., Chapter 9 of [24]), for some positive δ, the number of zeros of $\zeta(s)$ is the rectangle $\frac{3}{4} \leq \Re s \leq 1, 0 < \Im s \leq T$ is $O(T^{1-\delta})$. Thus for most T_0, $\zeta(s)$ is zero free in the region $\frac{3}{4} \leq \Re s \leq 1$, $T_0 - 1 \leq \Im s \leq T_0$. Taking such T_0 which is sufficiently large, we see that Theorem 1 follows from Theorem 2.

To prove a totally explicit zero-free region of type (1.1) for $|t| \geq 3$, we make use of classical type (de la Valée Poussin type) zero-free regions for smaller $|t|$. These take the form

$$1 - \beta \leq \frac{c}{\log |t|} \qquad (|t| \geq 3). \qquad (1.5)$$

Stechkin [23] proved (1.5) with $c = 1/9.646$ (he rounded this to $c = 9.65$ in his Theorem 2). Very tiny refinements were subsequently given by Rosser and Schoenfeld [21] and by Ramaré and Rumely [17]. With an explicit version of van der Corput's bound $|\zeta(1/2 + it)| \ll |t|^{1/6} \log |t|$ for $|t| \geq 3$, the methods of this paper produce a zero-free region

$$1 - \beta \leq \frac{1}{C_1 (\log |t| + 6 \log \log |t|) + C_2}, \qquad (|t| \geq 3), \qquad (1.6)$$

with $C_1 \approx 3.36$ and an explicit C_2. Better upper bounds are known for $|\zeta(1/2 + it)|$ for large t, the best being $O_\varepsilon(|t|^{89/570+\varepsilon})$ due to Huxley [11]. The implied constants are too large to improve the zero-free region, however. The zero-free region (1.6) also follows from Heath-Brown's methods with the same C_1 (and slightly larger C_3). In fact, the methods of this paper do not improve on Heath-Brown's methods when it comes to classical type zero-free regions for $\zeta(s)$ or zero-free regions for Dirichlet L-functions $L(s, \chi)$ when $|t|$ is small and the conductor of χ is large (e.g., those in [9]). Our methods do improve the Vinogradov-Korobov type zero-free regions for $L(s, \chi)$ when the conductor of χ is fixed and $|t|$ becomes large.

It is known [25] that all zeros with $|\Im \rho| \leq 5.45 \times 10^8$ in fact lie on the critical line. Still, at $t = 5.45 \times 10^8$, $6 \log \log t \approx 0.895 \log t$, so improving greatly on Stechkin's region for all $|t| \geq 3$ with (1.6) is not possible. Still, we can make a modest improvement using the bound

$$|\zeta(1/2 + it)| \leq \min\left(6t^{1/4} + 57, 3t^{1/6}\log t\right) \qquad (t \geq 3). \qquad (1.7)$$

proved by Cheng and Graham [4].

Theorem 3. *Let* $T_0 = 5.45 \times 10^8$ *and let*

$$J(t) = \min\left(\frac{1}{4}\log t + 1.8521, \frac{1}{6}\log t + \log\log t + \log 3\right). \qquad (1.8)$$

Then $\zeta(\beta + it) \neq 0$ *for* $t \geq T_0$ *and*

$$1 - \beta \leq \frac{0.04962 - \dfrac{0.0196}{J(t) + 1.15}}{J(t) + 0.685 + 0.155\log\log t}. \qquad (1.9)$$

We note that $J(t)$ is an increasing function of t, and $(J(t) + 0.685 + 0.155\log\log t)/\log t$ is a decreasing function of t. Therefore, we conclude as a corollary that

Theorem 4. *We have* $\zeta(\beta + it) \neq 0$ *for* $|t| \geq 3$ *and*

$$1 - \beta \leq \frac{1}{8.463\log|t|}.$$

Further verification that the zeros of $\zeta(s)$ for some range of $t > 5.45 \times 10^8$ would give an improved constant in Theorem 4, as would an improvement in the bound for $|\zeta(1/2 + it)|$ in the vicinity of $t = T_0$.

We now return to the problem of producing a totally explicit zero-free regions of Korobov-Vinogradov type. Taking $B = 4.45$, $A = 76.2$ (from [8]), we find that

$$1.1585\log A + 0.859 + 0.2327\log\left(\frac{B}{\log\log t}\right) + \left(\frac{1.313}{B^{4/3}} - \frac{2.188}{B^{1/3}}\right)\left(\frac{\log\log t}{\log t}\right)^{\frac{1}{3}}$$

$$\leq 6.22660 - 0.2327\log\log\log t - 1.1508\left(\frac{\log\log t}{\log t}\right)^{1/3}$$

$$\leq 5.6008.$$

Thus

$$M_1 \geq \min\left(0.05507, \frac{0.1652}{2.9997 + \frac{5.6008}{\log\log T_0}}\right). \qquad (1.10)$$

We take $T_0 = e^{54550}$, use Theorem 3 for $t \leq T_0 + 1$, and Theorem 2 plus (1.10) for larger t. This gives

Theorem 5. *The function* $\zeta(\beta + it)$ *is nonzero in the region*

$$1 - \beta \leq \frac{1}{57.54(\log |t|)^{2/3}(\log \log |t|)^{1/3}}, \quad |t| \geq 3.$$

2   The Zero Detector

Lemma 2.1. *Suppose* f *is the quotient of two entire functions of order* $< k$, *where* k *is a positive integer, and* $f(0) \neq 0$. *If* z *is neither a pole nor a zero of* f, *then*

$$\frac{f'(z)}{f(z)} = \sum_{|\rho| \leq 2|z|} \frac{(z/\rho)^{k-1}}{z - \rho} m_\rho + O_f\left(|z|^{k-1}\right),$$

$$\left|\log |f(z)|\right| \leq \left| \sum_{|\rho| \leq 2|z|} \log \left|(1 - z/\rho)e^{g(z/\rho)}\right| \right| + O_f\left(|z|^k\right),$$

where ρ *runs over the zeros and poles of* f *(with multiplicity),* $g(y) = y + \frac{1}{2}y^2 + \cdots + \frac{1}{k-1}y^{k-1}$, *and* m_ρ *is either 1 (if* ρ *is a zero of* f*) or* -1 *(if* ρ *is a pole of* f*). The implied constants depend on* f.

Proof. By theorems of Weierstrass and Hadamard ([22], Ch. VII, (2.13) and (10.1)),

$$f(z) = e^{f_1(z)} \prod_\rho \left[(1 - z/\rho)e^{g(z/\rho)}\right]^{m_\rho},$$

where f_1 is a polynomial of degree $\leq k$. Therefore, assuming that z is not a zero or pole of f, we have

$$\log |f(z)| = \Re f_1(z) + \sum_\rho m_\rho \left(\log |(1 - z/\rho)e^{g(z/\rho)}|\right),$$

$$\frac{f'(z)}{f(z)} = f_1'(z) + \sum_\rho m_\rho \left(\frac{1}{z - \rho} + \frac{1}{\rho} + \frac{z}{\rho^2} + \cdots + \frac{z^{k-2}}{\rho^{k-1}}\right).$$

Now suppose $|\rho| > 2|z|$. We then have

$$\left|\frac{1}{z - \rho} + \frac{1}{\rho} + \frac{z}{\rho^2} + \cdots + \frac{z^{k-2}}{\rho^{k-1}}\right| = \left|\frac{(z/\rho)^{k-1}}{z - \rho}\right| \leq 2\frac{|z|^{k-1}}{|\rho|^k}.$$

Since $\sum_\rho 1/|\rho|^k$ converges, the first part of the lemma follows. Similarly

$$\left|(1 - z/\rho)e^{g(z/\rho)}\right| \leq \exp\left\{\frac{2}{k}|\frac{z}{\rho}|^k\right\},$$

and the second part follows. $\qquad\square$

The next lemma is the main "zero detector". Instead of integrating around a small circle centered at $z = z_0$ (as in [9], Lemma 3.2), we integrate over two vertical lines.

Lemma 2.2. *Suppose f is the quotient of two entire functions of finite order, and does not have a zero or a pole at $z = z_0$ nor at $z = 0$. Then, for all $\eta > 0$ except for a set of Lebesgue measure 0 (the exceptional set may depend on f and z_0), we have*

$$- \Re \frac{f'(z_0)}{f(z_0)} = \frac{\pi}{2\eta} \sum_{|\Re(z_0-\rho)|\leq\eta} m_\rho \Re \cot\left(\frac{\pi(\rho - z_0)}{2\eta}\right)$$

$$+ \frac{1}{4\eta} \int_{-\infty}^{\infty} \frac{\log\left|f\left(z_0 - \eta + \frac{2\eta iu}{\pi}\right)\right| - \log\left|f\left(z_0 + \eta + \frac{2\eta iu}{\pi}\right)\right|}{\cosh^2 u} du,$$

where ρ runs over the zeros and poles of f (with multiplicity), and m_ρ is either 1 (if ρ is a zero of f) or -1 (if ρ is a pole of f).

Proof. We must exclude η for which the lines $\Re z = z_0 \pm \eta$ come "too close" to a zero or pole of f, since otherwise the above integral might not converge. By hypothesis, for some integer k, f is the quotient of two entire functions of order $< k$. We say a positive real number η is "good" if there is a positive number δ such that for every zero/pole ρ of f, $|\Re(\rho - z_0) \pm \eta| \geq \delta|\rho|^{-k}$. The number δ may depend on η. Since $\sum_\rho |\rho|^{-k}$ converges, the set of η for which $|\Re(\rho - z_0) \pm \eta| \leq \delta|\rho|^{-k}$ has measure $O(\delta)$ (here and throughout this proof, implied constants depend on f and z_0). Taking $\delta \to 0$ shows that the measure of "bad" η is 0.

Suppose now that η is "good" with an associated number δ. We may assume that $0 < \delta \leq 1$. Let T be a large real number such that $T \geq \eta$, $T \geq 2|z_0|$ and for all zeros/poles ρ of f, $|\Im(\rho - z_0) \pm T| \geq |\rho|^{-k}$. Since $\sum_\rho |\rho|^{-k}$ converges, the set of "bad" T has measure $O(1)$. Consider the contour $C = C_1 \cup C_2 \cup C_3 \cup C_4$, where the C_j are the line segments connecting the points $\eta - iT, \eta + iT, -\eta + iT, -\eta - iT, \eta - iT$, respectively. Let

$$I = I_1 + I_2 + I_3 + I_4, \quad I_j = \int_{C_j} \frac{f'(z + z_0)}{f(z + z_0)} h(z)\, dz,$$

where

$$h(z) = \frac{\pi}{2\eta} \cot\left(\frac{\pi z}{2\eta}\right).$$

By Cauchy's Residue Theorem,

$$I = \frac{f'(z_0)}{f(z_0)} + \sum_{\substack{|\Re(\rho-z_0)|\leq\eta \\ |\Im(\rho-z_0)|\leq T}} m_\rho h(\rho - z_0). \tag{2.1}$$

There is a holomorphic branch of $\log f(z + z_0)$ on C^*, the contour C cut at the point η. Applying integration by parts, and noting that $h(\eta) = 0$, we have

$$I = \lim_{\varepsilon \to 0^+} [h(z) \log f(z + z_0)]_{\eta+i\varepsilon}^{\eta-i\varepsilon} - \frac{1}{2\pi i} \int_{C^*} h'(z) \log f(z + z_0)\, dz$$

$$= -(J_1 + J_2 + J_3 + J_4),\qquad\qquad (2.2)$$

$$J_j = \frac{1}{2\pi i} \int_{C_j} h'(z) \log f(z + z_0)\, dz.$$

The number of zeros/poles ρ with $|\rho| \le x$ is $O(x^k)$, and $|\rho| \gg 1$ for every ρ. By our assumptions about T, when $z \in C$ we have $|z+z_0| \ll T$. Therefore, by Lemma 2.1 and our assumption about η,

$$\left| \frac{f'(z + z_0)}{f(z + z_0)} \right| \ll T^{k-1} + \sum_{|\rho| \le 2|z|} \frac{|(z + z_0)/\rho|^{k-1}}{|z + z_0 - \rho|}$$

$$\ll T^{k-1} + \frac{T^{k-1}}{|\rho|^{k-1}} \frac{|\rho|^k}{\delta} \ll \delta^{-1} T^k.$$

Likewise, using the second part of Lemma 2.1,

$$|\log |f(z + z_0)|| = O(T^{2k-1} + T^k \log(T\delta^{-1}))$$

for $z \in C$. Thus, there is a branch of $\log f(z + z_0)$ with

$$|\log f(z + z_0)| \ll T^{2k} \delta^{-1}.$$

This is important to the estimation of J_2 and J_4. Since

$$h'(z) = -\frac{\pi^2}{4\eta^2} \csc^2 \left(\frac{\pi z}{2\eta} \right),$$

we have $|h'(\eta \pm iT)| \ll \eta^{-2} e^{-\pi T/(2\eta)}$. Therefore, $|J_2| + |J_4| \to 0$ as $T \to \infty$. Parameterizing the line segments C_1 and C_3 with $z = \pm\eta + \frac{2\eta iu}{\pi}$ and taking real parts gives

$$\Re(J_1 + J_3) = \frac{1}{4\eta} \int_{-\frac{\pi T}{2\eta}}^{\frac{\pi T}{2\eta}} \frac{\log \left| f \left(z_0 - \eta + \frac{2\eta iu}{\pi} \right) \right| - \log \left| f \left(z_0 + \eta + \frac{2\eta iu}{\pi} \right) \right|}{\cosh^2 u}\, du.$$

Recalling (2.1) and (2.2), this proves the lemma upon letting $T \to \infty$. □

3 Bounds for $\zeta(s)$

Lemma 3.1. *Suppose $1 < \sigma \le 1.06$ and t is real. Then*

$$\frac{1}{\zeta(\sigma)} \le |\zeta(\sigma + it)| \le \zeta(\sigma) \le 0.6 + \frac{1}{\sigma - 1}$$

and

$$\left| -\frac{\zeta'}{\zeta}(\sigma + it) \right| < \frac{1}{\sigma - 1}.$$

Proof. For the first line of inequalities, we start with

$$|\zeta(\sigma + it)| \le \sum_{n=1}^{\infty} n^{-\sigma} = \zeta(\sigma)$$

and similarly

$$|\zeta(\sigma + it)|^{-1} = \left| \sum_{n=1}^{\infty} \mu(n) n^{-\sigma - it} \right| \le \sum_{n=1}^{\infty} n^{-\sigma} = \zeta(\sigma).$$

Next, since $x^{-\sigma}$ is convex and $e^{-y} \le 1 - y + \frac{1}{2}y^2$ for $0 \le y \le 1$, we have

$$\zeta(\sigma) \le 1 + \int_{3/2}^{\infty} \frac{du}{u^\sigma} = 1 + \frac{(3/2)^{-(\sigma-1)}}{\sigma - 1}$$

$$\le 1 + \frac{1}{\sigma - 1} - \log(1.5) + \frac{1}{2}(\sigma - 1)\log^2(1.5)$$

$$\le 0.6 + \frac{1}{\sigma - 1}.$$

In fact, near $\sigma = 1$ we have $\zeta(\sigma) = \frac{1}{\sigma-1} + \gamma + O(\sigma - 1)$, where $\gamma = 0.5772\cdots$ is the Euler-Mascheroni constant (see, e.g., [24], (2.1.16)). The last inequality in the lemma follows from $|-\frac{\zeta'}{\zeta}(\sigma + it)| \le -\frac{\zeta'}{\zeta}(\sigma)$ and

$$-\zeta'(\sigma) = \sum_{n=1}^{\infty} \left(\sum_{m \ge n+1} m^{-\sigma} \right) \log\left(\frac{n+1}{n} \right) < \sum_{n=1}^{\infty} \frac{n^{1-\sigma}}{\sigma - 1} \frac{1}{n} = \frac{\zeta(\sigma)}{\sigma - 1}. \quad \square$$

Lemma 3.2. *For real u,*

$$\left| \frac{\zeta'(-\frac{1}{2} + iu)}{\zeta(-\frac{1}{2} + iu)} \right| \le 4.62 + \frac{1}{2}\log(1 + u^2/9).$$

Proof. By the functional equation for $\zeta(s)$ (cf. [6], Ch. 9, (8)–(10)),

$$-\frac{\zeta'(w)}{\zeta(w)} = \frac{\zeta'(1-w)}{\zeta(1-w)} - 2\log\left(\frac{2\pi}{e}\right) - \sum_{n=1}^{\infty}\left(\frac{1}{w+2n} + \frac{1}{1-w+2n} - \frac{1}{n}\right).$$

Writing $w = -\frac{1}{2} + iu$, we obtain

$$\left|\frac{\zeta'(w)}{\zeta(w)}\right| \leq -\frac{\zeta'(3/2)}{\zeta(3/2)} + 1.676 + \sum_{n=1}^{\infty}\left|\frac{u^2 + n - 3/4 + 2iu}{n(4n^2 + 2n - 3/4 + u^2 + 2iu)}\right|$$

$$\leq 4.19 + \sum_{n=2}^{\infty}\frac{u^2 + 2|u| + n}{n(4n^2 + u^2)}$$

$$\leq 4.19 + (u^2 + 2|u|)\int_{3/2}^{\infty}\frac{dx}{x(4x^2 + u^2)} + \frac{1}{4}\sum_{n=2}^{\infty}\frac{1}{n^2}$$

$$= 4.19 + \frac{1}{4}(\pi^2/6 - 1) + (1/2 + 1/|u|)\log(1 + u^2/9)$$

$$\leq 4.62 + \frac{1}{2}\log(1 + u^2/9). \qquad \square$$

Lemma 3.3. *We have*

$$\sum_{\rho}\frac{1}{|\rho|^2} \leq 0.0463,$$

where the sum is over all of the non-trivial zeros of $\zeta(s)$.

Proof. By ([6], Ch. 9, (10) an (11)), we have

$$\sum_{\rho}\frac{\Re\rho}{|\rho|^2} = 1 + \frac{1}{2}\gamma - \frac{1}{2}\log(4\pi).$$

If $\zeta(\rho) = 0$ then $\zeta(1-\rho) = 0$, and the minimum of $|\Im\rho|$ is > 14.1. Thus

$$\sum_{\rho}\frac{1}{|\rho|^2} = \sum_{\rho}\left(\frac{\Re\rho}{|\rho|^2} + \frac{\Re\rho}{|1-\rho|^2}\right)$$

$$\leq \left(1 + \sqrt{1 + 1/14.1^2}\right)\sum_{\rho}\frac{\Re\rho}{|\rho|^2}$$

$$\leq 0.0463. \qquad \square$$

Lemma 3.4. *Let us fix $\sigma \in [\frac{1}{2}, 1)$, and suppose for all $t \geq 3$ we have*

$$|\zeta(\sigma + iy)| \leq Xt^Y(\log t)^Z \qquad (1 \leq |y| \leq t), \qquad (3.1)$$

where X, Y and Z are positive constants with $Y + Z \geq 0.1$. If $0 < a \leq \frac{1}{2}$, $t \geq 100$ and $\frac{1}{2} \leq \sigma \leq 1 - 1/t$, then

$$\int_{-\infty}^{\infty} \frac{\log|\zeta(\sigma + it + iau)|}{\cosh^2 u} \, du \leq 2(\log X + Y \log t + Z \log\log t).$$

Proof. First, there is no difficulty if $\zeta(\sigma + it + iau) = 0$ for some points along the path of integration. Since all zeros have finite order, the integral in the lemma always converges. When $-\frac{2t}{a} \leq u \leq \frac{-t-1}{a}$, (3.1) gives $|\zeta(\sigma + it + iau)| \leq Xt^Y(\log t)^Z$. For $\frac{-t-1}{a} \leq u \leq \frac{-t+3}{a}$, we use the identity ([24], (2.1.4))

$$\zeta(s) = \frac{1}{s-1} + \frac{1}{2} + s \int_1^\infty \frac{\lfloor x \rfloor - x + 1/2}{x^{s+1}} \, dx.$$

Writing $s = \sigma + it + iau$, it follows that $|s - 1| \geq 1/t$ and $|s| \leq \sqrt{10}$ and thus $\log|\zeta(s)| \leq \log(t + 4)$ for this range of u. For $u \geq \frac{3-t}{a}$, we use the inequalities $\log(1 + x) \leq x$ and $\log(1 + x) \leq x - \frac{1}{2}x^2 + \frac{1}{3}x^3$, both valid for all $x > -1$. Then

$$\log|\zeta(\sigma + it + iau)| \leq \log X + Y \log(t + au) + Z \log\log(t + au)$$
$$\leq \log(Xt^Y(\log t)^Z)$$
$$+ \left(Y + \frac{Z}{\log t}\right)\left(\frac{au}{t} - \frac{(au)^2}{2t^2} + \frac{(au)^3}{3t^3}\right).$$

Similarly, using $\log(1 + x) \leq x$, for $u \leq -\frac{2t}{a}$

$$\log|\zeta(\sigma + it + iau)| \leq \log(Xt^Y(\log t)^Z) + \left(Y + \frac{Z}{\log t}\right)\left(\frac{-au - 2t}{t}\right).$$

Combining these estimates together with $\int_{-\infty}^\infty (\cosh u)^{-2} \, du = 2$ yields

$$\int_{-\infty}^\infty \frac{\log|\zeta(\sigma + it + iau)|}{\cosh^2 u} \, du \leq 2(\log X + Y \log t + Z \log\log t) + E,$$

where

$$E = \frac{4\log(t + 4)}{a \cosh^2\left(\frac{3-t}{a}\right)}$$
$$+ \left(Y + \frac{Z}{\log t}\right)\left(\int_{-\infty}^{-\frac{2t}{a}} \frac{-au - 2t}{t \cosh^2 u} \, du + \int_{\frac{3-t}{a}}^\infty \frac{\frac{au}{t} - \frac{(au)^2}{2t^2} + \frac{(au)^3}{3t^3}}{\cosh^2 u} \, du\right).$$

Now $\frac{1}{4}e^{2|u|} \leq \cosh^2 u \leq e^{2|u|}$, $a \leq \frac{1}{2}$ and $t \geq 100$. Hence

$$ae^{(t-6)/a} \geq 2e^{2t-12}.$$

Therefore

$$E \leq \frac{16 \log(t+4)}{a e^{2(t-3)/a}}$$
$$+ \left(Y + \frac{Z}{\log t}\right) \left(\frac{4a}{t} e^{-4t/a} \int_0^\infty v e^{-2v} \, dv\right.$$
$$+ \int_{-\infty}^\infty \frac{\frac{au}{t} - \frac{(au)^2}{2t^2} + \frac{(au)^3}{3t^3}}{\cosh^2 u} \, du + \left. \int_{\frac{t-3}{a}}^\infty \frac{\frac{au}{t} + \frac{(au)^2}{2t^2} + \frac{(au)^3}{3t^3}}{\frac{1}{4} e^{2u}} \, du \right)$$
$$\leq \frac{32 \log(t+4)}{e^{t/a+2t-12}} + \left(Y + \frac{Z}{\log t}\right) \left(\frac{e^{-4t/a}}{t} - \frac{\pi^2 a^2}{12 t^2} + \frac{8a^3}{t^3} \int_{2t-6}^\infty u^3 e^{-2u} \, du\right)$$
$$\leq e^{-t/a} + \left(Y + \frac{Z}{\log t}\right) \left(e^{-4t/a} - \frac{\pi^2}{12} \frac{a^2}{t^2} + 48 a^3 e^{-4t+12}\right)$$
$$\leq e^{-t/a} - \frac{0.1}{\log t} \frac{a^2}{2t^2}$$
$$\leq 0. \qquad \qquad \square$$

4 Detecting Zeros of $\zeta(s)$

From now on, ρ will denote a zero of $\zeta(s)$ and in summations over the zeros, each zero is counted according to its multiplicity. Since $\zeta(s) = \overline{\zeta(\bar{s})}$, when proving zero-free regions we restrict our attention to the upper half plane.

Lemma 4.1. *Suppose* (1.2) *holds. Let* $s = \sigma + it$, $\eta > 0$, $\sigma - \eta \geq 1/2$, $1 \leq \sigma \leq 1 + \eta$ *and* $t \geq 100$. *If* S *is any subset of* $\{z : \sigma - \eta \leq \Re z \leq 1\}$, *then*

$$-\Re \frac{\zeta'(s)}{\zeta(s)} \leq - \sum_{\rho \in S, \zeta(\rho)=0} \Re \frac{\pi}{2\eta} \cot\left(\frac{\pi(s-\rho)}{2\eta}\right)$$
$$+ \frac{1}{2\eta} \left(\frac{2}{3} \log\log t + B(1 - \sigma + \eta)^{3/2} \log t + \log A\right)$$
$$- \frac{1}{4\eta} \int_{-\infty}^\infty \frac{\log |\zeta(s + \eta + 2\eta i u/\pi)|}{\cosh^2 u} \, du.$$

Proof. We apply Lemma 2.2 with $f = \zeta$ and $z_0 = s$, noting that $\zeta(0) \neq 0$, all zeros have real part < 1 and that $\Re \cot z \geq 0$ for $0 \leq \Re z \leq \frac{\pi}{2}$. Thus the right side in the conclusion of Lemma 2.2 is increased if we omit from the sum any subset of the zeros. Then we apply (1.2) and Lemma 3.4 (with $X = A$, $Y = B(1 - \sigma + \eta)^{3/2}$, $Z = 2/3$, $a = 2\eta/\pi$) to the integral over the line $\Re z = \sigma - \eta$. Note also that the integral on the right side in Lemma 4.1 always converges by Lemma 3.1. Therefore, if η is "bad" with respect to

Lemma 2.2, we can apply the above argument with a sequence of numbers η' tending to η from above.   □

We next require an upper bound on the number of zeros close to a point $1 + it$. Here $N(t, R)$ denotes the number of zeros ρ with $|1 + it - \rho| \leq R$.

Lemma 4.2. *Assume* (1.2) *holds with* $A > 1$ *and* $B > 0$. *Then, for* $0 < R \leq 1/4$, $t \geq 100$,

$$N(t, R) \leq 1.3478R^{3/2}B \log t + 0.49 + \frac{\log A - \log R + \frac{2}{3}\log\log t}{1.879}.$$

Proof. Apply Lemma 4.1 with $s = 1 + 0.6421R + it$, $\eta = 2.5R$ (so that $\sigma - \eta \geq \frac{1}{2}$) and $S = \{z : |1 + it - z| \leq R, \Re z \leq 1\}$. These parameters were chosen to minimize the first term on the right side of the inequality in the lemma. By Lemma 3.1, if v is real then

$$\left|\frac{\zeta'}{\zeta}(s + \eta + iv)\right| \leq \frac{1}{3.1421R},$$

$$|\zeta(s + \eta + iv)|^{-1} \leq \zeta(1 + 3.1421R) \leq 0.6 + \frac{1}{3.1421R}. \tag{4.1}$$

Next, in the region $U = \{z : \Re z \geq 0.6421, |z - 0.6421| \leq 1\}$, we prove

$$\Re \frac{\pi}{5} \cot\left(\frac{\pi z}{5}\right) \geq 0.3758. \tag{4.2}$$

By the maximum modulus principle, it suffices to prove (4.2) on the boundary of U. Using

$$\Re \cot(x + iy) = \frac{2\sin(2x)}{e^{2y} + e^{-2y} - 2\cos(2x)},$$

the minimum of $\Re \cot(x + iy)$ on the vertical segment $x = 0.6421\pi/5$, $|y| \leq \pi/5$ occurs at the endpoints. On the semicircular part of the boundary of U, we verified (4.2) by a short computation using the computer algebra package Maple. In particular, the relative minima on the boundary of U occur at $z = 1.6421$ and $z = 0.6421 \pm i$. Therefore, by (4.1), (4.2) and Lemma 4.1,

$$-\frac{1}{3.1421R} \leq -0.3758\frac{N(t, R)}{R}$$

$$+ \frac{1}{5R}\left(\frac{2}{3}\log\log t + (1.8579R)^{3/2}B \log t\right.$$

$$\left. + \log A + \log\left(0.6 + \frac{1}{3.1421R}\right)\right).$$

Since $\log(0.6 + \frac{1}{3.1421R}) \leq -\log(3.1421R) + 1.88526R \leq -\log R - 0.6735$, the lemma follows.   □

Remark. A qualitatively similar result may also be proved, in a similar way, from Lemma 2 of [HB1], or from Landau's lemma (§3.9 of [24]).

Lemma 4.3. *Suppose* $t \geq 10000$, $0 < v \leq 1/4$, *and* (1.2) *holds with* $A > 1$, $B > 0$. *Then*

$$\sum_{|1+it-\rho|\geq v} \frac{1}{|1+it-\rho|^2} \leq (6.132 + 5.392B(v^{-1/2} - 2))\log t + 13.5$$

$$- 8.5\log A + 4\log\log t + \frac{\frac{\log A - \log v + \frac{2}{3}\log\log t}{1.879} + 0.224 - N(t,v)}{v^2}.$$

Proof. Divide the zeros with $|1 + it - \rho| \geq v$ into three sets:

$$Z_1 = \{\rho : |\Im\rho - t| \geq 1\},$$

$$Z_2 = \{\rho \notin Z_1 : |1 + it - \rho| \geq \frac{1}{4} \text{ and } |it - \rho| \geq \frac{1}{4}\},$$

$$Z_3 = \{\rho : \rho \notin Z_2, \rho \notin Z_1 \text{ and } |1 + it - \rho| \geq v\}.$$

For $i = 1, 2, 3$, let S_i be the sum over $\rho \in Z_i$ of $|1 + it - \rho|^{-2}$. By Theorem 19 of [19], the number, $N(T)$, of nontrivial zeros of $\zeta(s)$ with imaginary part in $[0, T]$ satisfies

$$N(T) = \frac{T}{2\pi}\log\frac{T}{2\pi} - \frac{T}{2\pi} + \frac{7}{8} + Q(T), \qquad (4.3)$$

where

$$|Q(T)| \leq 0.137\log T + 0.443\log\log T + 1.588 \qquad (T \geq 2).$$

Since there are no zeros ρ with $|\Im\rho| \leq 14$,

$$S_1 \leq \int_{t+1}^{\infty}\frac{dN(u)}{(u-t)^2} + \int_{14}^{t-1}\frac{dN(u)}{(t-u)^2} + \int_{14}^{\infty}\frac{dN(u)}{(u+t)^2} = I_1 + I_2 + I_3.$$

Since $dN(u) = \frac{1}{2\pi}\log\frac{u}{2\pi} + dQ(u)$, $\log(t+x) \leq \log t + \frac{x}{t}$ and $\log\log(t+x) \leq \log\log t + \frac{x}{t\log t}$, we have

$$I_1 \leq \frac{1}{2\pi}\int_1^{\infty}\frac{\log(t+x) - \log 2\pi}{x^2}\,dx + |Q(t+1)| + 2\int_1^{\infty}\frac{|Q(t+x)|}{x^3}\,dx$$

$$= \frac{(1+\frac{1}{t})\log(1+t) - \log(2\pi)}{2\pi} + |Q(t+1)| + 2\int_1^{\infty}\frac{|Q(t+x)|}{x^3}\,dx$$

$$\leq 0.4332\log t + 0.886\log\log t + 2.884 + 2\int_1^{\infty}\frac{0.1851x/t}{x^3}\,dx$$

$$\leq 0.4332\log t + 0.886\log\log t + 2.885.$$

Similarly, noting that $Q(14) \geq 0$, we get

$$I_2 \leq \frac{1}{2\pi} \log\left(\frac{t}{2\pi}\right) + 2 \max_{14 \leq u \leq t-1} |Q(u)|$$
$$\leq 0.4332 \log t + 0.886 \log\log t + 2.884$$

and

$$I_3 \leq \frac{1}{2\pi} \int_{14}^{\infty} \frac{\log\left(\frac{u+t}{2\pi}\right)}{(u+t)^2} \, du + 2 \int_{14}^{\infty} \frac{|Q(u)|}{(u+t)^3} \, du \leq 0.00014.$$

Thus

$$S_1 \leq 0.8664 \log t + 1.772 \log\log t + 5.77. \tag{4.4}$$

Next let $N_2 = |Z_2|$ and $N_3 = |Z_3|$. By (4.3),

$$N_2 + N_3 = N(t+1) - N(t-1) - N(t,v)$$
$$\leq 0.59231 \log t + 0.886 \log\log t + 2.591 - N(t,v). \tag{4.5}$$

In the sum S_2, each zero on the critical line contributes ≤ 4 and each pair of zeros $\rho = \beta + i\gamma$, $\rho' = 1 - \beta + i\gamma$ with $\beta > 1/2$ contributes at most $4^2 + (4/3)^2$ to the sum. Therefore,

$$S_2 \leq \frac{80N_2}{9}.$$

For S_3, $N(t,1/4)$ of the zeros contribute at most $(4/3)^2$ each, since $N_3 + N(t,v) = 2N(t,1/4)$. By partial summation,

$$S_3 \leq \frac{16N(t,1/4)}{9} + \int_v^{1/4} \frac{dN(t,u)}{u^2}$$
$$= \frac{160}{9} N(t,1/4) - \frac{N(t,v)}{v^2} + 2 \int_v^{1/4} \frac{N(t,u)}{u^3} \, du$$
$$= \frac{80N_3}{9} + \left(\frac{80}{9} - \frac{1}{v^2}\right) N(t,v) + 2 \int_v^{1/4} \frac{N(t,u)}{u^3} \, du.$$

By Lemma 4.2,

$$2 \int_v^{1/4} \frac{N(t,u)}{u^3} \, du \leq \left(\frac{\log A + \frac{2}{3}\log\log t}{1.879} + 0.49\right)\left(v^{-2} - 16\right)$$
$$+ 5.3912 B\left(v^{-1/2} - 2\right) \log t$$
$$+ \frac{1}{1.879}\left(8 - 16\log 4 - \frac{1 + 2\log v}{2v^2}\right).$$

Therefore, using (4.5), we obtain

$$S_2 + S_3 \leq (5.2650 + 5.3912 B(v^{-1/2} - 2)) \log t + 2.2 \log \log t - 8.5 \log A$$
$$+ 7.65 + \frac{1}{v^2} \left(\frac{\log A - \log v + \frac{2}{3} \log \log t}{1.879} + 0.224 \right) - \frac{N(t, v)}{v^2}.$$

Combining this with (4.4) gives the lemma. $\qquad \square$

Lemma 4.4. *Suppose that* $\Re z \geq 0$ *and* $|z| \leq \pi/2$. *Then*

$$\Re \left(\cot z - \frac{1}{z} + \frac{4z}{\pi^2} \right) \geq 0.$$

Proof. By the maximum modulus principle it suffices to prove the inequality on the boundary of the region. On the vertical segment $z = iy$, $-\pi/2 \leq y \leq \pi/2$, the left side is zero. When $|z| = \pi/2$, $z = x + iy$ and $x \geq 0$, the left side is

$$\frac{2 \sin(2x)}{e^{2y} + e^{-2y} - 2\cos(2x)} - \frac{x}{x^2 + y^2} + \frac{4x}{\pi^2} = \frac{2 \sin(2x)}{e^{2y} + e^{-2y} - 2\cos(2x)} \geq 0.$$

This proves the lemma. $\qquad \square$

The next two lemmas are related to Heath-Brown's method for detecting zeros from [HB2]. These give bounds for a "mollified" sum, similar to Lemmas 5.1 and 5.2 of [HB2].

Lemma 4.5. *Suppose* f *is a non-negative real function which has continuous derivative on* $(0, \infty)$. *Suppose the Laplace transform*

$$F(z) = \int_0^\infty f(y) e^{-zy} \, dy$$

of f *is absolutely convergent for* $\Re z > 0$. *Let* $F_0(z) = F(z) - f(0)/z$ *and suppose*

$$|F_0(z)| \leq \frac{D}{|z|^2} \qquad (\Re z \geq 0, |z| \geq \eta), \tag{4.6}$$

where $0 < \eta \leq \frac{3}{2}$. *If* $\Re s > 1$ *and* $\Im s \geq 0$, *then*

$$K(s) := \sum_{n=1}^\infty \Lambda(n) n^{-s} f(\log n)$$
$$= -f(0) \frac{\zeta'(s)}{\zeta(s)} - \sum_\rho F_0(s - \rho) + F_0(s - 1) + E,$$

where $|E| \leq D(1.72 + \frac{1}{3} \log(1 + \Im s))$.

Proof. We follow the proof of Lemma 5.1 of [HB2]. Suppose $s = \sigma + it$ and $1 < \alpha < \sigma$. Define

$$I = \frac{1}{2\pi i} \int_{\alpha - i\infty}^{\alpha + i\infty} -\frac{\zeta'(w)}{\zeta(w)} F_0(s - w) \, dw.$$

Since $-\zeta'(w)/\zeta(w) = \sum_n \Lambda(n) n^{-w}$, the sum converging uniformly on $\Re w = \alpha$, we may integrate term by term. Thus $I = \sum_n \Lambda(n) J_n$, where

$$J_n = \frac{1}{2\pi i} \int_{\alpha - i\infty}^{\alpha + i\infty} n^{-w} F_0(s - w) \, dw = \frac{n^{-s}}{2\pi i} \int_{\sigma - \alpha - i\infty}^{\sigma - \alpha + i\infty} n^u F_0(u) \, du.$$

The integral on the right converges absolutely by (4.6). Since

$$F_0(z) = \frac{1}{z} \int_0^\infty e^{-zy} f'(y) \, dy,$$

we have

$$\begin{aligned} J_n &= \frac{n^{-s}}{2\pi i} \int_0^\infty f'(y) \int_{\sigma - \alpha - i\infty}^{\sigma - \alpha + i\infty} \frac{(ne^{-y})^u}{u} \, du \, dy \\ &= n^{-s} \int_0^{\log n} f'(y) \, dy = n^{-s} \left(f(\log n) - f(0) \right). \end{aligned}$$

Thus

$$I = K(s) + f(0) \frac{\zeta'(s)}{\zeta(s)}. \tag{4.7}$$

Moving the line of integration to $\Re w = -1/2$, we have

$$I = \frac{1}{2\pi i} \int_{-1/2 - i\infty}^{-1/2 + i\infty} -\frac{\zeta'(w)}{\zeta(w)} F_0(s - w) \, dw - \sum_\rho F_0(s - \rho) + F_0(s - 1). \tag{4.8}$$

By (4.6) and Lemma 3.2, the integral in (4.8) is $\leq \frac{D}{2\pi} I'$, where

$$\begin{aligned} I' &\leq \int_{-\infty}^\infty \frac{4.62 + \frac{1}{2} \log(1 + u^2/9)}{9/4 + (u - t)^2} \, du \\ &= 3.08\pi + \frac{1}{3} \int_{-\infty}^\infty \frac{\log(1 + (t/3 + v/2)^2)}{1 + v^2} \, dv \\ &\leq 3.08\pi + \frac{1}{3} \int_{-\infty}^\infty \frac{\log(1 + t^2) + \log(1 + v^2)}{1 + v^2} \, dv \\ &\leq 10.8 + \frac{2\pi \log(1 + t)}{3}. \end{aligned}$$

The lemma now follows from (4.7) and (4.8). $\square$

Remarks. Examples of functions f satisfying the conditions of Lemma 4.5 are those with compact support (say $[0, x_0]$) and with f'' continuous and bounded on $(0, x_0)$. These are the functions considered in [HB2]. To see that (4.6) holds, apply integration by parts twice, noting that $f(x_0) = f'(x_0) = 0$. This gives

$$F_0(z) = z^{-2} \left(f'(0^+) + \int_0^{x_0} e^{-zt} f''(t) \, dt \right).$$

Lemma 4.6. *Suppose* $0 < \eta \leq \frac{1}{2}$ *and* (1.2) *holds with* $A > 1$, $B > 0$. *Let* f *have compact support and satisfy* (4.6). *Suppose* $s = 1 + it$ *with* $t \geq 1000$. *Then*

$$\Re K(s) \leq - \sum_{|1 + it - \rho| \leq \eta} \Re \left\{ F(s - \rho) + f(0) \left(\frac{\pi}{2\eta} \cot \left(\frac{\pi(s - \rho)}{2\eta} \right) - \frac{1}{s - \rho} \right) \right\}$$

$$+ \frac{f(0)}{2\eta} \left[\frac{2 \log \log t}{3} + B \eta^{3/2} \log t + \log A \right.$$

$$\left. - \frac{1}{2} \int_{-\infty}^{\infty} \frac{\log |\zeta(s + \eta + \frac{2\eta u i}{\pi})|}{\cosh^2 u} \, du \right]$$

$$+ D \left(1.8 + \frac{\log t}{3} + \sum_{|1 + it - \rho| \geq \eta} \frac{1}{|1 + it - \rho|^2} \right).$$

In addition,

$$K(1) \leq F(0) + 1.8D.$$

Proof. Suppose that $\sigma > 1$. By Lemma 4.5,

$$K(\sigma) \leq -f(0) \frac{\zeta'(\sigma)}{\zeta(\sigma)} + F_0(\sigma - 1) + 1.72D + D \sum_\rho \frac{1}{|1 - \rho|^2}.$$

Since $\zeta(\rho) = 0$ implies $\zeta(1 - \rho) = 0$, we may replace $|1 - \rho|^2$ by $|\rho|^2$ in the last sum. Using Lemmas 3.1 and 3.3, we obtain

$$K(\sigma) \leq \frac{f(0)}{\sigma - 1} + F_0(\sigma - 1) + 1.8D$$

$$= F(\sigma - 1) + 1.8D. \tag{4.9}$$

When $t \geq 1000$ and $s = \sigma + it$, $\Re F_0(s - 1) \leq |F_0(s - 1)| \leq Dt^{-1} \leq 0.001D$. Also by (4.6),

$$\sum_{|1 + it - \rho| > \eta} |F_0(s - \rho)| \leq D \sum_{|1 + it - \rho| > \eta} \frac{1}{|1 + it - \rho|^2}.$$

Therefore, combining Lemma 4.1 (with $S = \{z : \Re z \leq 1, |1 + it - z| \leq \eta\}$) and Lemma 4.5 gives

$$\Re K(s) \leq -\sum_{|1+it-\rho|\leq\eta} \Re\left\{F(s-\rho) + f(0)\left(\frac{\pi}{2\eta}\cot\left(\frac{\pi(s-\rho)}{2\eta}\right) - \frac{1}{s-\rho}\right)\right\}$$

$$+ \frac{f(0)}{2\eta}\left[\frac{2}{3}\log\log t + B\eta^{3/2}\log t + \log A\right.$$

$$\left. -\frac{1}{2}\int_{-\infty}^{\infty}\frac{\log|\zeta(s+\eta+\frac{2\eta u i}{\pi})|}{\cosh^2 u}\,du\right]$$

$$+ D\left(1.8 + \frac{\log t}{3} + \sum_{|1+it-\rho|>\eta}\frac{1}{|1+it-\rho|^2}\right).$$

$$(4.10)$$

Since f has compact support, $K(s)$ and $F(s)$ are both entire functions. Also, on the right side of (4.10), $|\log|\zeta(\alpha + i\beta)|| \leq |\log\zeta(\alpha)|$ when $\alpha > 1$ (by Lemma 3.1). Thus we may let $\sigma \to 1^+$ in (4.9) and (4.10), and this proves the lemma. □

5   A Trigonometric Inequality

We use a trigonometric inequality that is very similar to what is used in standard treatments. For any real numbers a_1, a_2 we have

$$\sum_{j=0}^{4} b_j\cos(j\theta) = 8(\cos\theta + a_1)^2(\cos\theta + a_2)^2 \geq 0 \quad (\theta \in \mathbb{R}), \qquad (5.1)$$

where

$$b_4 = 1, \quad b_3 = 4(a_1 + a_2), \quad b_2 = 4(1 + a_1^2 + a_2^2 + 4a_1a_2),$$
$$b_1 = (a_1 + a_2)(12 + 16a_1a_2), \quad b_0 = b_2 - 1 + 8(a_1a_2)^2. \qquad (5.2)$$

Lemma 5.1. *Suppose a_1, a_2 are real numbers and define $b_0, \ldots, b_4$ by (5.2). Suppose that $\eta > 0$ and t_1, t_2 are real numbers. Then*

$$\int_{-\infty}^{\infty}\frac{1}{\cosh^2 u}\sum_{j=1}^{4} b_j\log|\zeta(1 + \eta + ijt_1 + iut_2)|\,du \geq -2b_0\log\zeta(1 + \eta).$$

Remark. Lemma 5.1 marks a departure from other treatments, where the bound $|\zeta(1+\eta+iw)| \geq \zeta(1+\eta)^{-1}$ is used at the outset (in the context of a different integral), which in our situation gives

$$I \geq -2(b_1 + \cdots + b_4) \log \zeta(1 + \eta).$$

The new idea is to combine the $\log |\zeta(\cdot)|$ terms using (5.1) to significantly reduce this part of the estimation. The idea in Lemma 6.1 accounts for the majority of the improvement over Heath-Brown's zero-free region. See also the remarks at the end of section 8.

Proof. Denote by I the integral in the lemma. We begin with the Euler product representation for $\zeta(s)$ in the form

$$\log |\zeta(s)| = -\Re \sum_p \log(1 - p^{-s}) = \Re \sum_{\substack{p \\ m \geq 1}} \frac{1}{m} p^{-ms} \quad (\Re s > 1). \tag{5.3}$$

Next, if $y \neq 0$,

$$U(y) := \int_{-\infty}^{\infty} \frac{e^{iyu}}{\cosh^2 u} \, du = \frac{\pi y}{\sinh(\pi y/2)} \geq 0, \tag{5.4}$$

which can be proved by contour integration. By (5.2), (5.3) and (5.4),

$$I = \sum_{p,m} \frac{1}{m} p^{-m(1+\eta)} \Re \left(\sum_{j=1}^{4} b_j p^{-ijmt_1} \int_{-\infty}^{\infty} \frac{p^{-imut_2}}{\cosh^2 u} \, du \right)$$

$$= \sum_{p,m} \frac{1}{m} p^{-m(1+\eta)} U(mt_2 \log p) \sum_{j=1}^{4} b_j \cos(jmt \log p)$$

$$\geq -b_0 \sum_{p,m} \frac{1}{m} p^{-m(1+\eta)} U(mt_2 \log p).$$

Since $U(y) \leq 2$ for all y, we obtain $I \geq -2b_0 \log \zeta(1 + \eta)$, as claimed. $\quad \square$

6   The Functions f, F and K

Suppose that $t \geq 10000$, $\zeta(\beta+it) = 0$ and λ is a number with $0 < \lambda \leq 1-\beta$ such that

$$\zeta(s) \neq 0 \quad (1 - \lambda < \Re s \leq 1, t - 1 \leq \Im s \leq 4t + 1). \tag{6.1}$$

Let f be a function with compact support, define F, F_0 and K as in Lemma 4.5, and assume that (4.6) holds. Let a_1, a_2 be real numbers and define $b_0, \ldots, b_4$ by (5.2). Put $b_5 = b_1 + b_2 + b_3 + b_4$. By (5.1),

$$\Re \sum_{j=0}^{4} b_j K(1 + ijt) = \sum_{n=1}^{\infty} \Lambda(n) n^{-1} f(\log n) \sum_{j=0}^{4} b_j \cos(jt \log n) \geq 0. \quad (6.2)$$

We next apply Lemma 4.6 with $s = 1$ and $s = 1 + ijt$ ($j = 1, 2, 3, 4$). Together with Lemma 5.1 (with $t_2 = \frac{2\eta}{\pi}$) and (6.2), this gives

$$
\begin{aligned}
0 \leq -\Re \sum_{\substack{1 \leq j \leq 4 \\ |1+ijt-\rho| \leq \eta}} b_j & \left(F(1 + ijt - \rho) \right. \\
& + f(0) \left(\frac{\pi}{2\eta} \cot \left(\frac{\pi(1 + ijt - \rho)}{2\eta} \right) - \frac{1}{1 + ijt - \rho} \right) \Bigg) \\
+ \frac{f(0)}{2\eta} & \left[b_5 \left(\frac{2}{3} L_2 + B\eta^{3/2} L_1 + \log A \right) + b_0 \log \zeta(1 + \eta) \right] \\
+ b_0 F(0) & + D \left(b_5 \left(1.8 + \frac{L_1}{3} \right) + 1.8 b_0 \right. \\
& + \sum_{j=1}^{4} b_j \sum_{|1+ijt-\rho| \geq \eta} \frac{1}{|1 + ijt - \rho|^2} \Bigg),
\end{aligned}
$$

$$(6.3)$$

where for brevity we write

$$L_1 = \log(4t + 1), \qquad L_2 = \log\log(4t + 1).$$

We choose a function f which is based on the functions given by Lemma 7.5 of [HB2]. Let θ be the unique solution of

$$\sin^2 \theta = \frac{b_1}{b_0}(1 - \theta \cot \theta), \quad 0 < \theta < \pi/2, \quad (6.4)$$

and define the real function

$$g(u) = \begin{cases} (\cos(u \tan \theta) - \cos \theta) \sec^2 \theta & |u| \leq \frac{\theta}{\tan \theta}, \\ 0 & \text{else.} \end{cases} \quad (6.5)$$

Set $w(u) = g * g(u)$ (the convolution square of g) for $u \geq 0$ and

$$W(z) = \int_0^{\infty} e^{-zu} w(u) \, du.$$

From (6.5) we deduce (cf. Lemma 7.1 of [HB2]) the identities

$$W(0) = 2\sec^2\theta(1 - \theta\cot\theta)^2,$$
$$W(-1) = 2\tan^2\theta + 3 - 3\theta(\tan\theta + \cot\theta),$$
$$w(0) = \sec^2\theta(\theta\tan\theta + 3\theta\cot\theta - 3).$$

(6.6)

Then we take (see (6.1))

$$f(u) = \lambda e^{\lambda u} w(\lambda u) \qquad (u \geq 0)$$

(6.7)

and

$$F(z) = \int_0^\infty e^{-zu} f(u)\, du = W\left(\frac{z}{\lambda} - 1\right).$$

(6.8)

For real y,

$$\Re W(iy) = 2\left(\int_0^\infty w(u)\cos(uy)\, du\right)^2 \geq 0.$$

Since $W(z) \to 0$ uniformly as $|z| \to \infty$ and $\Re z \geq 0$, it follows from the maximum modulus principle (applied to $e^{-W(z)}$) that

$$\Re W(z) \geq 0 \qquad (\Re z \geq 0).$$

(6.9)

7   An Inequality for the Real Part of a Zero

In this section, we take specific values for a_1 and a_2 and prove the following inequality.

Lemma 7.1. *Suppose $t \geq 10000$, $\zeta(\beta + it) = 0$ and (6.1) holds. Suppose further that (1.2) holds with $B > 0$ and $A > 6.5$, and that*

$$1 - \beta \leq \eta/2, \quad 0 < \lambda \leq \min\left(1 - \beta, \frac{1}{250}\eta\right).$$

(7.1)

Then

$$\frac{1}{\lambda}\left[0.16521 - 0.1876\left(\frac{1-\beta}{\lambda} - 1\right)\right]$$

$$\leq 1.471\frac{1-\beta}{\eta^2} + \frac{1}{2\eta}\left[\frac{666550}{200211}\left(\frac{2}{3}L_2 + B\eta^{3/2}L_1 + \log A\right) + \log\zeta(1+\eta)\right]$$

$$+ 3.683\lambda\Bigg[(6.466 + 5.392B(\eta^{-\frac{1}{2}} - 2))L_1$$

$$+ 4L_2 + \frac{\dfrac{\log(A/\eta) + \frac{2}{3}L_2}{1.879} + 0.224}{\eta^2}\Bigg].$$

Proof. A near optimal choice of parameters is $a_1 = 0.225$, $a_2 = 0.9$. By (5.2),

$$
\begin{aligned}
b_0 &= 10.01055 & b_3 &= 4.5, \\
b_1 &= 17.14500 & b_4 &= 1.0, \\
b_2 &= 10.68250 & b_5 &= 33.3275,
\end{aligned}
$$

and by (6.4) and (6.6),

$$\theta = 1.152214629976363048877\ldots, \quad w(0) = 6.82602968445295450905\ldots.$$

The function $W(z)$ has the explicit formula (found with the aid of Maple)

$$W(z) = \frac{w(0)}{z} + W_0(z), \tag{7.2}$$

where

$$W_0(z) = \frac{c_0\left(c_2((z+1)^2 e^{-2(\theta/\tan\theta)z} + z^2 - 1) + c_1 z + c_3 z^3\right)}{z^2(z^2 + \tan^2\theta)^2} \tag{7.3}$$

and

$$
\begin{aligned}
c_0 &= \frac{1}{\sin\theta\cos^3\theta} = 16.2983216223932350562\ldots \\
c_1 &= (1 + 2(\theta\cos\theta - \sin\theta)\cos\theta)\tan^4\theta = 16.28781036821666331825\ldots \\
c_2 &= \tan^3\theta\sin^2\theta = 9.4813169452950521682\ldots \\
c_3 &= (2 - 5\sin\theta\cos\theta + \theta + 4\theta\cos^2\theta)\tan^2\theta = 10.3924962150333624895\ldots
\end{aligned}
$$

If $R \geq 3$, (7.3) implies

$$|W_0(z)| \leq \frac{H(R)}{|z|^3} \qquad (\Re z \geq -1, |z| \geq R), \tag{7.4}$$

where

$$H(R) = \frac{c_0\left(c_2 \frac{(R+1)^2}{R^3}\left(e^{2\theta/\tan\theta} + 1\right) + \frac{c_1}{R^2} + c_3\right)}{\left(1 - \frac{\tan^2\theta}{R^2}\right)^2}.$$

By (6.7), (6.8) and (7.2),

$$
\begin{aligned}
F_0(z) &= F(z) - \frac{f(0)}{z} = W\left(\frac{z}{\lambda} - 1\right) - \frac{\lambda w(0)}{z} \\
&= W_0\left(\frac{z}{\lambda} - 1\right) + \frac{\lambda f(0)}{z(z - \lambda)}.
\end{aligned}
$$

Suppose $\Re z \geq 0$ and $|z| \geq (R+1)\lambda$. Writing $z' = \frac{z}{\lambda} - 1$, we have $\Re z' \geq -1$ and $|z'| \geq R$. Thus, by (6.7) and (7.4), we obtain

$$|F_0(z)| \leq \frac{H(R)\lambda^3}{|z - \lambda|^3} + \frac{w(0)\lambda^2}{|z(z - \lambda)|} \leq c_4 \frac{\lambda f(0)}{|z|^2},$$

where

$$c_4 = \frac{H(R)(R+1)^2}{R^3 w(0)} + 1 + 1/R. \tag{7.5}$$

Therefore, providing that $\eta \geq (R+1)\lambda$, (4.6) holds with

$$D = c_4 \lambda f(0). \tag{7.6}$$

Next, define

$$V_c(z) = cw(0)\left(\cot z - \frac{1}{z}\right) + W\left(\frac{z}{c} - 1\right).$$

By (6.7) and (6.8),

$$F(1 + ijt - \rho) + f(0)\left(\frac{\pi}{2\eta}\cot\left(\frac{\pi(1 + ijt - \rho)}{2\eta}\right) - \frac{1}{1 + ijt - \rho}\right) = V_c(z),$$

where $z = \frac{\pi}{2\eta}(1 + ijt - \rho)$ and $c = \frac{\pi\lambda}{2\eta}$. In order to bound the first double sum in (6.3) (leaving only the single term corresponding to $\rho = \beta + it$), we prove that for $0 < c \leq \frac{\pi}{2R+2}$,

$$\Re V_c(z) \geq -c_5 c^2 w(0) \qquad \left(\Re z \geq c, |z| \leq \frac{\pi}{2}\right), \tag{7.7}$$

where

$$c_5 = \frac{4}{\pi^2}\left(1 + \frac{(R+1)^2 H(R)}{w(0)R^3}\right) = \frac{4}{\pi^2}(c_4 - 1/R). \tag{7.8}$$

By the maximum modulus principle (applied to $e^{-V_c(z)}$), it suffices to prove (7.7) on the boundary of the region. First consider z satisfying $\Re z = c$, $|z| \leq \pi/2$. By Lemma 4.4 and (6.9),

$$\Re V_c(z) \geq cw(0)\Re\left(\cot z - \frac{1}{z}\right) \geq -\frac{4c^2 w(0)}{\pi^2}.$$

When $|z| = \pi/2$ and $x = \Re z \geq c$, we have $|z/c - 1| \geq R$, so by (7.4),

$|W_0(z/c - 1)| \leq H(R)|z/c - 1|^{-3}$. Thus, by (7.2) and Lemma 4.4,

$$\Re V_c(z) \geq -\frac{4cw(0)x}{\pi^2} + \frac{cw(0)(x-c)}{|z-c|^2} - \frac{H(R)c^3}{|z-c|^3}$$

$$\geq -\frac{4cw(0)x}{\pi^2} + \frac{cw(0)(x-c)}{(\pi/2)^2} - \frac{H(R)c^3}{(\pi/2-c)^3}$$

$$= c^2 w(0)\left(-\frac{4}{\pi^2} - \frac{H(R)c}{w(0)(\pi/2-c)^3}\right).$$

Noting that $c \leq \frac{\pi}{2R+2}$ completes the proof of (7.7). In fact, with more work one can prove that (7.7) holds with $c_5 = \frac{1}{3}$.

By (7.7), we have

$$-\Re \sum_{\substack{1 \leq j \leq 4 \\ |1+ijt-\rho| \leq \eta}} b_j \Bigg(F(1 + ijt - \rho)$$

$$+ f(0)\left(\frac{\pi}{2\eta}\cot\left(\frac{\pi(1+ijt-\rho)}{2\eta}\right) - \frac{1}{1+ijt-\rho}\right)\Bigg)$$

$$\leq -b_1 V_c(\frac{\pi}{2\eta}(1-\beta)) + c_5 c^2 w(0) \sum_{j=1}^{4} b_j N(jt,\eta).$$

Combining this last estimate with (6.3), (6.7), (7.6) and Lemma 4.3 gives

$$0 \leq b_0 F(0) - b_1 V_c\left(\frac{\pi}{2\eta}(1-\beta)\right) + \frac{\lambda f(0)}{\eta^2}\left(\frac{\pi^2 c_5}{4} - c_4\right) \sum_{j=1}^{4} b_j N(jt,\eta)$$

$$+ \frac{f(0)}{2\eta}\left[b_5\left(\frac{2}{3}L_2 + B\eta^{3/2}L_1 + \log A\right) + b_0 \log \zeta(1+\eta)\right]$$

$$+ c_4 \lambda f(0) b_5\left[1.8 + \frac{L_1}{3} + 1.8\frac{b_0}{b_5} + \left(6.132 + 5.392B(\eta^{-\frac{1}{2}} - 2)\right)L_1\right.$$

$$\left. + 13.5 - 8.5\log A + 4L_2 + \frac{1}{\eta^2}\left(\frac{\log A - \log \eta + \frac{2}{3}L_2}{1.879} + 0.224\right)\right].$$

$$(7.9)$$

The sum in (7.9) can be ignored because of (7.8). Also, by the lower bound on A we have

$$1.8 + 1.8\frac{b_0}{b_5} + 13.5 - 8.5\log A < 0. \qquad (7.10)$$

Put $R = 249$, and compute $H(249) \leq 171.8$ and $c_4 \leq 1.106$. Since $\cot x - \frac{1}{x} \geq -0.348x$ for $0 < x \leq \frac{\pi}{4}$ and $1 - \beta \leq \frac{1}{2}\eta$, we have

$$V_c\left(\frac{\pi}{2\eta}(1-\beta)\right) \geq F(1-\beta) - 0.348 f(0)\frac{\pi^2}{4\eta^2}(1-\beta). \qquad (7.11)$$

By (6.6), (6.7) and (6.8),

$$-\frac{b_1}{b_0}F(1-\beta)+F(0)$$

$$= -\left(\frac{b_1}{b_0}W\left(\frac{1-\beta}{\lambda}-1\right)-W(-1)\right)$$

$$= -\left(\frac{b_1}{b_0}W(0)-W(-1)\right)+\frac{b_1}{b_0}\left(W(0)-W\left(\frac{1-\beta}{\lambda}-1\right)\right)$$

$$= \frac{-f(0)\cos^2\theta}{\lambda}+\frac{b_1}{b_0}\left(W(0)-W\left(\frac{1-\beta}{\lambda}-1\right)\right).$$

$$(7.12)$$

Since $W(x)$ and $W'(x)$ are both decreasing, we have

$$W(0)-W\left(\frac{1-\beta}{\lambda}-1\right)\leq\left(\frac{1-\beta}{\lambda}-1\right)W'(0)\leq 0.7475\left(\frac{1-\beta}{\lambda}-1\right).$$

Thus, by (7.11) and (7.12),

$$F(0)-\frac{b_1}{b_0}V_c\left(\frac{\pi}{2\eta}(1-\beta)\right)\leq 0.348f(0)\frac{\pi^2}{4\eta^2}\frac{b_1}{b_0}(1-\beta)$$

$$+\frac{f(0)}{\lambda}\left(-\cos^2\theta+\frac{0.7475b_1}{b_0w(0)}\left(\frac{1-\beta}{\lambda}-1\right)\right).$$

$$(7.13)$$

Dividing both sides of (7.9) by $b_0f(0)$ and using (7.10), (7.13) and the numerical values of b_0, b_1, b_5 and θ completes the proof of the lemma. □

8   The Proof of Theorem 2

Suppose T_0 satisfies the hypotheses of Theorem 2 and let

$$M = \inf_{\substack{\zeta(\beta+it)=0 \\ t\geq T_0}} Z(\beta,t), \quad Z(\beta,t) := (1-\beta)(B\log t)^{\frac{2}{3}}(\log\log t)^{\frac{1}{3}}. \quad (8.1)$$

By the Korobov-Vinogradov theorem, $M > 0$. If $M \geq M_1$, then the theorem is immediate. Otherwise, suppose that $M < M_1 \leq 0.05507$. Then there is a zero $\beta+it$ of $\zeta(s)$ with $t \geq T_0$ and

$$Z(\beta,t) \in [M, M(1+\delta)], \quad \delta = \min\left(\frac{10^{-100}}{\log T_0}, \frac{M_1-M}{2M}\right).$$

By (8.1), (6.1) holds with

$$\lambda = ML_1^{-2/3}L_2^{-1/3}B^{-2/3}. \quad (8.2)$$

Again we make the abbreviations $L_1 = \log(4t+1)$, $L_2 = \log\log(4t+1)$. Define b_0, b_5 as in the previous section. We apply Lemma 7.1, taking

$$\eta = EB^{-\frac{2}{3}}\left(\frac{L_2}{L_1}\right)^{\frac{2}{3}}, \quad E = \left(\frac{4(1+b_0/b_5)}{3}\right)^{\frac{2}{3}} = \left(\frac{1733522}{999825}\right)^{\frac{2}{3}}. \quad (8.3)$$

The lower bound $\frac{\log T_0}{\log\log T_0} \geq \frac{1740}{B}$ ensures that $\eta \leq 0.01$ and

$$\lambda \leq 0.5507(BL_1)^{-\frac{2}{3}}L_2^{-\frac{1}{3}} \leq \frac{\eta}{250}.$$

The inequalities $T_0 \geq e^{30000}$ and $M_1 \leq 0.05507$ ensure that the other hypotheses of Lemma 7.1 are met. In addition,

$$\frac{1-\beta}{\lambda} - 1 \leq (1+\delta)\left(\frac{L_1}{\log t}\right)^{\frac{2}{3}}\left(\frac{L_2}{\log\log t}\right)^{\frac{1}{3}} - 1 \leq \frac{0.97}{\log T_0}. \quad (8.4)$$

By Lemma 3.1,

$$\log\zeta(1+\eta) \leq \log(1/\eta + 0.6) \leq \log(1/\eta) + 0.006. \quad (8.5)$$

We now apply Lemma 7.1, using the upper bounds for $(1-\beta)$ and λ on the right side of the conclusion. First, since $-\log\eta \approx \frac{2}{3}L_2$, we have by (8.3),

$$\frac{1}{2\eta}\left[\frac{b_5}{b_0}\left(\frac{2L_2}{3} + B\eta^{\frac{3}{2}}L_1\right) + \frac{2L_2}{3}\right] = \frac{b_5}{b_0}\left(1+\frac{b_0}{b_5}\right)^{\frac{1}{3}}\left(\frac{3B}{4}\right)^{\frac{2}{3}}L_1^{\frac{2}{3}}L_2^{\frac{1}{3}} \quad (8.6)$$
$$\leq 2.99968(BL_1)^{\frac{2}{3}}L_2^{\frac{1}{3}}.$$

This constitutes the main term as $t \to \infty$. Next, since $Z(\beta,t) \leq M_1$ and by the lower bound on T_0,

$$1.471\frac{1-\beta}{\eta^2} \leq 0.039B^{2/3}L_1^{2/3}L_2^{-5/3} \leq 0.0038B^{2/3}\left(\frac{L_1}{L_2}\right)^{2/3}. \quad (8.7)$$

Using (8.5), the remaining part of the second line in the conclusion of Lemma 7.1 is

$$\leq \frac{1}{2\eta}\left[\frac{b_5}{b_0}\log A - \log E + \frac{2}{3}\log(B/L_2) + 0.006\right]$$
$$\leq \frac{B^{\frac{2}{3}}}{2E}\left(\frac{L_1}{L_2}\right)^{\frac{2}{3}}\left[3.3293\log A - 0.3608 + \frac{2}{3}\log(B/L_2)\right] \quad (8.8)$$
$$\leq \left(\frac{BL_1}{L_2}\right)^{\frac{2}{3}}(1.1534\log A - 0.125 + 0.2310\log(B/L_2)).$$

By (8.2), (8.3), and $L_2 \leq 0.00035 L_1$, the third line in the conclusion of Lemma 7.1 is

$$\leq 0.2029 L_1^{-\frac{2}{3}} L_2^{-\frac{1}{3}} B^{-\frac{2}{3}} \left[\left(6.468 + \frac{5.392 B^{\frac{4}{3}}}{\sqrt{E}} \left(\frac{L_1}{L_2} \right)^{\frac{1}{3}} - 10.784 B \right) L_1 \right.$$
$$\left. + \frac{B^{\frac{4}{3}}}{E^2} \left(\frac{L_1}{L_2} \right)^{\frac{4}{3}} \left(\frac{\log A + \frac{4}{3} L_2 + \frac{2}{3} \log(B/L_2) - \log E}{1.879} + 0.224 \right) \right]$$
$$\leq \left(\frac{BL_1}{L_2} \right)^{\frac{2}{3}} \left[0.9798 + \frac{1.313 - 2.188 B}{B^{\frac{4}{3}}} \left(\frac{L_2}{L_1} \right)^{\frac{1}{3}} \right.$$
$$\left. + \frac{0.05185}{L_2} \left(\log A + \frac{2}{3} \log(B/L_2) + 0.05401 \right) \right]$$
$$\leq \left(\frac{BL_1}{L_2} \right)^{\frac{2}{3}} \left[\frac{1.313 - 2.188 B}{B^{4/3}} \left(\frac{L_2}{L_1} \right)^{\frac{1}{3}} + 0.0051 (\log A + \frac{2}{3} \log(\frac{B}{L_2})) + 0.9801 \right]$$
$$(8.9)$$

Combining (8.4)–(8.9) with Lemma 7.1 gives

$$\frac{1}{\lambda} \left(0.16521 - \frac{0.182}{\log T_0} \right) \leq (BL_1)^{\frac{2}{3}} L_2^{\frac{1}{3}} \left(2.99968 + \frac{X(t)}{L_2} \right).$$

By (8.2), this gives

$$M \geq \frac{0.16521 - 0.182/\log T_0}{2.99968 + X(t)/\log\log t} \geq M_1.$$

This concludes the proof of Theorem 2.

Remarks. Compared with the methods in [HB1], there are two improvements evident in (8.6). First, the factor $(3/4)^{2/3} \approx 0.82548$ replaces the factor $2^{-1/3} K_2 \approx 0.843445$ from ([10], p. 197). This improvement comes from integrating over two vertical lines (Lemma 2.2). The second and larger improvement is the factor $(1 + b_0/b_5)^{1/3}$, which is $2^{1/3}$ in the treatment of [10], and comes from combining the $\log |\zeta(\cdot)|$ terms in Lemma 5.1. Together these improve the bounds from [10] by about 17%.

9 The Proof of Theorem 3

Almost everything in Sections 2–6 is identical. In place of (1.2) we use the explicit form of the Van der Corput bound (1.7). We fix $\eta = \frac{1}{2}$, and the proof of Lemma 5.1 gives

$$\int_{-\infty}^{\infty} \frac{\sum_{j=1}^{4} b_j \log|\zeta\left(\frac{3}{2} + ijt + \frac{iu}{\pi}\right)|}{\cosh^2 u} \, du \leq b_0 \sum_{p,m} \frac{1}{m} p^{-\frac{3}{2}m} U\left(\frac{m}{\pi} \log p\right)$$

$$= 2b_0 \sum_{p,m} \frac{\log p}{p^{2m} - p^m} \tag{9.1}$$

$$= 2b_0 \sum_{n=2}^{\infty} \frac{\Lambda(n)}{n^2 - n}$$

$$\leq 1.702 b_0.$$

Let $T_0 = 545000000$ and suppose that $\zeta(\beta + it) = 0$ with $t \geq T_0$ (it is known that all zeros with $|t| < T_0$ have real part $\frac{1}{2}$ [25]). In place of Lemma 3.4 we use

Lemma 9.1. *If $t \geq T_0$, then*

$$I(t) = \int_{-\infty}^{\infty} \frac{\log|\zeta(1/2 + it + iu/\pi)|}{\cosh^2 u} \, du \leq 2J(t),$$

where $J(t)$ is given by (1.8).

Proof. From (1.7), $|\zeta(1/2 + it)| \leq 3t^{1/6} \log t$ for $t \geq 3$, so by Lemma 3.4, $I(t) \leq 2(\frac{1}{6} \log t + \log \log t + \log 3)$. Using the first inequality from (1.7), we have

$$I(y) \leq \int_{-\infty}^{\infty} \frac{\log(57 + 6(t + |u|)^{1/4})}{\cosh^2 u} \, du = 2 \int_{0}^{\infty} \frac{\log(57 + 6(t + u)^{1/4})}{\cosh^2 u} \, du.$$

When $0 \leq u \leq \log t$, the numerator is $\leq \log(6.37306 t^{1/4})$ and when $u > \log t$, the numerator is $\leq \log(6.4(e^u + u)^{1/4}) \leq u$ and the denominator is $\geq \frac{1}{4} e^{2u}$. Therefore,

$$I(t) \leq 2 \log(6.37306 t^{1/4}) + 8 \int_{\log t}^{\infty} u e^{-2u} \, du \leq \frac{\log t}{2} + 3.7042. \qquad \square$$

We make the assumption (6.1) as before and take the same values for a_1, a_2 (so $b_0, \ldots, b_4$, θ, w, f, F, W are the same as in section 7). The only change in (6.3) is that the term $\frac{2}{3} \log \log t + B\eta^{3/2} \log t + \log A$ is replaced by $J(t)$. Next, we follow the proof of Lemma 7.1. Using (4.3) (Rosser's theorem) as in the proof of Lemma 4.3, we obtain for $t \geq 10000$

$$\sum_{|1+ijt-\rho| \geq \frac{1}{2}} \frac{1}{|1 + ijt - \rho|^2}$$

$$\leq 3.2357 \log t + 5.316 \log \log t + 16.134 - 4N(t, 1/2). \tag{9.2}$$

Assume that

$$0 < \lambda \leq 1 - \beta \leq \frac{1}{160}. \tag{9.3}$$

Let $R = \frac{1}{2(1-\beta)} - 1 \geq 79$. By (9.3), $\eta \leq 80\lambda$. As in the proof of (7.5), we deduce that (4.6) holds with

$$D = c_4 \lambda f(0), \quad c_4 = \frac{H(79)(R+1)^2}{R^3 w(0)} + 1 + \frac{1}{R} \leq 1.35. \tag{9.4}$$

Also, (7.7) is replaced by

$$\Re V_c(z) \geq -c_5 c^2 w(0) = -c_5 \pi^2 \lambda f(0) \quad (\Re z \geq c, |z| \leq \pi/2), \tag{9.5}$$

valid for $0 < c \leq \pi(1-\beta)$ with

$$c_5 = \frac{4}{\pi^2} + \frac{\pi(1-\beta)H(79)}{w(0)(\pi/2 - \pi(1-\beta))^2}. \tag{9.6}$$

Analogously to (7.9), the inequalities (9.1), (9.2), (9.4), and (9.5) give

$$\begin{aligned}
0 \leq b_0 F(0) &- b_1 V_{\pi\lambda}(\pi(1-\beta)) + (\pi^2 c_5 - 4c_4)\lambda f(0) \sum_{j=1}^{4} b_j N\left(jt, \frac{1}{2}\right) \\
&+ f(0)\left(b_5 J(4t+1) + 0.851 b_0\right) \\
&+ 1.35\lambda f(0)\left[b_5\left(1.8 + \frac{L_1}{3}\right) + 1.8 b_0 \right. \\
&\qquad \left. + b_5(3.2357 L_1 + 5.316 L_2 + 16.134)\right].
\end{aligned} \tag{9.7}$$

As before we use $L_1 = \log(4t+1)$, $L_2 = \log\log(4t+1)$. By (9.4) and (9.6), $\pi^2 c_5 - 4c_4 = -4/R < 0$, so the sum in (9.7) can be ignored. By (9.3), $\cot x - 1/x \geq -0.3334x$ for $0 < x \leq \pi(1-\beta)$ and this gives

$$V_{\pi\lambda}(\pi(1-\beta)) \geq F(1-\beta) - 0.3334\pi^2(1-\beta)f(0).$$

By an argument similar to that leading to (7.13), we obtain

$$\begin{aligned}
F(0) - \frac{b_1}{b_0}V_{\pi\lambda}(\pi(1-\beta)) &\leq 0.3334\pi^2(1-\beta)\frac{b_1}{b_0}f(0) \\
&+ \frac{f(0)}{\lambda}\left(-\cos^2\theta + \frac{0.7475 b_1}{b_0 w(0)}\left(\frac{1-\beta}{\lambda} - 1\right)\right)
\end{aligned} \tag{9.8}$$

Combining (9.8) with (9.7) gives the following bound.

Lemma 9.2. *Suppose that* $\zeta(\beta + it) = 0$ *with* $t \geq 545000000$ *and* $1 - \beta \leq \frac{1}{160}$. *Let* λ *be a positive number satisfying* (6.1). *Then*

$$\frac{0.16521 - 0.1876(\frac{1-\beta}{\lambda} - 1)}{\lambda} \leq 5.646(1 - \beta) + \frac{b_5}{b_0}J(4t + 1) + 0.851 \quad (9.9)$$

$$+ 1.35\lambda\frac{b_5}{b_0}(3.5691L_1 + 5.316L_2 + 18.475).$$

To prove Theorem 3, first define

$$c_6 = c_6(t) = \frac{1}{J(t) + 1.15}. \quad (9.10)$$

For a zero $\beta + it$ of ζ with $t \geq T_0$, define $Y(\beta, t)$ by the equation

$$1 - \beta = \frac{0.04962 - 0.0196c_6(t)}{J(t) + Y(\beta, t)}.$$

By the Korobov-Vinogradov theorem, $Y(\beta, t) \to -\infty$ as $t \to \infty$. Let $M = \max_{t \geq T_0} Y(\beta, t)$. If $M \leq 1.15$, Theorem 3 follows. Otherwise, suppose $\beta + it$ is a zero with $Y(\beta, t) = M > 1.15$. Then (6.1) holds with

$$\lambda = \frac{0.04962 - 0.0196c_6(t)}{J(4t + 1) + M} \quad (9.11)$$

By (9.10) and (9.11),

$$\frac{1 - \beta}{\lambda} - 1 = \frac{J(4t + 1) + M}{J(t) + M} - 1 = \frac{J(4t + 1) - J(t)}{J(t) + M} \leq \frac{0.3466}{J(t) + M}. \quad (9.12)$$

Apply Lemma 9.2, multiplying both sides of (9.9) by $6b_0/b_5$. By (9.12), the left side is

$$\geq \frac{0.04962 - 0.0196c_6(t)}{\lambda} = J(4t + 1) + M.$$

Using $L_1 \leq \log t + \log 4 + \frac{1}{4t}$ and $L_2 \leq \log\log t + (\log 4 + \frac{1}{4t})/\log t$, we conclude that

$$M \leq 0.25562 + (1 - \beta)[1.696 + 1.35(3.5691L_1 + 5.316L_2 + 18.475)]$$

$$\leq 0.25562 + (1 - \beta)[33.812 + 4.8183\log t + 7.1766\log\log t]. \quad (9.13)$$

Also, by assumption $1 - \beta \leq (0.04962 - 0.0196c_6(t))c_6(t)$. Plugging this into (9.13), and using a short Maple computation, we find that

$$M \leq 0.685 + 0.155\log\log t.$$

In fact, $M \leq 1.7$ for all $t \geq T_0$, but the above bound suffices for our purposes. This completes the proof of Theorem 3.

Thanks. The author thanks D. R. Heath-Brown, D. Kutzarova, O. Ramaré and the referee for helpful suggestions and comments. The author also thanks Y. Cheng and S. W. Graham for preprints of their work.

References

[1] Y. Cheng, *An explicit zero-free region for the Riemann zeta-function*, Rocky Mountain J. Math. **30** (2000), 135–148.

[2] ———, *Explicit estimates on prime numbers*, preprint.

[3] ———, *On primes between consecutive cubes*, preprint.

[4] Y. Cheng and S. W. Graham, *Explicit estimates for the Riemann zeta function*, preprint.

[5] N. G. Chudakov, *On the functions $\zeta(s)$ and $\pi(x)$*, C. R. Acad. Sci. USSR, N.S. **21** (1938), 421–422.

[6] H. Davenport, *Multiplicative number theory*, 2nd ed., Graduate Texts in Mathematics, vol. 74, Springer-Verlag, New York, 1980.

[7] C.-J. de la Vallée Poussin, *Sur la fonction $\zeta(s)$ de Riemann et le nombres des nombres premiers inférieurs à une limite donnée*, Mém. Couronnés et Autres Mém., vol. 59, Publ. Acad. Roy. Sci. des lettres Beaux-Arts Belg., 1899–1900, pp. 1–74.

[8] K. Ford, *Vinogradov's integral and bounds for the Riemannn zeta function*, Proc. London Math. Soc., to appear.

[9] D. R. Heath-Brown, *Zero-free regions for Dirichlet L-functions and the least prime in an arithmetic progression*, Proc. London Math. Soc. **64** (1992), 265–338.

[10] ———, *Zero-free regions of $\zeta(s)$ and $L(s,\chi)$*, Proceedings of the Amalfi conference on analytic number theory (Maiori, 1989) (Salerno, Italy), 1992, pp. 195–200.

[11] M. N. Huxley, *Exponential sums and the Riemann zeta function. IV*, Proc. London Math. Soc. **66** (1993), 1–40.

[12] N. M. Korobov, *Estimates of trigonometric sums and their applications (Russian)*, Uspehi Mat. Nauk **13** (1958), 185–192.

[13] M. Kulas, *Refinement of an estimate for the Hurwitz zeta function in a neighbourhood of the line $\sigma = 1$*, Acta Arith. **89** (1999), 301–309.

[14] J. E. Littlewood, *Researches in the theory of the Riemann ζ-function*, Proc. London Math. Soc. **20** (1922), XXII–XXVIII, Records, Feb. 10, 1921.

[15] J. Pintz, *On the remainder of the prime number formula and the zeros of Riemann's zeta-function*, Number theory, Noordwijkerhout 1983, Lecture Notes in Mathematics, vol. 1068, Springer-Verlag, 1984, pp. 186–197.

[16] O. V. Popov, *A derivation of a modern bound for the zeros of the Riemann zeta function by the Hadamard method (Russian)*, Vestnik Moskov. Univ. Ser. I Mat. Mekh. (1994), 42–45, 96.

[17] O. Ramaré and R. Rumely, *Primes in arithmetic progressions*, Math. Comp. **65** (1996), 397–425.

[18] H.-E. Richert, *Zur Abschätzung der Riemannschen Zetafunktion in der Nähe der Vertikalen $\sigma = 1$*, Math. Ann. **169** (1967), 97–101.

[19] J. B. Rosser, *Explicit bounds for some functions of prime numbers*, Amer. J. Math. **63** (1941), 211–232.

[20] J. B. Rosser and L. Schoenfeld, *Approximate formulas for some functions of prime numbers*, Illinois J. Math. **6** (1962), 64–94.

[21] _____, *Sharper bounds for the Chebyshev functions $\theta(x)$ and $\Psi(x)$*, Math. Comp. **29** (1975), 243–269.

[22] S. Saks and A. Zygmund, *Analytic functions*, 2nd, enlarged ed., Polish Scientific Publishers, Warszawa, 1965, English translation.

[23] S. B. Stechkin, *The zeros of the Riemann zeta-function (Russian)*, Mat. Zametki **8** (1970), 419–429, English translation: Math. Notes **8** (1970), 706–711.

[24] E. C. Titchmarsh, *The theory of the Riemann zeta-function*, 2nd ed., Oxford University Press, 1986.

[25] J. van de Lune, J. J. te Riele, and D. T. Winter, *On the zeros of the Riemann zeta function in the critical strip. IV*, Math. Comp. **46** (1986), 667–681.

[26] I. M. Vinogradov, *A new estimate for $\zeta(1 + it)$ (Russian)*, Izv. Akad. Nauk SSSR, Ser. Mat. **22** (1958), 161–164.

Asymptotic Lower Bounds and Formulas for Diophantine Inequalities

D. Eric Freeman[1]

1   Introduction

In 1946, Davenport and Heilbronn [4] gave an important result which sparked the study of Diophantine inequalities. Suppose that k and s are positive integers with $k \geq 2$ and $s \geq 2^k + 1$, that ϵ is a positive real number and that one is given a diagonal form

$$F(\mathbf{x}) = \lambda_1 x_1^k + \lambda_2 x_2^k + \cdots + \lambda_s x_s^k$$

with nonzero coefficients λ_i which are not all in rational ratio, and which are not all of the same sign if k is even. They essentially showed that in this situation, there must exist a nonzero integral solution $\mathbf{x}$ of the Diophantine inequality

$$|F(\mathbf{x})| < \epsilon. \tag{1.1}$$

Now define $|\mathbf{x}| = \max_{1 \leq i \leq s} |x_i|$ for any real vector $\mathbf{x}$. Then one can ask the following question which is often asked in additive number theory: how many integral vectors $\mathbf{x}$ in the box $|\mathbf{x}| \leq P$ are solutions of the inequality (1.1)? One expects that the number of such solutions is asymptotic to some constant multiple of ϵP^{s-k}. In fact, in the course of their proof, Davenport and Heilbronn showed that the number of such solutions is greater than some constant multiple of P^{s-k}, but only for some sequence of positive real numbers P tending to infinity. In a recent paper, as discussed at the Millennial conference, we have shown that in fact the number of such solutions is greater than a constant multiple of P^{s-k} for all sufficiently large positive real numbers P. We have the following result, proved in [7].

Theorem 1. *There is an absolute real positive constant $\tilde{C}$ with the following property:*

[1]Supported by a National Science Foundation Postdoctoral Fellowship.

Suppose that k is an integer with $k \geq 3$, and that s is a positive integer satisfying $s \geq m_0(k)$, where we define

$$m_0(k) = \min\left(2^k + 1, k(\log k + \log\log k + 3) + \frac{\tilde{C}k\log\log k}{\log k}\right).$$

Suppose also that

$$F(\boldsymbol{x}) = \lambda_1 x_1^k + \lambda_2 x_2^k + \cdots + \lambda_s x_s^k$$

is a diagonal form with nonzero real coefficients $\lambda_1, \ldots, \lambda_s$ whose ratios are not all rational, and such that if k is even, then not all of the coefficients λ_i have the same sign. Suppose as well that ϵ is a positive real number. If P is a positive real number, then the number of solutions of

$$|F(\boldsymbol{x})| < \epsilon$$

where $\boldsymbol{x} = (x_1, \ldots, x_s)$ is an integral vector with

$$|\boldsymbol{x}| \leq P,$$

is

$$\gg P^{s-k}.$$

We note that here the constant in Vinogradov's notation depends at most on s, the coefficients of F, and ϵ.

Theorem 1 is proved in [7].

We observe first of all that one may certainly exclude the case $k = 2$ from our statement of the theorem, as a superior result has been obtained in this case both by S. G. Dani and S. Mozes, and, independently, M. Ratner. They have shown that for a non-degenerate real indefinite quadratic form $Q(\mathbf{x})$, which need not be diagonal, in at least three variables and whose coefficients are not all in rational ratio, an asymptotic lower bound of the expected type does indeed hold. (These results are both unpublished; see the comment on p. 95 of [5].) We also note that if all of these conditions are satisfied, and if additionally Q is such a form in at least four variables, with signature not equal to $(2, 2)$, then A. Eskin, G. Margulis and S. Mozes [5] have shown that an asymptotic formula of the expected type holds.

Our result was proved by adapting some of the very clever recent work of Bentkus and Götze [1]. When attacking Diophantine inequalities, one usually uses the Davenport-Heilbronn method to find an integral over the real line which gives a lower bound for the number of integral solutions $\mathbf{x}$ of an inequality $|F(\mathbf{x})| < \epsilon$ in a box $|\mathbf{x}| \leq P$. When using the Davenport-Heilbronn method, one considers the contribution to the integral from the region of the real line consisting of numbers with small absolute value, which we call the major arc, and generally shows that its contribution is $\gg P^{s-k}$. One usually also shows, at least for certain choices of P, that the contribution to the integral from the remaining regions of the real line is $O\left(P^{s-k-\delta}\right)$, where δ is some small positive number. In this manner, one can often see that there are $\gg P^{s-k}$ solutions for certain choices of P. One of the key ideas of the work of Bentkus and Götze is that one can show in some cases that the contribution of this remaining region is $o\left(P^{s-k}\right)$ for all large P. This is a weaker bound, but the fact that one get such a bound for all large P is extremely helpful: it allows one more flexibility in the use of the Davenport-Heilbronn method. In fact, there is reason to believe, as discussed in [7], that this might be the best possible error term one can obtain using this method. We note that, strictly speaking, Bentkus and Götze do not use the Davenport-Heilbronn method, but their approach is analogous.

We now turn to asymptotic formulas for Diophantine inequalities, referring the reader to [7] for more details on asymptotic lower bounds and the methods used, and for some of the history of related problems. At the Millennial conference, I spoke to Professor Brüdern about the question of finding an asymptotic formula for diagonal Diophantine inequalities. He suggested that one would only need to adjust the kernel functions one uses in the Davenport-Heilbronn method, and combine this with my work on asymptotic lower bounds. By building on this idea, I am able to present the following result.

Theorem 2. *Suppose that k is an integer with $k \geq 3$, and that s is a positive integer satisfying*

$$s \geq 2^k + 1. \tag{1.2}$$

Suppose also that

$$F(\boldsymbol{x}) = \lambda_1 x_1^k + \lambda_2 x_2^k + \cdots + \lambda_s x_s^k$$

is a diagonal form with nonzero real coefficients $\lambda_1, \ldots, \lambda_s$ whose ratios are not all rational, and such that if k is even, then not all of the coefficients λ_i have the same sign. Suppose as well that ϵ is a positive real number. Then for positive real numbers P, the number of solutions $N(P)$ of

$$|F(\boldsymbol{x})| < \epsilon, \tag{1.3}$$

where $\boldsymbol{x} = (x_1, \ldots, x_s)$ is an integral vector with

$$|\boldsymbol{x}| \leq P,$$

satisfies

$$N(P) = C(k, s, \lambda_1, \lambda_2, \ldots, \lambda_s)\epsilon P^{s-k} + o\left(P^{s-k}\right),$$

where $C(k, s, \lambda_1, \lambda_2, \ldots, \lambda_s)$ is a constant which depends at most on k and s and the coefficients λ_i of F, and where the implicit constant in the little "oh" notation depends at most on ϵ and k and s and the coefficients λ_i.

Recall that we have previously mentioned a result of Eskin, Margulis and Mozes which gives the expected asymptotic formula for suitable indefinite quadratic forms. Their result is much stronger than an analogue of Theorem 2 in the case $k = 2$ would be. For this reason and because it simplifies our proofs, we exclude the case $k = 2$ from our result. We also note that one could quite easily decrease the number s of variables that one requires, at least for large k, but for the sake of simplicity of exposition we do not give the details. But we take this opportunity to remark that by using the estimate (5.4) from the work of Ford [6], one can see that the requirement (1.2) can be replaced by the requirement $s \geq s_0(k)$, where $s_0(k)$ is a function which is asymptotic to $k^2 \log k$ for large k.

I note also that I have recently become aware that Bentkus and Götze [2] have announced a result on asymptotic formulas for certain Diophantine inequalities. For example, in their paper they address inequalities for positive definite diagonal forms, but not indefinite diagonal forms. Their results also require significantly larger bounds on the number of variables s required.

We give a proof of Theorem 2 in this paper. We note that the proof follows quite closely our previous work [7]; the present paper is not self-contained, and indeed a reading of it requires a copy of [7]. In fact, the only major differences which are necessary involve the kernel function and the treatment of the major arc.

We make one final observation. The techniques used in this paper could be employed to give asymptotic formulas for certain Diophantine inequalities of the type

$$\left|\lambda_1 x_1^k + \lambda_2 x_2^k + \cdots + \lambda_s x_s^k - M\right| < \epsilon,$$

for suitable real numbers M. Even the case in which k is even and all of the coefficients λ_i are positive can be treated in this way, assuming of course that M is in an interval $c_1 P^k \leq M \leq c_2 P^k$ for suitable choices of constants c_1 and c_2 which would depend on the coefficients λ_i.

I would like to thank Professor Brüdern for his comments in our conversation, which led to this paper. I thank the referee for useful remarks. I would also like to thank the organizers of the Millennial conference for an enjoyable meeting.

2 The Asymptotic Formula

2.1 The Setup

We now give a proof of Theorem 2. For the most part, we follow the method of proof in our earlier paper [7], making some adjustments along the way, so we do not always give all of the relevant details. We note that throughout the proof, implicit constants in the notations $o()$ and $\ll$ and $\gg$ may depend on s, k, the coefficients λ_i and also ϵ.

First, we note that as in the second section of [7], we may assume that $\frac{\lambda_1}{\lambda_2}$ is irrational and that λ_1 is negative while λ_2 is positive. Also we clearly need only prove the theorem when P is a large positive real number. As $[P]^{s-k} = P^{s-k} + O\left(P^{s-k-1}\right)$ for $P \geq 1$, it in fact suffices to assume that P is a large positive integer.

We now give a lemma about kernel functions which is very useful for our work. The lemma is simply a slight modification of Lemma 1 of Davenport [3], so we do not give many of the details of the proof. As is usual, we define $e(x) = e^{2\pi i x}$ for any real number x.

Lemma 1. *Fix a positive integer h. Fix positive real numbers a and b with $0 < a < b$. Then there is an even real function $K(\alpha)$ of the real variable α such that the function ψ defined by*

$$\psi(\theta) = \int_{\mathbb{R}} e(\theta \alpha) K(\alpha) d\alpha$$

satisfies

$$\psi(\theta) \begin{cases} \in [0, 1] & for \quad \theta \in \mathbb{R} \\ = 0 & for \quad |\theta| \geq b \\ = 1 & for \quad |\theta| \leq a. \end{cases} \tag{2.1}$$

Moreover, K satisfies the bound

$$K(\alpha) \ll_h \min\left(b, |\alpha|^{-1}, |\alpha|^{-h-1}(b-a)^{-h}\right), \tag{2.2}$$

where here the constant in Vinogradov's notation depends only on h.

Proof. We follow the proof of the aforementioned lemma of Davenport. We take $n = h$, and $\delta = \frac{b-a}{h(a+b)}$. Note that the required condition $0 < \delta < 1$ follows from the fact that we have $0 < a < b$.

Following Davenport's proof, we obtain

$$\psi_h(\theta) = 0 \text{ for } |\theta| \geq \frac{2b}{a+b} \quad \text{and} \quad \psi_h(\theta) = 1 \text{ for } |\theta| \leq \frac{2a}{a+b}.$$

Now we define $\psi(\theta) = \psi_h\left(\frac{2\theta}{a+b}\right)$ for all real θ. It follows immediately that

$$\psi(\theta) = 0 \text{ for } |\theta| \geq b \quad \text{and} \quad \psi(\theta) = 1 \text{ for } |\theta| \leq a.$$

After inserting our choices into the second (unlabeled) equation on page 86 of Davenport's paper and doing some calculations, one may see that if $K(\alpha)$ is defined by

$$K(\alpha) = (a+b)\frac{\sin(\pi(a+b)\alpha)}{(\pi(a+b)\alpha)}\left(\frac{\sin\left(\frac{\pi(b-a)\alpha}{h}\right)}{\left(\frac{\pi(b-a)\alpha}{h}\right)}\right)^h,$$

then we have $\psi(\theta) = \int_{\mathbb{R}} e(\theta\alpha)K(\alpha)d\alpha$. The bound (2.2) then follows. This completes the proof of Lemma 1.   □

We now apply this lemma to define two kernel functions. We first fix a positive function $T(P)$ to be chosen later. In fact, this function will be the same function as in [7], but chosen to satisfy one more condition. As in [7], the function $T(P)$ will tend to infinity as P tends to infinity. We define the function

$$R(P) = (T(P))^{\frac{1}{2}}. \tag{2.3}$$

We assume that P is large enough so that we have $R(P) \geq 2$. For both kernel functions, we shall choose $h = 1$. For the first function, we take

$$a = \epsilon - \frac{\epsilon}{R(P)} \quad \text{and} \quad b = \epsilon. \tag{2.4}$$

In this case, we call the kernel function K_- since we shall see that its Fourier transform gives a lower bound for the number of solutions of our inequality (1.3). On the other hand, for the second kernel function we take

$$a = \epsilon \quad \text{and} \quad b = \epsilon + \frac{\epsilon}{R(P)}. \tag{2.5}$$

In this case, we call the kernel function K_+ since we shall see that its Fourier transform gives instead an upper bound.

We note for future reference that from (2.2) we have

$$K_\pm(\alpha) \ll \min\left(1, |\alpha|^{-1}, |\alpha|^{-2}(R(P))\right). \tag{2.6}$$

Here by using $K_\pm$ we of course mean that the bound (2.6) holds for both K_- and K_+.

We may now proceed to the setup of the proof. In a manner similar to [7], we define

$$f(\alpha) = f(\alpha, P) = \sum_{|x| \leq P} e\left(\alpha x^k\right). \tag{2.7}$$

Then, for $1 \leq i \leq s$, we define

$$f_i(\alpha) = f_i(\alpha, P) = f(\lambda_i \alpha, P) = \sum_{|x| \leq P} e\left(\lambda_i \alpha x^k\right). \tag{2.8}$$

We note that here it was necessary to diverge slightly from [7]. In that paper, the summation was only over integers from 1 to P. We also define, as in [7],

$$S(\alpha) = S(\alpha, P) = |f_1(\alpha, P) f_2(\alpha, P)|. \tag{2.9}$$

Recall that we seek an asymptotic formula for the number $N(P)$ of solutions $\mathbf{x}$ of the inequality (1.3) with $|\mathbf{x}| \leq P$. By the definitions of K_- and K_+ and the identity (2.1), we can see that we have

$$\int_{\mathbb{R}} \left(\prod_{i=1}^{s} f_i(\alpha)\right) K_-(\alpha) d\alpha \leq N(P) \leq \int_{\mathbb{R}} \left(\prod_{i=1}^{s} f_i(\alpha)\right) K_+(\alpha) d\alpha. \tag{2.10}$$

As in [7], we define the major arcs, minor arcs and trivial arcs respectively by

$$\mathcal{M} = \left\{\alpha \in \mathbb{R} : |\alpha| \leq (\log P)^{\frac{1}{3}} P^{-k}\right\};$$

$$\mathfrak{m} = \left\{\alpha \in \mathbb{R} : (\log P)^{\frac{1}{3}} P^{-k} < |\alpha| \leq T(P)\right\};$$

$$\mathfrak{t} = \left\{\alpha \in \mathbb{R} : |\alpha| > T(P)\right\},$$

$$\tag{2.11}$$

where $T(P)$ is the function mentioned earlier, to be chosen later.

2.2 The Minor Arcs

Now Lemma 2 of [7] holds for our functions f_i, although with perhaps larger constants c_3 and c_4. To see this, note that if f_i satisfies $|f_i(\alpha, P)| \geq \gamma^{\frac{1}{2k}}$, then for sufficiently large P, one has

$$\left|\sum_{x=1}^{P} e\left(\lambda_i \alpha x^k\right)\right| \geq \frac{\gamma^{\frac{1}{2k}}}{3} P, \tag{2.12}$$

on noting that we have

$$\sum_{x=-P}^{-1} e\left(\lambda_i \alpha x^k\right) = \sum_{x=1}^{P} e\left(\lambda_i \alpha x^k\right) \text{ if } k \text{ is even, while}$$

$$\sum_{x=-P}^{-1} e\left(\lambda_i \alpha x^k\right) = \overline{\sum_{x=1}^{P} e\left(\lambda_i \alpha x^k\right)} \text{ if } k \text{ is odd.}$$

$$(2.13)$$

Thus one may apply the original lemma, whence Lemma 2 of [7] holds in our situation.

It is clear that Lemmas 3 and 4 and also 5 of [7] carry over to our situation as well. Now from equations (52) and (53) of [7] and trivial estimates, we can see that

$$\int_0^1 |f(\alpha)|^s d\alpha \ll_s P^{s-k} \tag{2.14}$$

for $s \geq 2^k$.

Technically, we should observe how one can derive this directly from Vaughan's bounds. To do so, one can split up the sum $f(\alpha)$ into a sum over x ranging from 1 to P, from $-P$ to -1, and also over $x = 0$. One can then use (2.13) to see that

$$\int_0^1 \left| \sum_{\substack{-P \leq x \leq P \\ x \neq 0}} e\left(\lambda_i \alpha x^k\right) \right|^s d\alpha = 2^s \int_0^1 \left| \sum_{x=1}^{P} e\left(\lambda_i \alpha x^k\right) \right|^s d\alpha.$$

By treating the integral on the left side of (2.14) as the number of solutions of a Diophantine equation for even s, one can see, after a little work, that the number of solutions with one of the variables equal to zero can be estimated satisfactorily using Hua's lemma, an application of Hölder's inequality, and trivial estimates. In this manner we have our analogue of Lemma 6 of [7], namely (2.14).

Our analogue of Lemma 8 of [7] is the following.

Lemma 2. *Suppose that we are in the setting of Theorem 2. Suppose as well that c and d are two real numbers with $c \leq d$ and $d - c \leq 1$. Fix i_0*

satisfying $1 \le i_0 \le s$. Then one has

$$\int_c^d \left(\prod_{\substack{1 \le i \le s \\ i \ne i_0}} |f_i(\alpha)| \right) |K_\pm(\alpha)| d\alpha$$

$$\ll P^{s-k-1} \min \left(1, (|c| + |d|)^{-1}, (|c| + |d|)^{-2} (R(P)) \right).$$

We note that Lemma 2 can be proved in exactly the same manner as Lemma 8 of [7], except that here we use (2.6) and the fact that one has $s - 1 \ge 2^k$.

Now we use Lemma 4 and 5 of [7] to obtain a certain function $T(P)$ for which

$$\sup_{(\log P)^{\frac{1}{3}} P^{-k} \le |\alpha| \le T(P)} S(\alpha, P) = o\left(P^2\right).$$

Define $E(P)$ to be a function satisfying $E(P) = o(1)$ for which one has

$$\sup_{(\log P)^{\frac{1}{3}} P^{-k} \le |\alpha| \le T(P)} S(\alpha, P) \le E(P) P^2. \tag{2.15}$$

Then it is clear that by replacing $T(P)$ by $\left(\min \left(T(P), e^{(E(P))^{-\frac{1}{4}}} \right) \right)$ for large P that (2.15) still holds for large P, that we still have $T(P) \to \infty$ as $P \to \infty$, and that we have

$$\log T(P) \le E(P)^{-\frac{1}{4}} \quad \text{for large } P. \tag{2.16}$$

By (2.15), we may write $\mathfrak{m} = \mathfrak{m}_1 \cup \mathfrak{m}_2$ where for $i = 1$ and $i = 2$, one has

$$f_i(\alpha, P) = O\left((E(P))^{\frac{1}{2}} P \right) \quad \text{for } \alpha \in \mathfrak{m}_i. \tag{2.17}$$

Now we consider the contribution of the minor arcs. Using the fact that the functions $K_\pm(\alpha)$ are even, for $i_0 = 1$ and 2 we have

$$\int_{\mathfrak{m}_{i_0}} \left(\prod_{1 \le i \le s} |f_i(\alpha)| \right) |K_\pm(\alpha)| d\alpha$$

$$\ll \left(\sup_{\alpha \in \mathfrak{m}_{i_0}} |f_{i_0}(\alpha)| \right) \int_{0 \le \alpha \le T(P)} \left(\prod_{\substack{1 \le i \le s \\ i \ne i_0}} |f_i(\alpha)| \right) |K_\pm(\alpha)| d\alpha.$$

$$\tag{2.18}$$

Using Lemma 2, we have

$$\int_0^{T(P)} \prod_{\substack{1 \le i \le s \\ i \ne i_0}} |f_i(\alpha)| \cdot |K_\pm(\alpha)| d\alpha \ll \sum_{n=0}^{\lceil T(P) \rceil} \int_n^{n+1} \prod_{\substack{1 \le i \le s \\ i \ne i_0}} |f_i(\alpha)| \cdot |K_\pm(\alpha)| d\alpha$$

$$\ll \left(\log \left(T(P) \right) \right) P^{s-k-1}.$$

Combining this bound with (2.17) and inserting into (2.18), we have

$$\int_{\mathfrak{m}_{i_0}} \left(\prod_{1 \le i \le s} |f_i(\alpha)| \right) |K_\pm(\alpha)| d\alpha \ll (E(P))^{\frac{1}{2}} \left(\log T(P) \right) P^{s-k}$$

$$\ll (E(P))^{\frac{1}{4}} P^{s-k},$$

by the condition (2.16), whence the contribution of the minor arcs to the integrals in (2.10) is $o\left(P^{s-k}\right)$, as we have $E(P) = o(1)$.

2.3   The Trivial Arcs

To handle the contribution from the trivial arcs, we can proceed as above, except that the process is slightly easier. As above, we obtain

$$\int_{\mathfrak{t}} \left(\prod_{1 \le i \le s} |f_i(\alpha)| \right) |K_\pm(\alpha)| d\alpha \ll \sum_{n=[T(P)]}^{\infty} n^{-2} R(P) P^{s-k}$$

$$\ll (T(P))^{-1} (T(P))^{\frac{1}{2}} P^{s-k},$$

by the definition (2.3), whence the contribution of the minor arcs to the integrals in (2.10) is $o\left(P^{s-k}\right)$, because $T(P) \to \infty$ as $P \to \infty$.

2.4   The Major Arc

Now we consider the contribution of the major arc to the integrals in (2.10). We show that for some constant $C = C(k, s, \lambda_1, \lambda_2, \ldots, \lambda_s)$, the contribution to each integral is equal to $CP^{s-k} + o\left(P^{s-k}\right)$. First, from our work on the minor arcs and the trivial arcs, we have

$$\int_{\mathbb{R}} \left(\prod_{i=1}^s f_i(\alpha) \right) K_\pm(\alpha) d\alpha = \int_{\mathcal{M}} \left(\prod_{i=1}^s f_i(\alpha) \right) K_\pm(\alpha) d\alpha + o\left(P^{s-k}\right).$$

It follows from (2.10) that there is a function $U(P)$ satisfying $U(P) = o\left(P^{s-k}\right)$ such that we have

$$\int_{\mathcal{M}} \prod_{i=1}^{s} f_i(\alpha) K_-(\alpha)d\alpha - U(P) \leq N(P) \leq \int_{\mathcal{M}} \prod_{i=1}^{s} f_i(\alpha) K_+(\alpha)d\alpha + U(P).$$

$$(2.19)$$

One can show in a standard way that

$$\int_{\mathcal{M}} \left(\prod_{i=1}^{s} f_i(\alpha) \right) K_\pm(\alpha)d\alpha$$
$$= \int_{[-P,P]^s} \int_{\mathbb{R}} e\left(\alpha \left(\lambda_1 x_1^k + \cdots + \lambda_s x_s^k \right) \right) K_\pm(\alpha)d\alpha d\mathbf{x} + o\left(P^{s-k} \right);$$

here we use the condition $s \geq k + 1$, which clearly holds. After a change of variable, and an application of the identity (2.1) while recalling the definitions of K_- and K_+, we see that the integral on the right side above can be rewritten in the form

$$P^s \int_{A\left(\epsilon \pm \frac{\epsilon}{R(P)}\right)} d\mathbf{y} + W(P),$$

where

$$A(\delta) = \left\{ \mathbf{y} \in [-1,1]^s : \left| \lambda_1 y_1^k + \cdots + \lambda_s y_s^k \right| < \delta P^{-k} \right\},$$

where $W(P)$ is an error term satisfying

$$W(P) \ll P^s \int_H d\mathbf{y},$$

$$(2.20)$$

and where H is the set of $\mathbf{y} \in [-1,1]^s$ which satisfy

$$\left(\epsilon - \frac{\epsilon}{R(P)} \right) P^{-k} \leq \left| \lambda_1 y_1^k + \cdots + \lambda_s y_s^k \right| < \left(\epsilon + \frac{\epsilon}{R(P)} \right) P^{-k}.$$

Combining the last two observations and using (2.19), we can see that we have

$$N(P) = P^s \int_{A(\epsilon)} d\mathbf{y} + O\left(W(P) + V(P) \right),$$

$$(2.21)$$

where $V(P)$ is a function satisfying $V(P) = o\left(P^{s-k} \right)$.

We now need a technical lemma.

Lemma 3. *Suppose that k and s are positive integers with $s \geq k + 1$. Suppose that for $1 \leq i \leq s$, the sets B_i are intervals and we have $\gamma_i \in$*

$\{-1, 1\}$. *Suppose also that L and M are real numbers with $L > 0$. Then one has*

$$\int_{Q(M)} d\boldsymbol{x} \ll L,$$

where

$$Q(M) = \left\{ \boldsymbol{x} \in \prod_{i=1}^{s} B_i : \left| \gamma_1 x_1^k + \cdots + \gamma_s x_s^k - M \right| < L \right\},$$

and where the constant in Vinogradov's notation depends only on s and k and the intervals B_i.

Proof. We first use Lemma 1 to choose a kernel function $K(\alpha)$, taking $h = 1$ and $a = L$ and $b = 2L$. Then by (2.2), we certainly have

$$K(\alpha) \ll L. \tag{2.22}$$

It also follows that the integral in the statement of the lemma is bounded above by

$$\int_{\mathbb{R}} \int_{\boldsymbol{x} \in \prod_{i=1}^{s} B_i} e\left(\alpha \left(\gamma_1 x_1^k + \cdots + \gamma_s x_s^k - M \right) \right) K(\alpha) d\boldsymbol{x} d\alpha.$$

By a standard estimate, one has

$$\int_{x_i \in B_i} e\left(\alpha \gamma_i x_i^k \right) \ll \min\left(1, |\alpha|^{-\frac{1}{k}} \right) \quad \text{for } 1 \le i \le s.$$

(See for example the comments right after [8], Lemma 2.8. In fact, some slight tweaking of this result is also needed, but this is straightforward.) It follows from (2.22) that we have

$$\int_{\mathbb{R}} \int_{\boldsymbol{x} \in \prod_{i=1}^{s} B_i} e\left(\alpha \left(\gamma_1 x_1^k + \cdots + \gamma_s x_s^k - M \right) \right) K(\alpha) d\boldsymbol{x} d\alpha$$

$$\ll L \left(\int_0^1 d\alpha + \int_1^\infty \alpha^{-\frac{s}{k}} d\alpha \right)$$

$$\ll L.$$

This completes the proof of Lemma 3.   □

Now we give another technical lemma, whose proof comprises the main work of this section.

Lemma 4. *Fix a positive integer k, and a positive integer s with $s \geq k+2$. For $1 \leq i \leq s$, suppose that a_i is a positive real number. If k is odd, suppose that $B_i = [-a_i, a_i]$ for $1 \leq i \leq s$, while if k is even, suppose that $B_i = [0, a_i]$ for $1 \leq i \leq s$. Suppose as well that $\gamma_i \in \{-1, 1\}$ for $1 \leq i \leq s$. Also, if k is even, suppose that for some i and j with $1 \leq i \neq j \leq s$, we have $\gamma_i = -\gamma_j$. Suppose as well that L is a positive real number which is sufficiently small in terms of s and k and the intervals B_i. Then there is a positive constant D depending only on s and k and the intervals B_i such that one has*

$$\int_{Q(0)} d\boldsymbol{x} = DL + O\left(L^{1+\frac{1}{k}}\right),$$

where $Q(0)$ is defined as in Lemma 3.

Proof. We start off by assuming without loss of generality that $\gamma_2 = -\gamma_3$ in the case that k is even. We now give a simple bound. Suppose that δ is a real number satisfying $0 < \delta < 1$, and that M is any real number. Then we have

$$\mu\left(\{x \in \mathbb{R} : |\pm x^k - M| \leq \delta\}\right) \ll \delta^{\frac{1}{k}} \tag{2.23}$$

for either choice of sign in front of x^k, where μ denotes standard Euclidean measure. This can be proved readily by, for example, considering the cases $|M| \leq 2\delta$ and $|M| > 2\delta$ separately.

Now, for sufficiently small L, we define the region

$$\mathcal{W} = \left\{(x_2, \ldots, x_s) \in \prod_{i=2}^{s} B_i : \left(-\gamma_1\gamma_2 x_2^k - \cdots - \gamma_1\gamma_s x_s^k\right) \in [-2L, 2L]\right\}.$$

By applying Lemma 3, noting that we have $s - 1 \geq k + 1$, we obtain

$$\mu\left(\mathcal{W}\right) \ll L. \tag{2.24}$$

We define the interval B_1' by

$$B_1' = \left[-a_1^k, a_1^k\right] \text{ if } k \text{ is odd} \quad \text{and} \quad B_1' = \left[0, a_1^k\right] \text{ if } k \text{ is even.}$$

Now, as either k is odd, or we have $B_i = [0, a_i]$ for $1 \leq i \leq s$, we can make the change of variable

$$u = x_1^k + \gamma_1\gamma_2 x_2^k + \cdots + \gamma_1\gamma_s x_s^k.$$

In this manner, we have

$$\int_{Q(0)} d\mathbf{x} = \frac{1}{k} \int_{-L}^{L} \phi(u)du, \tag{2.25}$$

where $\phi(u)$ is the function defined by

$$\phi(u) = \int_{\mathcal{R}_u} \left| u - \gamma_1 \gamma_2 x_2^k - \cdots - \gamma_1 \gamma_s x_s^k \right|^{\frac{1}{k}-1} dx_2 \dots dx_s,$$

and $\mathcal{R}_u$ is the region defined by

$$\mathcal{R}_u = \left\{ (x_2, \dots, x_s) \in \prod_{i=2}^s B_i : \left(u - \gamma_1 \gamma_2 x_2^k - \cdots - \gamma_1 \gamma_s x_s^k \right) \in B_1' \right\}.$$

Also define

$$g(u) = g(u, x_2, \dots, x_s) = \left| u - \gamma_1 \gamma_2 x_2^k - \cdots - \gamma_1 \gamma_s x_s^k \right|^{\frac{1}{k}-1}.$$

Now we consider $\phi(u) - \phi(0)$. Writing $\mathbf{x}' = (x_2, \dots, x_s)$ for convenience, we may see that one has

$$\phi(u) - \phi(0) \ll \int_{\mathcal{R}_0 \setminus \mathcal{W}} |g(u) - g(0)| d\mathbf{x}'$$
$$+ \int_{(\mathcal{R}_u \setminus (\mathcal{R}_0 \cup \mathcal{W})) \cup (\mathcal{R}_0 \setminus (\mathcal{R}_u \cup \mathcal{W})) \cup (\mathcal{W})} (|g(u)| + |g(0)|)\, d\mathbf{x}'.$$
$$(2.26)$$

We first look at the region $\mathcal{R}_u \setminus (\mathcal{R}_0 \cup \mathcal{W})$. We can see that there is some real number M such that for $|u| \leq L$, one has

$$\left| \gamma_1 \gamma_2 x_2^k + \cdots + \gamma_1 \gamma_s x_s^k - M \right| \leq |u| \leq L \quad \text{for } (x_2, \dots, x_s) \in \mathcal{R}_u \setminus \mathcal{R}_0.$$

By Lemma 3, noting that we have $s - 1 \geq k + 1$, we obtain

$$\mu\left(\mathcal{R}_u \setminus (\mathcal{R}_0 \cup \mathcal{W}) \right) \ll L.$$

Now observe that for $|u| \leq L$, we have

$$|g(u)| \leq L^{\frac{1}{k}-1} \quad \text{for} \quad (x_2, \dots, x_s) \in \left(\prod_{i=2}^s B_i \right) \setminus \mathcal{W}.$$

It follows for $|u| \leq L$ that

$$\int_{(\mathcal{R}_u \setminus (\mathcal{R}_0 \cup \mathcal{W}))} (|g(u)| + |g(0)|)\, d\mathbf{x}' \ll L^{\frac{1}{k}}. \qquad (2.27)$$

Similarly, we can obtain such a bound for the corresponding integral over $(\mathcal{R}_0 \setminus (\mathcal{R}_u \cup \mathcal{W}))$.

Now we consider the second integral in (2.26) over $\mathcal{W}$. For integers n, let $\mathcal{X}_{n,u}$ be the region

$$\left\{ (x_2, \ldots, x_s) \in \prod_{i=2}^{s} B_i : 2^{-n-1} \leq \left| u - \gamma_1\gamma_2 x_2^k - \cdots - \gamma_1\gamma_s x_s^k \right| < 2^{-n} \right\}.$$

Observe that by Lemma 3, one has $\mu(\mathcal{X}_{n,u}) \ll 2^{-n}$. For fixed u with $|u| \leq L$, we write the integral $\int_{\mathcal{W}} |g(u)| dx'$ as a sum of integrals over the regions $\mathcal{X}_{n,u} \cap \mathcal{W}$. One obtains

$$\int_{\mathcal{W}} |g(u)| dx' \ll \sum_{n=[-\log_2(3L)]}^{\infty} 2^{-n} 2^{\left(-n\left(\frac{1}{k}-1\right)\right)} \ll \sum_{n=[-\log_2(3L)]}^{\infty} 2^{-\frac{n}{k}} \ll L^{\frac{1}{k}}.$$

We apply this bound for general u with $|u| \leq L$ and for $u = 0$ to obtain

$$\int_{\mathcal{W}} (|g(u)| + |g(0)|) dx' \ll L^{\frac{1}{k}}. \tag{2.28}$$

We now turn to the first integral in (2.26). For $|u| \leq L$ and $(x_2, \ldots, x_s) \in (\prod_{i=2}^{s} B_i) \setminus \mathcal{W}$, we can see that 0 is not in the interval with endpoints

$$\left(u - \gamma_1\gamma_2 x_2^k - \cdots - \gamma_1\gamma_s x_s^k \right) \quad \text{and} \quad \left(-\gamma_1\gamma_2 x_2^k - \cdots - \gamma_1\gamma_s x_s^k \right).$$

Thus we may apply the mean value theorem to conclude that we have

$$g(u) - g(0) \ll_k |u| \cdot \left| -\gamma_1\gamma_2 x_2^k - \cdots - \gamma_1\gamma_s x_s^k \right|^{\frac{1}{k}-2}.$$

Splitting up the integration into integrals over the regions $\mathcal{X}_{n,0} \cap (\mathcal{R}_0 \setminus \mathcal{W})$, one obtains

$$\int_{\mathcal{R}_0 \setminus \mathcal{W}} |g(u) - g(0)| dx' \ll \sum_{n=-n_0}^{\lceil -\log_2 L \rceil} 2^{-n} 2^{\left(-n\left(\frac{1}{k}-2\right)\right)} |u|$$

$$\ll \sum_{n=-\infty}^{\lceil -\log_2 L \rceil} 2^{\left(-n\left(\frac{1}{k}-1\right)\right)} |u|$$

$$\ll L^{\frac{1}{k}-1} |u| \ll L^{\frac{1}{k}}$$

for $|u| \leq L$; here n_0 is some integer depending on s and k and the intervals B_i.

Inserting this last bound, the bound (2.27), and its analogue, and finally the bound (2.28) into (2.26), we have

$$\phi(u) - \phi(0) \ll L^{\frac{1}{k}} \quad \text{for} \quad |u| \leq L.$$

It follows that

$$\frac{1}{k}\int_{-L}^{L}\phi(u)du = \frac{1}{k}\int_{-L}^{L}\phi(0)du + O\left(L^{1+\frac{1}{k}}\right).$$

Combining this approximation with (2.25), we have

$$\int_{Q(0)}d\mathbf{x} = \frac{L}{k}\phi(0) + O\left(L^{1+\frac{1}{k}}\right). \tag{2.29}$$

Observe that $\phi(0)$ does not depend on L; it depends only on k, s, the numbers γ_i and the intervals B_i. Thus, to prove Lemma 4, it remains only to prove that $\phi(0)$ is positive.

But this can be seen easily. First note that, for $(x_2,\ldots,x_s) \in \mathcal{R}_0$, one has that $g(0)$ is greater than some constant which depends only on k, s, the numbers γ_i and the intervals B_i. If k is odd, then each B_i is of the form $[-a_i, a_i]$ and thus there is some nonempty open set around $\mathbf{0}$ in the region $\mathcal{R}_0$. If k is even, then by assumption the ratio $\frac{-\gamma_1\gamma_2}{-\gamma_1\gamma_3}$ is negative, whence we also have a nonempty open set contained in the region $\mathcal{R}_0$. Thus in both cases, $\mu(\mathcal{R}_0)$ is positive, whence $\phi(0)$ is positive. This completes the proof of Lemma 4. $\qquad\qquad\square$

Now we are in a position to complete our work on the major arc contribution. Recall the approximation (2.21) and the bound (2.20). By a change of variable in each integral, we can write

$$N(P) = P^s|\lambda_1\lambda_2\ldots\lambda_s|^{-\frac{1}{k}}\int_{Z_1}d\mathbf{x} + W_0(P), \tag{2.30}$$

where

$$Z_1 = \left\{\mathbf{x} \in \prod_{i=1}^{s}\left[-|\lambda_i|^{\frac{1}{k}},|\lambda_i|^{\frac{1}{k}}\right] : \left|\gamma_1 x_1^k + \cdots + \gamma_s x_s^k\right| < \epsilon P^{-k}\right\},$$

where $\gamma_i = \mathrm{sgn}(\lambda_i)$ and where $W_0(P)$ is an error term satisfying

$$W_0(P) \ll P^s\int_{Z_2}d\mathbf{x} + V(P), \tag{2.31}$$

where Z_2 is the set of $\mathbf{x} \in \prod_{i=1}^{s}\left[-|\lambda_i|^{\frac{1}{k}},|\lambda_i|^{\frac{1}{k}}\right]$ satisfying

$$\left(\epsilon - \frac{\epsilon}{R(P)}\right)P^{-k} \le \left|\gamma_1 x_1^k + \cdots + \gamma_s x_s^k\right| < \left(\epsilon + \frac{\epsilon}{R(P)}\right)P^{-k},$$

and where $V(P)$ is a function satisfying $V(P) = o\left(P^{s-k}\right)$.

We first consider the case in which k is odd. Afterwards, we briefly mention the slight adjustments needed in the case which k is even. We apply Lemma 4 to approximate the integral in (2.30) and the integral in (2.31). Note that we have $s \geq k+2$. We assume that P is sufficiently large in terms of ϵ, s, k and the coefficients λ_i. In this manner we ensure that the choice $L = \epsilon P^{-k}$ is sufficiently small for an application of the lemma. Thus by Lemma 4, we have

$$N(P) = C\epsilon P^{s-k} + O\left(P^s \left(\epsilon P^{-k}\right)^{1+\frac{1}{k}} + W_0(P)\right) \qquad (2.32)$$

for some constant C which depends only on s, k and the coefficients λ_i. Now by two applications of Lemma 4, we can approximate the integral in (2.31). In this manner, we obtain

$$W_0(P) \ll 2C\frac{\epsilon}{R(P)}P^{s-k} + P^s \left(\epsilon P^{-k}\right)^{1+\frac{1}{k}} + V(P), \qquad (2.33)$$

where $V(P) = o\left(P^{s-k}\right)$. We note that in fact this argument shows that $W_0(P)$ is asymptotic to the first term on the right side of (2.33), but we do not need this fact. Now recall from (2.3) that $R(P) = (T(P))^{\frac{1}{2}}$. Also recall that $T(P)$ tends to positive infinity as P goes to positive infinity. Thus it follows from (2.33) that one has

$$W_0(P) = o\left(P^{s-k}\right).$$

Combining this with the approximation (2.32) yields

$$N(P) = C\epsilon P^{s-k} + o\left(P^{s-k}\right), \qquad (2.34)$$

as desired. This completes the proof in the case in which k is odd.

Now, if k is even, then by symmetry we have

$$N(P) = (2P)^s |\lambda_1 \lambda_2 \ldots \lambda_s|^{-\frac{1}{k}} \int_{\substack{\mathbf{x} \in \prod_{i=1}^s \left[0,|\lambda_i|^{\frac{1}{k}}\right] \\ |\gamma_1 x_1^k + \cdots + \gamma_s x_s^k| < \epsilon P^{-k}}} d\mathbf{x} + W_0(P),$$

and a similar result holds for the integral in $W_0(P)$. With these observations in hand, the result (2.34) can be proved in much the same way as in the odd case. Thus the proof of Theorem 2 is complete.

References

[1] V. Bentkus and F. Götze, *Lattice point problems and distribution of values of quadratic forms*, Ann. of Math. (2) **150** (1999), 977–1027.

[2] _____, *Lattice points in multidimensional bodies*, Forum Math. **13** (2001), 149–225.

[3] H. Davenport, *Indefinite quadratic forms in many variables*, Mathematika **3** (1956), 81–101.

[4] H. Davenport and H. Heilbronn, *On indefinite quadratic forms in five variables*, J. London Math. Soc. **21** (1946), 185–193.

[5] A. Eskin, G. A. Margulis, and S. Mozes, *Upper bounds and asymptotics in a quantitative version of the Oppenheim conjecture*, Ann. of Math. (2) **147** (1998), 93–141.

[6] K. Ford, *New estimates for mean values of Weyl sums*, Intern. Math. Res. Notices **1995** (1995), 155–171.

[7] D. E. Freeman, *Asymptotic lower bounds for Diophantine inequalities*, to appear in Mathematika.

[8] R. C. Vaughan, *The Hardy-Littlewood method*, 2nd ed., Cambridge University Press, Cambridge, U.K., 1997.

Shifted and Shiftless Partition Identities

F. G. Garvan[1]

1 Introduction

Let S and T be sets of positive integers. Let a be a fixed positive integer. A *shifted partition identity* has the form

$$p(S, n) = p(T, n - a), \qquad \text{for all } n \geq a.$$

We will mainly consider the case $a = 1$. Assume $a = 1$. If S or T is finite then it is not hard to show that $S = T = \{1\}$. Andrews [4] found the following two non-trivial examples:

$$S = \{n: n \text{ odd or } n \equiv \pm 4, \pm 6, \pm 8, \pm 10 \pmod{32}\},$$
$$T = \{n: n \text{ odd or } n \equiv \pm 2, \pm 8, \pm 12, \pm 14 \pmod{32}\};$$

and

$$S = \{n: n \equiv \pm 1, \pm 4, \pm 5, \pm 6, \pm 7, \pm 9, \pm 10, \pm 11, \pm 13, \pm 15, \qquad (1.1)$$
$$\pm 16, \pm 19 \pmod{40}\},$$
$$T = \{n: n \equiv \pm 1, \pm 3, \pm 4, \pm 5, \pm 9, \pm 10, \pm 11, \pm 14, \pm 15, \pm 16,$$
$$\pm 17, \pm 19 \pmod{40}\}.$$

In these examples, each S and T is the union of arithmetic progressions modulo M for some M; namely $M = 32$ and $M = 40$. In fact, each is a union of 24 such arithmetic progressions. Later, Kalvade [10] found five more identities with $M = 42, 48$ and 60, each also involving the union of 24 arithmetic progressions. Through a computer search we have found a further 48 identities with $M = 40, 42, 46, 48, 54, 56, 60, 62, 66, 70$ and 72. All but two of these identities involve unions of 24 arithmetic progressions. The remaining two involve unions of 48 arithmetic progressions. In the present paper, we show how the theory of modular functions may be used to prove certain shifted partition identities.

---

[1]Research supported in part by the NSF under grant number DMS-9870052.

The generating function for $p(S, n)$ is an infinite product:

$$\sum_{n \geq 0} p(S, n) q^m = \prod_{n \in S} \frac{1}{(1 - q^n)},$$

where $|q| < 1$. For the case $a = 1$, we can write a shifted partition identity as an equivalent q-series identity:

$$\prod_{n \in S} \frac{1}{(1 - q^n)} - q \prod_{n \in T} \frac{1}{(1 - q^n)} = 1. \tag{1.2}$$

There is a simple proof of Andrews's mod 32 and 40 shifted partition identities using Jacobi's triple product identity [2, p. 21]:

$$\sum_{n=-\infty}^{\infty} (-1)^n z^n q^{n(n-1)/2} = \prod_{n=1}^{\infty} (1 - zq^{n-1})(1 - z^{-1}q^n)(1 - q^n). \tag{1.3}$$

Let $T(z, q)$ denote the left side. By splitting $T(z, q)$ into its even and odd parts (as a function of z) we easily find that

$$T(z, q) = T(-z^2 q, q^4) - zT(-z^{-2}q, q^4),$$

or

$$\frac{T(-z^2 q, q^4)}{T(z, q)} - z \frac{T(-z^{-2}q, q^4)}{T(z, q)} = 1. \tag{1.4}$$

The idea is to replace q by q^m, and let $z = q$ in (1.4) so that the numerator in each term simplifies to 1 after cancellation in the infinite products. This only occurs in two cases: $m = 4$ giving Andrews's mod 32 shifted identity, and $m = 5$ giving Andrews's mod 40 shifted identity. Other choices for m do not give shifted partition identities but variants in which the parts from some residue classes must be distinct. These types of identities have been considered by Alladi [1].

When comparing shifted identities, quite often it is helpful to write (1.2) in the form:

$$\prod_{n \in T \setminus S} (1 - q^n) - q \prod_{n \in S \setminus T} (1 - q^n) = \prod_{n \in S \cup T} (1 - q^n).$$

For example, Andrews's modulus $M = 40$ result involves

$$T \setminus S = \{n : n \equiv \pm 3, \pm 14, \pm 17 \pmod{40}\},$$
$$S \setminus T = \{n : n \equiv \pm 6, \pm 7, \pm 13 \pmod{40}\}.$$

There is another shifted result with $M = 40$ that Andrews missed. It involves

$$T \setminus S = \{n : n \equiv \pm 2, \pm 9, \pm 11, \pm 12 \pmod{40}\},$$
$$S \setminus T = \{n : n \equiv \pm 4, \pm 6, \pm 7, \pm 13 \pmod{40}\}.$$

We prove this identity using the theory of modular functions in section 4.

By considering the effect of modular transformations on shifted identities we were led to consider shiftless partition identities. Let a be a fixed positive integer, and let S, T be distinct sets of positive integers. A *shiftless partition identity* has the form:

$$p(S, T) = p(T, n), \qquad \text{for all } n \neq a.$$

Our simplest example is for modulus $M = 40$. Let

$$S = \{n : n \equiv \pm 1, \pm 2, \pm 5, \pm 6, \pm 7, \pm 8, \pm 9, \pm 11,$$
$$\pm 12, \pm 13, \pm 15, \pm 19 \pmod{40}\},$$
$$T = \{n : n \equiv \pm 1, \pm 3, \pm 4, \pm 5, \pm 6, \pm 7, \pm 8, \pm 13,$$
$$\pm 14, \pm 15, \pm 17, \pm 19 \pmod{40}\}.$$

Then

$$p(S, n) = p(T, n), \qquad \text{for all } n \neq 2. \tag{1.5}$$

This identity follows from our shifted result with $M = 40$ by applying a modular transformation. The details are given in section 4. Other shiftless identities exist for the moduli $M = 42, 46, 48, 54, 56, 60, 62, 66$ and 72. In section 5 we present some examples of shifted and shiftless identities. Details for these other identities will be left to a later paper.

2 Computer Search

Our new shifted partition identities were found via a computer search. The idea is two consider a finite analogue of (1.2). Let N be an integer $N > 2$. A pair of sets $[S, T]$ is called a *truncated ST-pair* $O(q^N)$ if

$$\prod_{n \in S} \frac{1}{(1 - q^n)} - q \prod_{n \in T} \frac{1}{(1 - q^n)} = 1 + O(q^N), \tag{2.1}$$

$$S \subset \{1, 2, 3, \ldots, N - 1\},$$

and

$$T \subset \{1, 2, 3, \ldots, N - 2\}.$$

It is clear that if S, T give a shifted partition theorem then they also give rise to a truncated ST-pair $O(q^N)$, for all N. For $N \geq 3$, let $\mathcal{T}_N$ be the set of truncated ST-pairs $O(q^N)$. For a given N, we may find all truncated ST-pairs by a boot-strapping method:

0.   Initialize: $\mathcal{T}_3 = \{[\{1\}, \{1\}]\}$

1.   Suppose we are given $\mathcal{T}_N$. For each $[S, T] \in \mathcal{T}_N$, we consider four possible $[S', T']$:

a.	$S' = S,$	$T' = T$
b.	$S' = S \cup \{N\},$	$T' = T$
c.	$S' = S,$	$T' = T \cup \{N - 1\}$
d.	$S' = S \cup \{N\},$	$T' = T \cup \{N - 1\}$

For each of these $[S', T']$, if

$$\prod_{n \in S'} \frac{1}{(1 - q^n)} - q \prod_{n \in T'} \frac{1}{(1 - q^n)} = 1 + 0(q^{N+1}), \qquad (2.2)$$

then we include $[S', T']$ in $\mathcal{T}_{N+1}$.

In this way we may construct the sets $\mathcal{T}_N$. The computation in (2.2) would be done using a computer algebra package like Maple. For small N, we may calculate by hand.

We give some examples.

$\mathcal{T}_3 = \{[\{1\}, \{1\}]\}$
$\mathcal{T}_4 = \{[\{1, 3\}, \{1, 2\}], [\{1\}, \{1\}]\}$
$\mathcal{T}_5 = \{[\{1, 3, 4\}, \{1, 2, 3\}], [\{1, 3\}, \{1, 2\}], [\{1, 4\}, \{1, 3\}], [\{1\}, \{1\}]\}$
$\mathcal{T}_6 = \{[\{1, 3, 4, 5\}, \{1, 2, 3\}], [\{1, 3, 5\}, \{1, 2\}], [\{1, 4, 5\}, \{1, 3, 4\}],$
$\qquad [\{1, 4\}, \{1, 3\}], [\{1, 5\}, \{1, 4\}], [\{1\}, \{1\}]\}$

We have calculated the $\mathcal{T}_N$ for $N \leq 37$. Let $t(N) = |\mathcal{T}_N|$, ie. the number of truncated ST-pairs $O(q^N)$. We have the following table:

n	$t(n)$	n	$t(n)$	n	$t(n)$	n	$t(n)$
3	1	12	96	21	2447	30	70456
4	2	13	138	22	3425	31	110214
5	4	14	197	23	4962	32	159686
6	6	15	300	24	6839	33	253265
7	11	16	431	25	10000	34	374385
8	15	17	636	26	13989	35	591648
9	26	18	893	27	21383	36	876405
10	41	19	1258	28	30781	37	1354888
11	67	20	1723	29	48292		

It appears that $t(n)$ grows exponentially. We do not have enough data to make a real conjecture.

All the known shifted partition identities involve sets of integers S and T that remain unchanged on multiplication by -1 (mod M) for a certain modulus M. Armed with $\mathcal{T}_{37}$, we searched for such shifted partition identities with even modulus up to $M = 74$.

3 Modular Functions

In this section we set up the necessary theory of modular functions to prove our shifted partition identities.

3.1 Background Theory

The necessary background theory of modular functions and modular forms may be found in [16], [17], [11] and [15]. Many of the results that we require are contained in [5], [6] and [7]. Let $\Gamma(1)$ denote the full modular group and as usual let

$$\Gamma_0(N) := \left\{ \begin{pmatrix} a & b \\ c & d \end{pmatrix} \in \Gamma(1) : c \equiv 0 \pmod{N} \right\}.$$

Let Γ be a subgroup of $\Gamma(1)$ with finite index. For a modular function f on Γ and a cusp ζ the order of f (mod Γ) at ζ is denoted by $\mathrm{Ord}\,(f; \zeta; \Gamma)$ and the *invariant order* of f at ζ is denoted by $\mathrm{ord}\,(f; \zeta)$. We have

$$\mathrm{Ord}\,(f; \zeta; \Gamma) = \kappa(\Gamma; \zeta)\,\mathrm{ord}\,(f; \zeta) \tag{3.1}$$

where $\kappa(\Gamma; \zeta)$ denotes the fan width of the cusp ζ (mod Γ), and

$$\text{ord}\,(f \mid A; \zeta) = \text{ord}\,(f; A\zeta), \tag{3.2}$$

for $A \in \Gamma(1)$. Here we use the usual stroke operator notation

$$(f \mid A)(\tau) := f(A\tau).$$

Any non-zero modular function f must satisfy the valence formula:

$$\sum_{s \in \mathcal{F}} \text{Ord}\,(f; s; \Gamma) = 0, \tag{3.3}$$

where $\mathcal{F}$ is a fundamental set of Γ and the sum is taken over s with non-zero order so that this is a finite sum. For $s \in \mathcal{H}$ the order is interpreted in the usual sense. See [16] for more details. We shall use the valence formula to prove certain modular function identities. This is a standard technique.

3.2   Theta Products and Eta Products

We need results from [7] on transformation formulae and multiplier systems for the Dedekind eta function

$$\eta(\tau) = \exp(\pi i/12) \prod_{n=1}^{\infty} (1 - \exp(2\pi i\tau)), \tag{3.4}$$

and the theta function

$$\vartheta_1(v \mid \tau) = -i \sum_{m=-\infty}^{\infty} (-1)^m \exp(\pi i\tau(m + \tfrac{1}{2})^2) \exp(2\pi iv(m + \tfrac{1}{2})),$$

where $\tau \in \mathcal{H}$ and $v \in \mathbb{C}$.

For $\begin{pmatrix} a & b \\ c & d \end{pmatrix} \in \Gamma(1)$, $\eta(\tau)$ and $\vartheta(\tau)$ have well-known transformation formulae

$$\eta\left(\frac{a\tau + b}{c\tau + d}\right) = \nu_\eta(A)\,\sqrt{c\tau + d}\,\eta(\tau), \tag{3.5}$$

and

$$\vartheta_1\left(\frac{v}{c\tau + d}\mid\frac{a\tau + b}{c\tau + d}\right) = \nu_{\vartheta_1}(A)\,\sqrt{c\tau + d}\,\exp(\pi icv^2/(c\tau + d))\,\vartheta_1(v \mid \tau).$$

Here $\nu_\eta(A)$, $\nu_{\vartheta_1}(A)$ are explicit 24-th and 8-th roots of unity, respectively. Formulae for these multipliers are given in [7, §2]. We have

$$\nu_{\vartheta_1}(A) = \nu_\eta^3(A). \tag{3.6}$$

Let $q = \exp(2\pi i\tau)$. Let N, ρ be integers, $N \geq 1$, $\rho \nmid N$. We define the theta function $\theta_{\rho;N}(\tau)$ by

$$\theta_{\rho;N}(\tau) := \sum_{m=-\infty}^{\infty} (-1)^m q^{\frac{1}{8N}(2Nm+2\rho-N)^2} \tag{3.7}$$

$$= q^{\frac{1}{8N}(N-2\rho)^2} \prod_{m=1}^{\infty} (1 - q^{Nm-\rho})(1 - q^{Nm-(N-\rho)})(1 - q^{Nm}),$$

by Jacobi's triple product identity [3, (7.1)]. Our function $\theta_{\rho;N}$ corresponds to Biagioli's [7, (2.8)] $f_{N,\rho}$. From the definition of $\theta_{\rho;N}$ we have

$$\theta_{\rho+N;N} = \theta_{-\rho;N} = \theta_{\rho;N}. \tag{3.8}$$

Biagioli [7, Lemma 2.1] gives the following transformation formula:

$$\theta_{\rho;N}(A\tau)$$
$$= (-1)^{b\rho+\lfloor \rho a/N \rfloor + \lfloor \rho/N \rfloor} \exp(\rho^2 \pi iab/N) \, \nu_{\vartheta_1}\left(^N A\right) \sqrt{c\tau + d}\,\theta_{a\rho;N}(\tau), \tag{3.9}$$

for $A = \begin{pmatrix} a & b \\ c & d \end{pmatrix} \in \Gamma_0(N)$. Here

$$^N A = \begin{pmatrix} a & Nb \\ c/N & d \end{pmatrix}.$$

Similarly, from (3.5) we find that

$$\eta(N\tau) \,|\, A = \nu_\eta \left(^N A\right) \sqrt{c\tau + d}\,\eta(N\tau). \tag{3.10}$$

For a set of integers $\boldsymbol{\rho} = \{\rho_1, \rho_2, \ldots, \rho_k\}$, where each $\rho_j \nmid N$, we define the following theta product

$$\theta_{\boldsymbol{\rho};N}(\tau) := \prod_{j=1}^{k} \theta_{\rho_j;N}(\tau).$$

From (3.9) we have the following transformation formula. For $A \in \Gamma_0(N)$,

$$\theta_{\boldsymbol{\rho};N}(A\tau)$$
$$= (-1)^{bL(\boldsymbol{\rho})+M(\boldsymbol{\rho},a)} \exp(Q(\boldsymbol{\rho})\pi iab/N) \, \nu_{\vartheta_1}^k \left(^N A\right) (c\tau + d)^{k/2} \, \theta_{a\boldsymbol{\rho};N}(\tau), \tag{3.11}$$

where

$$L(\boldsymbol{\rho}) := \sum_j \rho_j,$$

$$M(\boldsymbol{\rho}, x) := \sum_j \lfloor \rho_j x/N \rfloor + \lfloor \rho_j/N \rfloor,$$

$$Q(\boldsymbol{\rho}) := \sum_j \rho_j^2,$$

and where $a\boldsymbol{\rho} = \{a\rho_1, \ldots, a\rho_k\}$.

In view of (3.8) we define a relation on $\mathbb{Z}$. Let N be a fixed positive integer. For x, $y \in \mathbb{Z}$ we define $x \sim_N y$ if either $x \equiv y \pmod{N}$ or $x \equiv -y \pmod{N}$. This is an equivalence relation. If $x \sim_N y$ then

$$\theta_{x;N} = \theta_{y;N}, \tag{3.12}$$

by (3.8). This equivalence relation extends naturally to certain sets of integers. We consider finite sets of integers whose elements are inequivalent mod $\sim_N$. For any two such sets of integers $\boldsymbol{\rho}$ and $\boldsymbol{\sigma}$ we define $\boldsymbol{\rho} \sim_N \boldsymbol{\sigma}$ if each element ρ of $\boldsymbol{\rho}$ is equivalent to some element σ of $\boldsymbol{\sigma}$ modulo $\sim_N$ and vice versa. For example,

$$\{3, 9, 12, 15, 18, 21, 39, 45, 48, 51\} \sim_{40} \{1, 3, 5, 8, 9, 11, 12, 15, 18, 19\}$$

The analogue of (3.12) holds for theta products. Thus, if $\boldsymbol{\rho} \sim_N \boldsymbol{\sigma}$ then

$$\theta_{\boldsymbol{\rho};N} = \theta_{\boldsymbol{\sigma};N}, \tag{3.13}$$

There is a group action of $\mathbb{Z}_N^\times$ on equivalence classes $\boldsymbol{\rho}$. For $(a, N) = 1$, we define $a\,\boldsymbol{\rho}$ in the natural way by

$$a\,\boldsymbol{\rho} = \{a\rho \colon \rho \in \boldsymbol{\rho}\}. \tag{3.14}$$

Since $(a, N) = 1$, the $a\rho$ are distinct modulo $\sim_N$. In later sections we will need to compute the orbit of certain sets $\boldsymbol{\rho}$.

3.3 Orders at Cusps

We will need the following lemmas

Lemma 3.1 ([7]). *If $(r, s) = 1$, then the fan width of $\Gamma_0(N)$ at $\frac{r}{s}$ is*

$$\kappa\left(\Gamma_0(N); \frac{r}{s}\right) = \frac{N}{(N, s^2)}. \tag{3.15}$$

Lemma 3.2 ([9]). *Let S_N be the set of integer pairs (c, a) satisfying*

(0) $(1, 0) \in S_N;$

(1) $c > 1$, $c \mid N$, $1 \le a < c$, $\gcd(c, a) = 1$, *and;*

(2) *If* (c, a), $(c, a') \in S_N$ *and* $a' \equiv a \pmod{\gcd(c, N/c)}$, *then* $a = a'$.

Then the set

$$\left\{ \frac{a}{c} : (c, a) \in S_N \right\}$$

is a complete set of inequivalent cusps for $\Gamma_0(N)$.

We will need results on the orders at cusps of eta-products and theta-functions. Newman [14] has found necessary and sufficient conditions under which an eta-product is a modular function on $\Gamma_0(N)$. Ligozat [12] has computed the order of a general eta-product at the cusps of $\Gamma_0(N)$. We need the behavior of the modular form $\eta(N\tau)$ near each cusp of $\Gamma_0(N)$. The following result follows from [9, Lemma 3.5].

Lemma 3.3 ([12]). *The order at the cusp* $s = \frac{b}{c}$ *(assuming* $(b, c) = 1$*) of the eta function* $\eta(N\tau)$ *is*

$$\operatorname{ord}(\eta(N\tau); s) = \frac{c^2}{24N}. \tag{3.16}$$

Biagoli [7] has computed the order of the theta function $\theta_{\rho;N}$ at any cusp.

Lemma 3.4 ([7, Lemma 3.2, p.285]). *The order at the cusp* $s = \frac{b}{c}$ *(assuming* $(b, c) = 1$*) of the theta function* $\theta_{\rho;N}(\tau)$ *(defined above and assuming* $\rho \nmid N$*) is*

$$\operatorname{ord}(\theta_{\rho;N}(\tau); s) = \frac{e^2}{2N} \left(\frac{b\rho}{e} - \left\lfloor \frac{b\rho}{e} \right\rfloor - \frac{1}{2} \right)^2, \tag{3.17}$$

where $e = (N, c)$ *and* $\lfloor \ \rfloor$ *is the greatest integer function.*

4 Proof of a Shifted Partition Identity with $M = 40$

There are two shifted partition identities with modulus $M = 40$ and $a = 1$. The first one is the identity (1.1) found by Andrews. We found the second one via a computer search. Let

$$\begin{aligned} S = \{n: &\ \pm 1, \pm 3, \pm 4, \pm 5, \pm 6, \pm 7, \pm 13, \pm 15, \pm 16, \pm 17, \qquad (4.1) \\ &\ \pm 18, \pm 19 \pmod{40}\}, \\ T = \{n: &\ \pm 1, \pm 2, \pm 3, \pm 5, \pm 9, \pm 11, \pm 12, \pm 15, \pm 16, \pm 17, \\ &\ \pm 18, \pm 19 \pmod{40}\}. \end{aligned}$$

Then
$$p(S, n) = p(T, n - 1),$$

for all $n \geq 1$. In this section we prove (4.1) in detail. We show how this identity is equivalent to other shifted partition identities and how it leads to a shiftless partition identity (1.5). The proofs of our other shifted partition identities are analogous. Some of the details are given in the next section.

We identify S with the set of "positive" residue classes mod 40:

$$S \leftrightarrow \rho_1 := \{1, 3, 4, 5, 6, 7, 13, 15, 16, 17, 18, 19\},$$

and T with another set of "positive" residue class mod 40:

$$T \leftrightarrow \sigma_1 := \{1, 2, 3, 5, 9, 11, 12, 15, 16, 17, 18, 19\}.$$

From (3.4), (3.7), we have

$$\sum_{n \geq 0} p(S, n) q^n = \frac{\eta^{12}(40\tau)}{\theta_{\rho_1;40}(\tau)},$$

and

$$q \sum_{n \geq 0} p(T, n) q^n = \frac{\eta^{12}(40\tau)}{\theta_{\sigma_1;40}(\tau)}.$$

Hence, our shifted partition identity is equivalent to showing that

$$\left(\frac{1}{\theta_{\rho_1;40}(\tau)} - \frac{1}{\theta_{\sigma_1;40}(\tau)} \right) \eta^{12}(40\tau) = 1. \tag{4.2}$$

In view of (3.11) we calculate the set of theta functions $\theta_{a\rho_1;40}$ for $(a, 40) = 1$. This reduces to calculating the orbit of the equivalence class ρ_1 modulo $\sim_{40}$ under multiplication by the group $\mathbb{Z}_{40}^{\times}$. A calculation shows that this orbit is

$$\{\rho_1, \rho_3, \rho_7, \rho_9\},$$

where

$$\rho_3 := \{1, 3, 5, 8, 9, 11, 12, 14, 15, 17, 18, 19\} \sim_{40} 3\rho_1,$$
$$\rho_7 := \{1, 2, 5, 6, 7, 8, 9, 11, 12, 13, 15, 19\} \sim_{40} 7\rho_1,$$
$$\rho_9 := \{2, 3, 4, 5, 7, 9, 11, 13, 14, 15, 16, 17\} \sim_{40} 9\rho_1.$$

We may extend the definition of ρ_j to all $j \in \mathbb{Z}_{40}^{\times}$, by

$$\rho_m = \rho_{-m}, \qquad \rho_{m+20} = \rho_m.$$

In this way, we find that

$$a\rho_j \sim_{40} \rho_{aj},$$

for all $a, j \in \mathbb{Z}_{40}^\times$. Similarly, we find that the orbit of the equivalence class σ_1 modulo $\sim_{40}$ under multiplication by the group $\mathbb{Z}_{40}^\times$ is

$$\{\sigma_1, \sigma_3, \sigma_7, \sigma_9\},$$

where

$$\sigma_3 := \{3, 4, 5, 6, 7, 8, 9, 11, 13, 14, 15, 17\} \sim_{40} 3\sigma_1,$$
$$\sigma_7 := \{1, 3, 4, 5, 6, 7, 8, 13, 14, 15, 17, 19\} \sim_{40} 7\sigma_1,$$
$$\sigma_9 := \{1, 2, 5, 7, 9, 11, 12, 13, 15, 16, 18, 19\} \sim_{40} 9\sigma_1,$$

and we may extend the definition of σ_j in analogous way so that

$$a\sigma_j \sim_{40} \sigma_{aj},$$

for all $a, j \in \mathbb{Z}_{40}^\times$.

We calculate $\theta_{\rho;40}(A\tau)$, for $A \in \Gamma_0(40)$, and $\rho = \rho_j, \sigma_j$. A calculation shows that

$$Q(\rho_j) \equiv Q(\sigma_j) \equiv 0 \pmod{80}, \tag{4.3}$$

$$L(\rho_j) \equiv L(\sigma_j) \equiv 0 \pmod{2}, \tag{4.4}$$

and

$$M(\rho_j, x) \equiv M(\sigma_j, x) \pmod{2}, \tag{4.5}$$

for $j = 1, 3, 7$ and 9 and $(x, 10) = 1$. Assuming $(x, 10) = 1$,

$$M(\rho_1, x) \equiv 1 \pmod{2}, \tag{4.6}$$

if and only if $x \equiv \pm 3 \pmod{20}$;

$$M(\rho_3, x) \equiv 1 \pmod{2}, \tag{4.7}$$

if and only if $x \not\equiv \pm 1 \pmod{20}$;

$$M(\rho_7, x) \equiv 1 \pmod{2}, \tag{4.8}$$

if and only if $x \equiv \pm 9 \pmod{20}$; and

$$M(\rho_9, x) \equiv 1 \pmod{2}, \tag{4.9}$$

if and only if $x \equiv \pm 7 \pmod{20}$.

We define four functions:

$$f_1(\tau) := \left(\frac{1}{\theta_{\rho_1}(\tau)} - \frac{1}{\theta_{\sigma_1}(\tau)} \right) \eta^{12}(40\tau), \tag{4.10}$$

$$f_3(\tau) := \left(\frac{1}{\theta_{\sigma_3}(\tau)} - \frac{1}{\theta_{\rho_3}(\tau)} \right) \eta^{12}(40\tau), \tag{4.11}$$

$$f_7(\tau) := \left(\frac{1}{\theta_{\rho_7}(\tau)} - \frac{1}{\theta_{\sigma_7}(\tau)} \right) \eta^{12}(40\tau), \tag{4.12}$$

$$f_9(\tau) := \left(\frac{1}{\theta_{\rho_9}(\tau)} - \frac{1}{\theta_{\sigma_9}(\tau)} \right) \eta^{12}(40\tau). \tag{4.13}$$

We note that $\{f_1, f_3, f_7, f_9\}$ is the orbit of f_1 by the group $\Gamma_0(40)$. We have

Theorem 4.1. *For* $A = \left(\begin{smallmatrix} a & b \\ c & d \end{smallmatrix} \right) \in \Gamma_0(40)$,

$$f_k(A\tau) = f_{ak}(\tau), \tag{4.14}$$

where $\pm ak$ *is reduced* (mod 20).

Proof. The functions f_3, f_7, and f_9 are obtained by applying a modular transformation to f_1. In fact, from (3.6), (3.10), (3.11) and (4.3)–(4.6) we have

$$f_1 \,|\, A = f_a, \tag{4.15}$$

for $A = \left(\begin{smallmatrix} a & b \\ c & d \end{smallmatrix} \right) \in \Gamma_0(40)$. In the subscript of f_a we have reduced $\pm a$ (mod 20). For $k \in \{1, 3, 7, 9\}$, choose a fixed matrix $A_k = \left(\begin{smallmatrix} k & * \\ * & * \end{smallmatrix} \right) \in \Gamma_0(40)$, so that

$$f_k = f_1 \,|\, A_k.$$

For any $A = \left(\begin{smallmatrix} a & b \\ c & d \end{smallmatrix} \right) \in \Gamma_0(40)$, and $k \in \{1, 3, 7, 9\}$,

$$\begin{aligned} f_k \,|\, A &= (f_1 \,|\, A_k) \,|\, A \\ &= f_1 \,|\, (A_k A) \\ &= f_{ak}, \end{aligned}$$

since $A_k A \equiv \left(\begin{smallmatrix} ak & * \\ * & * \end{smallmatrix} \right)$ (mod 40). $\qquad\qquad\qquad\qquad\qquad \square$

It is well-known that the theta functions θ_ρ, θ_σ and the eta function $\eta^{12}(40\tau)$, are meromorphic at all cusps $A\infty$, $A \in \Gamma(1)$. This together with the above theorem gives the

Corollary 4.1. *Any symmetric polynomial in* f_1, f_3, f_7, f_9 *is a modular function of* $\Gamma_0(40)$.

Now we are in a position to complete the proof of our shifted partition identity (4.1). This identity is equivalent to showing that

$$f_1 \equiv 1. \tag{4.16}$$

We define $F(\tau)$ by

$$F := (f_1 - 1)(f_3 - 1)(f_7 - 1)(f_9 - 1). \tag{4.17}$$

Our result is equivalent to showing that $F \equiv 0$. Clearly, if (4.16) holds, then $F \equiv 0$. Conversely, if $F \equiv 0$, then $f_j \equiv 0$, for some j. But $\Gamma_0(40)$ acts transitively on the $\{f_k\}$, and so

$$f_1 = f_3 = f_7 = f_9 \equiv 1. \tag{4.18}$$

Let $\mathcal{F}$ be a fundamental set for $\Gamma_0(40)$. By the valence formula (3.3), either $F \equiv 0$ or

$$\sum_{s \in \mathcal{F}} \mathrm{Ord}\,(F; s; \Gamma_0(40)) = 0. \tag{4.19}$$

A set of inequivalent cusps for $\Gamma_0(40)$ is

$$\left\{ \frac{0}{1}, \frac{1}{40}, \frac{1}{20}, \frac{1}{10}, \frac{1}{8}, \frac{1}{5}, \frac{1}{4}, \frac{1}{2} \right\} \tag{4.20}$$

From the definition of the f_j we find that

$$\mathrm{ord}\,(f_j; s) \geq \min\{-\mathrm{ord}\,(\theta_{\rho_1}; s), -\mathrm{ord}\,(\theta_{\sigma_1}; s)\} + 12\,\mathrm{ord}\,(\eta(40\tau), s).$$

Using (3.1), (3.15), (3.16) and (3.17) this gives lower bounds for $\mathrm{Ord}\,(f_j; s; \Gamma_0(40))$ at each cusp s of $\Gamma_0(40)$, and $j = 1, 3, 7, 9$. We denote this lower bound by $\ell(j, s)$. We have

$$\mathrm{Ord}\,(f_j; s; \Gamma_0(40)) \geq \ell(j, s), \tag{4.21}$$

where

$$\ell(j, 0) = -1, \quad \text{for all } j,$$
$$\ell(7, 1/40) = -2,$$
$$\ell(1, 1/20) = -1$$

and

$$\ell(j, s) = 0,$$

for all other (j, s). Since F has no poles in the complex upper half plane, we have

$$\sum_{s \neq i\infty} \mathrm{Ord}\,(F; s; \Gamma_0(40)) \geq -5.$$

We only need calculate the first couple of terms in the q-expansion of each f_j. The lowest power in the q-expansion gives the order at the cusp $i\infty$, which is equivalent to $1/40 \bmod \Gamma_0(40)$. Now,

$$f_1 = \frac{1}{(1-q)(1-q^3)\cdots} - q\frac{1}{(1-q)(1-q^2)\cdots}$$
$$= 1 + O(q^3),$$

$$f_3 = \frac{1}{(1-q^3)\cdots} - q^3\frac{1}{(1-q)\cdots}$$
$$= 1 + O(q^4),$$

$$f_7 = \frac{1}{q^2(1-q)(1-q^2)\cdots} - \frac{1}{q^2(1-q)(1-q^3)\cdots}$$
$$= O(1),$$

and

$$f_9 = \frac{1}{(1-q^2)(1-q^3)\cdots} - q^2\frac{1}{(1-q)(1-q^2)\cdots}$$
$$= 1 + O(q^3).$$

Hence,

$$\mathrm{Ord}\,(F, i\infty, \Gamma_0(40)) \geq 3 + 4 + 0 + 3 = 10,$$

and

$$\sum_{s\in\mathcal{F}} \mathrm{Ord}\,(F; s; \Gamma_0(40)) \geq 5.$$

Hence,

$$F \equiv 0,$$

and our result follows.

We see that our shifted partition identity (4.1) is equivalent to the four identities given in (4.18). We collect these together into the following

Theorem 4.2.

(i)   Let $S \equiv \pm\{1, 3, 4, 5, 6, 7, 13, 15, 16, 17, 18, 19\}$ (mod 40), and $T \equiv \pm\{1, 2, 3, 5, 9, 11, 12, 15, 16, 17, 18, 19\}$ (mod 40). Then $p(S, n) = p(T, n-1)$, for all $n \neq 0$.

(ii) Let $S \equiv \pm\{3, 4, 5, 6, 7, 8, 9, 11, 13, 14, 15, 17\}$ (mod 40), and $T \equiv \pm\{1, 3, 5, 8, 9, 11, 12, 14, 15, 17, 18, 19\}$ (mod 40). Then $p(S, n) = p(T, n-3)$, for all $n \neq 0$.

(iii) *Let* $S \equiv \pm \{1, 2, 5, 6, 7, 8, 9, 11, 12, 13, 15, 19\}$ (mod 40), *and*
$T \equiv \pm \{1, 3, 4, 5, 6, 7, 8, 13, 14, 15, 17, 19\}$ (mod 40). *Then*
$p(S, n) = p(T, n)$, *for all* $n \neq 2$.

(iv) *Let* $S \equiv \pm \{2, 3, 4, 5, 7, 9, 11, 13, 14, 15, 16, 17\}$ (mod 40), *and*
$T \equiv \pm \{1, 2, 5, 7, 9, 11, 12, 13, 15, 16, 18, 19\}$ (mod 40). *Then*
$p(S, n) = p(T, n - 2)$, *for all* $n \neq 0$.

Part (i) is our shifted partition identity (4.1). Part (iii) is equivalent to the identity

$$f_7 \equiv 1,$$

and is our shiftless partition identity (1.5).

5 Other Shifted and Shiftless Partition Identities

In this section we present some other shifted and shiftless partition identities. Details of proofs of these and other identities we have found will be presented in a later paper.

(i) Let $S \equiv \pm \{1, 4, 5, 7, 8, 9, 11, 12, 13, 15, 16, 19\}$ (mod 42), and
$T \equiv \pm \{1, 3, 4, 7, 8, 11, 12, 13, 15, 17, 19, 20\}$ (mod 42). Then
$p(S, n) = p(T, n - 1)$, for all $n \neq 0$.

(ii) Let $S \equiv \pm \{1, 2, 3, 4, 5, 6, 7, 8, 11, 13, 15, 17\}$ (mod 42), and
$T \equiv \pm \{1, 2, 3, 4, 5, 6, 7, 9, 10, 11, 17, 19\}$ (mod 42). Then
$p(S, n) = p(T, n)$, for all $n \neq 8$.

(iii) Let $S \equiv \pm \{1, 3, 4, 5, 7, 13, 15, 16, 17, 18, 19, 22\}$ (mod 46), and
$T \equiv \pm \{1, 2, 3, 7, 11, 12, 15, 16, 17, 19, 20, 21\}$ (mod 46). Then
$p(S, n) = p(T, n - 1)$, for all $n \neq 0$.

(iv) Let $S \equiv \pm \{1, 2, 3, 5, 6, 9, 10, 11, 13, 14, 17, 21\}$ (mod 46), and
$T \equiv \pm \{1, 2, 3, 5, 7, 8, 9, 11, 12, 15, 20, 21\}$ (mod 46). Then
$p(S, n) = p(T, n)$, for all $n \neq 6$.

(v) Let $S \equiv \pm \{1, 5, 6, 8, 9, 10, 11, 13, 15, 19, 20, 23\}$ (mod 48), and
$T \equiv \pm \{1, 4, 5, 7, 9, 10, 15, 17, 18, 19, 20, 23\}$ (mod 48). Then
$p(S, n) = p(T, n - 1)$, for all $n \neq 0$.

(vi) Let $S \equiv \pm \{1, 2, 3, 4, 5, 6, 11, 13, 19, 20, 21, 23\}$ (mod 48), and
$T \equiv \pm \{1, 2, 3, 4, 5, 7, 8, 17, 18, 19, 21, 23\}$ (mod 48). Then
$p(S, n) = p(T, n)$, for all $n \neq 6$.

(vii)   Let $S \equiv \pm\{1, 3, 5, 7, 11, 12, 13, 20, 22, 23, 24, 25\}$ (mod 54), and
$T \equiv \pm\{1, 2, 5, 9, 11, 12, 17, 19, 22, 23, 24, 25\}$ (mod 54). Then
$p(S, n) = p(T, n - 1)$, for all $n \neq 0$.

(viii) Let $S \equiv \pm\{1, 2, 3, 4, 5, 7, 9, 11, 16, 17, 24, 25\}$ (mod 54), and
$T \equiv \pm\{1, 2, 3, 4, 5, 7, 9, 12, 13, 20, 23, 25\}$ (mod 54). Then
$p(S, n) = p(T, n)$, for all $n \neq 11$.

(ix)   Let $S \equiv \pm\{1, 3, 4, 5, 7, 18, 21, 22, 23, 24, 25, 27\}$ (mod 56), and
$T \equiv \pm\{1, 2, 3, 7, 11, 17, 20, 21, 24, 25, 26, 27\}$ (mod 56). Then
$p(S, n) = p(T, n - 1)$, for all $n \neq 0$.

(x)   Let $S \equiv \pm\{1, 5, 7, 8, 10, 12, 13, 15, 18, 21, 23, 27\}$ (mod 56), and
$T \equiv \pm\{2, 3, 5, 7, 8, 13, 15, 20, 21, 22, 23, 25\}$ (mod 56). Then
$p(S, n) = p(T, n)$, for all $n \neq 1$.

(xi)   Let $S \equiv \pm\{1, 7, 8, 10, 11, 13, 14, 17, 19, 20, 23, 29\}$ (mod 60), and
$T \equiv \pm\{1, 6, 7, 9, 10, 13, 17, 20, 21, 23, 28, 29\}$ (mod 60). Then
$p(S, n) = p(T, n - 1)$, for all $n \neq 0$.

(xii)   Let $S \equiv \pm\{1, 2, 7, 10, 11, 13, 16, 17, 19, 20, 23, 29\}$ (mod 60), and
$T \equiv \pm\{1, 3, 4, 10, 11, 13, 17, 18, 19, 20, 27, 29\}$ (mod 60). Then
$p(S, n) = p(T, n)$, for all $n \neq 2$.

(xiii) Let $S \equiv \pm\{1, 3, 5, 7, 8, 19, 21, 23, 25, 26, 28, 30\}$ (mod 62), and
$T \equiv \pm\{1, 2, 5, 7, 13, 18, 21, 23, 24, 27, 29, 30\}$ (mod 62). Then
$p(S, n) = p(T, n - 1)$, for all $n \neq 0$.

(xiv)   Let $S \equiv \pm\{1, 3, 5, 7, 9, 13, 15, 16, 21, 22, 24, 28\}$ (mod 62), and
$T \equiv \pm\{1, 3, 6, 7, 8, 10, 15, 19, 21, 23, 25, 28\}$ (mod 62). Then
$p(S, n) = p(T, n)$, for all $n \neq 5$.

(xv)   Let $S \equiv \pm\{1, 10, 11, 12, 13, 14, 15, 16, 19, 21, 23, 25\}$ (mod 66), and
$T \equiv \pm\{1, 9, 10, 11, 12, 13, 14, 15, 25, 29, 31, 32\}$ (mod 66). Then
$p(S, n) = p(T, n - 1)$, for all $n \neq 0$.

(xvi)   Let $S \equiv \pm\{1, 4, 5, 6, 7, 9, 11, 13, 16, 21, 23, 28\}$ (mod 66), and
$T \equiv \pm\{1, 4, 5, 6, 7, 9, 11, 14, 16, 17, 27, 29\}$ (mod 66). Then
$p(S, n) = p(T, n)$, for all $n \neq 13$.

(xvii) Let $S \equiv \pm\{1, 3, 4, 5, 6, 7, 9, 11, 13, 14, 15, 16, 17, 19, 23, 24, 25,$
$26, 27, 28, 29, 31, 33, 34\}$ (mod 70), and

$T \equiv \pm \{1, 2, 3, 5, 8, 9, 11, 12, 13, 14, 15, 17, 18, 19, 21, 22, 23, 25,$
$27, 28, 29, 31, 32, 33\}$ (mod 70). Then
$p(S, n) = p(T, n - 1)$, for all $n \neq 0$.

(xix) Let $S \equiv \pm \{1, 3, 5, 7, 8, 26, 28, 29, 30, 31, 33, 35\}$ (mod 72), and
$T \equiv \pm \{1, 2, 5, 7, 13, 23, 28, 29, 31, 32, 34, 35\}$ (mod 72). Then
$p(S, n) = p(T, n - 1)$, for all $n \neq 0$.

(xx) Let $S \equiv \pm \{1, 4, 5, 6, 11, 14, 15, 21, 25, 31, 32, 35\}$ (mod 72), and
$T \equiv \pm \{1, 4, 5, 7, 10, 11, 16, 25, 26, 29, 31, 35\}$ (mod 72). Then
$p(S, n) = p(T, n)$, for all $n \neq 6$.

(xxi) Let $S \equiv \pm \{1, 3, 5, 7, 8, 9, 10, 11, 12, 13, 14, 16, 17, 18, 19, 23, 25,$
$27, 29, 31, 32, 33, 34, 35\}$ (mod 72), and
$T \equiv \pm \{1, 2, 5, 7, 8, 9, 11, 12, 13, 15, 16, 17, 18, 19, 21, 22, 23, 25, 26,$
$27, 29, 31, 32, 35\}$ (mod 72). Then
$p(S, n) = p(T, n - 1)$, for all $n \neq 0$.

References

[1] K. Alladi, *The quintuple product identity and shifted partition functions*, J. Comput. Appl. Math. **68** (1996), 3–13.

[2] G. E. Andrews, *The theory of partitions*, Encyclopedia of Mathematics and its Applications, vol. 2, Addison-Wesley Publishing Co., Reading, Mass.-London-Amsterdam, 1976.

[3] _____, *q-series: their development and application in analysis, number theory, combinatorics, physics, and computer algebra*, CBMS Regional Conference Series in Mathematics (Washington, D.C.), vol. 66, Conference Board of the Mathematical Sciences, 1986.

[4] _____, *Further problems on partitions*, Amer. Math. Monthly **94** (1987), 437–439.

[5] B. C. Berndt, A. J. Biagioli, and J. M. Purtilo, *Ramanujan's "mixed" modular equations*, J. Ramanujan Math. Soc. **1** (1986), 46–70.

[6] _____, *Ramanujan's modular equations of "large" prime degree*, J. Indian Math. Soc. (N.S.) **51** (1987), 75–110.

[7] A. J. F. Biagioli, *A proof of some identities of Ramanujan using modular forms*, Glasgow Math. J. **31** (1989), 271–295.

[8]   F. Garvan, *A q-product tutorial for a q-series maple package*, Sém. Lothar. Combin. **42** (1999), Art. B24d, 27 pp. (electronic).

[9]   K. Harada, *Modular functions, modular forms and finite groups*, Lecture notes, The Ohio State University, 1987.

[10] A. Kalvade, *Equality of shifted partition functions*, J. Indian Math. Soc. (N.S.) **54** (1989), 155–164.

[11] M. I. Knopp, *Modular functions in analytic number theory*, Markham Publishing Co., Chicago, Ill., 1970.

[12] G. Ligozat, *Courbes modulaires de genre 1*, Bull. Soc. Math. France, Mém. 43, Supplément au Bull. Soc. Math. (Paris, France), no. 3, vol. 103, Société Mathématique de France, 1975.

[13] I. G. Macdonald, *Affine root systems and Dedekind's η-function*, Invent. Math. **15** (1972), 91–143.

[14] M. Newman, *Construction and application of a class of modular functions II*, Proc. London Math. Soc. **9** (1959), 373–387.

[15] H. Rademacher, *Topics in analytic number theory*, Springer-Verlag, New York, 1973.

[16] R. A. Rankin, *Modular forms and functions*, Cambridge University Press, Cambridge, 1978.

[17] B. Schoeneberg, *Elliptic modular functions: an introduction*, Springer-Verlag, New York, 1974.

[18] D. Stanton, *Sign variations of the Macdonald identities*, SIAM J. Math. Anal. **17** (1986), 1454–1460.

[19] ———, *An elementary approach to the Macdonald identities*, q-series and partitions (Minneapolis, MN, 1988), IMA Vol. Math. Appl., vol. 18, Springer, New York, 1989, pp. 139–149.

Deformations of Pseudorepresentations Coming from Reducible Representations

Emiliano Gómez

1 Introduction

This is a study of the deformation theory of (Wiles) pseudorepresentations. In many cases these have the same universal deformation ring as the 2-dimensional, odd representations they come from (deformations of such representations have been studied by Boston and Mazur, whose ideas we use). When they come from reducible representations, however, the two deformation rings might be different. We consider the factorization of a pseudorepresentation through an appropriate quotient, resulting in an easier deformation problem from which we can obtain information about the universal deformation ring. Finally, we present a concrete example.

2 Basic Definitions

Let G be a profinite group with a specified element $\mathbf{c} \in G$ of order 2, and let B be a ring with $2 \in B^*$. We define pseudorepresentations as in [3], based on the work of Hida [4] and Wiles [11]. However, it is worth mentioning that Taylor has a different definition [10], for which some deformation theory has already been developed. See [7] and [8]. We remark that the (Wiles) pseudorepresentations we now define always yield (2-dimensional) Taylor pseudorepresentations, simply by adding the maps A and D below.

Definition 2.1. *A pseudorepresentation* π *from* G *to* B *($\pi \colon G \to B$) is a triple of continuous maps* $A, D \colon G \to B$ *and* $X \colon G \times G \to B$ *such that for all* σ, τ, γ, η *in* G:

$$A(\sigma\tau) = A(\sigma)A(\tau) + X(\sigma, \tau)$$
$$D(\sigma\tau) = D(\sigma)D(\tau) + X(\tau, \sigma)$$
$$X(\sigma\tau, \gamma) = A(\sigma)X(\tau, \gamma) + D(\tau)X(\sigma, \gamma)$$
$$X(\sigma, \tau\gamma) = A(\gamma)X(\sigma, \tau) + D(\tau)X(\sigma, \gamma)$$
$$A(1) = D(1) = 1; \qquad A(\mathbf{c}) = 1, \quad D(\mathbf{c}) = -1$$

$$X(\sigma, 1) = X(1, \sigma) = X(\sigma, \mathbf{c}) = X(\mathbf{c}, \sigma) = 0$$
$$X(\sigma, \tau)X(\gamma, \eta) = X(\sigma, \eta)X(\gamma, \tau).$$

If $\rho\colon G \to \mathrm{GL}_2(B)$ is an odd representation, meaning a continuous homomorphism with $\rho(\mathbf{c}) = \begin{pmatrix} 1 & 0 \\ 0 & -1 \end{pmatrix}$, and if we let $\rho(\sigma) = \begin{pmatrix} a(\sigma) & b(\sigma) \\ c(\sigma) & d(\sigma) \end{pmatrix}$, then $\pi = (A, D, X)$ given by

$$A(\sigma) = a(\sigma), \quad D(\sigma) = d(\sigma) \quad \text{and} \quad X(\sigma, \tau) = b(\sigma)c(\tau)$$

is a pseudorepresentation from G to B. We say π *comes from* ρ. We will reserve the letter π for pseudorepresentations and ρ for representations.

It is an easy fact that if X is identically zero or $X(h_1, h_2) \in B^*$ for some h_1, h_2, then π comes from an odd representation $\rho\colon G \to \mathrm{GL}_2(B)$. In particular, this is always the case if $B = k$, a field.

Definition 2.2. *The* kernel *of* π *is*

$$\ker \pi = \{\sigma \in G \mid A(\sigma) = 1 = D(\sigma), \ X(\sigma, w) = 0 = X(w, \sigma) \ \forall \ w \in G\}.$$

One checks that $\ker \pi \lhd G$. If $M \lhd G$, $M \subset \ker \pi$, then we can factor π through G/M. The resulting pseudorepresentation $\tilde{\pi}\colon G/M \to B$ is called the *factorization* of π through G/M. We will use a tilde to indicate the factorization of a pseudorepresentation through a certain quotient group.

3 Deformation Theory

Let k be a finite field, $\mathrm{char}\, k = p \neq 2$, and G a group as above and satisfying Mazur's Φ_p condition: for any open subgroup $G_0 \subset G$, there are only finitely many continuous homomorphisms $G_0 \to \mathbf{F}_p$. See [5] for equivalent ways of stating this property.

A (pseudo)representation into k will be called *residual*, and we will use the notation $\bar{\pi}$ or $\bar{\rho}$.

Let $\mathcal{C}$ be the category of "coefficient rings:" its objects are complete noetherian local rings with residue field k, and the morphisms ($\mathcal{C}$-morphisms) are homomorphisms of complete local rings inducing the identity on residue fields.

Definition 3.1. *A* deformation *of* $\bar{\pi} = (\bar{A}, \bar{D}, \bar{X})$ *into* $B \in \mathcal{C}$ *is a pseudorepresentation* $\pi = (A, D, X)$ *from* G *to* B *which, when composed with the natural reduction map* $r\colon B \to k$, *yields* $\bar{\pi}$. *That is,* $r \circ A = \bar{A}$, $r \circ D = \bar{D}$ *and* $r \circ X = \bar{X}$.

Definition 3.2. *A* deformation *of $\bar{\rho}$ into $B \in C$ is a strict equivalence class of lifts of $\bar{\rho}$ to B. Two lifts are strictly equivalent if conjugate by an element of $\Gamma_2(B) = \ker\left(GL_2(B) \to GL_2(k)\right)$.*

The deformation functor $F_{\bar{\pi}}: C \to Sets$ (see [9]) is defined by

$$F_{\bar{\pi}}(B) = \text{set of deformations of } \bar{\pi} \text{ into } B \in C.$$

Similarly,

$$F_{\bar{\rho}}(B) = \text{set of deformations of } \bar{\rho} \text{ into } B \in C.$$

Definition 3.3. *The* tangent space *of a functor $F: C \to Sets$ is $t_F = F(k[\varepsilon])$. Here $k[\varepsilon]$ denotes the ring of dual numbers over k ($\varepsilon^2 = 0$). We denote the tangent spaces of $F_{\bar{\pi}}$ and $F_{\bar{\rho}}$ by $t_{\bar{\pi}}$ and $t_{\bar{\rho}}$, respectively.*

One can show as in [9] that $t_{\bar{\pi}}$ and $t_{\bar{\rho}}$ are vector spaces over k. In fact they are finite-dimensional, and it is easy to give their vector space structure (see [6]).

Using Schlessinger's criteria for the representability of this type of functors, Coleman and Mazur [3], [5] have shown:

Proposition 3.1 (Mazur). *If $\bar{\rho}: G \to GL_2(k)$ is absolutely irreducible, then there exist a universal deformation ring $R_{\bar{\rho}} = R(G, k, \bar{\rho}) \in C$ and a universal deformation $\rho_u: G \to GL_2(R_{\bar{\rho}})$. If $\bar{\rho}$ is not absolutely irreducible, then at least it has a hull.*

Let us explain this. If ρ_u is the universal deformation of $\bar{\rho}$, then for any deformation $\rho: G \to GL_2(B)$ of $\bar{\rho}$ into any $B \in C$, there exists a *unique* C-morphism $\psi: R_{\bar{\rho}} \to B$ such that the induced map $GL_2(R_{\bar{\rho}}) \to GL_2(B)$ (which we also call ψ abusing notation) brings ρ_u to ρ:

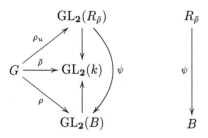

In this case $R_{\bar{\rho}}$ is unique up to canonical isomorphism. If $\bar{\rho}$ only has a hull, then the morphism $\psi: R_{\bar{\rho}} \to B$ need not be unique, and $R_{\bar{\rho}}$ is determined only up to non-canonical isomorphism. In this case we call $R_{\bar{\rho}}$ the *versal* deformation ring of $\bar{\rho}$, and ρ_u the *versal* deformation.

For pseudorepresentations the situation is nicer, which is one good reason to study them:

Proposition 3.2 (Coleman, Mazur). *If $\bar{\pi}\colon G \to k$ is any residual pseudorepresentation, then it has a universal deformation ring $R_{\bar{\pi}} = R(G, k, \bar{\pi}) \in \mathcal{C}$ and a universal deformation $\pi_u\colon G \to R_{\bar{\pi}}$.*

This means that for any deformation $\pi\colon G \to B$ of $\bar{\pi}$ into any $B \in \mathcal{C}$, there exists a *unique* $\mathcal{C}$-morphism ψ such that $\psi \circ \pi_u = \pi$:

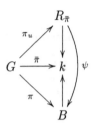

The ring $R_{\bar{\pi}}$ is unique up to canonical isomorphism.

We know from Section 2 that $\bar{\pi}$ comes from a representation $\bar{\rho}$. There is a natural map $\phi\colon R_{\bar{\pi}} \to R_{\bar{\rho}}$:

$$
\begin{array}{ccc}
G & \xrightarrow{\ \pi_u\ } & R_{\bar{\pi}} \\
 & {}_{\pi}\searrow & \downarrow{\scriptstyle\phi} \\
 & & R_{\bar{\rho}}
\end{array}
$$

In this picture $\pi\colon G \to R_{\bar{\rho}}$ is the pseudorepresentation coming from $\rho_u\colon G \to \mathrm{GL}_2(R_{\bar{\rho}})$, the "versal" deformation of $\bar{\rho}$.

Lemma 3.3. *If $\bar{\rho}$ is absolutely irreducible, then $\phi\colon R_{\bar{\pi}} \xrightarrow{\sim} R_{\bar{\rho}}$.*

This is an easy exercise in category theory.

We now turn our attention to the case where $\bar{\rho}$ is not absolutely irreducible. Otherwise (according to the lemma), looking for $R_{\bar{\pi}}$ is basically the same as looking for $R_{\bar{\rho}}$. Now, if $\bar{X}$ is not identically zero, then $\bar{\rho}$ is easily seen to be absolutely irreducible. Therefore, from now on we look only at the case where $\bar{X}$ is identically zero. We also restrict our setting to a situation which is of interest from a number-theoretic perspective.

4   The Case Where $\bar{X}$ is Identically Zero

Let k be a finite field, $\operatorname{char} k = p \neq 2$ and let $G_{\mathbf{Q},S}$ be the Galois group of the maximal extension $\mathbf{Q}_S$ of $\mathbf{Q}$ (in a fixed algebraic closure) which is unramified outside a finite set of primes S containing p and ∞. As noted in [5], $G_{\mathbf{Q},S}$ satisfies Mazur's Φ_p condition.

Let $\bar{\rho}\colon G_{\mathbf{Q},S} \to \mathrm{GL}_2(k)$ be an odd, reducible representation. Say $\bar{\rho}(\sigma) = \begin{pmatrix} \bar{a}(\sigma) & \bar{b}(\sigma) \\ 0 & \bar{d}(\sigma) \end{pmatrix}$. Let $\bar{\pi} = (\bar{A}, \bar{D}, \bar{X})$ be its associated residual pseudorepresentation. Then $\bar{X}(\sigma, \tau) = 0 \; \forall \, \sigma, \tau \in G_{\mathbf{Q},S}$. Equivalently, we can start with a residual pseudorepresentation $\bar{\pi}\colon G_{\mathbf{Q},S} \to k$ with $\bar{X}$ identically zero and let $\bar{\rho}\colon G_{\mathbf{Q},S} \to \mathrm{GL}_2(k)$ be an odd representation such that $\bar{\pi}$ comes from $\bar{\rho}$.

Lemma 4.1. $\ker \bar{\pi} / \ker \bar{\rho}$ *is a finite, elementary p-abelian group.*

Proof. $(\ker \bar{\pi}\colon \ker \bar{\rho}) < \infty$ because $(G_{\mathbf{Q},S}\colon \ker \bar{\rho}) < \infty$, $\mathrm{Im}\, \bar{\rho}$ being finite. If $\sigma \in \ker \bar{\pi}$, then $\sigma^p \in \ker \bar{\rho}$ because $\mathrm{char}\, k = p$. Finally, the restriction of $\bar{\rho}$ to $\ker \bar{\pi}$ maps into an abelian group. $\qquad\square$

Let $N_{\bar{\pi}}$ be the characteristic subgroup of $\ker \bar{\pi}$ such that $\ker \bar{\pi}/N_{\bar{\pi}}$ is the maximal pro-p quotient of $\ker \bar{\pi}$. Similarly, let $N_{\bar{\rho}}$ be the characteristic subgroup of $\ker \bar{\rho}$ such that $\ker \bar{\rho}/N_{\bar{\rho}}$ is the maximal pro-p quotient of $\ker \bar{\rho}$.

Lemma 4.2. $N_{\bar{\pi}} = N_{\bar{\rho}}$.

This follows from the lemma we just proved. Call this group N.

Definition 4.1. *The quotient group $G = G_{\mathbf{Q},S}/N$ is called the p-completion of $G_{\mathbf{Q},S}$ with respect to $\bar{\pi}$ (or $\bar{\rho}$).*

Our strategy for finding the universal deformation of $\bar{\pi}$ is to factor through G, following Boston and Mazur [1], [2], [5]. The resulting deformation problem is easier because we can get a hold on the structure of G. For representations, all deformations factor through G. For pseudorepresentations, we only have a weaker result, as will be seen below: all $\pi \in t_{\bar{\pi}}$ factor through G. This yields the result that $\dim_k t_{\bar{\pi}} = \dim_k t_{\bar{\bar{\pi}}}$, which can be helpful for computing $R_{\bar{\pi}}$ from $R_{\bar{\bar{\pi}}}$.

Proposition 4.3. *For all $\pi \in t_{\bar{\pi}}$, $N \subset \ker \pi \subset \ker \bar{\pi}$.*

Proof. Computations using the axioms of a pseudorepresentation yield a lemma: if $\sigma \in \ker \bar{\pi}$, then $\sigma^p \in \ker \pi$. Therefore if $(\ker \bar{\pi})^p$ is the subgroup of $\ker \bar{\pi}$ topologically generated by the p-th powers, then we have $N \subset (\ker \bar{\pi})^p \subset \ker \pi$. $\qquad\square$

Corollary 4.4. *For all $\pi \in t_{\bar{\pi}}$, π factors through $G = G_{\mathbf{Q},S}/N$.*

Proposition 4.5. $\dim_k t_{\bar{\pi}} = \dim_k t_{\bar{\bar{\pi}}}$.

Proof. Take a basis for $t_{\bar{\bar{\pi}}}$ and compose each basis element with the quotient homomorphism $\varphi\colon G_{\mathbf{Q},S} \to G$. The result is a basis for $t_{\bar{\pi}}$. It is easy to show its elements are linearly independent, and they span all of $t_{\bar{\pi}}$ because of the corollary. $\qquad\square$

5   An Example

Let p be an odd prime and $\bar{\rho}\colon G_{\mathbf{Q},S_0} \to \mathrm{GL}_2(\mathbf{F_p})$ an odd representation with image $\left\{ \begin{pmatrix} 1 & 0 \\ 0 & 1 \end{pmatrix}, \begin{pmatrix} 1 & 0 \\ 0 & -1 \end{pmatrix} \right\}$. Let $\bar{\pi}\colon G_{\mathbf{Q},S_0} \to \mathbf{F_p}$ be the associated residual pseudorepresentation. We see that $\ker \bar{\pi} = \ker \bar{\rho}$. Call this group G_K, and let K be its fixed field. Then K is an imaginary quadratic number field. We assume that S_0 contains p and the primes that ramify in the extension $K/\mathbf{Q}$. Let S be the set of primes of K lying above the primes in $S_0 - \{\infty\}$.

We need to borrow a definition from a paper by Boston and Mazur [2]. Let E denote the group of units in the ring of integers of K, and for any nonarchimedean place ν, let E_ν be the group of units of the ring of integers of K_ν (the completion of K with respect to ν). For any field F, let $\mu_p(F)$ be the group of p-th roots of unity in F. Finally, for any group W, let $\bar{W}$ denote its p-Frattini quotient, i.e., its maximal elementary p-abelian quotient.

Definition 5.1. *The pair (K,S) is* neat *for p if the following conditions hold:*

(a)   The map $\bar{E} \to \bigoplus_{\nu \in S} \bar{E}_\nu$ is injective. That is, any global unit which is locally a p-th power for all $\nu \in S$ is globally a p-th power.

(b)   The class number h_K of K is prime to p.

(c)   The map $\mu_p(K) \to \bigoplus_{\nu \in S} \mu_p(K_\nu)$ is surjective (and so it is an isomorphism).

As an example, let $p = 3$ and $K = \mathbf{Q}(i)$, where $i^2 = -1$. $K/\mathbf{Q}$ is ramified at 2 and ∞. Let $S_0 = \{2,3,\infty\}$ and let $\bar{\rho}\colon G_{\mathbf{Q},S_0} \to \mathrm{GL}_2(\mathbf{F_3})$ have image as above and kernel G_K, the group with fixed field K. Then S is the set of primes of K lying above 2 and 3. In this example, $\bar{E}$ is trivial (since E has four elements) and $h_K = 1$. Also, it is easy to check that the $\mu_3(K_\nu)$ are trivial as well. Hence (K,S) is neat for $p = 3$.

In what follows, we assume that (K,S) is neat for p. Let L be the maximal pro-p extension of K unramified outside S. We have a tower of fields and their corresponding Galois groups:

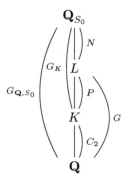

Boston (using results of Koch) shows that $G = P \rtimes C_2$, where P is free pro-p on two generators $\tilde{x}_1$ and $\tilde{x}_2$, the first of which is fixed and the second inverted by the non-trivial element $\tilde{x}$ of C_2.

Proposition 5.1. $R_{\tilde{\pi}} \cong \mathbf{Z}_p[[Y_1, Y_2, Y_3]]$.

Proof. We use the axioms of a pseudorepresentation and the structure of G to show that all deformations must have certain properties. The most general such deformation is the universal deformation of $\tilde{\pi}$, and it is given by $\tilde{\pi}_u = (\tilde{A}_u, \tilde{D}_u, \tilde{X}_u)$, where

$$\tilde{A}_u(\tilde{x}) = 1, \quad \tilde{D}_u(\tilde{x}) = -1,$$
$$\tilde{A}_u(\tilde{x}_1) = 1 + Y_1, \quad \tilde{D}_u(\tilde{x}_1) = 1 + Y_2,$$
$$\tilde{A}_u(\tilde{x}_2) = \sqrt{1 + Y_3} = \tilde{D}_u(\tilde{x}_2), \quad \tilde{X}_u(\tilde{x}_2, \tilde{x}_2) = Y_3.$$

$\square$

Lemma 5.2. $\dim_{\mathbf{F}_p} t_{\tilde{\pi}} = 3$.

Proof. For $i = 1, 2, 3$ define the $\mathcal{C}$-morphisms $f_i \colon R_{\tilde{\pi}} \to \mathbf{F}_p[\varepsilon]$ by $f_i \colon Y_i \mapsto \varepsilon$, $Y_j \mapsto 0$ for $j \neq i$, and by reducing coefficients modulo p. Let $\tilde{\pi}_i = f_i \circ \tilde{\pi}_u$. Then one checks that $\{\tilde{\pi}_1, \tilde{\pi}_2, \tilde{\pi}_3\}$ is a basis for $t_{\tilde{\pi}}$. $\square$

By Proposition 4.5, $\dim_{\mathbf{F}_p} t_{\bar{\pi}} = 3$ also. From Schlessinger's work, $R_{\bar{\pi}}$ is a quotient of $\mathbf{Z}_p[[Y_1, Y_2, Y_3]]$.

Proposition 5.3. $R_{\bar{\pi}} \cong R_{\tilde{\pi}} = \mathbf{Z}_p[[Y_1, Y_2, Y_3]]$.

Proof. Say $R_{\bar{\pi}} = \mathbf{Z}_p[[Y_1, Y_2, Y_3]]/I$. We have a diagram

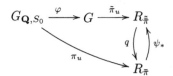

where q is the natural quotient homomorphism and ψ_* the (unique) $\mathcal{C}$-morphism coming from the universal property of π_u. One checks from the knowledge of $\bar{\pi}_u$ that the image of ψ_* contains the subring $\mathbf{Z}[Y_1, Y_2, Y_3]$. This is a dense subring, so $\operatorname{Im}\psi_*$ is dense in $R_{\bar{\bar{\pi}}}$. But ψ_* is continuous ($\mathcal{C}$-morphisms must be continuous) and $R_{\bar{\pi}}$ is compact, so that $\operatorname{Im}\psi_*$ is compact and, therefore, closed. Since it is dense, it must be all of $R_{\bar{\bar{\pi}}}$. That is, ψ_* is surjective. Therefore the map $\psi_* \circ q\colon R_{\bar{\bar{\pi}}} \to R_{\bar{\bar{\pi}}}$ is surjective. Since any surjective endomorphism of a noetherian ring is actually an isomorphism, $\psi_* \circ q$ must also be injective. This implies that q is injective, which means that $I = \{0\}$. Therefore $R_{\bar{\bar{\pi}}} \cong R_{\bar{\pi}} = \mathbf{Z_p}[[Y_1, Y_2, Y_3]]$. $\qquad\square$

As a final comment, we can also compute the "versal" deformation ring of $\bar{\rho}$. It is $R_{\bar{\rho}} \cong \mathbf{Z_p}[[T_1, T_2, T_3, T_4]]$, and the natural map $\phi\colon R_{\bar{\pi}} \to R_{\bar{\rho}}$ takes $Y_1 \mapsto T_1$, $Y_2 \mapsto T_2$ and $Y_3 \mapsto T_3 T_4$.

References

[1] N. Boston, *Explicit deformation of Galois representations*, Invent. Math. **103** (1991), 181–196.

[2] N. Boston and B. Mazur, *Explicit universal deformations of Galois representations*, Algebraic Number Theory – in honor of K. Iwasawa (Coates, Greenberg, Mazur, , and Satake, eds.), Advanced Studies in Pure Mathematics, vol. 17, Academic Press, 1989, pp. 1–21.

[3] R. Coleman and B. Mazur, *The eigencurve*, Galois Representations in Arithmetic Algebraic Geometry (Scholl and Taylor, eds.), Cambridge University Press, 1998, pp. 1–113.

[4] H. Hida, *Nearly ordinary Hecke algebras and Galois representations of several variables*, JAMI Inaugural Conference Proceedings, 1990, (supplement to) Amer. J. Math., pp. 115–134.

[5] B. Mazur, *Deforming Galois representations*, Galois Groups over $\mathbf{Q}$ (Ihara, Ribet, and Serre, eds.), MSRI Publications, vol. 16, Springer-Verlag, 1989, pp. 385–437.

[6] ———, *An introduction to the deformation theory of Galois representations*, Modular Forms and Fermat's Last Theorem (Cornell, Silverman, and Stevens, eds.), Springer-Verlag, 1997, pp. 243–311.

[7] L. Nyssen, *Pseudo-représentations*, Math. Ann. **306** (1996), 257–283.

[8] R. Rouquier, *Caractérisation des caractères et pseudo-caractères*, J. Algebra **180** (1996), 571–586.

[9] M. Schlessinger, *Functors on Artin rings,* Trans. Amer. Math.Soc. **130** (1968), 208–222.

[10] R. Taylor, *Galois representations associated to Siegel modular forms of low weight,* Duke Math. J. **63** (1991), no. 2, 281–332.

[11] A. Wiles, *On ordinary λ-adic representations associated to modular forms,* Invent. Math. **94** (1988), 529–573.

The Sum of Multiplicative Functions Arising in Selberg's Sieve

George Greaves

1 Introduction

Following a notation frequently used in this subject, let $\mathcal{A}$ be a finite sequence of integers, X an associated real number and ω a multiplicative function such that $0 \le \omega(p) < p$ for primes p. Define "remainders" $r_{\mathcal{A}}(d)$ by the equation

$$\sum_{\substack{a \in \mathcal{A} \\ a \equiv 0, \, \bmod \, d}} 1 = \frac{X \, \omega(d)}{d} + r_{\mathcal{A}}(d). \tag{1.1}$$

The sieve method deals with the number of $a \in \mathcal{A}$ divisible by no "sifting" prime p with $p < z$:

$$S(\mathcal{A}, z) = \sum_{\substack{a \in \mathcal{A} \\ (a, P(z)) = 1}} 1, \quad \text{where} \quad P(z) = \prod_{p < z; \, p \in \mathcal{P}} p,$$

$\mathcal{P}$ being a specified set of primes. If we make $\omega(p) = 0$ when $p \notin \mathcal{P}$, then the condition $p \in \mathcal{P}$ will be implicit in the sequel.

For squarefree d write

$$g(d) = \prod_{p \mid d} \frac{\omega(p)}{p - \omega(p)}. \tag{1.2}$$

Then Selberg's upper bound sieve estimate takes the form

$$S(\mathcal{A}, z) \le \frac{X}{G_z(x)} + \frac{1}{\lambda^2(1)} \sum_{d_1 < x} \sum_{d_2 < x} \lambda(d_1) \lambda(d_2) r_{\mathcal{A}}([d_1, d_2]), \tag{1.3}$$

in which $[\cdot, \cdot]$ denotes a least common multiple. An optimisation procedure leads to a specification of the numbers $\lambda(d)$ in which

$$\frac{\lambda(d)\omega(d)}{d} = C \sum_{\substack{n < x \\ n \equiv 0, \, \bmod \, d}} g(n),$$

103

it being usual to choose C such that $\lambda(1) = 1$. Then

$$G_z(x) = \sum_{\substack{d|P(z) \\ d<x}} g(d) = \sum_{d<x} g_z(d), \tag{1.4}$$

where

$$g_z(d) = \begin{cases} g(d) & \text{if } d \mid P(z) \\ 0 & \text{otherwise.} \end{cases}$$

In practice the choice of x is restricted, depending on X, by the requirement that the sum over d_1 and d_2 in (1.3) should not be too large. In favourable situations a choice $x = \sqrt{X}/\log X$ might be viable. It is the estimation of the sum (1.4) that we discuss in this article.

A specimen application. An illustrative example of the type of sequence $\mathcal{A}$ to which this process might be applied is

$$\mathcal{A} = \Big\{ (n - a_1)(n - a_2) \dots (n - a_k) : 1 \leq n \leq X \Big\},$$

where the numbers a_i are distinct integers. Here the estimate (1.1) follows by estimating the contribution, for each i, from those n with $1 \leq n < X$ and $n \equiv a_i$, mod d. This gives (1.1), with $w(p)$ the number of distinct residue classes, mod p, occupied by the numbers a_i, and $|r_A(d)| \leq w(d)$. Thus $w(p) = k$, except for a finite number of primes dividing some $a_i - a_j$ for which $w(p)$ is smaller.

In such a situation it could be important that all results are uniform in the numbers a_i. For example, in an elementary approach to the Goldbach problem one might sift the sequence $n(N - n)$, where $1 \leq n \leq N$ and N is an even integer. Then (1.1) would follow with $X = N$ and $r_A(d)$ suitably small, and, while $w(p) = 2$ for all save finitely many p, we would have $w(p) = 1$ whenever $p \mid X$, which might have a serious effect on the X-dependence, and thus on the x-dependence, of the sum (1.4).

In this example, as in many others, the value of $w(p)$ is rather smaller at the "bad" primes, so the function $g(p)$ defined by (1.2) has the property that there exist constants $\kappa > 0$ and $A > 1$ such that

$$\sum_{u \leq p < v} g(p) \log p - \kappa \log \frac{v}{u} \leq A \quad \text{when} \quad 2 \leq u < v. \tag{1.5}$$

Here the constant κ is referred to as the *dimension* or *sifting density* of the function w. The subsidiary constant A is assumed to exceed 1 for convenience, so that certain O-terms can subsequently be written as $O(A)$.

Readers acquainted with the sieve literature will be familiar with the fact that in the combinatorial sieve methods of Brun, Rosser and Iwaniec [6] a one-sided condition on $\omega(p)$, similar to (1.5), fits in to the theory in a very natural way. In Selberg's method the situation is slightly less clear, because to obtain an upper bound in (1.3) one requires a lower bound for the sum $G_z(x)$ defined by (1.4) and (1.2), and this is not obtainable from (1.5). However, what is obtainable is a lower bound for the quotient $G_z(x)/G_z(\infty)$. Here we use a natural extension of the notation (1.4) whereby

$$G_z(\infty) = \sum_{d|P(z)} g(d) = \prod_{p<z}\left(1 + g(p)\right) = \prod_{p<z}\left(1 - \frac{\omega(p)}{p}\right)^{-1}. \qquad (1.6)$$

Again, this is a factor whose appearance in a sieve situation is entirely natural.

A method appearing in the literature for obtaining the required lower bound proceeds by initially assuming a two-sided bound in which the quantity on the left of (1.5) is also bounded below by a number L, which may depend on x. On this hypothesis one can obtain a lower bound (depending on L) for the sum $G_z(x)$. In the resulting lower bound for the quotient $G_z(x)/G_z(\infty)$ a certain "topping up" procedure on the function $\omega(p)$ can be used. This procedure was first suggested by W. B. Jurkat in a lecture in Urbana in 1973, and was developed further in [7]. In this approach the numbers $\omega(p)$ are selectively increased until the number L is replaced by one that depends only on the constants κ, A of the problem. It is important that the quotient $G_z(x)/G_z(\infty)$ decreases as this process proceeds.

A more direct approach to this question, in which no such number L need be mentioned, is also possible. The author and M. N. Huxley [4] showed how to deal with the question in this style in the case $z = x$. We need not state this result explicitly here, because it is just the case $s = 1$ of Theorem 1, to be enunciated below.

It is important, more particularly in instances where Selberg's upper bound is used as a weapon in a lower bound sieve method, to obtain good estimates in the case $z < x$. Write $x = z^s$, so that we will be concerned with the case $s > 1$. We will show

$$G_z(z^s) \geq G_z(\infty)\left(\sigma(s) + O\left(\frac{\eta(s)}{\log z}\right)\right), \qquad (1.7)$$

where σ, to be described, is a moderately recondite function satisfying $\sigma(s) \to 1$ as $s \to \infty$, and the expression $\eta(s)$ describes the s-dependence of the error estimate. Actually $\sigma(s) = 1 + O\left(e^{-s\log s}\right)$, but this fact will not be used in this paper.

The previous literature. A standard method establishes such a result recursively for $s \leq N$, using an induction on the integer N. Unless conducted with extraordinary delicacy such arguments are likely to lead to an increase in the error term $\eta(s)$ as N increases. In this fashion Halberstam and Richert [5] obtained a version of (1.7) in which the estimate for the error term was $\eta(s) \ll s^{2\kappa+1}$.

On the other hand when s is large one can proceed in a simpler way, by using the well-known device attributed to R. A. Rankin. In this way one arrives at the following lemma (see §7 in [8], or cf. Lemma 4.1 in [5]).

Lemma 1. *Let B be any constant such that*

$$B \geq B(z) := \frac{1}{\log z} \sum_{p<z} \frac{\omega(p) \log p}{p}.$$

Then

$$\frac{G_z(z^s)}{G_z(\infty)} \geq 1 - \exp(-\psi_B(s)),$$

where

$$\psi_B(s) = \max\left\{0, s\log\frac{s}{B} - s + B\right\}.$$

Since $g(p) \geq \omega(p)/p$ in (1.2), the property (1.5) guarantees the existence of such a number B, for example $\kappa + A/\log z$. On the other hand Lemma 1 does not depend on any assumption like (1.5), but is valid, for each z, with B equal to the number $B(z)$ itself.

We will not use Lemma 1 in this paper, except in the trivial case where

$$G_z(z^s) = G_z(\infty) \quad \text{if} \quad s > \log P(z)/\log z, \tag{1.8}$$

so that $z^s > P(z)$. Indeed, this equation will only be required when s is sufficiently large.

In the situation just summarised the entry $\eta(s)$ in (1.7) initially increases like $s^{2\kappa}$, but when $s > \log\log z$ assumes the much more satisfactory form $e^{-s\log s+O(s)}$. It would be bizarre if this were the true state of affairs, and indeed it is possible to obtain a substantial improvement, as in Theorems 1 and 2 below.

The function σ. This expression, which plays a key rôle in our theorems, appears in the paper [1], though not using the same notation. The notation σ follows [5], except that we write $\sigma(s)$ where $\sigma(2s)$ appears in [5].

For $s > 0$ this function will satisfy the integral equation

$$s\,\sigma(s) = \int_0^s \sigma(u)\,du + \kappa \int_{s-1}^s \sigma(u)\,du, \tag{1.9}$$

with $\sigma(s) = 0$ if $s < 0$. In differentiated form this gives

$$s\,\sigma'(s) = \kappa\big(\sigma(s) - \sigma(s-1)\big), \tag{1.10}$$

for all real s. The equation (1.10) can be rewritten as

$$\frac{d}{ds}\left(\frac{\sigma(s)}{s^\kappa}\right) + \frac{\kappa\,\sigma(s-1)}{s^{\kappa+1}} = 0.$$

Then (1.10) has a solution continuous at all real s for which

$$\sigma(s) = Cs^\kappa \quad \text{when} \quad 0 \le s \le 1; \qquad C = e^{-\gamma\kappa}/\Gamma(\kappa+1). \tag{1.11}$$

For larger s the expression $\sigma(s)$ can be obtained on $(0, N]$ by induction on the integer N, using successive integrations. It was shown in [1] that when C is as specified the object $\sigma(s)$ thus constructed increases with s and tends to 1 as $s \to \infty$.

Statement of results. Theorem 1 states a result of the type that one might expect to be true, following the preceding discussion. Theorem 2 is the relatively weak corollary of Theorem 1 that follows by observing $\psi_B(s) > 0$ in Lemma 1. It is, however, equally useful in situations where our concern is with $\sup_s R_z(s)$, taken over a set that includes small values of s. Examples of such situations might include the lower bound sieve method considered in [1] and in Chapter 7 of [5], and in its subsequent developments in [3].

Theorem 1. *Assume that g satisfies* (1.5), *where $\kappa \ge 1$. Define R_z by*

$$\frac{G_z(z^s)}{G_z(\infty)} = \sigma(s) - \frac{R_z(s)}{\log z},$$

where σ is as in (1.9). *Then, if $z \ge 2$ and $s \ge 1$,*

$$R_z(s) \le A^\kappa e^{-\psi_B(s)+O(B)},$$

with B and ψ_B are as in Lemma 1. The implied constant may depend upon κ.

Theorem 2. *In Theorem 1, there is a positive constant c such that*

$$R_z(s) \le cA^\kappa,$$

when $z \ge 2$ and $s \ge 1$.

A direct proof of Theorem 2 is also possible, independently of Theorem 1, and will be supplied in this article. It will become clear at what stage a factor $\exp\left(-\psi_B(s)\right)$ is abandoned as being merely $O(1)$. A proof of Theorem 1 involves paying attention to this s-dependence throughout, and in particular strengthening Lemma 3. This proof will appear in a book on Sieves in Number Theory that is in course of preparation by the present author, to be published by Springer-Verlag, and which may be expected to be in print at about the time when this paper appears.

2   Integral Equations

We collect some relevant results and methods relating to integral equations satisfied by the expressions $G_z(x)$ and $\sigma(s)$.

The integral equation for $G_z(s)$.   In connection with (1.5) denote

$$\eta(v) = \sum_{2 \le p < v} g(p) \log p - \kappa \log v,$$

so that (1.5) implies $\eta(v) \le A$. It is now the case that $G_z(x)$ satisfies the integral equation

$$G_z(x) \log x = \int_1^x G_z(t)\,\frac{dt}{t} + \kappa \int_{x/z}^x G_z(t)\,\frac{dt}{t} + \delta_z(x), \qquad (2.1)$$

when $z \le x$, where $\delta_z(x)$ is of the shape

$$\delta_z(x) = \sum_{n < x} g_z(n) E_{z,n}\left(\frac{x}{n}\right) \quad \text{with} \quad E_{z,n}(t) \le \eta\big(\min\{t, z\}\big) \le A. \quad (2.2)$$

The equation (2.1) appears in Chapter 5, Eq. (3.5) of [5], except that the expression given for δ is different because those authors work with an assumption equivalent to bounding the quantity η from below as well as from above. On the other hand the case $z = x$ of (2.1) with the estimate (2.2) for δ was derived in [4]. The central idea used in these references is, in a sum $\sum g(n) \log n$, to express $\log n$ as a sum of logarithms of primes.

Because the relation between the s-dependence of these two quantities is central to our theme, we will re-express G and δ by the change of variable

$$u = \frac{\log t}{\log z}, \quad s = \frac{\log x}{\log z}, \quad G_z(t) = F_z(u), \quad \delta_z(t) = \Delta_z(u), \qquad (2.3)$$

so that $G_z(x) = G_z(z^s) = F_z(s)$, for example. Then $\log z \, du = dt/t$, so that (2.1) assumes the form

$$s \, F_z(s) = \int_0^s F_z(u) \, du + \kappa \int_{s-1}^s F_z(u) \, du + \frac{1}{\log z} \Delta_z(s) \quad \text{if} \quad s \geq 1. \quad (2.4)$$

The relationship with the equation (1.9) satisfied by $\sigma(s)$ is transparent: (1.9) is obtained from (2.4) by replacing F_z by σ and ignoring the error term involving $\Delta_z(s)$.

The adjoint equation. This device was introduced in [2], and its use in the context of sieve methods was greatly developed in [6]. It is helpful in eliciting properties of functions defined by difference-differential equations such as (1.10). A short résumé of this technique is given here, because we plan to apply it also to the approximate integral equation (2.1) for the discontinuous function G_z.

The equation *adjoint* to the equation (1.10) for $\sigma(s)$ is

$$\frac{d}{ds}\big(s \, M(s)\big) + \kappa\big(M(s) - M(s+1)\big) = 0 \quad \text{when} \quad s > 0. \quad (2.5)$$

This equation can also be rewritten as

$$\frac{d}{ds}\big(s^{\kappa+1} M(s)\big) = \kappa s^\kappa M(s+1). \quad (2.6)$$

It is easy to verify that this has a solution in terms of Laplace transforms:

$$M(s) = \int_0^\infty e^{-sx} \phi(x) \, dx, \quad \text{with} \quad \phi(x) = \exp\left(\kappa \int_0^x \frac{1 - e^{-t}}{t} \, dt\right). \quad (2.7)$$

Then, since $\phi(0) = 1$,

$$s \, M(s) = \phi(0) + \int_0^\infty e^{-sx} \phi'(x) \, dx \to 1 \quad \text{as} \quad s \to \infty. \quad (2.8)$$

In fact $x \, \phi'(x) = \kappa\big(1 - e^{-x}\big)\phi(x)$, from which

$$\frac{d}{ds}\big(s \, M(s)\big) = -\int_0^\infty e^{-sx} \, x \, \phi'(x) \, dx = -\kappa \int_0^\infty e^{-sx}\big(1 - e^{-x}\big)\phi(x) \, dx$$

$$= -\kappa\big(M(s) - M(s+1)\big),$$

as asserted in (2.5).

The relation corresponding to (2.8) when $s \to 0$ follows on noting

$$M(s) = \int_0^\infty \exp\left(\kappa \int_0^x \frac{1 - e^{-t}}{t}\, dt - \kappa \int_1^x \frac{dt}{t}\right) x^\kappa e^{-sx}\, dx.$$

$$= \frac{1}{s^{\kappa+1}} \int_0^\infty \exp\left(\kappa \int_0^1 \frac{1 - e^{-t}}{t}\, dt - \kappa \int_1^{u/s} \frac{e^{-t}}{t}\, dt\right) u^\kappa e^{-u}\, du$$

$$\sim \frac{e^{\kappa\gamma}}{s^{\kappa+1}} \int_0^\infty u^\kappa e^{-u}\, du \quad \text{as} \quad s \to 0 + .$$

Thus

$$s^{\kappa+1} M(s) \to e^{\kappa\gamma} \Gamma(\kappa + 1) \quad \text{as} \quad s \to 0 + . \tag{2.9}$$

The connection between the equation (1.10) and its adjoint (2.5) involves the inner product

$$\langle Q, M \rangle(s) = s\, Q(s) M(s) - \kappa \int_{s-1}^s Q(u) M(u + 1)\, du. \tag{2.10}$$

This can be rewritten in the form

$$\langle Q, M \rangle(s) = Q(s) + \kappa \int_{s-1}^s \big(Q(s) - Q(u)\big) M(u + 1)\, du. \tag{2.11}$$

For integration of (2.5) gives

$$s\, M(s) - \kappa \int_{s-1}^s M(u + 1)\, du = 1,$$

the value of the constant on the right following from (2.8), and the equivalence of (2.10) and (2.11) follows.

The application to $\sigma(s)$. It is now the case that $\langle \sigma, M \rangle(s)$ is constant in $s > 0$. For (1.10) and (2.5) give

$$\frac{d}{ds}\big(s\, \sigma(s) M(s)\big) = \kappa\, \sigma(s) M(s + 1) - \kappa\, \sigma(s - 1) M(s).$$

But $\sigma(s)$ is as in (1.11) when $0 < s < 1$, so that

$$\langle \sigma, M \rangle(s) = \langle \sigma, M \rangle(1) = C M(1) - C\kappa \int_0^1 u^\kappa M(u + 1)\, du \tag{2.12}$$

$$= C \lim_{u \to 0+} u^{\kappa+1} M(u) = 1,$$

the last two steps following from (2.6) and (2.9). With (2.8) and (2.11) this gives that $\lim_{s \to \infty} \sigma(s)$, once known to exist, must take the value 1, as was shown (by a different method) in [1].

The application to $F_z(s)$. On comparison of (1.9) with (2.4) it appears that it is reasonable to expect an analogue for F_z of the relation (2.12) just described for σ. In fact, (2.4) gives

$$\frac{1}{\log z} \int_1^s M(u)\,d\Delta_z(u)$$
$$= \int_1^s M(u)\left(u\,dF_z(u) - \kappa\,F_z(u)\,du + \kappa\,F_z(u-1)\,du\right).$$

Since M satisfies (2.5) the first contribution on the right is

$$\int_1^s u\,M(u)\,dF_z(u)$$
$$= s\,M(s)F_z(s) - M(1)F_z(1) + \kappa \int_1^s F_z(u)\left(M(u) - M(u+1)\right)du.$$

Thus we obtain our analogue of (2.12),

$$\frac{1}{\log z} \int_1^s M(u)\,d\Delta_z(u) = s\,M(s)F_z(s) - M(1)F_z(1)$$
$$- \kappa \int_1^s \left(F_z(u)M(u+1) - F_z(u-1)M(u)\right)du,$$

that is to say

$$\frac{1}{\log z} \int_1^s M(u)\,d\Delta_z(u) = \langle F_z, M\rangle(s) + C_z,$$

where the inner product notation is as in (2.10), and where C_z is independent of s.

We can infer a more convenient presentation of C_z than the initial one. The fact (1.8) implies, in the current notation (2.3), that $F_z(s) = F_z(\infty)$ if s is large enough, in which $F_z(\infty)$ is just the number $G_z(\infty)$ appearing in Theorem 1. With the expression (2.11) of our inner product, this shows

$$\langle F_z, M\rangle(s) \to F_z(\infty) \quad \text{as} \quad s \to \infty,$$

because of (2.8). Hence

$$C_z = \frac{1}{\log z} \int_1^\infty M(u)\,d\Delta_z(u) - F_z(\infty),$$

so that

$$\langle F_z, M\rangle(s) = -\frac{1}{\log z} \int_s^\infty M(u)\,d\Delta_z(u) + F_z(\infty). \qquad (2.13)$$

Because of (2.3) the remainder term $R_z(s)$ in Theorems 1 and 2 satisfies

$$\frac{F_z(s)}{F_z(\infty)} = \sigma(s) - \frac{R_z(s)}{\log z}. \tag{2.14}$$

The inner product $\langle Q, M \rangle$ in (2.10) is linear in Q, so subtraction of (2.13) from (2.12) gives

$$\langle R_z, M \rangle(s) = \frac{1}{F_z(\infty)} \int_s^\infty M(u)\, d\Delta_z(u). \tag{2.15}$$

3   Proof of Theorem 2

The proof of this theorem uses the following three lemmas.

Lemma 2. *Suppose, when t and n are in an interval $[x,y)$, that $f'(t)$ is continuous, $f(t)$ is positive and decreasing, $g(n) \geq 0$, and $E(t,n) \leq A$. Then*

$$\int_{x<t<y} f(t)\, d \sum_{x \leq n < t} g(n)E(t,n) \leq A \sum_{x \leq n < y} f(n)g(n).$$

Since $f'(t) \leq 0$ the integral on the left is

$$f(y) \sum_{x \leq n < y} g(n)E(y,n) + \int_{x<t<y} \sum_{x \leq n < t} g(n)E(t,n)|f'(t)|\, dt$$

$$\leq A f(y) \sum_{x \leq n < y} g(n) + A \int_{x<t<y} \sum_{x \leq n < t} g(n)|f'(t)|\, dt$$

$$= A \int_{x<t<y} f(t)\, d \sum_{x \leq n < t} g(n) = A \sum_{x \leq n < y} f(n)g(n),$$

as required.

Lemma 3. *Assume that $Q(s)$ is continuous in $s > 0$, apart from possible jump discontinuities at which it may decrease. Suppose that $Q(s) < 0$ when $0 < s \leq 1$ and that $\langle Q, M \rangle(s) < 0$ when $s \geq 1$, where M is as in (2.7). Then $Q(s) \leq 0$ for all $s > 0$.*

If $Q(s) > 0$ for any $s > 0$, let s_1 be the infimum of these s. Then s_1 is not a discontinuity of Q, so that $Q(s_1) = 0$ and $Q(x) \leq 0$ when $x < s_1$. Furthermore $s_1 \geq 1$. But $M(x) > 0$ when $x > 0$, because of (2.7), so the expression (2.10) would then give $\langle Q, M \rangle(s_1) \geq 0$, a contradiction. Lemma 3 follows.

Lemma 4. *Assume the hypotheses of Theorem* 1. *Then when* $0 < s \le 1$

$$R_z(s) \le cA^\kappa,$$

for a certain c *depending only on* $\kappa \ge 1$.

It is at this point that we need prior knowledge of Theorem 2 in the case $s = 1$. In fact this is known in the sharper form

$$\frac{G_z(z)}{G_z(\infty)} \ge \sigma(1) + O\left(\frac{A}{\log z}\right).$$

This could be obtained from the hypothesis (1.5) using the method of [7] referred to in the introduction, but has more recently been given a more direct proof in [4].

The behaviour of $F_z(s)$ and $R_z(s)$ when $0 < s < 1$ now follows. Write $w = z^u$. When $0 < u \le 1$ we obtain

$$F_z(u) = \sum_{n < z^u} g(n) = F_w(w) \ge F_w(\infty)\left(C + O\left(\frac{A}{\log z^u}\right)\right),$$

where $C = \sigma(1)$ is the constant in (1.11). A standard procedure involving partial summation (cf. p. 53 of [5]) shows that the hypothesis (1.5) of Theorem 1 implies

$$\prod_{z^u \le p < z}\left(1 - \frac{\omega(p)}{p}\right)^{-1} \le \left(\frac{\log z}{\log z^u}\right)^\kappa\left(1 + O\left(\frac{A}{\log z^u}\right)\right) \quad \text{if} \quad \log z^u > A,$$

the product on the left being just $F_z(\infty)/F_w(\infty)$, from (1.6) and (2.3). Thus

$$\frac{F_z(u)}{F_z(\infty)} \ge Cu^\kappa\left(1 + O\left(\frac{A}{u\log z}\right)\right) \tag{3.1}$$

$$= \sigma(u) + O\left(\frac{Au^{\kappa-1}}{\log z}\right) \quad \text{if} \quad 1 \ge u > A/\log z.$$

When $u \le A/\log z$ the A-dependence of the O-terms in this argument is much weaker, but the trivial estimate $F_z(\infty) \ge 0$ shows

$$\frac{F_z(u)}{F_z(\infty)} \ge \sigma(u) + O(u^\kappa) = \sigma(u) + O\left(\frac{A^\kappa}{\log^\kappa z}\right) \quad \text{when} \quad 0 \le u \le A/\log z,$$

since $\sigma(u) = O(u^\kappa)$ for these u. With (3.1) this gives the result in Lemma 4, because $\kappa \ge 1$.

Theorem 2 completed. Reverting to the original variables via (2.3) and invoking (2.2) shows that the integral in (2.15) is

$$\int_s^\infty M(u)\, d\Delta_z(u) = \int_x^\infty M\left(\frac{\log t}{\log z}\right) d\delta_z(t)$$

$$= \int_x^\infty M\left(\frac{\log t}{\log z}\right) d\sum_{n<t} g_z(n) E_{z,n}\left(\frac{t}{n}\right),$$

in which $x = z^s$ as in (2.3). The condition on E in Lemma 2 is satisfied because of (2.2), and (2.7) shows that $M(s)$ decreases, so Lemma 2 gives

$$\int_s^\infty M(u)\, d\Delta_z(u) \le A \sum_{x \le n} M\left(\frac{\log n}{\log z}\right) g_z(n) \le A\, M(1) \sum_{x \le n} g_z(n) \quad (3.2)$$

$$\le A\, M(1) G_z(\infty) = A\, M(1) F_z(\infty),$$

where we estimated the x-dependence of the last sum in a completely trivial way. The equation (2.15) now gives

$$\langle R_z, M \rangle(s) \le A\, M(1).$$

Now define

$$Q_z(s) = R_z(s) - cA^\kappa,$$

where $c > M(1)$ and c is also large enough for Lemma 4 to apply. Reference to (2.11) shows, because $A \ge 1$ and $\kappa \ge 1$,

$$\langle Q_z, M \rangle(s) \le 0.$$

But (2.14) and (2.3) show that $Q_z(s)$ differs only by a continuous function from a negative multiple of the sum $G_z(z^s)$ given in (1.4), so that $Q_z(s)$ decreases at its discontinuities in the way required in Lemma 3. Furthermore, Lemma 4 shows that $Q_z(s) < 0$ when $0 < s < 1$. Now Lemma 3 gives $Q_z(s) \le 0$ whenever $s > 0$, so that $R_z(s) < cA^\kappa$. This gives Theorem 2, since $R_z(s)$ is as in (2.14).

References

[1] N. C. Ankeny and H. Onishi, *The general sieve*, Acta Arith. **10** (1964), 31–62.

[2] N. G. de Bruijn, *On the number of uncancelled elements in the sieve of Eratosthenes*, Nederl. Akad. Wetensch. Proc. **53** (1950), 803–812.

[3] H. Diamond, H. Halberstam, and H.-E. Richert, *Combinatorial sieves of dimension exceeding one*, J. Number Theory **28** (1988), 306–346.

[4] G. Greaves and M. Huxley, *One-sided sifting density hypotheses in Selberg's sieve*, Proceedings of the Turku Symposium on Number Theory in memory of Kustaa Inkeri (1999), W. de Gruyter, 2001, pp. 105–114.

[5] H. Halberstam and H.-E. Richert, *Sieve methods*, Academic Press, London, 1974.

[6] H. Iwaniec, *Rosser's sieve*, Acta Arith. **36** (1980), 171–202.

[7] D. A. Rawsthorne, *Selberg's sieve estimate with a one-sided hypothesis*, Acta Arith. **49** (1982), 281–289.

[8] A. Selberg, *Lectures on sieves*, Collected Works, vol. 2, Springer-Verlag, 1991, pp. 65–247.

Siegel Modular Forms and Hecke Operators in Degree 2

James Lee Hafner and Lynne H. Walling[1]

1 Introduction

Let F be a degree 2, weight k Siegel modular form of level N which is an eigenfunction for all the Hecke operators. That is, for each prime p,

$$F|T(p) = \lambda(p)\, F, \qquad F|\widetilde{T}_1(p^2) = \lambda_1(p^2)\, F$$

where here $T(p)$ and $\widetilde{T}_1(p^2)$ denote the operators described in [7]. As a Fourier series,

$$F(\tau) = \sum_\Lambda c(\Lambda)\, e^*\{\Lambda\tau\}$$

where τ is in the Siegel upper half-space, Λ varies over all isometry classes of rank 2 lattices equipped with a positive, semi-definite quadratic form Q, and with T a matrix representing Q on Λ,

$$e^*\{\Lambda\tau\} = \sum_{G \in O(\Lambda)\backslash \mathrm{GL}_2(\mathbb{Z})} \exp(\pi i Tr(^t GTG\tau)). \qquad (1.1)$$

Here $O(\Lambda)$ is the orthogonal group of Λ (see O'Meara [10] for the precise definitions and results on lattices and quadratic forms that we use in this paper). We view Λ as a representative of the isometry class in which it lies. (Actually, when k is odd, we need to equip each isometry class with an orientation, and in the above sum, $G \in O^+(\Lambda)\backslash SL_2(\mathbb{Z})$; cf. [7].) Note that $c(\Lambda) = 0$ unless Λ is even integral.

In the above expansion, $c(\Lambda) = c(T)$ where $c(T)$ is a Fourier coefficient of F for some matrix T representing the quadratic form Q on Λ. The G in the sum in (1.1) represents a change of basis for the quadratic form on Λ. We call $c(\Lambda)$ a *lattice-Fourier coefficient* of F.

We have multiple goals in this paper. First, from this viewpoint of lattice-Fourier coefficients, we give explicit expressions for the action of the Hecke operators in some average sense (see (2.5) and (2.6)). Second, these

[1]Second author partially supported by Max-Planck Institute, Bonn, Germany.

formulas lead to relations between (average) lattice-Fourier coefficients and Hecke-eigenvalues for F (see Lemmas 2.4 and 2.5). Third, from these relations we compute a generating function for each family of average lattice-Fourier coefficients (locally) in terms of the eigenvalues and a maximal lattice in the family (Propositions 2.2 and 2.3). Fourth, as a consequence, we prove a multiplicativity or local-global property for these average lattice-Fourier coefficients (Theorem 2.1) as well as give explicit formulas for the local coefficients in terms of the Satake-parameters (Proposition 2.4 and Section 2.4.2). Fifth, we compute a factorization of the Koecher-Maaß series for F in terms of three zeta-functions (Theorem 3.1). The first two are the spinor zeta-function (already known as a factor, Andrianov [1, p. 84ff]) and the standard zeta-function (cf. Andrianov [2]). These two zeta-functions are completely determined by the Hecke-eigenvalues (and show no explicit connection with the coefficients of F). The third is a complicated zeta-function whose coefficients depend on the eigenvalues and the average lattice-Fourier coefficients for maximal lattices. Finally, we use the explicit formulas to give optimal (but relative) bounds for the average lattice-Fourier coefficients, whenever the Ramanujan-Petersson conjecture holds (Theorem 4.1).

Remark 1.1. Our results raise to the foreground fundamental questions concerning the extent to which the Hecke eigenvalues characterize a Siegel modular form of degree 2 (the so-called multiplicity one problem). All our theorems express the complete dependence between the eigenvalues and lattice-Fourier coefficients but only *relative to the coefficients for maximal lattices*. In other words, the eigenvalues and the maximal lattice coefficients seem to be completely independent. This is analogous to the half-integral weight case where there are limitations with respect to square-classes.

This "information barrier" between eigenvalues and maximal lattice coefficients raises these specific questions (among others):

- can there be a multiplicity-one theorem based solely on eigenvalues?

- can (optimal and absolute) bounds on Fourier coefficients be derived from Ramanujan-Petersson bounds on eigenvalues?

- are there other invariants (besides eigenvalues) which summarize the uniqueness of coefficients for maximal lattices?

Our results suggest that the answer to the first two questions is "no". The answer to the third question is probably "yes", but we do not know what form such invariants must take.

We now begin by specifying some notation for lattices.

Let Λ be an integer lattice equipped with a even integral quadratic form Q. For a prime p, we let $\Lambda_{(p)}$ denote the local lattice $\mathbb{Z}_p \otimes \Lambda$. By Λ^u we mean the lattice "scaled" by u, that is, the lattice Λ equipped with the quadratic form uQ. By $u\Lambda$ we mean the lattice multiplied by u, that is, each vector is scaled by u. So $u\Lambda \simeq \Lambda^{u^2}$. Both of these scalings may occur in the global or local context.

Throughout, we will assume that our lattices satisfy disc $\Lambda \neq 0$.

2 Hecke Actions

Theorem 4.2 from [7] shows that, for p prime,

$$\lambda(p)c(\Lambda) = p^{2k-3}c(\Lambda^{1/p}) + p^{k-2} \sum_{\{\Lambda:\, \Omega\}=(1,p)} c(\Omega^{1/p}) + c(\Lambda^p). \quad (2.1)$$

Similarly, Theorem 4.1 of [7] gives us

$$\lambda_1(p^2)c(\Lambda) = p^{2k-3} \sum_{\{\Lambda:\, \Omega\}=(1/p,1)} c(\Omega) + p^{k-2}v_p(\Lambda)c(\Lambda) + \sum_{\{\Lambda:\, \Omega\}=(1,p)} c(\Omega)$$
$$(2.2)$$

where $v_p(\Lambda)$ is the number of isotropic lines in $\Lambda/p\Lambda$ (relative to the quadratic form $\frac{1}{2}Q$).

The above expansion is not completely specified until we compute the expression v_p and determine what Ω's can appear in the above summations. We do that in the next section, after incorporating some additional averaging.

Remark 2.1. There are three operators which generate the local Hecke algebra: $T(p)$ and $\widetilde{T}_1(p^2)$ described above, and a third denoted by $\widetilde{T}_2(p^2)$. The third is algebraically dependent on the first two (see [7, Prop. 5.1]). Consequently, it provides no additional information about the relationship of coefficients and eigenvalues and is not used in this paper.

2.1 Average Lattice-Fourier Coefficients and Hecke Action

Observe that these Hecke actions for a prime p carry a lattice-Fourier coefficient for a lattice Λ into a weighted sum of lattice-Fourier coefficients for some other lattices that agree locally with Λ at all places different from p and which have specific structural differences locally at p. These differences are relatively hard to describe globally without some averaging to "smooth" out the variances. We will do this averaging over isometry classes and over families, a notion which we define next.

Definition 2.1. $\Omega \in \text{fam}\, \Lambda$ if Ω and Λ have the same signature, and, at each finite prime, Ω looks like Λ scaled by a unit. That is, $\Omega \in \text{fam}\, \Lambda$ if, for each finite prime p, there is a unit u in $\mathbb{Z}_{(p)}$ so that $\Omega \simeq \Lambda^u$ locally at p, and $\mathbb{R} \otimes \Omega \simeq \mathbb{R} \otimes \Lambda$.

In the sum in (2.1) (and analogously the sums in (2.2)), we sort the Ω into families and then into isometry classes. Thus

$$\sum_{\{\Lambda:\, \Omega\}=(1,p)} c(\Omega^{1/p}) = \sum_{\substack{\text{fam}\, \Omega_0 \\ \text{cls}\, \Omega \in \text{fam}\, \Omega_0}} c(\Omega^{1/p}) \#\{\Omega' \in \text{cls}\, \Omega: \{\Lambda: \Omega'\} = (1,p)\}.$$

(2.3)

Given isometries σ, σ', we have $\sigma\Omega = \sigma'\Omega$ if and only if $\sigma^{-1}\sigma' \in O(\Omega)$, the orthogonal group of Ω. If $\Omega' \in \text{cls}\, \Omega$ and $\{\Lambda: \Omega'\} = (1,p)$, then there exists an isometry σ such that $\{\Lambda: \sigma\Omega\} = (1,p)$. Thus, given such an Ω' there exist $o(\Omega)$ isometries σ such that $\Omega' = \sigma\Omega$. (Here $o(\Omega)$ denotes the order of the orthogonal group $O(\Omega)$. Since disc $\Omega \neq 0$, we know the lattices on $\mathbb{Q}\Omega$ are positive definite, so their orthogonal groups are finite.) Hence, with σ denoting an isometry,

$$\#\{\Omega' \in \text{cls}\, \Omega: \{\Lambda: \Omega'\} = (1,p)\} = \frac{\#\{\sigma: \{\Lambda: \sigma\Omega\} = (1,p)\}}{o(\Omega)}.$$

Consequently, we can average over families in (2.3) as follows:

$$\sum_{\text{cls}\, \Lambda \in \text{fam}\, \Lambda_0} \frac{1}{o(\Lambda)} \sum_{\{\Lambda:\, \Omega\}=(1,p)} c(\Omega^{1/p}) \qquad (2.4)$$

$$= \sum_{\substack{\text{fam}\, \Omega_0 \\ \text{cls}\, \Lambda \in \text{fam}\, \Lambda_0 \\ \text{cls}\, \Omega \in \text{fam}\, \Omega_0}} \frac{c(\Omega^{1/p})}{o(\Omega)} \frac{\#\{\sigma: \{\Lambda: \sigma\Omega\} = (1,p)\}}{o(\Lambda)}$$

$$= \sum_{\substack{\text{fam}\, \Omega_0 \\ \text{cls}\, \Omega \in \text{fam}\, \Omega_0}} \frac{c(\Omega^{1/p})}{o(\Omega)} \sum_{\text{cls}\, \Lambda \in \text{fam}\, \Lambda_0} \frac{\#\{\sigma: \{\Omega: \sigma p\Lambda\} = (1,p)\}}{o(\Lambda)}$$

$$= \sum_{\substack{\text{fam}\, \Omega_0 \\ \text{cls}\, \Omega \in \text{fam}\, \Omega_0}} \frac{c(\Omega^{1/p})}{o(\Omega)} \#\{\Lambda' \in \text{fam}\, \Lambda_0: \{\Omega: p\Lambda'\} = (1,p)\}$$

$$= \sum_{\substack{\text{fam}\, \Omega_0 \\ \text{cls}\, \Omega \in \text{fam}\, \Omega_0}} \frac{c(\Omega^{1/p})}{o(\Omega)} w_p(\Omega, p\Lambda),$$

where $w_p(\Lambda, \Omega)$ is the number of sublattices Ω' of Λ such that $\{\Lambda: \Omega'\} = (1,p)$ and such that $\Omega' \simeq \Omega^u$ for some p-unit u, that is,

$$w_p(\Lambda, \Omega) = \#\{\Omega' \in \text{fam}\, \Omega: \{\Lambda: \Omega'\} = (1,p)\}.$$

It is trivial from the definition that w_p depends only on the family of either of its variables (i.e., is invariant under scaling by units in either variable). Explicit values for this w_p are given in Lemma 2.2.

These calculations suggest that we can simplify the expression of the Hecke actions (i.e., reduce the explicit calculation to computing $w_p(\Lambda, \Omega)$) by averaging over isometry classes (with the right weighting) and also over families. We need a couple of definitions.

Definition 2.2. Define the *family mass* of a lattice as

$$\mu(\Lambda_0) = \sum_{\text{cls } \Lambda \in \text{fam } \Lambda_0} \frac{1}{o(\Lambda)}$$

and define the *average lattice-Fourier coefficient* of F by the expression

$$a(\Lambda_0) = \frac{1}{\mu(\Lambda_0)} \sum_{\text{cls } \Lambda \in \text{fam } \Lambda_0} \frac{c(\Lambda)}{o(\Lambda)}.$$

Remark 2.2. When F is in Maaß' Spezialchar, $a(\Lambda) = c(\Lambda)$. Consequently, in this case no averaging occurs and (most of) the results of the rest of this paper apply directly to the lattice-Fourier coefficients. It also follows that not all $a(\Lambda)$ are zero. Additionally (as we will see later), if the Koecher-Maaß series is not identically zero, then again not all $a(\Lambda)$ vanish.

We will not be using the family mass for a while, but we will need a simple lemma immediately.

Lemma 2.1. *If $\{\Lambda : \Omega\} = (1, p)$, then*

$$\mu(\Lambda) \, w_p(\Lambda, \Omega) = \mu(\Omega) \, w_p(\Omega, p\Lambda).$$

Proof. The proof is easy from the definitions if we observe that

$$\#\{\sigma : \{\Omega : \sigma p\Lambda\} = (1, p)\} = \#\{\sigma : \{\Lambda : \sigma\Omega\} = (1, p)\}.$$

$\square$

With this lemma, the expression in (2.4) then equals

$$\sum_{\substack{\text{fam } \Omega_0 \\ \text{cls } \Omega \text{ fam } \Omega_0}} \frac{c(\Omega^{1/p})}{o(\Omega)} w_p(\Omega_0, p\Lambda_0) = \frac{\mu(\Lambda_0)}{\mu(\Omega_0)} \sum_{\substack{\text{fam } \Omega_0 \\ \text{cls } \Omega \text{ fam } \Omega_0}} \frac{c(\Omega^{1/p})}{o(\Omega)} w_p(\Lambda_0, \Omega_0).$$

Consequently, we have from the above, (2.4), (2.1), and (2.2)

$$\lambda(p)a(\Lambda) = p^{2k-3}a(\Lambda^{1/p}) \tag{2.5}$$
$$+ p^{k-2} \sum_{\text{fam } \Omega} w_p(\Lambda, \Omega)a(\Omega^{1/p}) + a(\Lambda^p)$$

and

$$\lambda_1(p^2)a(\Lambda) = p^{2k-3} \sum_{\text{fam } \Omega} w_p(\Lambda, \Omega) a(\tfrac{1}{p}\Omega) \tag{2.6}$$

$$+ p^{k-2} v_p(\Lambda) a(\Lambda) + \sum_{\text{fam } \Omega} w_p(\Lambda, \Omega) a(\Omega)$$

These relations provide the most compact and explicit expressions for the Hecke actions in degree 2.

We conclude this section by giving the explicit values for the combinatorial expressions $w_p(\Lambda, \Omega)$ and $v_p(\Lambda)$.

Lemma 2.2. *Fix a prime $p \nmid N$ and suppose $\Lambda_{(q)} \simeq \Omega_{(q)}$ at each prime $q \neq p$. Let r and m be positive integers. We have*

$$w_p(\Lambda^{p^m}, \Omega^{p^m}) = w_p(\Lambda, \Omega).$$

For p odd, δ a non-square p-unit and ϵ an arbitrary p-unit, we have

$$w_p(\Lambda, \Omega) = \begin{cases} 2 & \text{if } \Lambda_{(p)} \simeq \langle 1, -1\rangle,\ \Omega_{(p)} \simeq p\langle 1, -1\rangle, \\ p-1 & \text{if } \Lambda_{(p)} \simeq \langle 1, -1\rangle,\ \Omega_{(p)} \simeq \langle 1\rangle \perp p^2\langle -1\rangle, \\ p+1 & \text{if } \Lambda_{(p)} \simeq \langle 1, -\delta\rangle,\ \Omega_{(p)} \simeq \langle 1\rangle \perp p^2\langle -\delta\rangle, \\ p & \text{if } \Lambda_{(p)} \simeq \langle 1\rangle \perp p^r\langle -\epsilon\rangle,\ \Omega_{(p)} \simeq \langle 1\rangle \perp p^{r+2}\langle -\epsilon\rangle, \\ 1 & \text{if } \Lambda_{(p)} \simeq \langle 1\rangle \perp p^r\langle -\epsilon\rangle,\ \Omega_{(p)} \simeq p^2\langle 1\rangle \perp p^r\langle -\epsilon\rangle, \\ 0 & \text{otherwise.} \end{cases}$$

For $p = 2$ and ϵ an arbitrary 2-unit, we have

$$w_2(\Lambda, \Omega) = \begin{cases} 2 & \text{if } \Lambda_{(2)} \simeq \begin{pmatrix} 0 & 1 \\ 1 & 0 \end{pmatrix},\ \Omega_{(2)} \simeq 2\begin{pmatrix} 0 & 1 \\ 1 & 0 \end{pmatrix}, \\ 1 & \text{if } \Lambda_{(2)} \simeq \begin{pmatrix} 0 & 1 \\ 1 & 0 \end{pmatrix},\ \Omega_{(2)} \simeq \langle 2, -2\rangle, \\ 3 & \text{if } \Lambda_{(2)} \simeq \begin{pmatrix} 2 & 1 \\ 1 & 2 \end{pmatrix},\ \Omega_{(2)} \simeq \langle 2, 6\rangle, \\ 1 & \text{if } \Lambda_{(2)} \simeq \langle 2, -2\rangle,\ \Omega_{(2)} \simeq 2^2\begin{pmatrix} 0 & 1 \\ 1 & 0 \end{pmatrix}, \\ 1 & \text{if } \Lambda_{(2)} \simeq \langle 2, 6\rangle,\ \Omega_{(2)} \simeq 2^2\begin{pmatrix} 2 & 1 \\ 1 & 2 \end{pmatrix}, \\ 1 & \text{if } \Lambda_{(2)} \simeq \langle 2, -2\epsilon\rangle,\ \Omega_{(2)} \simeq 2\langle 2, -2\epsilon\rangle,\ \epsilon = -1, 3, \\ 2 & \text{if } \Lambda_{(2)} \simeq \langle 2, -2\epsilon\rangle,\ \Omega_{(2)} \simeq \langle 2\rangle \perp 2^2\langle -2\epsilon\rangle, \\ 2 & \text{if } \Lambda_{(2)} \simeq \langle 2\rangle \perp 2^r\langle -2\epsilon\rangle,\ \Omega_{(2)} \simeq \langle 2\rangle \perp 2^{r+2}\langle -2\epsilon\rangle, \\ 1 & \text{if } \Lambda_{(2)} \simeq \langle 2\rangle \perp 2^r\langle -2\epsilon\rangle,\ \Omega_{(2)} \simeq 2^2\langle 2\rangle \perp 2^r\langle -2\epsilon\rangle, \\ 0 & \text{otherwise.} \end{cases}$$

Remark 2.3. Note that $w_p(\Lambda, \Omega) = 0$ unless $\det(\Omega) = p^2 \det(\Lambda)$. More precisely, either Ω is Λ scaled by p or one Jordan component is "shifted" by p^2. Additionally, no change is made to the square-class of the unit.

Proof. For $p \neq 2$ and given $\Lambda_{(p)}$, we need to find all sublattices $\Omega_{(p)}$ with $\{\Lambda : \Omega\} = (1, p)$ (up to scaling by units). Such $\Omega_{(p)}$ are the preimages in $\Lambda_{(p)}$ of lines in the space $\Lambda_{(p)}/p\Lambda_{(p)} \simeq \Lambda/p\Lambda$. The reader is referred to O'Meara [10, Sections 91C, 92:1-2, 93B] for related background.

We do the cases where $\Lambda_{(p)} \simeq \langle 1, -1 \rangle$ relative to some bases x, y and leave the rest to the reader as the computations are similar. In this case, the lines in $\Lambda/p\Lambda$ are generated by $\bar{x}, \bar{y}$ and $\bar{x} + \beta \bar{y}, \bar{y}$ for $\beta \not\equiv 0 \mod p$. The first two cases give us $\Omega_{(p)} = \mathbb{Z}_p x \oplus \mathbb{Z}_p(py)$ and $\Omega_{(p)} = \mathbb{Z}_p(px) \oplus \mathbb{Z}_p y$. Both of these are equivalent to $p\langle 1, -1 \rangle$ and this proves the first part of this case.

Now take $\beta \not\equiv 0 \mod p$. Then $\Omega_{(p)} = \mathbb{Z}_p(x + \beta y) \oplus \mathbb{Z}_p(py) \simeq \begin{pmatrix} 2\beta & p \\ p & 0 \end{pmatrix}$. But the latter is equivalent to $\langle 2\beta \rangle \perp p^2\langle -2\beta \rangle$. Scaling by a unit, this is $\langle 1 \rangle \perp p^2\langle -1 \rangle$. Consequently, this occurs $p - 1$ times. This completes the proof for this case. $\qquad\square$

Lemma 2.3. *For p odd, a non-square p-unit δ, a p-unit ϵ and integer $r \geq 1$, we have*

$$
v_p(\Lambda) = \begin{cases} 2 & \text{if } \Lambda_{(p)} \simeq \langle 1, -1 \rangle, \\ 0 & \text{if } \Lambda_{(p)} \simeq \langle 1, -\delta \rangle, \\ 1 & \text{if } \Lambda_{(p)} \simeq \langle 1 \rangle \perp p^r\langle -\epsilon \rangle, \\ p+1 & \text{otherwise.} \end{cases}
$$

Similarly, for $p = 2$, ϵ a 2-unit and integer $r \geq 0$, we have

$$
v_2(\Lambda) = \begin{cases} 2 & \text{if } \Lambda_{(2)} \simeq \begin{pmatrix} 0 & 1 \\ 1 & 0 \end{pmatrix}, \\ 0 & \text{if } \Lambda_{(2)} \simeq \begin{pmatrix} 2 & 1 \\ 1 & 2 \end{pmatrix}, \\ 1 & \text{if } \Lambda_{(2)} \simeq \langle 2 \rangle \perp 2^r\langle -2\epsilon \rangle, \\ 3 & \text{otherwise.} \end{cases}
$$

Proof. This is just a special case of Proposition 4.1 in [7]; see the remark following that result. $\qquad\square$

2.2   Explicit Hecke Actions and Family Types

For an odd prime p, a lattice family in degree 2 is determined by three parameters: (a) the square class of a local unit, (b) the *split p-structure* without scaling (this can also be referred to as the primitivity) and (c) the *scaling* by a power of p. This can be expressed by $\Lambda_{(p)} \simeq \Lambda_r^{p^m}$ where in Jordan decomposition $\Lambda_r = \Lambda_r(t, \epsilon) \simeq \langle 1 \rangle \perp p^{2r+t} \langle -\epsilon \rangle$, ϵ is the local unit, $t = 0, 1$ reflects the splitting in the maximal lattice, $2r + t$ reflects the total local splitting, r reflects (half of) the non-maximal extra splitting and m reflects the amount of local scaling.

Lemma 2.2 shows us that the Hecke action preserves the parameters (t, ϵ) and changes only r and m. That is, the lattices Ω such that $w_p(\Lambda, \Omega) \neq 0$ have the same (t, ϵ) values as Λ. (This explains the choice of parameters $2r + t$ for the split structure.) Consequently, we can partition the set of lattices according to these parameters. We call each partition a *family type*. There are three family types described by the terms hyperbolic, anisotropic and split (which characterize the maximal lattice Λ_0 in each family type). They can be parameterized via $\Lambda_0 = \Lambda_0(t, \epsilon)$ or via additional parameters u_0 and v_0 as follows:

Case 1 (hyperbolic): $t = 0$, $\epsilon = 1$ (a square), $u_0 = 2$, $v_0 = -1$;

Case 2 (anisotropic): $t = 0$, $\epsilon = \delta$ (a non-square), $u_0 = 0$, $v_0 = 1$;

Case 3 (split): $t = 1$, ϵ arbitrary, $u_0 = 1$, $v_0 = 0$, where

$$
\begin{aligned}
u_0 &= u_0(\Lambda_0) = 1 + \left(\frac{-\operatorname{disc} \Lambda_0}{p} \right), \\
v_0 &= 1 - u_0,
\end{aligned}
\tag{2.7}
$$

and the symbol $\left(\frac{*}{p} \right)$ is the Kronecker symbol.

For $p = 2$, there are four family types. One family type has a maximal lattice with Jordan decomposition $\Lambda_0 \simeq \begin{pmatrix} 0 & 1 \\ 1 & 0 \end{pmatrix}$, that is, unimodular hyperbolic. The other (unscaled) lattices in this family type are locally of the form $\Lambda_r \simeq \langle 2 \rangle \perp 2^{2r-2} \langle -2 \rangle$ for $r \geq 1$. The second family type has maximal lattice with Jordan decomposition $\Lambda_0 \simeq \begin{pmatrix} 2 & 1 \\ 1 & 2 \end{pmatrix}$, that is, unimodular, anisotropic, non-diagonal. The other (unscaled) lattices in this family are locally of the form $\Lambda_r \simeq \langle 2 \rangle \perp 2^{2r-2} \langle 6 \rangle$ for $r \geq 1$. The third family type contains the (unscaled) lattices of the form $\Lambda_r \simeq \langle 2 \rangle \perp 2^{2r} \langle -2\epsilon \rangle$, for $r \geq 0$ and $\epsilon = -1, 3$. The maximal lattice is 2-modular, diagonal and non-split isotropic. Finally, the fourth family type contains (unscaled) lattices of the form $\Lambda_r \simeq \langle 2 \rangle \perp 2^{2r+1} \langle -2\epsilon \rangle$, for $r \geq 0$ and ϵ arbitrary. The maximal lattice is split. With u_0 and v_0 as in (2.7), we can parameterize these family types as follows:

Case 1 (hyperbolic): $t = 0$, $\epsilon = 1$, $u_0 = 2$, $v_0 = -1$;

Case 2 (anisotropic): $t = 0$, $\epsilon = -3$, $u_0 = 1$, $v_0 = 0$;

Case 3a (2-modular): $t = 0$, $\epsilon = -1, 3$, $u_0 = 1$, $v_0 = 0$;

Case 3b (split): $t = 1$, ϵ arbitrary, $u_0 = 1$, $v_0 = 0$.

We will see that the latter two cases have combinatorially identical Hecke actions so that we combine these two into a single Case 3. Consequently, we can (loosely) refer to the three cases for $p = 2$ by the same terms and, conveniently, by the same parameter u_0 as for p odd.

In the following, we will use the notation

$$\Lambda_{(p)} \simeq \Lambda_r (u_0)^{p^m} \tag{2.8}$$

to indicate locally at p the family type (parameterized by u_0), the scaling (parameterized by m) and the extra splitting (parameterized by r).

Remark 2.4. As noted, the parameter u_0 provides a simple device for parameterizing the family types. Additionally, $u_0 = v_p(\Lambda)$ if $\Lambda_{(p)}$ is maximal (see Lemma 2.3). Finally, v_0 seems extraneous; it is a notational convenience so that many formulas become easier to write down and to digest.

We can now reformulate the Hecke actions in the form of recursion relations in the parameters r and m. Fix a prime p and assume that F is an eigenform for $T(p)$ and $\widetilde{T}_1(p^2)$, with eigenvalues $\lambda = \lambda(p)$ and $\lambda_1 = \lambda_1(p^2)$, respectively.

Since the Hecke action is local and depends only on the local family type, we define some notation which carries only the information relevant for the purposes at hand (though it suppresses a number of otherwise important data). For $\Lambda_{(p)} \simeq \Lambda_r^{p^m}$ with $r, m \geq 0$ (in the notation above for Λ_r), let

$$a(r, m) = a_p(r, m; u_0) = a(\Lambda), \tag{2.9}$$

where u_0 captures the family type of Λ at p and (r, m) set the extra scaling and component shift parameters at p. We will see later how this "local" notation relates globally.

Set $a(r, m) = 0$ if r or m is negative.

For $r \geq 0, m \geq -1$, define the auxiliary sequence

$$b(r,m) \quad = \quad u_r a(r,m) + a(r-1,m+1) + (p+v_r)a(r+1,m-1)$$

where

$$u_r \quad = \quad \begin{cases} u_0 & \text{if } r = 0 \\ 0 & \text{if } r \geq 1 \end{cases} \quad \text{and} \quad v_r = \begin{cases} v_0 & \text{if } r = 0 \\ 0 & \text{if } r \geq 1. \end{cases}$$

Set $b(r,m) = 0$ in all other cases. Note that $b(r,-1) = a(r-1,0)$ for $r \geq 0$.

Lemma 2.4. *Let p be a fixed prime not dividing the level N of F. The Hecke action of (2.5) is expressed for $r, m \geq 0$ by the relation*

$$\lambda a(r,m) = p^{2k-3}a(r,m-1) + p^{k-2}b(r,m) + a(r,m+1). \qquad (2.10)$$

Additionally, the Hecke action of (2.6) is expressed for $r, m \geq 0$ by the relation

$$\lambda_1 a(r,m) = p^{2k-3}b(r,m-1) + p^{k-2}v(r,m)a(r,m) + b(r,m+1), \qquad (2.11)$$

where

$$v(r,m) = \begin{cases} u_0 & \text{if } r = m = 0 \\ 1 & \text{if } r \geq 1, m = 0 \\ p+1 & \text{otherwise (i.e., } m \geq 1, r \geq 0). \end{cases}$$

Proof. The proof is essentially just a change of notation, together with Lemmas 2.2 and 2.3 and the observation that $v(r,m) = v_p(\Lambda)$. ☐

In the next lemma, we detail the action when $p|N$.

Lemma 2.5. *If $p|N$, then the Hecke action can be expressed by the relations*

$$\begin{aligned} \lambda a(r,m) &= a(r,m+1) \\ \lambda_1 a(r,m) &= b(r,m+1). \end{aligned} \qquad (2.12)$$

Proof. The proof is immediate from (2.5) and (2.6). ☐

This next proposition is our first step towards relating the average lattice-Fourier coefficients for a lattice to those of a maximal lattice, globally.

Proposition 2.1. *If Δ is a globally maximal lattice such that $a(\Delta) = 0$, then $a(\Lambda) = 0$ for every lattice Λ contained in Δ.*

Proof. A straight-forward double-induction on the relations of Lemma 2.4 and 2.5 can be used to show that $a(0,0) = 0$ implies $a(r,m) = 0$ for all $r, m \geq 0$. But $a(0,0) = a_p(0,0;u_0) = a(\Delta)$ for every p, when Δ is globally maximal. The result follows immediately. ☐

Our next goal is to derive from the two-term recurrence relations in Lemmas 2.4 and 2.5 a generating function for $a(r, m)$. We do this in a series of lemmas in the next few subsections.

2.2.1 Some Lemmas

By Proposition 2.1, we can assume without loss of generality that $a(0,0) = 1$. We revert to the explicit dependence on $a(0,0)$ in major statements of lemmas and propositions, for completeness. (Recall that $a(0,0)$ is an average lattice-Fourier coefficient for a lattice which is maximal at p, though the dependence on p (and the structure at other primes) is suppressed in the notation.)

For $m \in \mathbb{Z}$, let

$$A_m(x) = \sum_{r \geq 0} a(r, m) x^r$$

$$B_m(x) = \sum_{r \geq 0} b(r, m) x^r.$$

Lemma 2.6. *For any prime p, we have*

$$B_{-1}(x) = x A_0(x)$$
$$B_0(x) = x A_1(x) + u_0$$
$$B_1(x) = x A_2(x) + \frac{p}{x} A_0(x) - \frac{p}{x} + \beta$$

where

$$\beta = u_0 a(0,1) + v_0 a(1,0).$$

Proof. The first two formulas can be derived easily from the definitions, since $b(r,-1) = a(r-1,0)$ and $a(-1,0) = 0$ (for the first formula) and $b(r,0) = u_r a(r,0) + a(r-1,1)$ and $u_r = 0$ if $r \geq 1$ (for the second formula). The third formula is also straightforward but a bit more involved. We give the details. By the definition,

$$b(r,1) = u_r a(r,1) + a(r-1,2) + (p+v_r) a(r+1,0),$$

and $u_r = v_r = 0$ for $r \geq 1$. Consequenlty,

$$B_1(x) = u_0 a(0,1) + v_0 a(1,0) + \sum_{r \geq 1} a(r-1,2) x^r + p \sum_{r \geq 0} a(r+1,0) x^r$$
$$= u_0 a(0,1) + v_0 a(1,0) + x A_2(x) + \frac{p}{x} (A_0(x) - 1),$$

from which the result follows. $\qquad\square$

Lemma 2.7. *If $p \nmid N$, we have*

$$a(0,1) = (\lambda - u_0 p^{k-2})$$

$$a(1,0) = \frac{1}{p+v_0}((\lambda_1 - u_0 p^{k-2}) - u_0 a(0,1)).$$

Proof. These are both derived from special cases of (2.10) and (2.11) with $r = m = 0$ and the definitions. $\qquad\qquad\qquad\qquad\qquad\qquad\qquad\qquad\square$

Lemma 2.8. *Assume $p \nmid N$. For $m = 0, 1, 2$, there exist polynomials $N_m(x)$ (in x) of degree at most three such that*

$$A_m(x) = N_m(x)/G_0(x),$$

where

$$\begin{aligned}
G_0(x) &= 1 - (\lambda_1/p^{k-1} - 1/p - 1)p^{k-2}x \\
&\quad + (\lambda^2/p^{2k-3} - 2\lambda_1/p^{k-1} + 2/p)p^{2k-4}x^2 \\
&\quad - (\lambda_1/p^{k-1} - 1/p - 1)p^{3k-6}x^3 + p^{4k-8}x^4.
\end{aligned}$$

In particular,

$$N_0(x) = 1 + (p^{k-3}(p+v_0) - \beta/p)x + (\lambda u_0 + 2v_0 p^{k-2} - \beta)p^{k-3}x^2 + v_0 p^{3k-7}x^3.$$

Furthermore,

$$\lambda x N_0(x) + u_0 G_0(x) \equiv 0 \mod (1 + xp^{k-2}).$$

Proof. For the proof, we will derive three linear equations in the "variables" $A_m(x)$, for $m = 0, 1, 2$ and then apply Cramer's rule. Take $m = 0$ in (2.10), multiply by x^r and sum on $r \geq 0$. This gives the first equation

$$\lambda A_0(x) - (1 + xp^{k-2})A_1(x) + 0 \cdot A_2(x) = u_0 p^{k-2}. \qquad (2.13)$$

Next, take $m = 1$ in (2.10), multiply by x^{r+1} and sum on $r \geq 0$. This yields

$$\lambda x A_1(x) = p^{2k-3}x A_0(x) + p^{k-2}x B_1(x) + x A_2(x).$$

Now apply the third formula in Lemma 2.6 to get the second equation

$$p^{k-1}(1 + xp^{k-2})A_0(x) - \lambda x A_1(x) + x(1 + xp^{k-2})A_2(x) = p^{k-2}(p - \beta x).$$

Finally, take $m = 0$ in (2.11), multiply by x^{r+1} and sum on $r \geq 0$. This computation is a bit more involved. The immediate consequence of this step is the formula

$$\lambda_1 x A_0(x) = p^{2k-3}x B_{-1}(x) + p^{k-2}x \sum_{r \geq 0} v(r,0)a(r,0)x^r + x B_1(x).$$

The summation term here is

$$\sum_{r\geq 0} v(r,0)a(r,0)x^r = \sum_{r\geq 0} a(r,0)x^r - v_0 = A_0(x) - v_0.$$

Substitute this as well as the first and third formulas from Lemma 2.6 into the previous formula. This yields the third equation

$$(p - \lambda_1 x + xp^{k-2} + x^2 p^{2k-3})A_0(x) + 0 \cdot A_1(x) + x^2 A_2(x) = p + (v_0 p^{k-2} - \beta)x.$$

The result follows by solving the linear system for A_0, A_1, A_2. In particular, note that the coefficient matrix of the system is

$$\begin{pmatrix} \lambda & -(1 + xp^{k-2}) & 0 \\ p^{k-1}(1 + xp^{k-2}) & -\lambda x & x(1 + xp^{k-2}) \\ (p - \lambda_1 x + xp^{k-2} + x^2 p^{2k-3}) & 0 & x^2 \end{pmatrix}.$$

We observe that the last column is divisible by x. Its determinant is a polynomial in x of degree five, but with zero constant term and linear term $-px$. Call this denominator $-pxG_0(x)$ (so $G_0(x)$ has constant term 1 and degree 4). The constants in the system of equations is the vector

$$\begin{pmatrix} u_0 p^{k-2} \\ p^{k-2}(p - \beta x) \\ p + (v_0 p^{k-2} - \beta)x \end{pmatrix}.$$

It is easy to see that the numerator determinants are of degree at most 4. Also, the determinants for the numerator of A_0 and A_1 are divisible by x (as seen from their last column). The numerator determinant of A_2 vanishes at $x = 0$, so this is also divisible by x. Consequently, $N_m(x)$ $(m = 0, 1, 2)$ are polynomials in x, of degree at most 3. ($N_m(x)$ is the respective numerator determinant divided by $-px$.)

Additionally, it is easy to see that $G_0(x) \equiv \lambda^2 x^2/p \mod (1 + xp^{k-2})$ and $N_0(x) \equiv \lambda x^2 u_0 p^{k-3} \mod (1 + xp^{k-2})$. So the last statement in the lemma also holds.

Finally, the explicit expression for $G_0(x)$ and $N_0(x)$ come from direct calculation. $\qquad\square$

Define two generating functions:

$$
\begin{aligned}
A(x, y) &= \sum_{r,m\geq 0} a(r,m)x^r y^m \\
B(x, y) &= \sum_{r,m\geq 0} b(r, m-1)x^r y^m.
\end{aligned}
$$

Note that we start the B series with second index equal to -1 because $b(r,-1) = a(r-1,0)$ which is not zero for $r \geq 1$.

The next two lemmas will contain linear relations between these two generating functions from which we can compute $A(x,y)$ directly.

Lemma 2.9. *If $p \nmid N$, we have*

$$(p^{2k-3}y^2 - \lambda y + 1)A(x,y) + p^{k-2}B(x,y) = (1 + xp^{k-2})A_0(x).$$

Proof. Multiply (2.10) by $x^r y^{m+1}$ and sum on $r, m \geq 0$. We get

$$\begin{aligned}
\lambda y A(x,y) &= p^{2k-3} \sum_{r,m \geq 0} a(r, m-1)x^r y^{m+1} \\
&\quad + p^{k-2} \sum_{r,m \geq 0} b(r,m)x^r y^{m+1} \\
&\quad + \sum_{r,m \geq 0} a(r, m+1)x^r y^{m+1}.
\end{aligned}$$

The first sum on the right is easily seen to equal $y^2 A(x,y)$. The second sum is $B(x,y) - B_{-1}(x) = B(x,y) - xA_0(x)$ by Lemma 2.6. The third sum is $A(x,y) - A_0(x)$. Collecting terms we prove the lemma. $\qquad\square$

Lemma 2.10. *If $p \nmid N$, there exists a polynomial $N_0^*(x,y)$ of degree 4 in x and 2 in y so that*

$$(p^{k-2}(p+1) - \lambda_1)y^2 A(x,y) + (p^{2k-3}y^2 + 1)B(x,y) = \frac{N_0^*(x,y)}{G_0(x)}.$$

Furthermore, the coefficient of x^4 in $N_0^*(x,y)$ is $v_0 p^{3k-7}(1 + y^2 p^{2k-3})$.

Proof. Multiply (2.11) by $x^r y^{m+2}$ and sum on $r, m \geq 0$. We get

$$\begin{aligned}
\lambda_1 y^2 A(x,y) &= p^{2k-3}y^2 B(x,y) \\
&\quad + p^{k-2}y^2 \sum_{r,m \geq 0} v(r,m)a(r,m)x^r y^m \\
&\quad + \sum_{r,m \geq 0} b(r, m+1)x^r y^{m+2}.
\end{aligned}$$

The last sum is $B(x,y) - B_{-1}(x) - yB_0(x) = B(x,y) - (xA_0(x) + yxA_1(x) + u_0y)$ by Lemma 2.6. The second sum is (since $v(r,m) = p+1$ for $m \geq 1$)

$$\begin{aligned}
(p+1)A(x,y) + \sum_{r,m \geq 0}(v(r,0) - p - 1)a(r,0)x^r \\
= (p+1)A(x,y) - p\sum_{r \geq 1}a(r,0)x^r - (v_0 + p) \\
= (p+1)A(x,y) - pA_0(x) - v_0.
\end{aligned}$$

Equation (2.13) computed in the proof of Lemma 2.8 has

$$(1 + xp^{k-2})A_1(x) = \lambda A_0(x) - u_0 p^{k-2}.$$

Collecting terms, we derive the relation

$$(p^{k-2}(p+1) - \lambda_1)y^2 A(x,y) + (p^{2k-3}y^2 + 1)B(x,y) \qquad (2.14)$$

$$= (p^{k-1}y^2 + x + \frac{xy\lambda}{1 + xp^{k-2}})A_0(x) + v_0 p^{k-2}y^2 + \frac{u_0 y}{1 + xp^{k-2}}.$$

It remains to show that the right hand side of (2.14) has the form specified in the statement of the lemma. Collecting terms, we write (2.14) in the form

$$(p^{k-2}y^2 + x)A_0(x) + v_0 p^{k-2}y^2 + \frac{y}{1 + xp^{k-2}}(x\lambda A_0(x) + u_0).$$

By Lemma 2.8, the last part of this expression is

$$\frac{y}{(1 + xp^{k-2})G_0(x)}(x\lambda N_0(x) + G_0(x)u_0) = y\frac{N^*(x)}{G_0(x)}$$

for some polynomial $N^*(x)$ of degree at most 3. We conclude the proof by putting

$$N_0^*(x,y) = (p^{k-2}y^2 + x)N_0(x) + v_0 p^{k-2}y^2 G_0(x) + yN^*(x)$$

and observing that the coefficient of x^4 in this expression is

$$v_0 p^{3k-7} + v_0 y^2 p^{5k-10} = v_0 p^{3k-7}(1 + y^2 p^{2k-3})$$

which is derived from the leading coefficients of N_0 and G_0 in Lemma 2.8.

$\square$

2.2.2 The Generating Function, $p \nmid N$

We are now in a position to compute the generating function for $a(r, m)$.

Proposition 2.2. *If $p \nmid N$, there exists a polynomial $N_p(x, y; u_0)$ of degree 3 in x and degree 2 in y such that*

$$\sum_{r,m \geq 0} \frac{a_p(r, m; u_0)}{p^{r(k-2)+m(k-3/2)}} x^r y^m = a_p(0, 0; u_0) \frac{N_p(x, y; u_0)}{G_p(x)H_p(y)}$$

where with

$$\lambda_0 = \lambda/p^{k-3/2}$$
$$\lambda_2 = \lambda_1/p^{k-1} + 1 - 1/p$$

we have

$$G_p(x) = 1 + (2 - \lambda_2)x + (2 + \lambda_0^2 - 2\lambda_2)x^2 + (2 - \lambda_2)x^3 + x^4$$
$$H_p(y) = 1 - \lambda_0 y + \lambda_2 y^2 - \lambda_0 y^3 + y^4.$$

Remark 2.5. This result was first proved by Zagier (private communication) with $k = 2$ in our Case 1 above.

Remark 2.6. Note that the denominator factors G_p and H_p depend only on the eigenvalues and not on the family type. That dependence appears explicitly in N_p and implicitly in $a(0,0)$.

Proof. Without loss of generality, we assume that $a_p(0,0;u_0) = 1$. The left hand side of the main result in this proposition is just $A(x/p^{k-2}, y/p^{k-3/2})$, so our first goal is to find $A(x,y)$ from the two linear equations in Lemmas 2.9 and 2.10. The denominator determinant is

$$\begin{vmatrix} (p^{2k-3}y^2 - \lambda y + 1) & p^{k-2} \\ (p^{k-2}(p+1) - \lambda_1)y^2 & (p^{2k-3}y^2 + 1) \end{vmatrix} = \begin{vmatrix} (y'^2 - \lambda_0 y' + 1) & p^{k-2} \\ p^{-k+2}(2 - \lambda_2)y'^2 & (y'^2 + 1) \end{vmatrix}$$

where $y' = yp^{k-3/2}$. This is easily seen to equal $H_p(y') = H_p(yp^{k-3/2})$ by a computation.

The numerator determinant for $A(x,y)$ is

$$\begin{vmatrix} (1 + xp^{k-2})N_0(x)/G_0(x) & p^{k-2} \\ N_0^*(x,y)/G_0(x) & (p^{2k-3}y^2 + 1) \end{vmatrix}$$

$$= \frac{1}{G_0(x)} \begin{vmatrix} (1 + xp^{k-2})N_0(x) & p^{k-2} \\ N_0^*(x,y) & (p^{2k-3}y^2 + 1) \end{vmatrix}.$$

The last determinant is a polynomial in x and y of degree 2 in y and at most degree 4 in x. But, the coefficient of x^4 in this expression is zero (using Lemmas 2.8 and 2.10), so that the actual degree is at most 3.

Observing that $G_p(x) = G_0(xp^{k-2})$, we complete the proof.   □

Remark 2.7. The specific form of the polynomial $N_p(x, y; u_0)$ in the proposition is not very important (mostly because it is so complicated). However, we give it here for completeness:

$$
\begin{aligned}
N_p(x, y; u_0) = 1 &- \left[\frac{u_0}{p^{1/2}}\right] y - \left[\frac{v_0}{p}\right] y^2 \\
&+ \left[\frac{(u_0^2 + 3v_0 - v_0\lambda_2) - u_0\lambda_0 p^{1/2} + p}{p + v_0}\right] x \\
&+ \left[\frac{u_0(\lambda_2 - 1) - \lambda_0 p^{1/2}}{p^{1/2}}\right] xy \\
&+ \left[\frac{v_0(\lambda_2 - 2) + p}{p}\right] xy^2 \\
&+ \left[\frac{v_0^2 + u_0 v_0 \lambda_0 p^{1/2} + (u_0^2 + 3v_0 - v_0\lambda_2)p}{p(p + v_0)}\right] x^2 \\
&+ \left[\frac{u_0 v_0(\lambda_2 - \lambda_0^2 - 1) + (v_0\lambda_2 - u_0^2 - 2v_0)\lambda_0 p^{1/2} + u_0(\lambda_2 - 1)p}{p^{1/2}(p + v_0)}\right] x^2 y \\
&+ \left[\frac{v_0^2(2\lambda_2 - \lambda_0^2 - 2) + (u_0^2 + v_0(1 + \lambda_2 - \lambda_0^2))p - u_0\lambda_0 p^{3/2} + p^2}{p(p + v_0)}\right] x^2 y^2 \\
&+ \left[\frac{v_0}{p}\right] x^3 \\
&+ \left[\frac{-(v_0\lambda_0 + u_0 p^{1/2})}{p}\right] x^3 y \\
&+ \left[\frac{v_0^2(\lambda_2 - 1) + u_0 v_0 \lambda_0 p^{1/2} + (u_0^2 + v_0)p}{p(p + v_0)}\right] x^3 y^2 .
\end{aligned}
$$

In none of the cases (that is, for particular choices of u_0 and $v_0 = 1 - u_0$) does it appear that this expression gets significantly simpler. For example, it does not factor (symbolically). However, $N_p(0, p^{1/2}y; u_0)$ has very simple forms in each case: $(1 - y)^2$, $1 - y^2$, and $1 - y$, respectively. On the other hand, $N_p(x, 0; u_0)$ is cubic, irreducible in the hyperbolic case, a product of a linear and quadratic in the anisotropic case and is (curiously) only a quadratic in the split case.

2.2.3 The Generating Function, $p|N$

In this section, we use the two-term recurrence relations in Lemma 2.5 to derive a generating function for the sequence $a(r, m)$ in this special case.

It is easy to see that $a(r,m) = \lambda^m a(r,0)$ so that

$$A(x,y) = \frac{A_0(x)}{(1-\lambda y)}.$$

It remains to compute $A_0(x)$. From (2.12) with $m = 0$ and Lemma 2.6, we get

$$\lambda_1 A_0(x) = B_1(x) = xA_2(x) + \frac{p}{x}A_0(x) - \frac{p}{x} + \beta.$$

But $A_2(x) = \lambda^2 A_0(x)$, so

$$A_0(x) = \frac{p - \beta x}{p - \lambda_1 x + \lambda^2 x^2}.$$

Also, $\beta = u_0 a(0,1) + v_0 a(1,0)$ and $a(0,1) = \lambda a(0,0) = \lambda$ and

$$a(1,0) = \frac{\lambda_1 - u_0 \lambda}{p + v_0},$$

which is easily seen from (2.12) with $r = m = 0$. Thus, reinserting the explicit dependence on $a(0,0)$,

$$A(x,y) = a(0,0)\frac{p(p+v_0) - x(p\lambda u_0 + v_0\lambda_1)}{(p+v_0)(p-\lambda_1 x + \lambda^2 x^2)(1-\lambda y)}. \qquad (2.15)$$

We reformulate this in the next proposition in a form which is notationally consistent with the case $p \nmid N$.

Proposition 2.3. *If $p|N$, we have*

$$\sum_{r,m \geq 0} \frac{a_p(r,m;u_0)}{p^{r(k-2)+m(k-3/2)}} x^r y^m = a_p(0,0;u_0)\frac{N_p(x,y;u_0)}{G_p(x)H_p(y)},$$

where with

$$\begin{aligned}\lambda_0 &= \lambda/p^{k-3/2}\\ \lambda_2 &= \lambda_1/p^{k-1} + 1 - 1/p\end{aligned}$$

we have

$$\begin{aligned}N_p(x,y;u_0) &= 1 - \left[\frac{(\lambda_2 - 1 + 1/p)v_0 + \lambda_0\sqrt{p}u_0}{(p+v_0)}\right]x\\ G_p(x) &= 1 - (\lambda_2 - 1 + 1/p)x + \lambda_0^2 x^2\\ H_p(y) &= 1 - \lambda_0 y.\end{aligned}$$

Note that, in contrast to the formula in Proposition 2.2, the numerator $N_p(x,y;u_0)$ is independent of y. Additionally, the factor $G_p(x)$ is quadratic even though the factor $H_p(y)$ is linear and neither is monic (as a side-effect of our uniform normalization in x and y). The proof of this proposition is immediate from (2.15).

2.3 A Local-global Property

In this section, we prove a local-global property for the average lattice-Fourier coefficients of Hecke eigenforms. This theorem also shows that the average lattice-Fourier coefficients for a Hecke eigenform are completely determined by the coefficients on maximal lattices and on the eigenvalues. This is a weak form of a multiplicity one theorem and should be compared with results of Breulmann-Kohnen [6], the recent result of Scharlau-Walling [14] and others. We will see other applications of this notion later.

First we introduce additional notation. In the expansions of the expressions $N_p(x, y; u_0)/(G_p(x)H_p(y))$ in Propositions 2.2 and 2.3, let $\lambda_p(r, m; u_0)$ be the coefficient of $x^r y^m$, so that

$$\sum_{r,m \geq 0} \lambda_p(r, m; u_0) x^r y^m = \frac{N_p(x, y; u_0)}{G_p(x)H_p(y)}, \qquad (2.16)$$

or

$$\lambda_p(r, m; u_0) = \frac{a_p(r, m; u_0)/a_p(0, 0; u_0)}{p^{r(k-2)+m(k-3/2)}}$$

(when $a_p(0, 0; u_0) \neq 0$).

Theorem 2.1. *Let Λ be a (global) lattice and let Δ be a maximal lattice containing Λ. Then*

$$a(\Lambda) = a(\Delta) \prod_p \lambda_p(r_p, m_p; u_0) p^{r_p(k-2)+m_p(k-3/2)}. \qquad (2.17)$$

Here the triple (r_p, m_p, u_0) is determined by $\Lambda_{(p)} \simeq \Lambda_{r_p}(u_0)^{p^{m_p}}$, in the notation of (2.8).

Remark 2.8. The product in the above expression is actually finite. The only factors that are non-trivial (i.e., not equal to one) are for those primes p at which Λ is not maximal.

Remark 2.9. This theorem provides a refinement of Proposition 2.1.

Proof. The proof is straightforward by induction on the number of primes at which Λ is not maximal. Assume Λ is maximal, then the theorem holds as $\lambda_p(0, 0; u_0) = 1$ for all p. Assume Λ is not maximal at n primes. Then pick any such prime p. We use the results of Propositions 2.2 or 2.3 (depending on whether $p \nmid N$ or $p|N$, respectively) to express

$$a(\Lambda) = a(\Lambda_p)\lambda_p(r_p, m_p; u_0) p^{r_p(k-2)+m_p(k-3/2)} \qquad (2.18)$$

where (in the notation of those propositions) $a(\Lambda_p) = a_p(0, 0; u_0)$ for Λ_p isometric to Λ at all primes not equal to p and maximal at p and $\lambda_p(r_p, m_p; u_0)$

and u_0 are defined in (2.16) and (2.7), respectively. Now Λ_p has $n-1$
primes at which it is not maximal. By the induction hypothesis, this is
can be factored as in the statement of the theorem. This completes the
proof.   □

Remark 2.10. As seen in the proof, the expressions $\lambda_p(r_p, m_p; u_0)$ are, by
Propositions 2.2 and 2.3, polynomials in the eigenvalues $\lambda(p)$ and $\lambda_1(p^2)$.
The explicit formulation depends on the family type (as expressed by the
parameter u_0) and by the extra splitting and scaling (r_p, m_p). This justifies
the statements made prior to the theorem.

2.4 Explicit Formulas via Partial Fractions

With the explicit generating functions given above, we can use the theory of
partial fractions to derive explicit formulas for the coefficients $\lambda_p(r, m; u_0)$
of the generating functions. These explicit formulas will be stated in terms
of the roots of the polynomial $H_p(y)$. We will see that these roots are
connected to the Satake parameters (Satake [13]). (This is clear because
of the relationship to the spinor zeta-function, but we make the connection
explicit below.) However, the formulation of the results seem to be easiest
and cleanest to state, not directly in terms of the Satake parameters, but
in terms of these roots. In the next section we deal with the case where
$p \nmid N$. We follow that with the case where $p | N$.

2.4.1 *Expansion for $p \nmid N$*

Both $H_p(y)$ and $G_p(x)$ are symmetric polynomials (that is, $y^4 H_p(1/y) =
H_p(y)$). This means that if a is a root, then so is $1/a$. Let $a_1, a_2, 1/a_1, 1/a_2$
be the roots of $H_p(y)$; we can then express λ_0 and λ_2 as rational functions
in a_1, a_2:

$$\begin{aligned}
\lambda_0 &= a_1 + a_2 + 1/a_1 + 1/a_2, \\
\lambda_2 &= a_1 a_2 + a_1/a_2 + a_2/a_1 + 1/(a_1 a_2) + 2.
\end{aligned}$$

We find by simple algebra that the roots of $G_p(x)$ are $a_1 a_2, a_1/a_2, 1/(a_1 a_2)$
and a_2/a_1. In other words,

$$\begin{aligned}
H_p(y) &= (1 - y/a_1)(1 - y/a_2)(1 - a_1 y)(1 - a_2 y) \\
G_p(x) &= (1 - x/(a_1 a_2))(1 - a_2 x/a_1)(1 - a_1 a_2 x)(1 - a_1 x/a_2).
\end{aligned}$$

By the same token, we can express $N_p(x, y; u_0)$ strictly in terms of these
parameters and $q = p^{1/2}$ (but we don't write out that expansion here).

We've chosen the notation for the roots in a convenient way for what follows. However, it turns out that the a_1, a_2 are related to the Satake parameters because H_p is essentially the local factor of the Spinor zeta-function (Satake [13], Shimura [15] and Andrianov [1, p. 66ff] and the references therein). If $\alpha_0, \alpha_1, \alpha_2$ are the Satake parameters, then one formulation of the relationship is $\alpha_0 = (1/a_1)p^{k-3/2}, \alpha_1 = a_1/a_2, \alpha_2 = a_1 a_2$. This means that $G_p(x) = (1 - \alpha_1 x)(1 - \alpha_2 x)(1 - x/\alpha_1)(1 - x/\alpha_2)$. It follows that G_p is exactly the local factor of the standard zeta-function (Andrianov [2]).

For $p|N$, H_p is linear and G_p can be factored as $G_p(x) = (1-\lambda_0 b_1 x)(1-\lambda_0 x/b_1)$ where $b_1+1/b_1 = (\lambda_2-1+1/p)/\lambda_0$. Again there are two parameters defining the roots of G_p and H_p, namely, b_1 and λ_0.

The more interesting fact now is that we can expand the quotient $N_p/(G_p \cdot H_p)$ in a partial fraction decomposition with denominators consisting of one linear factor from G_p and one from H_p, with constant coefficients (for fixed a_1, a_2, p). Begin with the following:

Genericity condition. *For all primes p, all of the roots of G_p or H_p are simple. Equivalently, for $p \nmid N$, $a_1 \neq \pm a_2^{\pm 1}$ and $a_1, a_2 \neq \pm 1$. For $p|N$, $b_1 \neq \pm 1$.*

Remark 2.11. This condition is only a technical convenience in order to avoid excessive cases and notation. It is possible to perform all the necessary calculations in the cases where the assumption is false; we leave that to the reader, however.

Now, we set some additional notation by setting

$$
\begin{aligned}
q &= \sqrt{p} \\
P(q, u, v; u_0) &= (q - u)(q - v)(q + v_0 u)(q + v_0 v) \\
U(x, y, a_1, a_2) &= (1 - a_1 a_2 x)(1 - a_1 y) \\
D(a_1, a_2) &= (a_1 a_2)^{-2}(1 - a_1^2)(1 - a_2^2)(1 - a_1 a_2)(a_2 - a_1) \\
&= (a_1 - \frac{1}{a_1})(a_2 - \frac{1}{a_2})(\sqrt{\frac{a_1}{a_2}} - \sqrt{\frac{a_2}{a_1}})(\sqrt{a_1 a_2} - \frac{1}{\sqrt{a_1 a_2}}).
\end{aligned}
$$

Recall that $u_0 = 1 - v_0$ and that $\lambda_p(r, m; u_0)$ is the coefficient of $x^r y^m$ in $N_p(x, y; u_0)/(G_p(x)H_p(y))$.

Let W_0 be the group of actions on a_1, a_2 defined by $W_0 = \langle \sigma_0 \rangle \times \langle \sigma_1 \rangle \times \langle \sigma_2 \rangle$ where σ_0 permutes a_1 and a_2, and σ_i ($i = 1, 2$) inverts a_i. (This is actually a part of the Weyl group when viewed as acting on the Satake parameters.) Then, the partial fraction expansion of the generating function is given by

$$\frac{N_p(x, y; u_0)}{G_p(x)H_p(y)}$$

$$= \frac{1}{D(a_1, a_2)q^2(q^2 + v_0)} \sum_{\sigma \in W_0} \operatorname{sgn}(\sigma) \left(\frac{a_1^2 a_2 P(q, a_1^{-1}, a_2^{-1}; u_0)}{U(x, y, a_1, a_2)} \right)^\sigma.$$

Remark 2.12. In general, a partial fraction expansion with denominator $G_p(x)H_p(y)$ has 16 terms (four from G_p times four from H_p). However, with the particular numerator $N_p(x, y; u_0)$ we have, only eight terms survive. Although it is not entirely clear why this should be the case (computationally), it seems that it must be the case because W_0 has order eight.

Note that with the exception of the leading factor depending on v_0, only the expression $P(q, u, v; u_0)$ depends on the family type. For completeness, we write down $P(q, u, v; u_0)$ for each family type since they are quite simple on specialization:

Case 1 (hyperbolic): $P(q, u, v; 2) = (q - u)^2(q - v)^2.$

Case 2 (anisotropic): $P(q, u, v; 0) = (q^2 - u^2)(q^2 - v^2).$

Case 3 (split): $P(q, u, v; 1) = q^2(q - u)(q - v).$

This expansion into partial fractions immediately implies the following formula under the Genericity Condition.

$$\lambda_p(r, m; u_0) \tag{2.19}$$

$$= \frac{1}{D(a_1, a_2)q^2(q^2 + v_0)} \sum_{\sigma \in W_0} \operatorname{sgn}(\sigma) \left(P(q, a_1^{-1}, a_2^{-1}; u_0)\, a_1^{r+m+2} a_2^{r+1} \right)^\sigma.$$

This formulation is still not in the most convenient form. We next want to interchange the summation on σ with the terms in $P(q, u, v; u_0)$. To this end, we let

$$\rho(r, m) = \sum_{\sigma \in W_0} \operatorname{sgn}(\sigma)(a_1^{r+m+2} a_2^{r+1})^\sigma \tag{2.20}$$

$$= (a_1^{m+r+2} a_2^{r+1} - a_1^{r+1} a_2^{m+r+2})$$
$$- (a_1^{m+r+2} a_2^{-r-1} - a_1^{r+1} a_2^{-m-r-2})$$
$$- (a_1^{-m-r-2} a_2^{r+1} - a_1^{-r-1} a_2^{m+r+2})$$
$$+ (a_1^{-m-r-2} a_2^{-r-1} - a_1^{-r-1} a_2^{-m-r-2}).$$

This does not depend on the initial conditions, but is strictly a function of the parameters a_1, a_2.

Expanding $P(q, u, v; u_0)$ and collecting terms involving $\rho(r, m)$, we easily see that

$$\sum_{\sigma \in W_0} \text{sgn}(\sigma) \left(P(q, a_1^{-1}, a_2^{-1}; u_0) \, a_1^{r+m+2} a_2^{r+1} \right)^\sigma$$

$$= P(q, U, V; u_0) \circ \rho(r, m)$$

$$= q^2 \left[q^2 \rho(r, m) - u_0 q \rho(r, m-1) - v_0 \rho(r, m-2) \right] \qquad (2.21)$$

$$- u_0 q \left[q^2 \rho(r-1, m+1) - u_0 q \rho(r-1, m) - v_0 \rho(r-1, m-1) \right]$$

$$- v_0 \left[q^2 \rho(r-2, m+2) - u_0 q \rho(r-2, m+1) + v_0 \rho(r-2, m) \right],$$

where U and V are the operators defined by

$$U \circ \rho(r, m) = \rho(r, m-1), \quad V \circ \rho(r, m) = \rho(r-1, m+1).$$

There is an interesting pattern in this expansion. The first parameter in ρ decreases and the second increases as one moves down the rows. Within one of these rows, only the second parameter changes (decreases). Additionally, the coefficients within each of these rows are $q^2, -u_0 q, -v_0$ (which are the coefficients of $N_p(0, y; u_0)$). Similarly, the coefficients between these rows are $q^2, -u_0 q, -v_0$. However, we failed to find any direct connection with these coefficients and $N_p(x, y; u_0)$, in general. Note that in Case 2, the second row and second columns do not appear. In Case 3, the third row and third column vanish.

We collect what we have so far in a proposition.

Proposition 2.4. *Assume the Genericity Condition. Then if* $p \nmid N$,

$$\lambda_p(r, m; u_0) = \frac{1}{D(a_1, a_2) q^2 (q^2 + v_0)} P(q, U, V; u_0) \circ \rho(r, m)$$

Remark 2.13. The formula in Proposition 2.4 is the degree 2 generalization of the classical formula from degree 1 for Ramanujan's τ-function:

$$\frac{\tau(p^m)}{p^{11m/2}} = \frac{a^{m+1} - a^{-m-1}}{a - a^{-1}}.$$

In our case, the denominator role is played by $D(a_1, a_2)$ (essentially). In this classical case, it is possible to explicitly divide the numerator by the denominator and therefore write $\tau(p^m)$ as a sum of powers of a^2. Furthermore, there is a formulation for $\tau(p^m)$ involving an angle θ, where $\tau(p)p^{-11/2} = 2\cos\theta$ which facilitates bounds on $\tau(p^m)$, particularly since $|a| = 1$. In Section 4, we do the analogous "division" in the degree 2 case, and provide the analogous bound (though we do not given the representation in terms of angular parameters).

2.4.2   Expansion for $p|N$

In Lemma 2.3), write

$$G_p(x) = (1 - \lambda_0 b_1 x)(1 - \lambda_0 x/b_1),$$

so that

$$b_1 + \frac{1}{b_1} = \frac{\lambda_2 - 1 + 1/p}{\lambda_0}.$$

Assume $b_1 \neq \pm 1$ (this is the Genericity Condition for this case). We can now write (after a moderately lengthy calculation)

$$\lambda_p(r, m; u_0) = \frac{\lambda_0^{m+r}}{D(b_1)(p + v_0)} \sum_{\sigma \in W_0} \operatorname{sgn}(\sigma)(P(q, b_1^{-1}; u_0) \, b_1^{r+1})^\sigma,$$

where W_0 is the group of two elements which acts on b_1 by inverse, and

$$\begin{aligned}
q &= \sqrt{p}, \\
D(b_1) &= b_1^{-1}(b_1^2 - 1) = b_1 - 1/b_1, \\
P(q, u; u_0) &= q^2 - u_0 u q - v_0 u^2 = (q - u)(q + v_0 u).
\end{aligned}$$

Set

$$\rho(r) = b_1^{r+1} - b_1^{-r-1} \qquad (2.22)$$

and we find that

$$\lambda_p(r, m; u_0) = \frac{\lambda_0^{m+r}}{D(b_1)(p + v_0)}(q^2 \rho(r) - u_0 q \rho(r - 1) - v_0 \rho(r - 2)).$$

Note that in the three family type cases, $P(q, u; u_0) = (q - u)^2$ (hyperbolic), $q^2 - u^2$ (anisotropic) and $q(q - u)$ (split). Additionally, note the explicit dependence on λ_0. These two remarks should be viewed also in the context of the observations preceding (2.19) and the statement of Proposition 2.4.

3   Koecher-Maaß Zeta-functions

The results above, in particular the explicit generating functions (Propositions 2.2 and 2.3) and relationship between a lattice and its maximal lattice (Theorem 2.1), allow us to express the Koecher-Maaß zeta-function for a Hecke-eigenform in a very special way. This is given in the next theorem.

Let $K_F(s)$ be the Koecher-Maaß zeta-function defined by

$$K_F(s) = \sum_{T > 0} \frac{c(T)}{\# \operatorname{aut}(T)} (\det T)^{-s},$$

where $c(T)$ are the Fourier coefficients of F (as functions of matrices). Collecting terms by summing over lattices, we can write this as

$$K_F(s) = \sum_\Lambda a(\Lambda)\mu(\Lambda)(\det \Lambda)^{-s},$$

where $a(\Lambda)$ is the average lattice-Fourier coefficient and $\mu(\Lambda)$ is the family mass as defined in Definition 2.2.

Theorem 3.1. *Let F be a Hecke eigenform of degree 2, weight k and level N. Then*

$$K_F(s) = Z_F(2s)Z_F^\times(2s) \sum_{\Delta \text{ max}} a(\Delta)\mu(\Delta)(\det \Delta)^{-s} N_\Delta(2s)$$

where the sum is over maximal lattices and

$$
\begin{aligned}
Z_F(s) &= \prod_p H_p(p^{k-3/2-s})^{-1}, \\
Z_F^\times(s) &= \prod_p G_p(p^{k-1-s})^{-1}, \\
N_\Delta(s) &= \prod_p N_p(p^{-s}; \Delta), \\
N_p(x; \Delta) &= \left(1 + \frac{v_0}{p}\right) N_p(xp^{k-1}, xp^{k-3/2}; u_0) \\
&\quad - \left(\frac{v_0}{p}\right) N_p(0, xp^{k-3/2}; u_0) G_p(xp^{k-1}).
\end{aligned}
$$

and furthermore $H_p(y)$, $G_p(x)$ and $N_p(x, y; u_0)$ are the expressions in the generating functions for the family type of $\Delta_{(p)}$ (parameterized by u_0; see Propositions 2.2 and 2.3).

Proof. Write $K_F(s)$ as

$$\sum_{\Delta \text{ max}} a(\Delta)\mu(\Delta)(\det \Delta)^{-s} \sum_{\Lambda \subseteq \Delta} \frac{a(\Lambda)}{a(\Delta)} \frac{\mu(\Lambda)}{\mu(\Delta)} \left(\frac{\det \Lambda}{\det \Delta}\right)^{-s}. \tag{3.1}$$

That is, sum first on maximal lattices and then on all lattices contained within the maximal lattice. Note that in the inner sum, we can assume $a(\Delta) \neq 0$ as otherwise that term is missing from the expression (this is a consequence of Proposition 2.1).

We claim that the inner sum in (3.1) equals

$$\prod_p \sum_{r,m \geq 0} \lambda_p(r, m; u_0)\mu_p(p^r; v_0)p^{r(k-2-2s)+m(k-3/2-2s)}. \tag{3.2}$$

where $u_0 = 1 - v_0$ parameterizes the family type of $\Delta_{(p)}$, λ_p is defined in (2.16), and μ_p is defined by $\mu_p(p^r; u_0) = (p + v_0)p^{r-1}$ for $r \geq 1$ and $\mu_p(1) = 1$.

To justify (3.2), we proceed as follows. For a prime p, write $\Lambda_{(p)} \simeq \Lambda_r(u_0)^{p^m}$. Then $\Delta_{(p)} \simeq \Lambda_0(u_0)$. Let $l = \prod_p p^r$ and $n = \prod_p p^m$. These two integers give us a convenient parameterization of Λ.

Given this parameterization, Theorem 2.1 gives us

$$\frac{a(\Lambda)}{a(\Delta)} = \prod_{p^r \| l, p^m \| n} \lambda_p(r, m; u_0) p^{r(k-2) + m(k-3/2)}.$$

Similarly,

$$\left(\frac{\det \Lambda}{\det \Delta}\right)^{-s} = \prod_{p^r \| l, p^m \| n} p^{-2s(r+m)}.$$

Additionally, from Lemmas 2.1 and 2.2 and a simple induction argument, it is easy to see that

$$\frac{\mu(\Lambda)}{\mu(\Delta)} = \prod_p \frac{\mu(\Lambda_r(u_0))}{\mu(\Lambda_0(u_0))} = \prod_{p^r \| l, p^m \| n} \mu_p(p^r; v_0)$$

Consequently, the inner sum in (3.1) equals

$$\sum_{l,n \geq 1} \prod_{p^r \| l, p^m \| n} \lambda_p(r, m; u_0) \mu_p(p^r; v_0) p^{r(k-2-2s) + m(k-3/2-2s)}.$$

This is easily seen to equal the expression in (3.2).

To complete the proof, we have to relate the expression in (3.2) to the generating functions in Propositions 2.2 and 2.3. The double sum on $r, m \geq 0$ is

$$\left(1 + \frac{v_0}{p}\right) \frac{N_p(x, y; u_0)}{G_p(x) H_p(y)} - \left(\frac{v_0}{p}\right) \frac{N_p(0, y; u_0)}{H_p(y)},$$

with $x = p^{-2s+k-1}$ and $y = p^{-2s+k-3/2}$ when $p \nmid N$ and $x = p^{-2s-1/2}\lambda$ and $y = p^{-2s}\lambda$ when $p | N$. This yields the result in the statement of the theorem. Note that only the numerator of the above expression depends on the local maximal type of Δ and that the denominator involving G_p and H_p is independent of Δ.   □

Remark 3.1. The function $Z_F(s)$ is the spinor zeta-function (cf. Andrianov [1]). The function $Z_F^\times(s - k + 1)$ is the standard zeta-function (cf. Andrianov [2]). This should be compared to the result of Andrianov [1,

p. 85] which, though more general in many respects, has only the spinor-zeta-function as a factor. Now $K_F(s)$, $Z_F(s)$ and $Z_F^\times(s)$ all have functional equations (see Imai [8], Andrianov [1], and Böcherer [4] or Andrianov-Kalinin [3], respectively). The interesting (and curious) observation is that as combined in the statement of the theorem, the functional equations are inconsistent ($s \mapsto k - s, k - 1 - s, k - 1/2 - s$, respectively). Breulmann-Kohnen [6] used the cited Andrianov result and a similar observation (with only two functional equations) to prove a multiplicity one theorem.

Remark 3.2. For a given maximal lattice Δ, the function $N_\Delta(s)$ is always an infinite Euler product. For a finite set of primes (those which divide the determinant of Δ, i.e., those where Δ is split), the local factor N_p will be of type Case 3. For asymptotically half the odd primes the local factor will be of type Case 1 and the other half of type Case 2. (For that finite set of primes that divide the level, the factors are a bit different but still involve the same three cases.) So, the "coefficient" of $a(\Delta)\mu(\Delta)(\det \Delta)^{-s}$ in the zeta-function is an Euler product that has structure analogous to a Dirichlet L-function. In a sense, we have expressed the Koecher-Maaß series in terms of a zeta-function whose coefficients are also zeta-functions.

Remark 3.3. The local factor $N_p(x; \Delta)$ in the above expression is a degree 6 polynomial in $x = p^{-2s}$, when $p \nmid N$ and $u_0 = 2, -1$, is degree 5 when $p \nmid N$ and $u_0 = 1$ and is quadratic if $p|N$.

4   A Strong Inequality

Our final goal is to provide a strong bound for the average lattice-Fourier coefficients assuming the eigenvalues satisfy the Ramanujan-Petersson conjecture and the Genericity Condition. In the next proposition, we explicitly divide $\rho(r, m)$ by the factor $D(a_1, a_2)$ and $\rho(r)$ by $D(b_1)$. Recall that $\rho(r, m)$ is defined in (2.20) and $\rho(r)$ in (2.22).

Proposition 4.1. *The term $\rho(r, m)$ is "divisible" by $D(a_1, a_2)$ for all r, m. More explicitly,*

$$
\frac{\rho(r, m)}{D(a_1, a_2)} = (a_1 a_2)^{-m-r} \sum_{j=0}^{r} \sum_{i=0}^{[m/2]+j} \epsilon(i + j)(a_1 a_2)^{2i+\delta(i+j)} a_1^{m+2(j-i)}
$$
$$
\times R(m + 2(j - i), a_2/a_1)
$$
$$
\times R(|m + 2(r - i - j) + 1| - 1, (a_1 a_2)^{\delta(i+j)}),
$$

where $R(v, x) = \sum_{l=0}^{v} x^l$, and

$$
\delta(i) = \begin{cases} -1 & \text{if } i \geq m/2 + r + 1 \\ 0 & \text{else} \end{cases}
$$

and

$$\epsilon(i) = \begin{cases} -1 & \text{if } i \geq m/2 + r + 1 \\ 1 & \text{else} \end{cases}$$

Similarly, $D(b_1)$ divides $\rho(r)$ in the form $\sum_{l=0}^{r} b_1^{2l-r} = b_1^{-r} R(r, b_1^2)$.

Proof. The proof of the last statement is trivial so we skip that and give the details of the proof for $\rho(r, m)$.

By a simple rearrangement and collection of terms, we easily see that

$$(a_1 a_2)^{m+r+2} \rho(r, m)$$
$$= a_2^{m+1}(1 - a_1^{2m+2r+4})(1 - a_2^{2r+2})$$
$$\quad - a_1^{m+1}(1 - a_1^{2r+2})(1 - a_2^{2m+2r+4})$$
$$= (1 - a_1^2)(1 - a_2^2) a_2^{m+1} \sum_{j=0}^{r} a_2^{2j} \sum_{i=0}^{m+r+1} a_1^{2i}$$
$$\quad - (1 - a_1^2)(1 - a_2^2) a_1^{m+1} \sum_{j=0}^{r} a_1^{2j} \sum_{i=0}^{m+r+1} a_2^{2i}$$
$$= (1 - a_1^2)(1 - a_2^2) \sum_{j=0}^{r} \sum_{i=0}^{m+r+1} (a_1^{2i} a_2^{m+2j+1} - a_1^{m+2j+1} a_2^{2i})$$
$$= (1 - a_1^2)(1 - a_2^2) \sum_{j=0}^{r} \sum_{i=0}^{m+r+1} \gamma(2i, m + 2j + 1),$$

where $\gamma(u, v) = a_1^u a_2^v - a_1^v a_2^u$. So, it suffices to show that the last double sum is divisible by $(a_2 - a_1)(1 - a_1 a_2)$ and find the quotient. Note that $\gamma(u, v)$ is already divisible by $(a_2 - a_1)$, but we ignore that fact for the moment.

Let $t = [m/2]$ and $S(r, m)$ be the double sum in the expression above. Write the inner sum on i (for fixed j) in two pieces. The first piece is for $0 \leq i \leq t + j$ and the second is for $t + j + 1 \leq i \leq m + r + 1$. So

$$S(r, m) = \sum_{j=0}^{r} \sum_{i=0}^{t+j} \gamma(2i, m + 2j + 1) + \sum_{j=0}^{r} \sum_{i=t+j+1}^{m+r+1} \gamma(2i, m + 2j + 1).$$

In the last double sum, first replace j by $r - j$ (invert the order of summation) and then replace i by $m+r+1-i$ (again invert). The resulting sum is on $0 \leq j \leq r$ and $0 \leq i \leq m - t + j$, and the summand is $\gamma(2m + 2r - 2i + 2, m + 2r - 2j + 1)$. If m is even, then $m - t = t$ so the upper limit on i is $t + j$ (as in the first sum above). If m is odd, then $m = 2t + 1$, so $m - t + j = t + j + 1$. For $i = t + j + 1$, the summand is zero, so we can ignore this term. Consequently, we can combine the two double sums to get

$$S(r, m) = \sum_{j=0}^{r} \sum_{i=0}^{t+j} [\gamma(2i, m + 2j + 1) + \gamma(2m + 2r - 2i + 2, m + 2r - 2j + 1)].$$

Now observe that $\gamma(v, u) = -\gamma(u, v)$ and $\gamma(u + w, v + w) = (a_1 a_2)^w \gamma(u, v)$. Applying this to the second term within the sum above, we can write the summand in square-brackets as

$$(1 - (a_1 a_2)^{m + 2(r - i - j) + 1}) \gamma(2i, m + 2j + 1)$$

or

$$(1 - (a_1 a_2)^{m + 2(r - i - j) + 1}) (a_1 a_2)^{2i} \gamma(0, m + 2(j - i) + 1).$$

Note that in the summation ranges, the second parameter in the last expression is always positive.

For $v + 1 \geq 1$, $\gamma(0, v + 1) = (a_2 - a_1) a_1^v R(v, a_2/a_1)$, with R defined in the statement of the proposition. Similarly, if $v + 1 \geq 1$, then

$$(1 - x^{v+1}) = (1 - x) R(v, x)$$

and if $v + 1 \leq -1$, then

$$(1 - x^{v+1}) = (1 - 1/x) R(|v + 1| - 1, 1/x) = -x^{-1}(1 - x) R(|v + 1| - 1, x^{-1}).$$

and if $v + 1 = 0$, this expression vanishes. To summarize this, we write

$$(1 - x^{v+1}) = \epsilon_v x^{\delta_v} (1 - x) R(|v + 1| - 1, x^{\delta_v}),$$

where $\epsilon_v = -1, \delta_v = -1$ if $v + 1 \leq -1$ and $\epsilon_v = 1, \delta_v = 0$, otherwise.

With this notation, we then can write

$$
\begin{aligned}
S(r, m) &= (a_2 - a_1)(1 - a_1 a_2) \sum_{j=0}^{r} \sum_{i=0}^{t+j} \epsilon(i + j)(a_1 a_2)^{2i + \delta(i+j)} a_1^{m + 2(j-i)} \\
&\quad \times R(m + 2(j - i), a_2/a_1) \\
&\quad \times R(|m + 2(r - i - j) + 1| - 1, (a_1 a_2)^{\delta(i+j)}).
\end{aligned}
$$

where $\delta(i) = \delta_{m+2r-2i}$ and $\epsilon(i) = \epsilon_{m+2r-2i}$, as defined in the statement of the proposition. This completes the proof. □

We can use this proposition to give explicit bounds for the average lattice-Fourier coefficient (actually the local factor in that coefficient) whenever the Ramanujan-Petersson conjecture holds (see Weissauer [16]).

Theorem 4.1. *Let F be an eigenform that satisfies both the Genericity Condition and the Ramanujan-Petersson conjecture. Let Δ be a maximal lattice containing a lattice Λ. Let p be a prime and write $\Lambda_{(p)} \simeq \Lambda_{r_p}(u_0)^{p^{m_p}}$. Write $l = \prod_p p^{r_p}$ and $n = \prod_p p^{m_p}$ and assume that $\gcd(ln, N) = 1$. Then for any $\epsilon > 0$,*

$$|a(\Lambda)| \ll_\epsilon |a(\Delta)| \, l^{k-2+\epsilon} n^{k-3/2+\epsilon}. \tag{4.1}$$

The implied constant depends only on ϵ and is computable.

Remark 4.1. A weaker form of this result can be stated in the form

$$|a(\Lambda)| \ll_{\Delta,\epsilon} l^{k-2+\epsilon} n^{k-3/2+\epsilon}.$$

Compare this to the conjectural bound of Resnikoff-Saldaña [12] (see also Böcherer-Raghavan [5], Raghavan [11] and Kohnen [9])

$$|a(\Lambda)| \ll_\epsilon (\det(\Lambda))^{k/2-3/4+\epsilon} \ll_{\Delta,\epsilon} (ln)^{k-3/2+\epsilon},$$

since $\det(\Lambda) = \det(\Delta)(ln)^2$. Thus, our theorem is in one sense stronger than the conjectured bounds in l-dependence, but weaker in that the result is relative to Δ. Note, however, that $\det(\Delta)$ is always square-free and may or may not have any prime factors in common with l and n.

This theorem then indicates two things. First, the optimal bound for Fourier coefficients is *not* a function of the determinant itself, but depends naturally (and more precisely) on the local structure of the lattices. Second, the independence of Hecke eigenvalues and Fourier coefficients for maximal lattices creates a fundamental barrier to methods which attempt to bound Fourier coefficients (in absolute terms) based on bounds for eigenvalues.

Remark 4.2. The dependence on ϵ in the inequality could be replaced by some powers of log and other explicit (well-known) arithmetic functions, if necessary. We state the results in the above form only for simplicity. Additionally, the gcd assumption is also only for simplicity, in order to avoid discussion of the cases where $p|N$.

Remark 4.3. A slightly stronger form of the inequality (without implied constants) can be given in the case that $\Lambda = \Delta^n$, that is, if $l = 1$.

Proof. From Proposition 4.1 we can easily derive a bound of the form

$$\left| \frac{\rho(r,m)}{D(a_1,a_2)} \right| \leq \sum_{j=0}^{r} \sum_{i=0}^{[m/2]+j} (m+2(j-i)+1)|m+2(r-i-j)+1| \ll (m+r)^4,$$

with a computable implied constant. From Proposition 2.4 and (2.21) we see that

$$|\lambda_p(r,m;u_0)| \ll (1-1/\sqrt{p})^{-4}(m+r)^4,$$

with the same implied constant. This together with Theorem 2.1 implies the result.   □

5 Open Questions and Future Work

We hope to investigate some of the following open questions in the future.

1. What can we say about other zeta-functions (e.g., Rankin-type) in this context? For example, do our results provide a second proof of Theorem 5.1.1 (for $n = 2$) in Andrianov [2]?

2. Is the "lattice-exponential" defined in (1.1) *naturally* a Whittaker function of some type?

3. Is there a more fundamental role for $\rho(r, m)$ other than as a building block for the average lattice-Fourier coefficients?

4. Is there a two-complex variable zeta-function, analytic and with functional equation in each variable, for which the Koecher-Maaß series is a one-variable specialization and which explains the apparent inconsistencies of the functional equations for the three zeta functions in Theorem 3.1?

5. Perhaps the most interesting question is: how far can these ideas be pushed in degree greater than two?

Acknowledgements. We'd like to thank Winfried Kohnen, Rudolph Scharlau and Tom Shemanske for many stimulating discussions on this and related topics. Additionally, we'd like to thank the referee for suggesting some changes which improved the readability of the paper.

References

[1] A. N. Andrianov, *Euler products corresponding to Siegel modular forms of genus 2*, Russian Math. Surveys **29** (1974), 45–116.

[2] _____, *The multiplicative arithmetic of Siegel modular forms*, Russian Math. Surveys **34** (1979), 75–148.

[3] A. N. Andrianov and A. N. Kalinin, *On the analytic properties of standard zeta functions of Siegel modular forms*, Math. USSR Sb. **35** (1979), 1–17.

[4] S. Böcherer, *Über die Funktionalgleichung automorpher L-Funktionen zur Siegelschen Modulgruppe*, J. Reine Angew. Math. **362** (1985), 146–168.

[5]   S. Böcherer and S. Raghavan, *On Fourier coefficients of Siegel modular forms*, J. Reine Angew. Math. **284** (1988), 80–101.

[6]   S. Breulmann and W. Kohnen, *Twisted Maaß-Koecher series and spinor zeta functions*, Nagoya Math. J. **155** (1999), 153–160.

[7]   J. L. Hafner and L. H. Walling, *Explicit action of Hecke operators on Siegel modular forms*, preprint, 2000.

[8]   K. Imai, *Generalization of Hecke's correspondence to Siegel modular forms*, Amer. J. Math. **102** (1980), 903–936.

[9]   W. Kohnen, *Fourier coefficients and Hecke eigenvalues*, Nagoya Math. J. **149** (1998), 83–92.

[10] O. T. O'Meara, *Introduction to Quadratic Forms*, Springer-Verlag, New York, 1973.

[11] S. Raghavan, *Estimation of Fourier coefficients of Siegel modular forms*, 1986, Math. Gottingensis **75**.

[12] H. L. Resnikoff and R. L. Saldaña, *Some properties of Fourier coefficients of Eisenstein series of degree two*, J. Reine Angew. Math. **265** (1974), 90–109.

[13] I. Satake, *Theory of spherical functions on reductive algebraic groups over p-adic fields*, Inst. Hautes Études Sci. Publ. Math. **18** (1963), 1–69.

[14] R. Scharlau and L. Walling, *A weak multiplicity-one theorem for Siegel modular forms*, preprint 1998, revised 2000.

[15] G. Shimura, *On modular correspondences for Sp(n,Z) and their congruence relations*, Proc. Nat. Acad. Sci. USA **49** (1963), 824–828.

[16] R. Weissauer, *The Ramanujan conjecture for genus two Siegel modular forms (an application of the trace formula)*, preprint, 1993.

One Hundred Years of Normal Numbers

Glyn Harman

Prologue

In this paper we survey the development of metric number theory in the twentieth century with an emphasis on Borel's notion of a *normal number*. The early work in this subject is best described as "analysis" or "probability", rather than "number theory". However, as the subject has developed through the twentieth century, arithmetical questions and methods have come to the fore. We will demonstrate this by placing Borel's normal number theorem in a wider context of the study of the arithmetical properties of the sequence $F(\alpha, n)$, $n = 1, 2, \ldots$, for families of functions F. We shall also survey other properties of a "typical" number and lay down a few challenges for the next century (or next millennium?).

1 Paris in 1900

Back in 1874, in Germany, Cantor had demonstrated that the continuum was uncountable. His ideas were resisted by some leading mathematicians at the time, notably Kronecker, but they found a more sympathetic audience in France where several influential figures had come to accept them by the turn of the century. Cantor's demonstration had set a limit on mankind's knowledge of numbers. It could not be possible to list all the real numbers with a description, even allowing for an infinite list. To embark on any analysis of the properties of real numbers would therefore require a system of classification: one would have to fit the reals into classes and describe the nature of those classes. As the 19th century drew to a close and the 20th century dawned, Borel and Lebesgue were laying the foundations of measure theory and a new definition of the integral. Measure theory gave researchers an excellent tool for an analysis of the continuum. A natural question to ask was: What is true for a typical number—in the sense that the exceptional set has measure zero? If you pick a number at random, what properties might you expect it to enjoy? The simplest analysis of the continuum is its division into rationals and irrationals. A "typical" number is irrational since the set of rationals has measure zero. We pause

149

to remark that it was Borel who first demonstrated that a countable set has measure zero; indeed, this was the motivating point in his foundation of measure theory (see [11, pp. 499–501]).

In 1898 Borel wrote his influential book "Leçons sur la théorie des fonctions" [6] in which he introduces the idea of measure and considers examples of uncountable sets of measure zero. On pages 41–42 he essentially proves the following result.

Let A be the set of real α for which there are infinitely many solutions in integers a and q to

$$\left| \alpha - \frac{a}{q} \right| < \frac{1}{q^4}.$$

Then A is an uncountable set with measure zero.

Putting this another way: A "typical number" satisfies

$$\left| \alpha - \frac{a}{q} \right| > \frac{C(\alpha)}{q^4} \tag{1.1}$$

for all a and q, although there are uncountably many exceptions.

This appears to be the first result in metric number theory in the literature. Borel's original proof from our vantage point may seem unnecessarily longwinded, until we remember that he was working on the foundation of a completely new theory. In 1898 he could easily have shown the best possible form of the above result with the right hand side of (1.1) replaced by $\psi(q)/q$ for any non-negative function $\psi(q)$ such that

$$\sum_{q=1}^{\infty} \psi(q)$$

converges. It looks obvious to us now, but Borel had not yet stated the result which is known to us at present (after Cantelli's removal of the hypothesis of independence in the convergence case) as the Borel–Cantelli lemma!

Many count the birth of metric number theory from Borel's paper [7] in which he considered the properties of 'typical' numbers from the viewpoint of their decimal or continued fraction expansion. Write a real number α to base b ($b \in \mathbb{Z}, b \geq 2$) as

$$\alpha = [\alpha] + \sum_{n=1}^{\infty} a_n b^{-n},$$

with

$$0 \leq a_n \leq b - 1, \qquad a_n < b - 1 \text{ infinitely often.}$$

Let
$$A(d, b, N) = |\{n \leq N : a_n = d\}|.$$
Then α is said to be *simply normal* to base b if, for $0 \leq d \leq b - 1$,
$$\lim_{N \to \infty} \frac{A(d, b, N)}{N} = \frac{1}{b},$$
and *entirely normal* to base b if α is simply normal to each base $b, b^2, b^3, \ldots$. Finally, call a number *absolutely normal* if it is simply normal to every base $b \geq 2$. Such a number is usually referred to as a *normal number*. Borel's original definitions were equivalent to those we have given here (see [17, 1–4]). He established the following result.

Theorem (Borel 1909). *Almost all real numbers are normal.*

Since a countable union of sets with measure zero itself has measure zero, it suffices to prove that, given any base b, almost all real numbers are simply normal to base b.

From Borel's result we see that a normal number is indeed a 'typical' number. We note that an equivalent definition of 'entirely normal to base b' is that
$$\lim_{N \to \infty} \frac{A(B_k, b, N)}{N} = \frac{1}{b^k}$$
for all $k \geq 1, B_k$, where B_k is a block of k digits to base b, and $A(B_k, b, N)$ denotes the number of times B_k occurs in the first N digits of α (see [17, 5–7]). This can be described more dramatically by pointing out that if your telephone number has 11 digits, say, and you pick a number α at random, then in the first N digits (N very large!) of the decimal expansion of α your telephone number should occur $N \cdot 10^{-11}$ times. Alternatively, since an individual's genetic code can be considered as a block of digits to base 4, a normal number contains the genetic code of every human being who has ever lived, or will live, with the correct asymptotic frequency. A similar argument shows that a normal number contains the sum of human knowledge: past, present and future. Of course, as with similar remarks (the large number of monkeys typing for many millennia to come up with the works of Shakespeare, for example), it would take a finite rational mind longer than the lifetime of the universe to find even one individual's genetic code in any given normal number!

2 Normal Numbers Through the Twentieth Century

The immediate interest after Borel's result was either to give constructions of normal numbers [26], or to give a quantitative estimate for the discrepancy

$$D(d, b, N) = A(d, b, N) - \frac{N}{b}$$

for almost all α. After earlier work by Hausdorff and Hardy and Littlewood, the definitive answer to this problem was supplied by Khintchine [23].

Theorem. *Suppose $b \in \mathbb{Z}, b \geq 2$. For almost all α, for $0 \leq d \leq b - 1$, we have*

$$\limsup_{N \to \infty} \frac{D(d, b, N)}{\sigma f(N)} = 1, \qquad \liminf_{N \to \infty} \frac{D(d, b, N)}{\sigma f(N)} = -1,$$

with

$$f(N) = (2N \log \log N)^{\frac{1}{2}}, \qquad \sigma^2 = b^{-1}(1 - b^{-1}).$$

Remark. The referee pointed out to the author that it follows that the set of limit points of $\frac{D(d,b,N)}{\sigma f(N)}$ is, with probability 1, precisely the interval $[-1, 1]$.

From our vantage point in the year 2000 the results of Borel and Khintchine could be considered "obvious". The a_n are independent variables taking the values 0 to $b - 1$ with equal probability. The quantity σ^2 represents the variance. However, it must be remembered that Borel's result is the first occurrence of what is now referred to as "the law of large numbers" after the work of Kolmogorov [25], and Khintchine's result is the first mention of the law of the iterated logarithm (see [5] for a survey of this topic).

Another more modern perspective on Borel's result is afforded by Ergodic Theory (see [4]). If $\{x\}$ represents the fractional part of x, then the mapping $T \colon \{x\} \to \{bx\}$ is measure preserving and ergodic on the interval $[0, 1)$ with respect to Lebesgue measure. The (pointwise) Ergodic Theorem (see [4, p. 13]) states that, for integrable functions f,

$$\frac{1}{N} \sum_{n=1}^{N} f(T^{n-1}\alpha) \longrightarrow \int_0^1 f(x) \, dx$$

for almost all α. Borel's theorem follows on taking

$$f(x) = \begin{cases} 1 & \text{if } \frac{d}{b} \leq x < \frac{d+1}{b} \\ 0 & \text{otherwise.} \end{cases}$$

One natural question arising from Borel's normal number theorem is the analysis of the set of non-normal numbers. Clearly no rational can be normal (although a rational can be simply normal to a given base, or finite set of bases, it cannot be entirely normal to any base). By modifying

Cantor's argument it is straightforward to show that the set of non-normal numbers is uncountable. With the introduction of Hausdorff dimension the technology became available to give a more precise definition of the exceptional set—it has dimension 1. From one point of view it is thus as large as it could be and still have measure zero. We pause briefly here to contrast the measure-theorist's and topologist's perspectives. The set of normal numbers is a set of the first Baire category (it is a countable union of nowhere dense sets), while the non-normal numbers form a set of the second category—and thus appear larger from this topological point of view.

One challenge has been to write down numbers which are entirely normal to a given base. Taking 10 as the natural example base, Champernowne [10] demonstrated that $0.12345678910111213141516171819202\ldots$ is entirely normal to base 10. He conjectured that one could replace the naturals in order with the primes to show that $0.235711131719232931374147347\ldots$ is entirely normal to base 10. Copeland and Erdős [12] proved this as a corollary to a more general result on decimals built up from a sufficiently dense sequence of integers. Davenport and Erdös [13] replaced the naturals with the values of a non-constant polynomial taking only non-negative values. The particular case that $0.149162536496481100121144\ldots$ is entirely normal to base 10 had been dealt with earlier by Besicovitch. We note that Davenport and Erdös used Weyl's inequality applied to

$$\frac{f(n)}{10^m}, \qquad f(x) \in \mathbb{Z}[x],$$

and so genuinely number-theoretic ideas were involved. These last three examples may at first sight appear to be rather surprising. It is clear that there must be large discrepancies for these numbers. However, as the length of the number added each time increases, the effect of these irregularities is masked to the extent that $D(10^n, b, N)/N \to 0$ as $N \to \infty$, even though infinitely often $D(10, b, N) \gg N/\log N$ (compare [35]). In the opposite direction see [27] for the construction of numbers with minimal discrepancy.

Various authors have investigated sets of numbers which are entirely normal to one given base (or set of bases), but not to another base or set of bases. Note that if α is entirely normal to base b_1, it is entirely normal to base b_2 where $b_2^r = b_1^s, r, s \in \mathbb{Z}^+$. For results on the dimension of such sets see [29], [32] (and the earlier work of Cassels [9] and Schmidt [36]). Brown, Moran and Pearce [8] considered writing every real β as $\beta_1 + \beta_2$, where each β_j is entirely normal to base b_1, but not to base b_2 (where $\log b_1 / \log b_2 \notin \mathbb{Q}$). Wagner [39] constructed rings of numbers which, with 0 excepted, are normal to one base, but not to another. Schmidt [37] and

Pollington [31] established hybrid results concerning α not normal to any given base and satisfying Diophantine approximation conditions.

Other authors have checked the decimal expansions of the best known irrationals such as $\sqrt{2}, e, \pi$, to see how closely they resemble normal numbers. This leads to a great challenge for mathematicians of the coming century (or millennium!):

Challenge. *Determine whether or not $\sqrt{2}, e$ or π are normal.*

Here are some weaker questions that still seem far too hard at present.

Challenge. *Find one base to which $\sqrt{2}$ is simply normal, or prove that no such base exists.*

Challenge. *Find a base $b > 2$ for which each of $0, \ldots, b-1$ occurs infinitely often in the base b expansion of $\sqrt{2}, e, \pi$.*

The result of this last challenge is trivial, of course, if $b = 2$. We note that Mahler [28] gave an approximation to the above:

For any given irrational α, positive integer N, and base $b > 2$ there exists an integer M such that every block of length N occurs infinitely often in the expansion of $M\alpha$ to base b.

We close this section with just one simple case which illustrates the difficulty of the problem. Suppose there is no 1 in the base 3 expansion of $\sqrt{2}$ from some point onwards. Then there exists m such that, upon writing $\beta = 3^m \sqrt{2}$, we have either

$$\{3^n\beta\} < \frac{1}{3} \quad \text{or} \quad \{3^n\beta\} > \frac{2}{3}$$

for all $n \geq 1$. Could this be possible? The set of β for which the above holds has Hausdorff dimension $\log 2/\log 3$, of course, for it is essentially Cantor's ternary set.

3 Normal Numbers Generalized

In this section we reformulate the normal number problem in a way that admits generalization, yet which still has a satisfactory solution. Let $f(n, \alpha)$ be a function of a positive integer variable n and a real variable α. The question is to investigate the arithmetical properties of $F(n, \alpha) = [f(n, \alpha)]$ for almost all α. Normal numbers correspond to taking $f(n, \alpha) = b^n\alpha$ and investigating the property $P(\alpha, b)$, that $F(n, \alpha)$ is equidistributed in the residue classes (mod b) as $n \to \infty$. Cast in this light, normal numbers are those for which $P(\alpha, b)$ is true for all $b \geq 2$. Now let $f(n, \alpha) = g^n\alpha$ ($g \in \mathbb{Z}$,

$g \geq 2$), and let $P(\alpha, b, g)$ be the property that $F(n, \alpha)$ is equidistributed in the residue classes (mod b) as $n \to \infty$. Using the fact that a countable union of sets with measure zero itself has measure zero, we need only investigate $P(\alpha, b, g)$ for any given pair (b, g). It is not difficult to show that $P(\alpha, b, g)$ is true for almost all α. This could be proved directly, or one could use

$$\left\{\frac{\alpha g^n}{b}\right\} \text{ uniformly distributed (mod 1)} \Rightarrow P(\alpha, b, g) \text{ true.}$$

Thus almost all α satisfy a more stringent condition than normality:

$$\lim_{N\to\infty} \max_{0\leq a\leq b-1} \left| \frac{1}{N} \sum_{\substack{[g^n\alpha]\equiv a(\bmod b) \\ n\leq N}} 1 - \frac{1}{b} \right| = 0,$$

for every pair b, g.

It should be noted that all we have said so far could be described as *hitting a fixed target*. Given an infinite sequence $\mathcal{C} = c_1, \ldots, c_n, \ldots$ of integers from 0 to $b - 1$ write $P(\mathcal{C}, g, b, \alpha)$ for the property that

$$\frac{1}{N} |\{n \leq N : [g^n\alpha] \equiv c_n \pmod{b}\}| \to \frac{1}{b}.$$

This could be described as a *moving target problem* (the terminology was suggested by [21]). Now we cannot have a result true for all possible $\mathcal{C}$ (since there are uncountably many!), but if we are given one sequence $\mathcal{C}(b, g)$ for each pair (b, g), then, for almost all α, $P(\mathcal{C}, g, b, \alpha)$ is true for every pair (b, g) with $b, g \geq 2$. This follows since we can give an asymptotic formula, valid for almost all α, for the number of solutions to

$$\left\{\frac{g^n\alpha - c_n}{b}\right\} < \frac{1}{b}$$

(see [30]).

We now give a much wider generalization of the normal number concept. Let a_n be any increasing sequence of positive reals with $a_{n+1} > a_n + n^{-1}(\log n)^{1+\delta}$ for some $\delta > 0$. Let $f(n, \alpha) = a_n\alpha$ with $P(\alpha, b)$ the property that

$$\lim_{N\to\infty} \max_{0\leq a\leq b-1} \left| \frac{1}{N} \sum_{\substack{n\leq N \\ F(n,\alpha)\equiv a \pmod{b}}} 1 - \frac{1}{b} \right| = 0.$$

Then $P(\alpha, b)$ is true for almost all α. This follows from [17, Theorem 5.8*]. Given a countable set of sequences a_n we could thus let $P(\alpha)$ be the property that $P(\alpha, b)$ is true for all $b \geq 2$, and for each of the given sequences. We then have that $P(\alpha, b)$ is true for almost all α. One can turn this into a "moving target" problem also, but at present the slightly faster growth criterion $a_{n+1} > a_n + n^\sigma$ with $\sigma > -1$ is required (see [17, Theorem 7.5]).

Since $\{b^n \alpha\}$ being uniformly distributed (mod 1) is equivalent to α being entirely normal to base b, it has become customary to define normality to a non-integer base $b > 1$ by using the criterion that $\{b^n \alpha\}$ should be uniformly distributed (mod 1). It follows from our previous discussion that given any countable set S of bases (for example $\mathbb{Q} \cap (1, \infty)$) almost all α are normal to every given base in S. The subject of uniform distribution gives many other characterisations of what should be expected of a 'typical' number (see [17, Chapter 5]). Write

$$D_N(a_n) = \sup_{\mathcal{I} \subset [0,1)} \left| \sum_{\substack{\{a_n\} \in \mathcal{I} \\ n \leq N}} 1 - \lambda(\mathcal{I})N \right|,$$

where $\mathcal{I}$ denotes an interval, and λ is used for Lebesgue measure. Khintchine [22] established the following result using the metric theory of continued fractions.

Theorem. *For almost all α we have*

$$D_N(n\alpha) \ll (\log N)G(\log\log 9N)$$

for any positive non-decreasing function G such that

$$\sum_{n=1}^{\infty} \frac{1}{G(n)}$$

converges. On the other hand, for almost all α we have

$$\limsup_{N \to \infty} \frac{D_N(n\alpha)}{(\log N)H(\log\log N)} > 0$$

where $H(x)$ is any positive non-decreasing function such that

$$\sum_{n=1}^{\infty} \frac{1}{H(n)}$$

diverges.

This gives a very precise result on the discrepancy of $n\alpha$ for almost all α. If a_n is an increasing sequence of integers then it is known that

$$D_N(a_n\alpha) = O\left(N^{\frac{1}{2}}(\log N)^{\frac{3}{2}+\epsilon}\right)$$

for almost all α [1], and that the exponent of the logarithm cannot be reduced below $\frac{1}{2}$ [2]. It is very frustrating that we are so close to knowing the exact order of magnitude for this problem, but there remains the $\log N$ gap.

Challenge. *Characterise the slowest growing functions $F(N)$ such that, for any increasing sequence of integers a_n,*

$$D_N(a_n\alpha) \ll F(N) \text{ for almost all } \alpha.$$

4 Prime Values of $\mathbf{F(n, \alpha)}$

In the previous section we considered the distribution of $F(n, \alpha)$ in arithmetic progressions. The property of integer sequences of greatest interest to number-theorists is usually whether they infinitely often take prime values. The author considered this question in [16] (see also [17, Chapter 8]). Among the results we proved was the following.

Theorem. *Let a_n be an increasing sequence of positive reals. Then $[a_n\alpha]$ is infinitely often prime, or only finitely often prime, for almost all α according to whether*

$$\sum_{n=1}^{\infty} \frac{R(n)}{\log a_n}$$

diverges or converges, respectively. Here

$$R(n) = \left(\sum_{\substack{m \le n \\ |a_m - a_n| \le 1}} 1\right)^{-1}.$$

A limited quantitative version of this result can be proved [18]. If $a_{n+1} \ge a_n + 1$, $P(\alpha, N) = |\{n \le N : [a_n\alpha] \text{ prime}\}|$, and

$$V(N) = \sum_{n=1}^{N} \frac{1}{\log a_n},$$

then

$$\limsup_{N \to \infty} \frac{P(\alpha, N)}{V(N)} \ge 1, \qquad \liminf_{N \to \infty} \frac{P(\alpha, N)}{V(N)} \le 1.$$

We might expect

$$\lim_{N \to \infty} \frac{P(\alpha, N)}{V(N)} = 1,$$

and this can be proved if a_n is a lacunary sequence. In particular, we called a number α *prime normal* to base b if

$$\frac{1}{\log N} |\{n \leq N : [b^n \alpha] \text{ prime}\}| \to \frac{1}{\log b} \text{ as } N \to \infty.$$

Our result showed that almost all $\alpha > 0$ are prime normal to every integer base $b \geq 2$. Hence, not only do all blocks of integers show up with the expected frequency in the base b expansion of a "typical" number, but also the number of primes occurring in the first N places obeys the expected asymptotic law. As would be expected, much more number-theoretic machinery is required to prove a result of this type.

5 Diophantine Approximation to a Typical Number

We now return to Borel's other main result in his 1909 paper which gave a characterisation of a typical number through the growth of the partial quotients in its continued fraction expansion. We write

$$\alpha = [a_0; a_1, \ldots, a_n, \ldots]$$

in the usual continued fraction notation [15], [34].

Theorem (Borel 1909). *Let $\psi(n) > 0$ be a decreasing function of n and let*

$$S = \sum_{n=1}^{\infty} \psi(n).$$

Then, if S diverges, for almost all α

$$\limsup_{n \to \infty} a_n \psi(n) = \infty,$$

while, if S converges, for almost all α,

$$\lim_{n \to \infty} a_n \psi(n) = 0.$$

This completely describes the behaviour of the large values of the partial quotients of a typical number. The behaviour of the average value of a_n was given later by the following result.

Theorem (Khintchine 1935). *For almost all α we have*

$$(a_1 \ldots a_n)^{\frac{1}{n}} \to \prod_{r=1}^{\infty} \left(1 + \frac{1}{r(r+2)}\right)^{\frac{\log r}{\log 2}} \quad \text{as } n \to \infty.$$

Indeed, Khintchine proved a more general result which also gives

$$\lim_{N \to \infty} \frac{1}{N} \sum_{\substack{n \le N \\ a_n = r}} 1 = \frac{\log\left(1 + \frac{1}{r(r+2)}\right)}{\log 2}$$

for almost all α. It can be shown [20] that, given a finite set of positive integers $\{r_1, \ldots, r_k\}$, the limit

$$\lim_{N \to \infty} \frac{1}{N} |\{n \le N : a_{n+j} = r_j, 1 \le j \le k\}|$$

exists and is constant for almost all α. This is a consequence of the pointwise ergodic theorem. Thus, for almost all α, not only does a given block of digits (to base b) occur in its base b expansion with a fixed frequency, the same block also occurs as consecutive numbers among the partial quotients with a certain fixed frequency also. The referee remarks that most limit theorems, such as the Law of the Iterated Logarithm, or the Central Limit Theorem, carry over from the independent case as the random variables involved are ψ-mixing with an exponential mixing rate.

The above results lead naturally on to a discussion of Diophantine approximation. We are now concerned with the fractional rather than integer parts of a sequence. Let

$$f_1(n, \alpha) = \min_{m \in \mathbb{Z}} |n\alpha - m|,$$

$$f_2(n, \alpha) = \min_{\substack{m \in \mathbb{Z} \\ (m,n)=1}} |n\alpha - m|,$$

and suppose that $\psi(n)$ is a sequence of positive reals. Let $P_j(\alpha, \psi)$ denote the property that $f_j(n, \alpha) < \psi(n)$ infinitely often. Then Khintchine proved the following result.

Theorem (Khintchine 1924). *If $n\psi(n)$ is monotonically decreasing then $P_1(\alpha, \psi)$ holds for almost all α or almost no α according to whether the series*

$$\sum_{n=1}^{\infty} \psi(n)$$

diverges or converges, respectively.

The convergence case is very easy: it could have been proved by Borel in 1898. The divergence case requires more work and has been the starting point for much further work (see [17, Chapters 2–4]). Khintchine's original proof depended on Borel's work on continued fractions. The condition that $n\psi(n)$ should be decreasing can be relaxed to $\psi(n)$ decreasing [14]. We note that if $\psi(n)$ is decreasing then

$$P_2(\alpha, \psi) \text{ true} \Rightarrow P_1(\alpha, \psi) \quad \text{true},$$

and, if $\alpha \notin \mathbb{Q}$,

$$P_1(\alpha, \psi) \text{ true} \Rightarrow P_2(\alpha, \psi) \quad \text{true}.$$

If our earlier problems were 'fixed target' or 'moving target', then results of the above type can be considered 'shrinking target'. One can, of course, consider targets which are moving and shrinking at once:

$$|n\alpha - m - \beta_n| < \psi(n)$$

(see [17, Chapter 3]). Having said this though, questions involving $P_2(\alpha, \psi)$ are the more natural, and the most important unsolved problem in this area is as follows.

Conjecture (Duffin and Schaeffer 1941). *Let $\psi(n)$ be any non-negative function of an integer variable. Then*

$$\sum_{n=1}^{\infty} \psi(n) \frac{\phi(n)}{n} = \infty \Rightarrow P_2(\alpha, \psi) \text{ true}.$$

We note that if the series converges then $P_2(\alpha, \psi)$ is false for almost all α as a simple consequence of the Borel–Cantelli lemma. This conjecture has stood for nearly 60 years and has been proved subject to various conditions on ψ (see [17, Chapters 2–4]). Will the next 100 years see a proof (or counterexample)?

The author has considered Khintchine's theorem with both variables restricted to sets of number theoretic interest [17, Chapter 6]. As an example of what can be obtained we cite the following.

Theorem. *Let $\psi(n)$ be monotonic decreasing. Then the inequality*

$$|\alpha p - q| < \psi(p), \quad p, q \text{ primes},$$

has infinitely many solutions or only finitely many solutions for almost all α according to whether the series

$$\sum_{n=2}^{\infty} \frac{\psi(n)}{\log^2 n}$$

diverges or converges, respectively.

6   Higher Dimensions

All the problems considered so far can be extended to $\alpha \in \mathbb{R}^k$. Usually the extra freedom of more variables makes the problem easier. For example, the Duffin and Schaeffer conjecture is known to be true in two or more dimensions [33]. The more interesting—and more demanding!—problems in higher dimensions concern restricting attention to a submanifold and working with the appropriate measure. Diophantine approximation problems from this perspective are covered in the recent tract of Bernik and Dodson [3]. This is a very active area of research. One goal likely to be achieved in the next few years is an analogue of Khintchine's theorem for all 'reasonable' manifolds. It should be noted (see Sprindzuk's theorem, for example [38]) that the convergence case is no longer trivial for such problems.

Very recently the author [19] has considered the arithmetical structure of the set of the integer parts of points on a curve. Among the results that can be proved we mention the following (where $R(N)$ has the same meaning as in §4.).

Theorem. *Let $b_1 = 1$, $b_j > (20/19)b_{j-1}$ $(j = 2, \ldots, k)$. Let a_n be an increasing sequence of positive reals. Then $[(\alpha a_n)^{b_1}], \ldots, [(\alpha a_n)^{b_k}]$ are infinitely often simultaneously prime for almost all positive α or almost no α according to whether the series*

$$\sum_{n=1}^{\infty} \frac{R(n)}{(\log a_n)^k}$$

diverges or converges respectively.

Of course, many of the classical conjectures concerning prime numbers (for example, prime twins) can be expressed in terms of finding prime points on lines or curves.

7   The Crucial Feature of the Proofs

We finish by outlining the most significant step in the proof of many of the results outlined above. Say $P(\alpha, n)$ denotes a property as discussed above, while $P(\alpha)$ indicates the property asymptotically holds the expected number of times (or alternatively at least that it holds infinitely often). To prove $P(\alpha)$ for almost all α we usually need first to restrict attention to an interval of finite length, say $\mathcal{I}$, and to consider

$$\lambda(\mathcal{B}_n \cap \mathcal{B}_m),$$

where
$$\mathcal{B}_n = \{\alpha \in \mathcal{I} \colon P(\alpha, n) \text{ true}\}.$$

For Borel's normal number theorem we can take $\mathcal{I} = [0, 1)$ and the sets $\mathcal{B}_n$ are independent, so that

$$\lambda(\mathcal{B}_m \cap \mathcal{B}_n) = \lambda(\mathcal{B}_m)\lambda(\mathcal{B}_n).$$

In general, to get results involving an asymptotic formula, we take $\mathcal{I}$ to have unit length and must show that

$$\lambda(\mathcal{B}_m \cap \mathcal{B}_n) = \lambda(\mathcal{B}_n)\lambda(\mathcal{B}_m) + g(m, n)$$

where

$$\sum_{1 \le m, n \le N} g(m, n) = O\left(\frac{S(N)^2}{\log^{3+\epsilon} S(N)}\right) \quad \text{with} \quad S(N) = \sum_{n=1}^{N} \lambda(\mathcal{B}_n);$$

see [17, Theorem 1.5]. In some problems one can show that

$$\lambda(\mathcal{B}_m \cap \mathcal{B}_n) = \lambda(\mathcal{B}_m)\lambda(\mathcal{B}_n)\left(1 + O\left(q^{-|m-n|}\right)\right).$$

[34, p. 159] where $0 < q < 1$, which suffices. In general, it is the 'm near n' terms which are problematical ('near' meaning 'in absolute value' or 'gcd(m, n) large', depending on the context).

In order to show only that a property holds infinitely often for almost all α it suffices to prove that

$$\lambda(\mathcal{I})\lambda(\mathcal{B}_m \cap \mathcal{B}_n) \le K\lambda(\mathcal{B}_m)\lambda(\mathcal{B}_n)$$

for *all* open intervals $\mathcal{I}$ with K an absolute constant. In fact, in many cases it can be shown that a "zero-one" law holds, in which case the above inequality need only be proved for *one* open interval $\mathcal{I}$.

These overlap estimates often involve obtaining an upper bound for the number of solutions to inequalities of the form

$$\left|\frac{m}{n} - \frac{r}{s}\right| < \eta,$$

with various restrictions on the integer variables m, n, r and s. This is the part of the proof which requires the most number-theoretic input. The reader will see its relation to the equation

$$ms - nr = 1$$

which has been at the heart of number theory from Euclid onwards. Sometimes the inequalities take the form

$$|\theta r - s| < A$$

where $\theta = a_n/a_m$, with a_n a given sequence. The proof of the appropriate upper bound then branches in different directions depending on Diophantine approximations to θ. It may be that there are "too many" potential solutions for certain θ: the idea is to show that this happens for few pairs m, n.

8 Conclusion

In the last one hundred years we have come a long way in characterising the arithmetical properties of almost all real numbers, yet several difficult challenges remain. Those presented at the end of Section 2 seem especially difficult! After presenting this talk Andrew Granville asked if I could exhibit an irrational absolutely abnormal number: that is a number which is not entirely normal to any base (clearly no rational can be entirely normal to any base). Greg Martin quickly provided one nice construction. Carl Pomerance thought that

$$\sum_{n=1}^{\infty} \frac{1}{(n!)^{n!}}$$

should work—here is another challenge for the reader!

Acknowledgements. I would like to thank the referee and my colleagues at Royal Holloway—Eira Scourfield and Gar de Barra—for their comments.

References

[1] R. C. Baker, *Metric number theory and the large sieve*, J. London Math. Soc. **24** (1981), 34–40.

[2] I. Berkes and W. Philipp, *The size of trigonometric and Walsh series and uniform distribution mod* 1, J. London Math. Soc. **50** (1994), 454–463.

[3] V. I. Bernik and M. M Dodson, *Metric Diophantine approximation on manifolds*, Cambridge Tracts in Mathematics, vol. 137, Cambridge University Press, Cambridge, 1999.

[4] P. Billingsley, *Ergodic theory and information*, Wiley, New York, 1965.

[5] N. H. Bingham, *Variants on the law of the iterated logarithm*, Bull. London Math. Soc. **18** (1986), 433–467.

[6] E. Borel, *Leçons sur la théorie des fonctions*, Gauthier-Villars, Paris, 1898.

[7] ——, *Les probabilités dénombrables et leurs applications arithmétiques*, Rend. Circ. Math. Palermo **27** (1909), 247–271.

[8] W. Brown, W. Moran, and C. E. M. Pearce, *Riesz products and normal numbers*, J. London Math. Soc. **32** (1985), 12–18.

[9] J. W. S. Cassels, *On a problem of Steinhaus about normal numbers*, Collect. Math. **7** (1959), 95–101.

[10] D. G. Champernowne, *The construction of decimals normal in the scale of ten*, J. London Math. Soc. **8** (1933), 254–260.

[11] E. F. Collingwood, *Emile Borel*, J. London Math. Soc. **34** (1959), 488–512.

[12] A. H. Copeland and P. Erdös, *Note on normal numbers*, Bull. Amer. Math. Soc. **52** (1946), 857–860.

[13] H. Davenport and P. Erdös, *Note on normal decimals*, Canad. J. Math. **4** (1952), 58–63.

[14] R. J. Duffin and A. C. Schaeffer, *Khintchine's problem in metric Diophantine approximation*, Duke Math. J. **8** (1941), 243–255.

[15] G. H. Hardy and E. M. Wright, *An introduction to the theory of numbers*, Clarendon Press, Oxford, 1979.

[16] G. Harman, *Metrical theorems on prime values of the integer parts of real sequences*, Proc. London Math. Soc. **75** (1997), 481–496.

[17] ——, *Metric number theory*, Clarendon Press, Oxford, 1998.

[18] ——, *Variants of the second Borel–Cantelli lemma and their applications in metric number theory*, Number Theory, HBA, New Delhi, 2000, pp. 121–140.

[19] ——, *Metrical theorems on prime values of the integer parts of real sequences. II*, J. London Math. Soc. (2) **64** (2001), 287–298.

[20] G. Harman and K. C. Wong, *A note on the metric theory of continued fractions*, Amer. Math. Monthly **107** (2000), 834–837.

[21] R. Hill and S. L. Velani, *The shrinking target problem for matrix transformations of tori*, J. London Math. Soc. **60** (1999), 381–398.

[22] A. Khintchine, *Einige Sätze über Kettenbrüche mit Anwendungen auf die Theorie der Diophantischen Approximationen*, Math. Ann. **92** (1924), 115–125.

[23] ———, *Über einen Satz der Wahrscheinlichkeitsrechnung*, Fund. Math. **6** (1924), 9–20.

[24] ———, *Metrische Kettenbrücheprobleme*, Comp. Math. **1** (1935), 361–382.

[25] A. N. Kolmogorov, *Über die Summen durch den Zufall bestimmten unabhängiger Grössen*, Math. Ann. **99** (1928), 309–319.

[26] H. Lebesgue, *Sur certains démonstrations d'existence*, Bull. Soc. Math. France **45** (1917), 132–144.

[27] M. B. Levin, *On the discrepancy estimate of normal numbers*, Acta Arith. **88** (1999), 99–111.

[28] K. Mahler, *Arithmetical properties of the digits of the multiples of an irrational number*, Bull. Austral. Math. Soc. **8** (1973), 191–203.

[29] K. Nagasaka, *La dimension de Hausdorff de certaines ensembles dans* [0, 1], Proc. Japan Acad. Ser. A Math. Sci. **54** (1979), 109–112.

[30] W. Philipp, *Some metrical theorems in number theory II*, Duke Math. J. **37** (1970), 447–458.

[31] A. D. Pollington, *The Hausdorff dimension of a set of nonnormal well approximable numbers*, Number Theory, Carbondale (1979), Lecture Notes in Math. 751, Springer-Verlag, Berlin, 1979, pp. 256–264.

[32] ———, *The Hausdorff dimension of a set of normal numbers*, Pacific J. Math. **95** (1981), 193–204.

[33] A. D. Pollington and R. C. Vaughan, *The k-dimensional Duffin and Schaeffer conjecture*, Mathematika **37** (1990), 190–200.

[34] A. M. Rockett and P. Szüsz, *Continued fractions*, World Scientific, Singapore, 1992.

[35] J. Schiffer, *Discrepancy of normal numbers*, Acta Arith. **47** (1986), 175–186.

[36] W. M. Schmidt, *On normal numbers*, Pacific J. Math. **10** (1960), 661–672.

[37] ———, *On badly approximable numbers*, Mathematika **12** (1965), 10–20.

[38] V. G. Sprindzuk, *A proof of Mahler's conjecture on the measure of the set of S-numbers (in Russian)*, Izv. Akad. Nauk SSSR, Ser. Mat. **29** (1965), 379–436.

[39] G. Wagner, *On rings of numbers which are normal to one base but non-normal to another*, J. Number Theory **54** (1995), 211–231.

A Reciprocity Relation Between Some Cyclotomic Integers

Charles Helou

1 Introduction

Let p, q be two prime numbers, ζ_p, ζ_q primitive p-th, q-th roots of unity in $\mathbb{C}$ and $\lambda_p = 1 - \zeta_p, \lambda_q = 1 - \zeta_q$ the distinguished prime elements of the cyclotomic fields $\mathbb{Q}(\zeta_p), \mathbb{Q}(\zeta_q)$ above p, q, respectively. The elements $1 - \lambda_p^n$ (n a positive integer) are well known to form a topological basis for the group of principal units in the λ_p-adic completion of $\mathbb{Q}(\zeta_p)$ ([1], p. 247). But little is known of their multiplicative properties as global objects, i.e., of their factorization in $\mathbb{Q}(\zeta_p)$. It turns out that there is a reciprocity relation between the factorization of $1 - \lambda_p^q$ in $\mathbb{Q}(\zeta_p)$ and that of $1 - \lambda_q^p$ in $\mathbb{Q}(\zeta_q)$ (Corollary 11, Proposition 15). In particular, the latter is prime if and only if the former is (Corollary 12). There are also simple expressions for the norm $N_p(1 - \lambda_p^n)$ of $1 - \lambda_p^n$ in $\mathbb{Q}(\zeta_p)|\mathbb{Q}$ for some small values of n (formulas (3.2), (3.4)–(3.8)). For distinct prime numbers p, q, the integers $N_p(1 - \lambda_p^q)$ generalize the Mersenne numbers $2^p - 1 = N_p(1 - \lambda_p^2)$ and share some of their properties (Proposition 14); numerical evidence seems to suggest that they are always squarefree. They satisfy the relation $N_p(1 - \lambda_p^q) = (-1)^{p+q} N_q(1 - \lambda_q^p)$ (Corollary 2), which is a special case of a more general one valid for any positive integers replacing p, q (Proposition 1).

2 General Properties

For any positive integer n, let ζ_n be a primitive n-th root of unity in $\mathbb{C}$, $K_n = \mathbb{Q}(\zeta_n)$, $O_n = \mathbb{Z}[\zeta_n]$, N_n the norm map in $K_n|\mathbb{Q}$, $\lambda_n = 1 - \zeta_n$ and ϕ_n the n-th cyclotomic polynomial, i.e., the irreducible polynomial of ζ_n over $\mathbb{Q}$. The Galois group $\mathrm{Gal}(K_n|\mathbb{Q})$, of $K_n|\mathbb{Q}$, consists of the automorphisms $\sigma_k^{(n)}$ of K_n defined by $\sigma_k^{(n)}(\zeta_n) = \zeta_n^k$, for $k \pmod{n}$ in the group $(\mathbb{Z}/n\mathbb{Z})^*$ of invertible residue classes $\pmod{n}$, so that $\mathrm{Gal}(K_n|\mathbb{Q}) \simeq (\mathbb{Z}/n\mathbb{Z})^*$.

Proposition 1. *For any positive integers m and n, we have*

$$N_m(\phi_n(\lambda_m)) = N_n(\phi_m(\lambda_n)).$$

Proof. Let R_m^* (resp. R_n^*) be a set of representatives in $\mathbb{Z}$ of the residue classes in $(\mathbb{Z}/m\mathbb{Z})^*$ (resp. $(\mathbb{Z}/n\mathbb{Z})^*$). In what follows, j (resp. k) ranges through R_m^* (resp. R_n^*). We have $\phi_n(X) = \prod_k (X - \zeta_n^k)$ and ϕ_n has rational coefficients, so that $\sigma_j^{(m)}(\phi_n(\lambda_m)) = \phi_n(1 - \zeta_m^j)$. Hence $N_m(\phi_n(\lambda_m)) = \prod_j \sigma_j^{(m)}(\phi_n(\lambda_m)) = \prod_{j,k}(1 - \zeta_m^j - \zeta_n^k)$. Similarly, exchanging m and n, we get $N_n(\phi_m(\lambda_n)) = \prod_{k,j}(1 - \zeta_n^k - \zeta_m^j)$. Hence the equality. $\square$

Corollary 2. *For any prime numbers p and q, we have*

$$N_p(1 - \lambda_p^q) = (-1)^{p+q} N_q(1 - \lambda_q^p).$$

Proof. We have $\phi_q(X) = \frac{1-X^q}{1-X}$, so that $\phi_q(\lambda_p) = \frac{1-\lambda_p^q}{\zeta_p}$. Moreover $N_p(\zeta_p) = \zeta_p^{\frac{p(p-1)}{2}} = (-1)^{p-1}$. Hence $N_p(\phi_q(\lambda_p)) = (-1)^{p-1} N_p(1 - \lambda_p^q)$. Similarly $N_q(\phi_p(\lambda_q)) = (-1)^{q-1} N_q(1 - \lambda_q^p)$. The result thus follows from Proposition 1. $\square$

Corollary 3. *For any prime number p and any integer $k \geq 0$, we have*

$$N_p(1 + \lambda_p^{2^k}) = (-1)^{2^k} N_{2^{k+1}}(1 - \lambda_{2^{k+1}}^p).$$

Proof. By Proposition 1, $N_p(\phi_{2^{k+1}}(\lambda_p)) = N_{2^{k+1}}(\phi_p(\lambda_{2^{k+1}}))$. Moreover $\phi_{2^{k+1}}(X) = 1 + X^{2^k}$ and $\phi_p(\lambda_{2^{k+1}}) = \frac{1-\lambda_{2^{k+1}}^p}{\zeta_{2^{k+1}}}$. The conclusion then follows by noting that $N_{2^{k+1}}(\zeta_{2^{k+1}}) = \zeta_{2^{k+1}}^{2^k} = (-1)^{2^k}$. $\square$

Corollary 4. *For any integer $k \geq 1$, we have*

$$2^{2^k} + 1 = N_{2^{k+1}}(2 - \zeta_{2^{k+1}}).$$

Thus $F_k = 2^{2^k} + 1$ is a Fermat prime if and only if $2 - \zeta_{2^{k+1}}$ is a prime element in $O_{2^{k+1}}$; and in this case, the prime F_k splits completely in $K_{2^{k+1}}$.

Proof. We apply Corollary 3 with $p = 2$ and $k \geq 1$, noting that $K_2 = \mathbb{Q}$ and $\lambda_2 = 2$, which gives $2^{2^k} + 1 = N_{2^{k+1}}(1 - \lambda_{2^{k+1}}^2)$. Moreover $1 - \lambda_{2^{k+1}}^2 = (1 - \lambda_{2^{k+1}})(1 + \lambda_{2^{k+1}}) = \zeta_{2^{k+1}}(2 - \zeta_{2^{k+1}})$ and, by the proof of Corollary 3, $N_{2^{k+1}}(\zeta_{2^{k+1}}) = 1$. Hence the equality.

In general, we have a factorization $(2 - \zeta_{2^{k+1}})O_{2^{k+1}} = \prod_{i=1}^r \mathfrak{p}_i^{n_i}$, with distinct prime ideals $\mathfrak{p}_i$ of $O_{2^{k+1}}$, lying above prime numbers p_i, having residue degrees f_i and positive exponents n_i ($1 \leq i \leq r$). Taking the absolute norms, we get $F_k = \prod_{i=1}^r p_i^{f_i n_i}$. Thus, if F_k is prime then $r = f_1 = n_1 = 1$ and $(2 - \zeta_{2^{k+1}})O_{2^{k+1}} = \mathfrak{p}_1$ is a prime ideal of $O_{2^{k+1}}$ of residue degree 1 above F_k. Therefore, in the Galois extension $K_{2^{k+1}}|\mathbb{Q}$, the unramified rational prime F_k has all the prime ideals above it (conjugates of $\mathfrak{p}_1$) of residue degree 1; hence F_k splits completely. Conversely, if $(2 - \zeta_{2^{k+1}})O_{2^{k+1}} = \mathfrak{p}$ is a prime ideal of $O_{2^{k+1}}$, lying above a prime number p and having residue degree f, then, upon taking absolute norms, we get $F_k = 2^{2^k} + 1 = p^f$. Hence $2^{2^k} \equiv -1 \pmod{p}$, which implies that the order of 2 (mod p) is 2^{k+1}. Therefore $p \equiv 1 \pmod{2^{k+1}}$, and thus p splits completely in $K_{2^{k+1}}$ ([3], p. 14). Hence $f = 1$ and $F_k = p$ is a prime number. $\qquad\square$

Corollary 5. *For any prime number p, we have*

$$2^p - 1 = N_p(2 - \zeta_p).$$

Thus $M_p = 2^p - 1$ is a Mersenne prime if and only if $2 - \zeta_p$ is a prime element in O_p; and in this case, the prime M_p splits completely in K_p.

Proof. An application of Corollary 3, with $k = 0$, immediately yields the equality.

If M_p is a prime number then an argument identical to the one in the proof of Corollary 4 implies that $2 - \zeta_p$ is a prime of O_p and that M_p splits completely in K_p. Conversely, if $(2 - \zeta_p)O_p = \mathfrak{q}$ is a prime ideal of O_p, lying above a prime number q and having residue degree f, then, upon taking absolute norms, we get $M_p = 2^p - 1 = q^f$. Hence $2^p \equiv 1 \pmod{q}$, which implies that the order of 2 (mod q) is p. Therefore $q \equiv 1 \pmod{p}$, and thus q splits completely in K_p. Hence $f = 1$ and $M_p = q$ is a prime number. $\qquad\square$

Proposition 6. *For any prime number p and any positive integer n, we have*

$$N_p(1 - \lambda_p^n) \equiv 1 \pmod{p^{h(n,p)}},$$

where $h(n,p)$ is the least integer $\geq \frac{n}{p-1}$.

Proof. The only prime ideal of O_p dividing p is $\lambda_p O_p$. Therefore, for any σ in $G = \mathrm{Gal}(K_p|\mathbb{Q})$, $\sigma(\lambda_p)O_p = \lambda_p O_p$. Hence $N_p(1 - \lambda_p^n) = \prod_{\sigma \in G}(1 - \sigma(\lambda_p)^n) \equiv 1 \pmod{\lambda_p^n}$. Since the left-hand side lies in $\mathbb{Z}$ and the ramification index of λ_p in $K_p|\mathbb{Q}$ is $p - 1$, this congruence means, in terms of the p-adic valuation v_p of $\mathbb{Q}$, that $(p - 1)v_p(N_p(1 - \lambda_p^n) - 1) \geq n$. Hence the result. $\qquad\square$

Proposition 6 and Corollary 2 then yield

Corollary 7. *For any distinct odd prime numbers p, q, we have*

$$N_p(1 - \lambda_p^q) = N_q(1 - \lambda_q^p) \equiv 1 \pmod{p^{[\frac{q}{p-1}]+1} q^{[\frac{p}{q-1}]+1}}),$$

where, for a real number x, $[x]$ is the largest integer $\leq x$.

Remark 8. More generally, arguing as in the proof of Proposition 6, we obtain, for any prime number p and any positive integers n, r,

$$N_{p^r}(1 - \lambda_{p^r}^n) \equiv 1 \pmod{p^{h(n,p^r)}},$$

where $h(n, p^r)$ is the least integer $\geq \frac{n}{\varphi(p^r)}$ and φ is Euler's function.

Also, for any prime numbers p, q and any positive integers r, s,

$$N_{p^r}(\phi_{q^s}(\lambda_{p^r})) \equiv 1 \pmod{p^{h(q^{s-1},p^r)}},$$

and when $p \neq q$, using Proposition 1, we further obtain

$$N_{p^r}(\phi_{q^s}(\lambda_{p^r})) = N_{q^s}(\phi_{p^r}(\lambda_{q^s})) \equiv 1 \pmod{p^{h(q^{s-1},p^r)} q^{h(p^{r-1},q^s)}}.$$

Proposition 9. *Let p, q and l be any prime numbers. If l divides $N_p(1 - \lambda_p^q) = \pm N_q(1 - \lambda_q^p)$, then the order of l modulo p is equal to the order of l modulo q.*

Proof. By Proposition 6, $l \neq p, q$. Let f_p (resp. f_q) be the order of l modulo p (resp. q). All the prime ideals $\mathfrak{l}$ of O_p dividing l have the same residue degree f_p. Moreover, any such $\mathfrak{l}$ divides the product of the conjugates of $1 - \lambda_p^q$ and thus it divides one of them. So there is a prime ideal $\mathfrak{l}_p$ of O_p dividing l and $1 - \lambda_p^q$. Hence the order of λ_p modulo $\mathfrak{l}_p$ divides q, and it is not 1 since $1 - \lambda_p$ is a unit. Therefore, in the residue field $O_p/\mathfrak{l}_p$, the order q of the element λ_p divides the order of the multiplicative group, which is $N\mathfrak{l}_p - 1 = l^{f_p} - 1$. Thus $l^{f_p} \equiv 1 \pmod{q}$, i.e., f_q divides f_p. Similarly, f_p divides f_q. Hence $f_p = f_q$. $\square$

Corollary 10. *For any prime numbers p, q and l, with l dividing $N_p(1 - \lambda_p^q) = \pm N_q(1 - \lambda_q^p)$, we have*

$$\sum_{\mathfrak{l}|l} v_{\mathfrak{l}}(1 - \lambda_p^q) = \sum_{\mathfrak{l}'|l} v_{\mathfrak{l}'}(1 - \lambda_q^p),$$

where $\mathfrak{l}$ (resp. $\mathfrak{l}'$) ranges through the prime ideals of O_p (resp. O_q) dividing l, and $v_{\mathfrak{l}}$ is the $\mathfrak{l}$-adic valuation of K_p (similarly for $v_{\mathfrak{l}'}$ in K_q).

Proof. We have $(1 - \lambda_p^q)O_p = \mathfrak{a} \prod_{\mathfrak{l}|l} \mathfrak{l}^{n_{\mathfrak{l}}}$, where $\mathfrak{a}$ is an ideal of O_p whose absolute norm $N\mathfrak{a}$ is prime to l and $n_{\mathfrak{l}} = v_{\mathfrak{l}}(1 - \lambda_p^q)$. Hence $|N_p(1 - \lambda_p^q)| = (N\mathfrak{a}) \prod_{\mathfrak{l}|l} l^{f n_{\mathfrak{l}}}$, where f is the order of l (mod p) and (mod q). A similar relation holds with p and q exchanged and $\mathfrak{l}'$ replacing $\mathfrak{l}$. Hence the l-adic valuation of $N_p(1 - \lambda_p^q)$ (and of $N_q(1 - \lambda_q^p)$) is given by

$$v_l(N_p(1 - \lambda_p^q)) = f \sum_{\mathfrak{l}|l} v_{\mathfrak{l}}(1 - \lambda_p^q) = f \sum_{\mathfrak{l}'|l} v_{\mathfrak{l}'}(1 - \lambda_q^p). \qquad (2.1)$$

The result follows. $\square$

Corollary 11. *Let p, q be any prime numbers. Assume that $N_p(1 - \lambda_p^q) = \pm \prod_{i=1}^r l_i^{f_i}$, where the l_i are distinct prime numbers and, for $1 \leq i \leq r$, f_i is the order of l_i (mod p) and (mod q). Then for every i, there is exactly one prime ideal $\mathfrak{l}_i$ of O_p dividing l_i and $1 - \lambda_p^q$ (resp. $\mathfrak{l}_i'$ of O_q dividing l_i and $1 - \lambda_q^p$). And we have*

$$(1 - \lambda_p^q)O_p = \prod_{i=1}^r \mathfrak{l}_i, \qquad (1 - \lambda_q^p)O_q = \prod_{i=1}^r \mathfrak{l}_i'.$$

Proof. It follows from (2.1) that $f_i = f_i \sum_{\mathfrak{l}|l_i} v_{\mathfrak{l}}(1 - \lambda_p^q) = f_i \sum_{\mathfrak{l}'|l_i} v_{\mathfrak{l}'}(1 - \lambda_q^p)$. Therefore the two sums in these relations are equal to 1. Hence the existence and uniqueness of $\mathfrak{l}_i$ (resp. $\mathfrak{l}_i'$) dividing l_i and $1 - \lambda_p^q$ (resp. l_i and $1 - \lambda_q^p$), which furthermore satisfies $v_{\mathfrak{l}_i}(1 - \lambda_p^q) = 1$ (resp. $v_{\mathfrak{l}_i'}(1 - \lambda_q^p) = 1$). The result follows. $\qquad \square$

In particular, we have

Corollary 12. *For any prime numbers p, q, the element $1 - \lambda_p^q$ is prime in O_p if and only if $1 - \lambda_q^p$ is prime in O_q.*

Remark 13. In all the numerical examples examined, for $p, q \leq 37$ and $p \neq q$, we found that $N_p(1 - \lambda_p^q)$ is square-free. This implies that the assumption in Corollary 11 is satisfied and that all positive prime factors l of $N_p(1 - \lambda_p^q)$ are $\equiv 1$ (mod pq).

3   Some Simple Cases

The above results can be sharpened when one of the primes p, q is 3. We first need a formula for $N_p(1 - \lambda_p^3)$. More generally, it is possible to give an expression for $N_p(1 - \lambda_p^n)$ for some small values of n. In the sequel, let p be a prime ≥ 3, n an integer ≥ 1 and $\zeta_p = e^{\frac{2\pi i}{p}}$.

We start with $n = 3$ and $p \geq 5$. By Corollary 2,

$$N_p(1 - \lambda_p^3) = N_3(1 - \lambda_3^p) = 1 - 2\Re(\lambda_3^p) + |\lambda_3^p|^2, \qquad (3.1)$$

where $\Re(z)$ is the real part of a complex number z. Moreover, since $\lambda_3^2 = -3\zeta_3 = 3e^{-\frac{\pi i}{3}}$ and $\Re(\lambda_3) > 0$, we have $\lambda_3 = \sqrt{3}e^{-\frac{\pi i}{6}}$. Hence $\lambda_3^p = \sqrt{3}^p e^{-\frac{p\pi i}{6}}$, so that $|\lambda_3^p|^2 = 3^p$ and $\Re(\lambda_3^p) = \sqrt{3}^p \cos\frac{p\pi}{6}$. Furthermore, either $p \equiv \pm 1$ (mod 12) and $\cos\frac{p\pi}{6} = \frac{\sqrt{3}}{2}$ or $p \equiv \pm 5$ (mod 12) and $\cos\frac{p\pi}{6} = -\frac{\sqrt{3}}{2}$. On the other hand, the Legendre symbol $\left(\frac{3}{p}\right) = 1$ or -1

according as $p \equiv \pm 1$ or ± 5 (mod 12), respectively ([2], p. 55). Therefore $2\Re(\lambda_3^p) = \left(\frac{3}{p}\right)\sqrt{3}^{p+1}$. Substituting this into (3.1), we get

$$N_p(1 - \lambda_p^3) = 1 - \left(\frac{3}{p}\right) 3^{\frac{p+1}{2}} + 3^p \qquad (p \geq 5). \qquad (3.2)$$

Proposition 14. *For any prime numbers p, l with $p \geq 5$ and l dividing $N_p(1 - \lambda_p^3)$, we have $l \equiv 1$ (mod 3p).*

Proof. Let $x = 3^{(p-1)/2}$. Then by (3.2), $N_p(1 - \lambda_p^3) = 3x^2 - 3\left(\frac{3}{p}\right)x + 1 \equiv 0$ (mod l). Hence the discriminant of this quadratic congruence, which is equal to -3, is a square (mod l), i.e., by the quadratic reciprocity law, $1 = \left(\frac{-3}{l}\right) = (-1)^{(l-1)/2}\left(\frac{3}{l}\right) = \left(\frac{l}{3}\right)$. Thus $l \equiv 1$ (mod 3). Therefore, by Proposition 9, we also have $l \equiv 1$ (mod p). Hence the result. ☐

Proposition 15. *Let p be a prime ≥ 5 and $N_p(1 - \lambda_p^3) = \prod_{i=1}^{r} l_i^{m_i}$, where the l_i are distinct prime numbers and the m_i are positive integers ($1 \leq i \leq r$). For every i, there is exactly one prime ideal $\mathfrak{l}_i$ of O_p dividing l_i and $1 - \lambda_p^3$ (resp. $\mathfrak{l}'_i$ of O_3 dividing l_i and $1 - \lambda_3^p$); and we have*

$$(1 - \lambda_p^3)O_p = \prod_{i=1}^{r} \mathfrak{l}_i^{m_i}, \qquad (1 - \lambda_3^p)O_3 = \prod_{i=1}^{r} \mathfrak{l}_i'^{m_i}.$$

Proof. By (2.1) and Proposition 14, for every i,

$$m_i = \sum_{\mathfrak{l}|l_i} v_{\mathfrak{l}}(1 - \lambda_p^3) = \sum_{\mathfrak{l}'|l_i} v_{\mathfrak{l}'}(1 - \lambda_3^p). \qquad (3.3)$$

Hence there is at least one prime ideal $\mathfrak{l}$ of O_p dividing l_i and $1 - \lambda_p^3$ (resp. $\mathfrak{l}'$ of O_3 dividing l_i and $1 - \lambda_3^p$). Since, by Proposition 14, $l_i \equiv 1$ (mod 3p), then l_i, $\mathfrak{l}$ and $\mathfrak{l}'$ split completely in K_{3p} as products of conjugate prime ideals. The Galois group $\text{Gal}(K_{3p}|\mathbb{Q})$ consists of the automorphisms $\sigma_{i,j}$ of K_{3p}, defined by $\sigma_{i,j}(\zeta_3) = \zeta_3^i$ and $\sigma_{i,j}(\zeta_p) = \zeta_p^j$ ($1 \leq i \leq 2, 1 \leq j \leq p - 1$), whose restrictions $\sigma_j^{(p)}$ to K_p (resp. $\sigma_i^{(3)}$ to K_3) make up $\text{Gal}(K_p|\mathbb{Q})$ (resp. $\text{Gal}(K_3|\mathbb{Q})$).

Assume that $\mathfrak{l}$ (resp. $\mathfrak{l}'$) is not unique to divide l_i and $1 - \lambda_p^3$ (resp. l_i and $1 - \lambda_3^p$), i.e., that for some $2 \leq j \leq p - 1$, $\sigma_j^{(p)}(\mathfrak{l})$ divides $1 - \lambda_p^3$ (resp. $\sigma_j^{(3)}(\mathfrak{l}')$ divides $1 - \lambda_3^p$) too. Then $\mathfrak{l}$ divides $1 - \lambda_p^3$ and $1 - \sigma_k^{(p)}(\lambda_p)^3$ where $2 \leq k \leq p - 1$ is the inverse of j (mod p) (resp. $\mathfrak{l}'$ divides $1 - \lambda_3^p$ and $1 - \sigma_2^{(3)}(\lambda_3)^p$). We have $\mathfrak{l}O_{3p} = \mathfrak{L}\sigma_{2,1}(\mathfrak{L})$ for some prime ideal $\mathfrak{L}$ of O_{3p}, and since it divides $1 - \lambda_p^3 = \zeta_p(1 - \zeta_3\lambda_p)(1 - \zeta_3^2\lambda_p)$, then $\mathfrak{L}$ divides

$1 - \zeta_3 \lambda_p$ or $1 - \zeta_3^2 \lambda_p$, and, provided we replace $\mathcal{L}$ by $\sigma_{2,1}(\mathcal{L})$ if necessary, we may assume that $\mathcal{L} | (1 - \zeta_3 \lambda_p)$. Moreover, since $\mathfrak{l}$ divides $1 - \sigma_k^{(p)}(\lambda_p)^3 = \zeta_p^k (1 - \zeta_3 \sigma_k^{(p)}(\lambda_p))(1 - \zeta_3^2 \sigma_k^{(p)}(\lambda_p))$, then $\mathcal{L}$ divides $1 - \zeta_3^i \sigma_k^{(p)}(\lambda_p)$, with $i = 1$ or 2. If $i = 1$, then $\mathcal{L}$ divides $(1 - \zeta_3 \lambda_p) - (1 - \zeta_3 \sigma_k^{(p)}(\lambda_p)) = \zeta_3 \zeta_p \sigma_{k-1}^{(p)}(\lambda_p)$ which is an associate of λ_p. This contradicts $\mathcal{L} | (1 - \zeta_3 \lambda_p)$. If $i = 2$, then $\mathcal{L}$ divides $\zeta_3^2 (1 - \zeta_3 \lambda_p) + (1 - \zeta_3^2 \sigma_k^{(p)}(\lambda_p)) = \zeta_p (1 + \zeta_3^2 \zeta_p^{k-1}) = \zeta_p (1 + \zeta_{3p}^{2p+3(k-1)})$ which is an associate of a conjugate of the cyclotomic unit $1 + \zeta_{3p}$ ([3], p. 144), therefore a unit itself. This contradicts that $\mathcal{L}$ is prime. On the other hand, the assumption that $\mathfrak{l}'$ divides $1 - \lambda_p^p$ and $1 - \sigma_2^{(3)}(\lambda_3)^p$ implies that $\mathfrak{l}'$ divides their difference $\lambda_3^p - \sigma_2^{(3)}(\lambda_3)^p = \lambda_3^p(1 - (1 + \zeta_3)^p) = -\zeta_3^p \lambda_3^p$. This contradicts $\mathfrak{l}' | (1 - \lambda_3^p)$.

Therefore there is only one prime ideal $\mathfrak{l}_i$ of O_p dividing l_i and $1 - \lambda_p^3$ (resp. $\mathfrak{l}'_i$ of O_3 dividing l_i and $1 - \lambda_3^p$). Thus, by (3.3), $m_i = v_{\mathfrak{l}_i}(1 - \lambda_p^3) = v_{\mathfrak{l}'_i}(1 - \lambda_3^p)$, for $1 \leq i \leq r$. Hence the result. $\qquad\square$

Remark 16. In the numerical examples explored, for all primes $5 \leq p \leq 127$, we found that $N_p(1 - \lambda_p^3)$ is square-free.

We now give a few more examples of formulas for $N_p(1 - \lambda_p^n)$, with a prime $p \geq 3$.

If $n = 2$ then it follows from Corollary 5 that

$$N_p(1 - \lambda_p^2) = N_p(1 + \lambda_p) = 2^p - 1 \qquad (p \geq 3). \qquad (3.4)$$

If $n = 4$, then $N_p(1 - \lambda_p^4) = N_p(1 - \lambda_p^2)N_p(1 + \lambda_p^2)$ and, by Corollary 3, $N_p(1 + \lambda_p^2) = N_4(1 - \lambda_4^p)$. We also have $\lambda_4^p = (1 - i)^p = \sqrt{2}^p e^{-\frac{p\pi}{4}i}$ and thus, as in (3.1) above, $N_4(1 - \lambda_4^p) = 1 - 2\sqrt{2}^p \cos\frac{p\pi}{4} + 2^p$. Moreover, $\cos\frac{p\pi}{4} = \frac{\sqrt{2}}{2}$ or $-\frac{\sqrt{2}}{2}$, according as $p \equiv \pm 1$ or $\pm 3 \pmod 8$. But also, the Legendre symbol $\left(\frac{2}{p}\right)$ equals 1 or -1, respectively, in these cases ([2], p. 53). Hence $\cos\frac{p\pi}{4} = \left(\frac{2}{p}\right)\frac{\sqrt{2}}{2}$. It follows that

$$N_p(1 + \lambda_p^2) = 1 - \left(\frac{2}{p}\right)2^{\frac{p+1}{2}} + 2^p \qquad (p \geq 3) \qquad (3.5)$$

and, in view of (3.4),

$$N_p(1 - \lambda_p^4) = (2^p - 1)\left(1 - \left(\frac{2}{p}\right)2^{\frac{p+1}{2}} + 2^p\right) \qquad (p \geq 3). \qquad (3.6)$$

If $n = 6$ then $N_p(1 - \lambda_p^6) = N_p(1 - \lambda_p^3)N_p(1 + \lambda_p^3)$. Moreover $1 + \lambda_p^3 = (1 + \lambda_p)(1 - \lambda_p + \lambda_p^2) = (1 + \lambda_p)\frac{1+\zeta_p^3}{1+\zeta_p}$, in which $1 + \zeta_p$ and $1 + \zeta_p^3$ are (cyclotomic) units if $p \geq 5$ and thus of norm 1. Hence, by Corollary 5,

$$N_p(1 + \lambda_p^3) = N_p(1 + \lambda_p) = 2^p - 1 \qquad (p \geq 5). \qquad (3.7)$$

Therefore, by (3.2) and (3.7), we have

$$N_p(1 - \lambda_p^6) = (2^p - 1)\left(1 - \left(\frac{3}{p}\right) 3^{\frac{p+1}{2}} + 3^p\right) \qquad (p \geq 5). \qquad (3.8)$$

Acknowledgements.   I would like to thank the referee whose suggestions helped improve an earlier version of the paper.

References

[1] H. Hasse, *Number theory*, Springer, Berlin, 1980.

[2] K. Ireland and M. Rosen, *A classical introduction to modern number theory*, 2d ed., Springer, New York, 1990.

[3] L. Washington, *Introduction to cyclotomic fields*, 2d ed., Springer, New York, 1997.

On the Spectrum of the Transfer Operator for Continued Fractions with Restricted Partial Quotients

Doug Hensley

1 Introduction

Let X_0 be a random variable with uniform density on $[0,1]$, and let $T\colon x \to 1/x - \lfloor x \rfloor$. In a letter to Laplace, Gauss stated that the event $T^n X_0 < a$ has asymptotic probability $\log_2(1 + a)$ for $0 \le a \le 1$. It seems reasonable to conclude that Gauss was aware, at any rate, that the probability density function $g(t) := 1/((1+t)\log 2)$ is invariant under T: If a random variable X has density g, then so does TX. The reason for this invariance is that for $0 \le a < a + h \le 1$, TX lies between a and $a + h$ if and only if there exists $k \ge 1$, so that X lies between $1/(k + a + h)$ and $1/(k + a)$. Thus

$$\text{prob}[a \le TX \le a + h] = \sum_{k=1}^{\infty} \text{prob}[1/(k + a + h) \le X \le 1/(k + a)]$$

Taking limits as $h \to 0$ gives, for an arbitrary probability density function f for X, the corresponding density Gf for TX:

$$G[f](t) = \sum_{k=1}^{\infty} (k+t)^{-2} f(1/(k+t)) \qquad (1.1)$$

Clearly $Gg = g$.

The first published proof of the conjecture that for initial density 1, $T^r X$ has density tending to g, is due to Kuz'min [15] in 1928. Subsequent work has focused on the question of which initial probability distributions converge, and at what rate, to a distribution with density g, and on what happens when we restrict our continued fractions to those in which all v_k come from a restricted subset M of the positive integers. For these purposes, one must have information not only about G, but about related operators $G_{M,s}$ where $s = \sigma + i\tau$, $\sigma > 0$, and

$$G_{M,s}\colon f(t) \to \sum_{k \in M} (k+t)^{-s} f(1/(k+t))$$

175

The operators $G_{M,s}$ seem to hold most of the information needed for applications. Given the right information about the operators, one may determine the limiting distribution of $T^r X$, or when $M \neq \mathbb{N}$, the limiting conditional distribution given that all a_i so far have been in M. One may determine a pair of natural measures on the continued fraction Cantor set $E_M := \{x : x = [a_1, a_2, \dots] : \forall i, a_i \in M\}$ that take the place of the usual measure and the Gauss measure with density g for the case $M = \mathbb{N}$. The Hausdorff dimension of E_M depends on the place where the spectral radius of $G_{M,s}$ crosses from greater than one, to less. [It can jump from infinity to less than one, though for the most common and interesting cases of M this radius passes smoothly through 1 and the dimension is simply half the value of s so that this radius is equal to one.] One may determine the asymptotic distribution of various sums associated to the a_i. In short, knowledge of $G_{M,s}$ is useful. For examples of applications of such results, see [6], [7], [11], [12], [10], [22]. The derivatives with respect to s of the leading eigenvalue are also important; the first and second derivatives encode key asymptotic data; see [11], [22]. The leading eigenvalue of $G_{M,s}$ is in general an analytic function of s; this is the topic of a forthcoming work by the author.

Babenko [2] showed that $G = G_{\mathbb{N},2}$, acting on a suitably, and carefully, chosen space of functions defined not just on the unit interval but on a half-plane in $\mathbb{C}$, has an isomorphic sister operator K, acting on a Hilbert space of functions defined on $(0, \infty)$ of the form $K[\phi](x) = \int_0^\infty \phi(y) k(x, y) \, dy$ where the kernel k is symmetric. From his work it is known that the operator $G = G_{\mathbb{N},2}$ has an eigenvalue-eigenvector expansion, with only zero in the continuous spectrum, and that the eigenvalues decline to zero at a roughly exponential rate. Later work by Mayer and Roepstorff [20], [21] has refined the results and techniques. Here we extend most of this to the case of general M and the operators $G_{M,\sigma}$ with $s = \sigma$ real. Although we don't get an isomorphism, $G_{M,\sigma}$ has a sister operator $K_{M,\sigma}$ that lives on a Hilbert space, is nuclear of order zero, trace class, and Hermitian. There are Hilbert spaces H_α and H_M and bounded operators $K_{M,\sigma}$ and $S_{M,\sigma}$ so that $S_{M,\sigma} : H_M \to H_\alpha$, $K_{M,\sigma} : H_M \to H_M$, and $S_{M,\sigma} K_{M,\sigma} = G_{M,\sigma} S_{M,\sigma}$.

There are two main results. The first deals with the general case of complex s for which $\sigma = \Re(s) > 0$. For a certain (reasonably natural) class of Hilbert spaces H_α of power series, the operator $G_{M,s}$ acting on H_α is compact. The spectrum and the corresponding set of eigenvectors is the same for all choices of α. It consists of a sequence of isolated nonzero eigenvalues, with the only accumulation point at zero. These eigenvalues, arranged in decreasing order of magnitude, satisfy the inequalities

$$|\lambda_1 \lambda_2 \dots \lambda_n| \leq D^n (2/3)^{n^2/2}$$

where $D = D[M, s]$ depends on M and, continuously, on s. As a corollary,

$|\lambda_n| \leq D[M,s](2/3)^n$. The other main result develops a basis for H_M involving Laguerre polynomials that elucidates the structure of K and gives an alternate proof for real σ that $|\lambda_n| \ll (2/3)^n$, again with $G_{M,\sigma}$ acting on some H_α. In this case the eigenvalues are all real.

For all sets M of at least two positive integers, and for all $\sigma > 0$ for which $\sum_{k \in M} k^{-\sigma} < \infty$, the eigenvalues and corresponding eigenfunctions of

$$G_{M,\sigma}: f(t) \to \sum_{k \in M} (k+t)^{-\sigma} f(1/(k+t))$$

are real, and zero is the only point in the continuous spectrum. Moreover, the spectral radius $\lambda_M(\sigma)$ of $G_{M,\sigma}$ is a positive eigenvalue of $G_{M,\sigma}$ of algebraic multiplicity one. The unique (normalized to $g(0) = 1$) corresponding eigenfunction $g = g_{M,\sigma}(t)$ is also positive on $[0,1]$.

2   Hilbert Spaces of Power Series

The Hilbert spaces H_α can be defined for arbitrary $\alpha > 0$, but in the application, we shall require $\alpha > 1$. Let H_α denote the set of all complex analytic functions on the disk $|z-1| < \sqrt{\alpha}$ so that if $f(z) = \sum_{n=0}^{\infty} c_n(z-1)^n$ then $\sum_{n=0}^{\infty} |c_n|^2 \alpha^n < \infty$, equipped with the norm

$$\|f\|_\alpha^2 = \sum_{n=0}^{\infty} |c_n|^2 \alpha^n = \lim_{r \to \sqrt{\alpha}} \oint_{|z-1|=r} |f^2(z)| \, ds.$$

This is clearly isomorphic, with the natural map, to the space of all sequences $a = (a_0, a_1, a_2, \dots)$ so that $\sum |a_n|^2 \alpha^n$ converges, with the norm $\|a\|_\alpha^2 = \sum |a_n|^2 \alpha^n$, and we shall identify this sequence space with H_α.

Lemma 2.1. *For $\alpha > 1$, $G_{M,s}$ maps H_α into $H_{9/4}$.*

Proof. Suppose $\alpha > 1$ and choose ϵ, $0 < \epsilon < \min[1/10, 1/2 - 1/(1 + \sqrt{\alpha})]$. The disk $|z-1| < (3/2) + \epsilon$ is mapped by $z \to 1/(k+z)$ onto the open disk of diameter $(1/(k + 5/2 + \epsilon), 1/(k - 1/2 - \epsilon))$, and the union of these, for $k \in M$, is a subset of the disk $|z-1| < 1 + 5\epsilon$ which in turn is a subset of the disk $|z-1| < 3/2$. Thus

$$G_{M,s}[f](z) = \sum_{k \in M} (k+z)^{-s} f(1/(k+z))$$

converges uniformly on

$$|z-1| \leq \frac{3+\epsilon}{2}$$

to an element of $H_{(9/4)+\epsilon}$.   □

Our next lemma says that there is an 'infinite matrix' representation of $G_{M,s}$ with respect to the 'basis' $\{(z-1)^n, n \geq 0\}$.

Lemma 2.2. *If*

$$f(z) = \sum_{n=0}^{\infty} a_n(z-1)^n \in H_\alpha$$

then

$$G_{M,s}[f](z) = \sum_{m=0}^{\infty} b_m(z-1)^m$$

where

$$b_m = \sum_{n=0}^{\infty} \gamma_{M,s}[m,n] a_n$$

and

$$\gamma_{M,s}[m,n] = \frac{1}{2\pi i} \oint_{|z-1|=3/2} (z-1)^{-m-1} \sum_{k \in M} (k+z)^{-s} \left(\frac{1}{k+z}-1\right)^m dz.$$

Proof. We have already seen that $G_{M,s}[f] \in H_{(9/4)+\epsilon}$ for some $\epsilon > 0$, so that there exist b_0, b_1, ... for which $G_{M,s}[f](z) = \sum_{m=0}^{\infty} b_m(z-1)^m$. What remains to be shown is that the announced values for b_m are correct, and that the sums and integrals involved converge.

We begin by noting that

$$b_m = \frac{1}{m!} \frac{d^m}{dz^m}|_{z=1} \left(G_{M,s}[f]\right)(z)$$

$$= \frac{1}{2\pi i} \oint_C (z-1)^{-m-1} \sum_{k \in M} (k+z)^{-s} f\left(\frac{1}{k+z}\right) dz$$

where C is the circle $|z-1| = 3/2$, and this integral expands to

$$\frac{1}{2\pi i} \oint_C (z-1)^{-m-1} \sum_{k \in M} (k+z)^{-s} \sum_{n=0}^{\infty} a_n \left(\frac{1}{k+z}-1\right)^n dz.$$

The order of summation and integration is arbitrary here because the integral and sum are absolutely convergent, by Cauchy's inequality,

$$\sum_{n=0}^{\infty} |a_n| \left|\frac{1}{k+z}-1\right|^n \ll \left(\sum_{n=0}^{\infty} |a_n|^2 \alpha^n\right)^{1/2},$$

so that

$$b_m = \sum_{n=0}^{\infty} a_n \frac{1}{2\pi i} \oint_C \sum_{k \in M} (k+z)^{-s}(z-1)^{-m-1} \left(\frac{1}{k+z} - 1\right)^n dz$$

as claimed.   □

Our next lemma says that $G_{M,s}$ is a compact operator. Definitions of this concept differ from author to author but the difference is in appearance only. One definition, the one we shall use here, is that there be a sequence of finite dimensional operators that converge in norm to the compact operator. The alternate definition terms an operator L on a Hilbert space H compact, if there exist orthonormal sequences (f_n) and (g_n) and positive real numbers (ρ_n) with $\rho_n \to 0$ so that $L = \sum \rho_n \langle \cdot, f_n \rangle g_n$ [19]. In [8, p. 28], it is shown that a compact operator, in the first sense, has such an expansion, there termed the Schmidt expansion; the ρ_n are precisely the so-called s-numbers of L, and will be defined and used later in this section.

The proof that $G_{M,s}$ is compact provided $s = \sigma + i\tau$ and $\zeta_M(s) := \sum_{k \in M} k^{-\sigma} < \infty$, depends on estimates for $|\gamma_{M,s}[m,n]|$ which we give as a lemma.

Lemma 2.3. *If $M \in \mathbb{Z}^+$ with $|M| \geq 2$ (including infinite M), and if $\sum_{k \in M} k^{-\sigma} < \infty$, then for all $m, n \geq 0$,*

$$|\gamma_{M,s}[m,n]| \leq 2^{\sigma}(3/2)^{-m}\zeta_M(s).$$

Proof. Let $C = C[M, s] := 2^{\sigma}\zeta_M(s)$. (For the moment, M and s are fixed so we drop the subscripts.) We must show $|\gamma[m,n]| \leq C(3/2)^{-m}$. We calculate:

$$|\gamma[m,n]|$$

$$\leq \frac{1}{2\pi} \max_{|z-1|=3/2} (3/2)^{-m} \sum_{k \in M} |k+z|^{-\sigma} |1 - 1/(k+z)|^n$$

$$\leq (3/2)^{-m} \sum_{k \in M} (k - 1/2)^{-\sigma} \max[|1 - 1/(k+5/2)|^n, |1 - 1/(k-1/2)|^n]$$

$$\leq C(3/2)^{-m}.$$

This proves the lemma. Note that $(3/2)$ is best possible in this proof for with a larger radius the factor involving powers of n could not be safely replaced with 1.   □

Now for $f(z) = \sum_{n=0}^{\infty} a_n(z-1)^n$, we set

$$G_N[f] = G_{M,s,N}[f] = \sum_{m=0}^{N-1} (z-1)^m \sum_{m=0}^{\infty} \gamma[m,n]a_n.$$

For $\|f\|_\alpha = 1$ we then have

$$\|(G - G_n)f\| \le \sum_{m=N}^{\infty} (3/2)^m \left| \sum_{n=0}^{\infty} \gamma[m,n]a_n \right|^2$$

$$\le \sum_{m=N}^{\infty} (3/2)^m C^2 \left(\sum_{n=0}^{\infty} |a_n|^2 \alpha^n \right)^{1/2} \left(\sum_{n=N}^{\infty} \alpha^{-n} \right)^{1/2}$$

$$\le 3(3/2)^N C^2 \sqrt{\frac{1}{1 - \alpha^{-1}}}.$$

Since the range of G_N is contained in the set of polynomials of degree less than N, this shows that G is approximated in norm by a sequence of finite dimensional operators and so is compact.

Remark. G itself is not a finite dimensional operator, because the polynomials z^n have linearly independent images under G: if in

$$\sum_{n=0}^{N} c_n \sum_{k \in M} (k + z)^{-n-s}$$

not all c_n are zero, then this sum has a singularity at $z = -\min_{k \in M} k$ and is thus not identically zero.

The spectrum of a compact operator on an infinite dimensional Hilbert space consists of a countable [finite or infinite] set of isolated eigenvalues, each of finite algebraic multiplicity, and zero, which is in the continuous spectrum. Here $G_{M,s}: H_\alpha \to H_\alpha$ provided $1 < \alpha \le 9/4$. In general, the spectrum of an operator may depend on the space on which the operator acts. Not this time.

Lemma 2.4. *The spectrum of $G_{M,s}$ is independent of α.*

Proof. Suppose $\lambda \ne 0$ belongs to the spectrum of $G_{M,s}$ seen as an operator on H_α, $1 < \alpha \le 9/4$. Then there exists $f \in H_\alpha$ so that $(G_{M,s} - \lambda I)[f] = 0$. Thus $f = \lambda^{-1} G_{M,s} f \in H_{9/4}$ so that λ is also an eigenvalue of $G_{M,s}$ in the context of $H_{9/4}$. Conversely, if λ is an eigenvalue of $G_{M,s}$ in this context, then there exists a nonzero $f \in H_{9/4}$ so that $G_{M,s}f = \lambda f$. Since $H_{9/4} \subset H_\alpha$, this selfsame f serves as an eigenfunction for λ in H_α, and this completes the proof. $\square$

This brings us to the question of estimating the eigenvalues of $G_{M,\sigma}$. Briefly, most of them are small, as would be the case if they declined exponentially. The main idea is that the compact operator $G_{M,s}$ crams any unit cube of dimension N into a Hilbert cube which is, in most directions, quite narrow, and this is inconsistent with the existence of N relatively large eigenvalues.

The machinery to make this idea work is developed in [8]. In Chapter 2, the "s-numbers" of a compact operator A on a Hilbert space H are described.

If $A: H \to H$ is compact, then the s-numbers of A are the eigenvalues, in decreasing order and taking multiplicity into account, of the self-adjoint operator $S = (A^*A)^{1/2}$. We cite two results.

Lemma 2.5. *If P is an orthogonal projection of H onto an n-dimensional subspace of H, then $s_{n+1}(A) \leq \sup_{\|x\|=1} \|Ax - PAx\|$.*

This is a corollary to Theorem 2.2, p. 31, of [8].

Lemma 2.6. *If $\lambda_1, \lambda_2, \ldots \lambda_n$ and $s_1, s_2, \ldots s_n$ are the first n eigenvalues and s-numbers of A respectively, then $|\lambda_1 \lambda_2 \ldots \lambda_n| \leq s_1 s_2 \ldots s_n$.*

This is Lemma 3.3, pages 35 and 36 of [8], and is due originally to H. Weyl.

We take $\alpha = 1 + 1/n$ and apply these results to $A = G_{M,\sigma}$ acting on H_α with $P = P_n[f] = \sum_{k=0}^{n-1} a_k(z-1)^k$ when $f(z) = \sum_0^\infty a_k(z-1)^k$. For $\|f\|_\alpha = 1$, with $f = \sum_0^\infty a_k(z-1)^k$, we calculate

$$\|G[f] - P_n G[f]\|_\alpha = \|\sum_{k=n}^\infty b_k(z-1)^k\| \text{ where } b_k = \sum_{j=0}^\infty \gamma[k,j]a_j.$$

Now

$$|b_k| \leq C(2/3)^k \sum_{j=0}^\infty |a_j|$$

$$\leq C(2/3)^k \left(\sum_0^\infty |a_j|^2 \alpha^n j\right)^{1/2} \left(\sum_0^\infty \alpha^{-j}\right)^{1/2}$$

$$= C(2/3)^k \sqrt{n+1}.$$

Thus

$$\|G - P_n G\|_\alpha \leq C\sqrt{n+1} \left(\sum_{k=n}^\infty (2/3)^{2k} \alpha^k\right)^{1/2}$$

$$= C(2/3)^n (1+1/n)^{n/2} (1 - (4/9)(1+1/n))^{-1/2} \leq 6C(2/3)^n.$$

From this the result we have been aiming at follows (where we again make explicit the dependence on M and s):

Theorem 2.7.

$$|\lambda_1[M,s]\lambda_2[M,s]\ldots\lambda_n[M,s]| \le (6C[M,s])^n(2/3)^{n(n+1)/2}$$

and so

$$|\lambda_n[M,s]| \le 6C[M,s](2/3)^{(n+1)/2}.$$

This representation of G and g is computationally accessible, most readily in the case of finite M. It turns out to be computationally more efficient to use series in powers of $(z - 1/2)$ rather than in powers of $(z - 1)$.

$$G_{M,s}\left[\sum_0^n a_k(z-1/2)^k\right] = \sum_0^n b_k(z-1/2)^k + O((z-1/2)^{n+1})$$
$$\Leftrightarrow \mathcal{G}_{M,s}[(a_0,a_1,\ldots a_n)] = (b_0,b_1,\ldots b_n).$$

We can also get eigenvectors and eigenvalues of the matrix. But here, caution is in order. Although the leading eigenvalue of a positive operator is relatively stable under perturbation of the operator, small perturbations of G loom larger in comparison to small eigenvalues. This topic has been developed further just recently in [7].

3   Positive Operators

In this section and in the next, we assume that $s = \sigma$ is real. The Perron-Frobenius theorem for finite $k \times k$ matrices says that if L is a matrix with non-negative entries and with the property that for some $n > 0$, L^n has all entries positive, then there is a unique eigenvector p of L so that all entries of p are positive, the corresponding eigenvalue λ is positive and has geometric multiplicity one (that is, the space of all vectors q so that for some $n > 0, (L - \lambda I)^n q = 0$ has dimension one and consists merely of the scalar multiples of p), and the rest of the spectrum of L lies inside a circle about the origin of radius less than λ. Thus uniformly on R^k,

$$(\lambda^{-1}L)^n v = c_v p + O((1-\epsilon)^n)\|v\|,$$

and the mapping $M: v \to c_v p$ is a projection. The restriction of the linear operator $v \to Lv$ to the null space of M has spectral radius less than λ.

Krasnoselskii has shown that the same sort of result holds for positive linear operators under suitable definitions and hypotheses. In a Banach space B, a *proper cone* P is a set P of vectors closed under addition and multiplication by non-negative scalars, and $P \cap -P = \emptyset$. If $B = P - P$, then the cone is called *reproducing*. Suppose the cone also has nonempty interior.

A compact operator L is *positive* if $LP \subseteq P$. It is u_0-positive if for all $p \neq 0$ in P there exist positive real numbers a and b and a positive integer n so that $bu_0 < L^n p < au_0$, where $u < v$ means that $v - u \in P$.

The cone P of elements $f \in H_\alpha$ so that $f(t) \geq 0$ for all $0 \leq t \leq 1$ is a proper, reproducing cone in the sense of Krasnoselskii. It has non-empty interior; the function $u_0(t) := 1$ lies in the interior of the cone. The operator $G_{M,s}$ is u_0-positive with respect to the cone P because for every $f \neq 0$ in P there exist real numbers $0 < a < b$, and a positive integer n, so that $b \leq G^n[f] \leq a$. For although f may have isolated zeros on $[0,1]$, the zeros of $G^n[f]$ are the set of all z so that for all $v \in M^n$, every number x of the form $x[v, z] = [v_1, v_2, \ldots v_{n-1}, v_n + z]$ is a zero of f itself. Thus if $G^r[f]$ had k zeros, f must have had at least $|M|^r k$ zeros. Finally, G is compact as we have seen.

With all this in place, Krasnoselskii's Perron-Frobenius theorem applies and we have that $G_{M,\sigma}$ has an isolated largest eigenvalue λ which is positive and simple, and a corresponding positive eigenfunction $g = g_{M,\sigma}$, and the rest of H_α tends to zero exponentially under iteration of $\lambda^{-1}G$. That is, the spectral radius of the $\lambda^{-1}G$ restricted to the set of $f \in H_M$ for which $(\lambda^{-1}G)^n f \to 0$ is less than 1. Clearly this largest eigenvalue λ is our old friend $\lambda_M(\sigma)$.

4   A Hilbert Space Structure for G when s = σ is Real

Our second space H_M depends on M. The idea of connecting G to a setting like the one presented here goes back to Babenko, who worked this kind of structure out for the case $M = \mathbb{Z}^+$, and to Mayer [19]. The existence of a Hilbert space structure, and the consequent reality of the eigenvalues of G acting on a suitably chosen space, for $G_{M,\sigma}$ and for the more general case in which the set M from which the a_r must be taken varies periodically in r, has been established by Vallée in [23]. What is new here is the 'matrix expansion' of the associated operator $K_{M,\sigma}$ on this 'hidden Hilbert space' in terms of Laguerre polynomials. This expansion affords a second proof that the eigenvalues of $G_{M,\sigma}$ decay at least as fast as $(2/3)^n$. We take

$$H_M := \left\{ \phi \colon [0, \infty) \to \mathbb{C} \colon f \text{ is measurable and } \int_0^\infty |\phi^2(x)| \sum_{k \in M} e^{-kx} \, dx < \infty \right\} \tag{4.1}$$

Equipped with the norm

$$\|\phi\|_M^2 := \int_0^\infty |\phi^2(x)| \sum_{k \in M} e^{-kx} \, dx, \tag{4.2}$$

H_M is a Hilbert space. The point of introducing H_M is that there is a symmetric operator $K = K_{M,\sigma}$ on H_M that replicates the spectrum and the action of $G_{M,\sigma}$ on H_α. This time the proofs work only for $\alpha < 9/4$, so we assume from now on that $1 < \alpha < 9/4$. We also assume that we have a fixed M and real σ in mind, and the operators K and S to be defined will depend, but only implicitly, on these two parameters.

Let $x! := \Gamma(x+1)$, which, in the case of non-negative integers, agrees with the usual definition of $x!$. We also extend the notation of binomial coefficients; $\binom{x}{y} := \frac{x!}{y!(x-y)!}$. Recall that $\zeta_M(\sigma) = \sum_{k \in M} k^{-\sigma}$ and let $e_M(x) := \sum_{k \in M} e^{-kx}$. In all this, we lean heavily on the approach taken by Mayer [19], but using the spaces H_α in place of a certain space of analytic functions he used, which was tied closely to the choice $M = \mathbb{Z}^+$. (We must pay somehow for generality in M.)

Let $S = S_{M,\sigma} \colon H_M \to H_\alpha$ be the operator given by

$$S \colon \phi \to \int_0^\infty e^{-xz} \phi(x) x^{(\sigma-1)/2} e_M(x)\, dx. \qquad (4.3)$$

Lemma 4.1. *S maps H_M into H_α and is a bounded linear operator.*

Proof. By the Cauchy-Schwarz inequality, if $z = u + iv$ with $u > -1/2$ then

$$|S_{M,\sigma}[\phi](z)| \le \left(\int_0^\infty x^{\sigma-1} e^{-2ux} e_M(x)\, dx \right)^{1/2} \|\phi\|_M \qquad (4.4)$$

$$< (\sigma - 1)!^{1/2}((1 + 2u)^{-\sigma} + 1)^{1/2} \left(\zeta_M(\sigma) \right)^{1/2} \|\phi\|_M,$$

which is finite. To bound $\|S_{M,\sigma}[\phi]\|_B$ we need bounds for the various derivatives, evaluated at 1, of $S_{M,\sigma}[\phi]$. Using the same inequality, if

$$S_{M,\sigma}[\phi] = f = \sum_{n=0}^\infty a_n (z-1)^n,$$

then using the facts that $\Gamma(\sigma + 2j) \ll (1 + \sigma^{-1})(1 + 2j)^\sigma \Gamma(2j + 1)$ and that $\binom{2j}{j} \ll 2^{2j}$, we have

$$|a_j| \le \frac{1}{j!} \left(\int_0^\infty x^{s-1+2j} e^{-2x} e_M(x)\, dx \right)^{1/2} \|\phi\|_M \qquad (4.5)$$

$$= \frac{\Gamma(\sigma + 2j)^{1/2}}{j!} \left(\sum_{k \in M} (k+2)^{-\sigma-2j} \right)^{1/2} \|\phi\|_M$$

$$\le j^{O(1)} (2/3)^j \left(\zeta_M(s) \right)^{1/2} \|\phi\|_M.$$

Thus

$$\|S[\phi]\|_\alpha / \|\phi\|_M \ll_{M,\sigma} \left(\sum_{j=0}^{\infty} j^{O(1)} (2/3)^{2j} \alpha^j \right)^{1/2}, \qquad (4.6)$$

which proves that S takes H_M into H_α and is bounded as claimed.   □

The bound tends to infinity as $\sigma \to 0$, as $\alpha \to 9/4$, or as $\zeta_M(s) \to \infty$; happily these cases have not arisen in applications. We now proceed with a string of lemmas.

Lemma 4.2. $S = S_{M,\sigma}$ *is nonsingular, that is, if $\phi \in H_M \neq 0$ then $S[\phi] \in H_\alpha \neq 0$.*

Proof. $S[\phi]$ is the classical Laplace transform of $x^{(\sigma-1)/2} e_M(x) \phi$. If ϕ is not the zero function in H_M (that is, it is not the case that ϕ is zero almost everywhere), then this other function is likewise not zero, so that its Laplace transform is not identically zero.   □

Our next lemma says that if f is a linear combination of functions of the form $(1 + \theta z)^{-\sigma}$ then $G[f] \in \text{Range}[S]$. The classical Gauss-Kuz'min question has to do with the convergence of $G_{\mathbb{Z}+,2}^r[1]$ and the constant function 1 is covered by this lemma. Furthermore, it so happens that the eigenfunction $g_{M,\sigma}$ corresponding to $\lambda_M(\sigma)$ from Section 3 above has this form, so $g_{M,\sigma}$ is in the range of S.

Lemma 4.3. *If $f(z) = \int_0^\infty (1 + \theta z)^{-\sigma} d\mu(\theta)$ where μ is a finite signed measure on $[0, \infty)$ then $G[f]$ has the same form, and*

$$S\left(\frac{1}{\Gamma(\sigma)} \int_0^\infty x^{(\sigma-1)/2} e^{-\theta x} d\mu(\theta) \right) = G_{M,\sigma}[f].$$

Proof. It is sufficient to establish the result for the case that μ is a probability measure on $[0, \infty)$. So suppose $f(z) = \int_0^\infty (1 + \theta z)^{-\sigma} d\mu(\theta)$. Then

$$G[f](z) = \int_0^\infty \sum_{k \in M} (k + \theta)^{-\sigma} (1 + z/(k + \theta))^{-\sigma} d\mu(\theta) \qquad (4.7)$$

$$= \int_0^\infty (1 + \theta z)^{-\sigma} d\nu(\theta),$$

where ν is the measure given by

$$\nu[0, \theta) = \sum_{k \in M} \int_{\max[1/\theta - k, 0]}^\infty (k + \phi)^{-\sigma} d\mu(\phi).$$

On the other hand,

$$S\left[\frac{1}{\Gamma(\sigma)}\int_0^\infty x^{(\sigma-1)/2}e^{-\theta x}\,d\mu(\theta)\right](z) \tag{4.8}$$

$$= \frac{1}{\Gamma(\sigma)}\int_{\theta=0}^\infty \sum_{k\in M}(k+\theta)^{-\sigma}(1+z/(k+\theta))^{-\sigma}\,d\mu(\theta). \tag{4.9}$$

A routine calculation shows that $\phi_f(x) := \int_0^\infty x^{(\sigma-1)/2}e^{-\theta x}\,d\mu(\theta)$ is indeed in H_M which completes the proof. $\square$

Our interest in H_M, as previously noted, is that there is a sister operator $K = K_{M,\sigma}$ on H_M so that $S_{M,\sigma}K_{M,\sigma} = G_{M,\sigma}S_{M,\sigma}$. We now discuss K. The tools are Bessel functions and Laguerre polynomials.

The Bessel J functions of order $\alpha > -1$ are defined by

$$J_\alpha(u) := \sum_{n=0}^\infty \frac{(-1)^n(u/2)^{2n+\alpha}}{n!\Gamma(n+\alpha+1)}.$$

The Bessel I functions are defined by the same series as the Bessel J functions, except with no alternating signs; they will arise later in this section. The J_α are linked to the generalized Laguerre polynomials

$$L_n^\alpha(u) := \sum_{j=0}^n \frac{(-1)^j u^j}{j!}\binom{n+\alpha}{j+\alpha} \tag{4.10}$$

by the identity

$$J_\alpha(2\sqrt{vw}) = (vw)^{\alpha/2}\sum_{k=0}^\infty L_k^\alpha(v)\frac{w^k e^{-w}}{\Gamma(k+\alpha+1)}. \tag{4.11}$$

Now we are ready to define K:

$$K_{M,\sigma}[\phi](w) := \int_{v=0}^\infty J_{\sigma-1}(2\sqrt{vw})\phi(v)e_M(v)\,dv. \tag{4.12}$$

That is, K is an integral operator for which the kernel is the symmetric function $J_{\sigma-1}(2\sqrt{vw})$ and for which the measure with respect to which the integration is done is $d\mu_M(v) = \sum_{k\in M}e^{-vk}\,dv$. Ultimately we show that K is a compact symmetric operator on H_M with the concomitant nice spectral properties: an orthogonal system of eigenfunctions, eigenvalues real, and as in the case of finite dimensional operators, nothing more to the story. But first we must show that K is well defined.

Proposition 4.4. *If $\phi \in H_M$ and if $\zeta_M(\sigma) = \sum_{k \in M} k^{-\sigma} < \infty$, then for $x > 0$, $K[\phi](x)$ is defined. Moreover, the function $f = K[\phi]$ satisfies $\|f\|_B \ll \zeta_M(\sigma)\|\phi\|_M$.*

Proof. By the Cauchy-Schwartz inequality,

$$|K[\phi](x)| \leq \|\phi\|_M \int_0^\infty \left(J_{\sigma-1}^2(2\sqrt{xy})e_M(y)\right)^{1/2} dy. \qquad (4.13)$$

Now

$$\int_0^\infty J_{\sigma-1}^2(2\sqrt{xy})e^{-ky} dy = k^{-1}e^{-2x/k}I_{\sigma-1}(2x/k). \qquad (4.14)$$

This is most easily checked using Mathematica [24]. A similar identity may be found in [1, formula 2.45W]; it reads

$$\int_0^\infty J_\alpha(xz)J_\alpha(yz)e^{-t^2z^2}z\,dz = \frac{1}{t^2}\exp(-(x^2+y^2)/(4t^2))I_\alpha\left(\frac{xy}{2t^2}\right),$$

but this has a typographical error in that the first factor $1/t^2$ on the right hand side should read $1/(2t^2)$. With this repaired, the identity used here is equivalent after a change of variable to the case $x = y$ of 2.45. Thus

$$|K[\phi](x)| \leq \left(\sum_{k \in M} k^{-1}e^{-2x/k}I_{\sigma-1}(2x/k)\right)^{1/2}\|\phi\|_M. \qquad (4.15)$$

Now $I_{\sigma-1}(2u) \leq u^{\sigma-1}e^{2\sqrt{u}}$, so

$$|K[\phi](x)| \leq \left(\sum_{k \in M} k^{-1}e^{-2x/k}(x/k)^{\sigma-1}e^{2\sqrt{x/k}}\right)^{1/2}\|\phi\|_M$$
$$\leq e^{1/4}x^{(\sigma-1)/2}\zeta_M(\sigma)^{1/2}\|\phi\|_M.$$

Thus

$$\|K\phi\|^2 \leq \|\phi\|^2 e^{1/2}\int_0^\infty x^{\sigma-1}\zeta_M(\sigma)e_M(x)\,dx$$
$$= \Gamma(\sigma)e^{1/2}\zeta_M(\sigma)^2\|\phi\|_M^2$$
$$\Rightarrow \|K\phi\|_M \leq e^{1/4}\sqrt{\Gamma(\sigma)}\zeta_M(\sigma)\|\phi\|_M.$$

Thus $K\colon H_M \to H_M$ and K is a bounded linear operator.

The claim that $SK = GS$ is verified by a straightforward evaluation of both sides of the equality, renaming v and x with x and v on the GS side, expanding $J_{\sigma-1}$ as a series, and integrating term by term.

The upshot is this: K is a compact, nuclear-of-order-zero operator, with an expansion in terms of Laguerre polynomials. $\qquad \square$

Theorem 4.5. *If $\sigma > 0$, $K = K_{M,\sigma}$, and $\phi \in H_M$, then*

$$K[\phi] = \sum_{k=0}^{\infty} \langle u_k, \phi \rangle e_k$$

where

$$u_k(w) = w^{(\sigma-1)/2} L_k^{s-1}(w) \ \ and \ \ e_k(w) = \frac{w^{k+(\sigma-1)/2} e^{-w}}{\Gamma(k+\sigma)}.$$

Furthermore,

$$\|u_k\|^2 \ll \left(k^{\sigma} + \frac{k^2 \Gamma(\sigma)}{\sigma^2} \zeta_M(\sigma) \right)$$

and

$$\|e_k\|^2 \ll (2/3)^{2k}.$$

Finally,

$$S^{-1} G[(z-1)^n] = \frac{x^{(\sigma-1)/2} n!}{\Gamma(n+\sigma)} L_n^{\sigma-1}(x)$$

Proof. First we establish the identity for $J_{\sigma-1}(2\sqrt{vw})$ claimed above.

$$J_{\sigma-1}(2\sqrt{vw}) = (vw)^{(\sigma-1)/2} \sum_{k=0}^{\infty} \frac{w^k e^{-w}}{\Gamma(k+\sigma)} L_k^{\sigma-1}(v) \qquad (4.16)$$

because if $f(v,w) = (vw)^{-(\sigma-1)/2} J_{(\sigma-1)}(2\sqrt{vw})$ then

$$
\begin{aligned}
f(v,w) &= \sum_{n=0}^{\infty} \frac{(-1)^n v^n w^n}{n! \Gamma(n+\sigma)} \\
&= \sum_{n=0}^{\infty} \frac{(-1)^n v^n w^n}{n! \Gamma(n+\sigma)} \sum_{l=0}^{\infty} \frac{w^l e^{-w}}{l!} \\
&= \sum_{n=0}^{\infty} \sum_{k=n}^{\infty} \frac{(-1)^n v^n w^k e^{-w}}{n! \Gamma(n+\sigma)(k-n)!} \\
&= \sum_{k=0}^{\infty} \sum_{n=0}^{k} \frac{(-1)^n v^n}{n!} \binom{k+\sigma-1}{n+\sigma-1} \frac{w^k e^{-w}}{\Gamma(k+\sigma)} = \sum_{k=0}^{\infty} \frac{w^k e^{-w}}{\Gamma(k+\sigma)} L_k^{\sigma-1}(v).
\end{aligned}
$$

Thus

$$K[\phi](w) = \int_0^\infty \phi(v) J_{\sigma-1}(2\sqrt{vw}) \sum_{j \in M} e^{-jv}\, dv$$

$$= \int_0^\infty \phi(v)(vw)^{(\sigma-1)/2} \sum_{k=0}^\infty L_k^{(\sigma-1)}(v) \frac{w^k e^{-w}}{\Gamma(k+\sigma)} \sum_{j \in M} e^{-jv}\, dv$$

$$= \sum_{k=0}^\infty \frac{w^{k+(\sigma-1)/2} e^{-w}}{\Gamma(k+\sigma)} \langle v^{(\sigma-1)/2} L_k^{\sigma-1}(v), \phi \rangle = \sum_{k=0}^\infty \langle u_k, \phi \rangle e_k.$$

For the bound on $\|e_k\|$, we calculate

$$\|e_k\|_M^2 = \int_0^\infty \frac{x^{2k+\sigma-1} e^{-2x}}{\Gamma(k+\sigma)^2} \sum_{j \in M} e^{-jx}\, dx$$

$$= \frac{\Gamma(2k+\sigma)}{\Gamma(k+\sigma)^2} \sum_{j \in M} (j+2)^{-\sigma-2k}.$$

For the case $k = 0$, this simplifies to $\|e_0\|^2 \le (\Gamma(\sigma))^{-1} \sum_{j \in M} j^{-\sigma}$. For $k \ge 1$, our bound simplifies to

$$\|e_k\| \le \sum_{l=3}^\infty l^{-2-2k} \frac{\Gamma(\sigma+2k)}{\Gamma(\sigma+k)^2},$$

since the sum here is $O(3^{-2k})$ while the ratio of Γ function values is $\ll (2e)^{\sigma-1} 2^{2k}$ by Stirling's formula.

To bound $\|u_k\|$, we take $\alpha = \sigma - 1$ in an identity given in Askey [1, p. 15]

$$\int_0^\infty x^\alpha (L_k^\alpha)^2(x) e^{-x}\, dx = \Gamma(k+\alpha+1)/k!. \qquad (4.17)$$

Next, we calculate that

$$\|u_k\|_M^2 = \sum_{j \in M} \int_0^\infty x^{\sigma-1} (L_k^{\sigma-1})^2(x) e^{-jx}\, dx. \qquad (4.18)$$

The term corresponding to $j = 1$, which will be present if $1 \in M$, is thus $\Gamma(k+\sigma)/k!$. We now consider the terms corresponding to $k = 1$ and the contribution to those terms coming first from the range $0 \le x \le 1/k$ and second from $x > 1/k$.

The contribution coming from $x \le 1/k$ is bounded above by

$$\sum_{j \in M, j > 1} \int_0^{1/k} x^{\sigma-1} \left(\sum_{l=0}^k \frac{x^l}{l!} \binom{k+\sigma-1}{\sigma} \right)^2 e^{-jx}\, dx.$$

Now for $x \le 1/k$,

$$\sum_{l=0}^{k} \binom{k+\sigma-1}{\sigma} \ll \frac{k}{\sigma}.$$

Thus for $x \le 1/k$,

$$\sum_{j \in M, j > 1} \int_0^{1/k} x^{\sigma-1} \left(\sum_{l=0}^{k} \frac{x^l}{l!} \binom{k+\sigma-1}{\sigma} \right)^2 e^{-jx} \, dx \qquad (4.19)$$

$$\ll \frac{k^2}{\sigma^2} \sum_{j \in M, j > 1} \int_0^{1/k} x^{\sigma-1} e^{-jx} \, dx \ll \frac{k^2 \Gamma(\sigma)}{\sigma^2} \sum_{j \in M} j^{-(\sigma-1)}.$$

For the contribution due to $x > 1/k$, and from $j > 1$, we calculate that

$$\int_{1/k}^{\infty} x^{\sigma-1} \left(L_k^{\sigma-1}(x) \right)^2 e^{-x} \sum_{j \in M, j > 1} e^{-(j-1)x} \, dx \qquad (4.20)$$

$$\le k \int_{1/k}^{\infty} x^{\sigma-1} \left(L_k^{\sigma-1}(x) \right)^2 e^{-x} \, dx$$

$$< k \int_0^{\infty} x^{\sigma-1} \left(L_k^{\sigma-1}(x) \right)^2 e^{-x} \, dx = \frac{\Gamma(k+\sigma)}{(k-1)!}$$

which is $\ll k^\sigma$. $\square$

In a sense, the formula

$$K[\phi] = \sum_{k=0}^{\infty} \langle u_k, \phi \rangle e_k$$

gives us a matrix representation of K. Even though K acts on a Hilbert space rather than a finite dimensional space, and even though the 'matrix' is with respect to not one basis but two, we learn a lot. Grothendieck has worked all this out in greater generality, but in our Hilbert space setting matters are simpler. A compact symmetric operator has an orthogonal eigenvalue-eigenvector expansion just as in the finite dimensional case: $K\phi = \sum_{n=1}^{\infty} \lambda_n \gamma_n$, with $\|\gamma_n\| = 1$, $\langle \gamma_j, \gamma_k \rangle = 0$ for $j \ne k$, and with $\lambda_n \to 0$. Now observe that the image under K of the unit ball is a sort of ellipsoid, with the property that any subspace of dimension n meets the ellipsoid in an ellipsoid of minimal diameter $2|\lambda_n|$. On the other hand, the image under K of the span of $\{u_1, u_2, \ldots u_{n-1}\}$ is an $n-1$ dimensional subspace of H, and if we write $K = K_{n-1} + R_{n-1}$ with $K_{n-1}\phi = \sum_{j=1}^{n-1} \langle u_j, \phi \rangle e_j$, then $\|R_{n-1}\| \ll (2/3)^n$. Thus, if B is the unit ball in H, then the minimal diameter of $K_n B$ is $\ll (2/3)^n$. Therefore, $\lambda_n \ll (2/3)^n$. We now claim that the spectra of G as on operator on H_α and K on H_M are identical. In both cases, we are dealing with compact operators so the spectrum is pure point spectrum with the exception of an accumulation point at zero. In neither case is zero itself an eigenvalue. Now if $K\phi = \lambda\phi$, let $f = S\phi \in H_2$. Then

$$Gf = GS\phi = SK\phi = S\lambda\phi = \lambda S\phi = \lambda f$$

so that λ is an eigenvalue of G as well. If $Gf = \lambda f$, let $\phi = S^{-1}Gf \in H$. Then

$$K\phi = KS^{-1}Gf = S^{-1}G^2 f = S^{-1}\lambda Gf = \lambda S^{-1}Gf = \lambda\phi$$

so that λ is an eigenvalue of K as well. This completes the proof that G and K have the same spectrum. We leave to the reader the verification that there is no instance of $(G - \zeta I)f_1 = f_2$, $(G - \zeta I)f_2 = f_3$, ... $(G - \zeta I)f_r = 0$. (Since K is symmetric, nothing of the sort can happen with K.) From this it follows that all eigenvalues of $G_{M,\sigma}$ are real. $\square$

There is some numerical evidence in support of the conjecture that the eigenvalues decline exponentially (we already know that $\lambda_n \ll (2/3)^n$), that they alternate in sign, and that the geometric multiplicity of each eigenvalue is one. (This multiplicity is the dimension of the space of all f so that $(G - \lambda I)^k f = 0$ for some $k > 0$, but under the circumstances, there can be no nilpotent action in G since there is none in K, so that if $(G - \lambda I)^k f = 0$ then $(G - \lambda I)f = 0$. Thus, the geometric multiplicity of an eigenvalue is simply the dimension of the eigenspace corresponding to the eigenvalue in question. In particular, recent work of Flajolet and Vallée [7] reports strong numerical evidence that for the successive eigenvalues of $G_{\mathbb{Z}^+,2}$, the ratio λ_{n+1}/λ_n tends to $-((\sqrt{5}-1)/2)^2)$. Bolstering the numerical evidence is the fact that G is the sum of operators $G_k : f \to (k + z)^{-\sigma}f(1/(k + z))$ each of which has a geometric sequence spectrum [18]. In the case at hand, the leading eigenvalue is $(1 + \theta/k)^{-\sigma}$, where $\theta = (-k + \sqrt{k^2 + 4})/2$ is the positive solution to $\theta = 1/(k + \theta)$, and the ratio of consecutive eigenvalues is $-\theta^2$, from which it would seem that this one component of G controls the behavior of the deep eigenvalues.

References

[1]   R. Askey, *Orthogonal polynomials and special functions*, Regional Conference Series in Applied Mathematics 21 (Philadelphia), SIAM, 1975.

[2]   K. I. Babenko, *On a problem of Gauss*, Soviet Mathematical Doklady **19** (1978), 136–140.

[3] R. Bumby, *Hausdorff dimension of sets arising in number theory*, New York Number Theory Seminar, Lecture Notes in Mathematics, vol. 1135, Springer, NY, 1985, pp. 1–8.

[4] T. W. Cusick, *Continuants with bounded digits I*, Mathematica **24** (1977), 166–172.

[5] ———, *Continuants with bounded digits II*, Mathematika **25** (1978), 107–109.

[6] P. Flajolet and B. Vallée, *Continued fraction algorithms, functional operators, and structure constants*, Theoret. Comput. Sci. **194** (1998), 1–34.

[7] ———, *Continued fractions, comparison algorithms, and fine structure constants*, Canadian Mathematical Society Conference Proceedings, vol. 27, 2000, pp. 53–82.

[8] I. C. Gohberg and M. G. Kreĭn, *Introduction to the theory of linear nonselfadjoint operators*, Translations of Mathematical Monographs, vol. 18, AMS, Providence, 1969.

[9] I. J. Good, *The fractional dimensional theory of continued fractions*, Math. Proc. Cambridge Philos. Soc. **37** (1941), 199–228.

[10] D. Hensley, *Continued fraction Cantor sets, Hausdorff dimension, and functional analysis*, J. Number Theory **40** (1992), 336–358.

[11] ———, *The number of steps in the Euclidean algorithm*, J. Number Theory **49** (1994), 142–182.

[12] ———, *A polynomial time algorithm for the Hausdorff dimension of continued fraction Cantor sets*, J. Number Theory **58** (1996), 9–45.

[13] K. E. Hirst, *Continued fractions with sequences of partial quotients*, Proc. Am. Math. Soc. **38** (1973), 221–227.

[14] H.-K. Hwang, *Théorèmes limites pour les structures combinatoires et les fonctions arithmetiques*, Ph.D. thesis, École Polytechnique, 1994.

[15] R.O. Kuz'min, *A problem of Gauss*, Dokl. Akad. Nauk. SSR A (1928), 375–380.

[16] D. Mauldin and M. Urbanski, *Dimensions and measures in infinite iterated function systems*, Proc. London Math. Soc. **73** (1996), 105–154.

[17] _____, *Conformal iterated function systems with applications to the geometry of continued fractions*, Trans. Amer. Math. Soc. **351** (1999), 4995–5025.

[18] D. Mayer, *On composition operators on Banach spaces of holomorphic functions*, J. Funct. Anal. **35** (1980), 191–206.

[19] _____, *Continued fractions and related transformations*, Ergodic Theory, Symbolic Dynamics and Hyperbolic Spaces, chapter 7 (M. Keane T. Bedford and C. Series, eds.), Oxford, 1991.

[20] D. Mayer and G. Roepstorff, *On the relaxation time of Gauss' continued fraction map. I. the Hilbert space approach*, J. Statist. Phys. **47** (1987), 149–171.

[21] _____, *II. The Banach space approach (transfer operator approach)*, J. Statist. Phys. **50** (1988), 331–344.

[22] B. Vallée, *Opérateurs de Ruelle-Mayer généralisés et analyse des algorithmes d'Euclide et de Gauss*, Rapport de Recherche de l'Université de Caen, Les Cahiers du GREYC # 4, Acta Arithmetica, 1995.

[23] _____, *Fractions continues á contraintes périodiques*, J. Number Theory **72** (1998), 183–235.

[24] Wolfram Research, *Mathematica*, Champaign IL.

On Theorems of
Barban-Davenport-Halberstam Type

C. Hooley

1 Foreword

In this survey all equations and notation should be interpreted in the manner usual in the analytical theory of numbers. Beyond this encompassing comment it would be pedantic in a work of this type to expand, save to say that the precise meaning of each statement is easily supplied by the reader in the knowledge that x is usually regarded as tending to infinity, that the constants implied by the O-notation are independent of k and Q, and that A denotes a positive number, often arbitrarily large, that is not necessarily the same at each occurrence.

2 Introduction

The subject of primes in arithmetical progression is of great interest both because it forms an important facet in the theory of prime numbers and because of its potential applications to a host of problems in additive and multiplicative number theory. Undoubtedly primarily initiated for the former reason, the study of this subject especially involves the behaviour of entities that are associated with the sum[1]

$$\theta(x; a, k) = \sum_{\substack{p \leq x \\ p \equiv a \bmod k}} \log p \qquad ((a, k) = 1) \qquad (2.1)$$

and the formal remainder term

$$E(x; a, k) = \theta(x; a, k) - x/\phi(k). \qquad (2.2)$$

---

[1]Note that we persevere in our practice of using a notation in which a, k appear in θ and other functional symbols in the reverse of the usual order. This idiosyncrasy, started more by accident than design, has the advantage that k always follows a as in the underlying residue class $a \bmod k$.

Confining descriptions to (2.1) and (2.2) for technical reasons and appreciating they apply equally well to cognate items such as

$$\psi(x; a, k) = \sum_{\substack{n \leq x \\ n \equiv a \bmod k}} \Lambda(n),$$

we can summarize the main state of the subject reached before the Second World War by enunciating the unconditional estimate

$$E(x; a, k) = O(x \log^{-A} x) \qquad ((a, k) = 1) \tag{2.3}$$

for any positive constant A, which was subject to the improved

$$E(x; a, k) = O\left(x e^{-A\sqrt{\log x}}\right)$$

provided that k were not the modulus of a Dirichlet's L-function having an exceptional 'Siegel-zero'.

Although both the fruit and inspiration of substantial development in the understanding of the zeta function and the L-functions, these results were disappointing in that they only gave substantial information when the common difference of the arithmetical progression was small compared with its length x—so much so that many relevant problems in number theory remained out of reach with the welcome exception of the Goldbach three primes conjecture settled by Vinogradov. In compensation, however, Titchmarsh [37] showed in 1930 on the extended Riemann hypothesis (E.R.H.) that

$$E(x; a, k) = O\left(x^{1/2} \log^2 x\right) \qquad ((a, k) = 1) \tag{2.4}$$

with the implication that we are provided with substantial conditional information for moduli k almost, but not quite, as large as $x^{1/2}$. The impact of this was that a conditional mechanism was supplied that brought about the resolution of some important conjectures when it was accompanied, if necessary, by an ancillary supporting apparatus. For example, in the latter part of his paper, Titchmarsh settled the divisor sum problem associated with

$$\sum_{p < n} d(n - p)$$

with the aid of Brun's sieve method, while later in 1957 the author conditionally proved the asymptotic formula conjectured by Hardy and Littlewood for the number of representations of a large n as the sum of two squares and a prime by creating an altogether more elaborate structure for coping with the situations where (2.4) did not suffice.

Even on E.R.H. the results yet mentioned are seriously deficient because they do not reflect our expectation that $\theta(x; a, k)$ should be asymptotic to $x/\phi(k)$ for values of k greater than $x^{1/2}$. But this shortcoming was partially ameliorated on E.R.H. by Turán in 1937, whose result was later slightly improved by Montgomery [31] to yield

$$\sum_{\substack{0 \le a \le k \\ (a,k)=1}} E^2(x; a, k) = O(x \log^4 x) \qquad (k \le x) \qquad (2.5)$$

by a simplification in the method; thus, in mean-square average,

$$E(x; a, k) = O\left(\frac{x^{1/2} \log^2 x}{\phi^{1/2}(k)}\right), \qquad (2.6)$$

which conclusion not only accords with what might have been forecast on probability grounds but also indicates in some sense that an asymptotic formula for $\theta(x; a, k)$ remains true for k as large as $x^{1-\epsilon}$. Somewhat surprisingly no applications of this potentially rewarding theorem seem to have been published, though we should perhaps mention that while studying the Hardy-Littlewood problem on small values of $p - p'$ we ascertained in 1956 with the aid of a close analogue that infinitely often on E.R.H. we have

$$0 < p - p' < (.466 \cdots + \epsilon) \log p \qquad (2.7)$$

and thus improved on previous work of Rankin. We should observe also that (2.5) is so strong that it actually implies the Titchmarsh estimate (2.4) for individual $E(x; a, k)$.

In post-war years the theory received a vital transfusion that had its genesis in Linnik's important 1941 paper on his large sieve. Conceived as its name suggests to deal with certain sieving problems in which many residues for each appropriate congruential modulus are removed from sequences, Linnik's method was developed and reinterpreted to such an extent that powerful new procedures emerged that far outran its original purpose and that were seen to have applications to prime number theory. Thence, following the work of Linnik and Renyi in particular, the progress of the method accelerated in the 1960s and culminated after the work of Roth in Bombieri's [3] version of the so called Bombieri-Vinogradov theorem[2] to the effect that, for any positive constant A_1, there exists another positive constant A_2 such that

$$\sum_{k \le Q} \max_{(a,k)=1} \max_{1 \le y \le x} |E(y; a, k)| = O\left(\frac{x}{\log^{A_1} x}\right) \qquad (2.8)$$

---

[2] Vinogradov's result was rather weaker and did not depend on the large sieve.

when

$$Q = x^{1/2} \log^{-A_2} x.$$

But, although denominated through the title of large sieve, the procedures behind these developments do not constitute sieve methods in themselves but depend on sums such as

$$S(\theta) = S(x, \theta) = \sum_{n \leq x} e^{2\pi i n \theta},$$

and their moments

$$\sum_{k \leq Q} \sum_{\substack{0 < h \leq k \\ (h,k)=1}} |S(h/k)|^2, \qquad (2.9)$$

inequalities for which have implications for sieving problems as well as the wider theory described.

The Bombieri-Vinogradov theorem, as a moment's reflection will reveal, shows that almost always up to values of k nearly as large as $x^{1/2}$ we have $\theta(x; a, k) \sim x/\phi(k)$ for $(a, k) = 1$ and that we therefore have a workable unconditional surrogate for the conditional bound (2.4) in the problems that had previously depended on it. In particular, superseding Linnik's unconditional treatment of the Hardy-Littlewood problem on sums of two squares and prime, we can simply substitute the use of (2.8) for (2.4) in our earlier paper referred to above, as was shown by Elliott and Halberstam or by us in our tract [17]; also, in somewhat parallel manner, Bombieri and Davenport obtained the bound (2.7) without any hypothesis.

This is as far as we take the story about individual values of $\theta(x; a, k)$ or their involvement in the sums in (2.8) because the first part of the review we now reach will concern the parallel developments in our knowledge of the sum in (2.5) that stem from, or have been motivated by, estimates for the sum (2.9). We could not have veered from this direction earlier because we shall need the results already quoted and because we wished to set the survey in a proper context.

So far we have deliberately confined our focus on the particular sequence of prime numbers, since this is the subject on which we weight our concentration and by which we are then moved to widen our thoughts to other sequences. Such sequences may either be familiar ones such as that of the square-free numbers for which analogues of (2.3) are known for small moduli or, more generally, simply ones for which such analogues are postulated, the emphasis of the enquiries being what can then be discovered about the distribution on average of the sequences among arithmetical progressions with large common differences.

3   The Barban-Davenport-Halberstam Theorem

At about the time of the promulgations of the Bombieri-Vinogradov theorem, Barban ([1], [2]) on the one hand and Davenport and Halberstam [6] on the other initiated a complementary advance in prime number theory by proving versions of a theorem[3] that stands in the same relationship to (2.8) as did on E.R.H. the conditional (2.5) to the conditional (2.4). Although the demonstrations—also based on bounds for (2.9)—are much simpler even than the later more compact derivations of the Bombieri-Vinogradov theorem, we shall see that this result in its own sphere of relevance is almost as influential in its applications as the theorem it parallels.

In sketching the principles used by Barban, Davenport, and Halberstam, we follow Gallagher's [9] improved treatment and statement of their work. We begin with his form of the large sieve inequality for (2.9) that states that

$$\sum_{k \leq Q} \sum_{\substack{0 < h \leq k \\ (h,k)=1}} |S(h/k)|^2 \leq (\pi x + Q^2) \sum_{n \leq x} |a_n|^2, \tag{3.1}$$

which is roughly tantamount to certain earlier versions and which was derived in his characteristically elegant manner (this is the most accessible form of the inequality but is not best possible; Montgomery and Vaughan have shown that the multiplier π can be removed, although this is of no significance here). From this, since for any primitive character χ_q^* mod q, we have

$$\chi_q^*(m) = \frac{1}{\tau\left(\overline{\chi}_q^*\right)} \sum_{0 < h \leq q} \overline{\chi}_q^*(h) e^{2\pi i m h/q}$$

with the usual definition of $\tau(\chi_q^*)$, it follows that the sums

$$T(\chi_q^*) = \sum_{n \leq x} a_n \chi_q^*(n) = \frac{1}{\tau\left(\overline{\chi}_q^*\right)} \sum_{n \leq x} a_n \sum_{0 < h \leq q} \overline{\chi}_q^*(h) e^{2\pi i n h/q}$$

$$= \frac{1}{\tau\left(\overline{\chi}_q^*\right)} \sum_{0 < h \leq q} \overline{\chi}_q^*(h) S(h/q)$$

---

[3]But note that this will not imply the truth of (2.8), whereas, as already stated, (2.5) implied (2.4).

are subject to the inequality

$$\sum_{q \leq Q} \frac{q}{\phi(q)} \sum_{\chi^*} |T(\chi_q^*)|^2 = \sum_{q \leq Q} \frac{q}{|\tau(\overline{\chi}_q^*)|^2 \phi(q)} \sum_{\chi_q^*} \left| \sum_{0 < h \leq q} \overline{\chi}_q^*(h) S(h/q) \right|^2$$

$$\leq \sum_{q \leq Q} \frac{1}{\phi(q)} \sum_{\chi_q} \left| \sum_{0 < h \leq q} \overline{\chi}_q(n) S(h/q) \right|^2$$

$$= \sum_{q \leq Q} \sum_{\substack{0 < h \leq q \\ (h,q)=1}} |S(h/q)|^2$$

$$\leq (\pi x + Q^2) \sum_{n \leq x} |a_n|^2.$$

Hence, on choosing a_n to be $\log n$ or 0 according as n is a prime or otherwise, we obtain,

$$\sum_{q \leq Q} \frac{q}{\phi(q)} \sum_{\chi_q^*} |\theta(x, \chi_q^*)|^2 = O\left((x + Q^2) x \log x\right) \tag{3.2}$$

with the usual meaning attached to the sums $\theta(x, \chi)$.

To benefit from this inequality, let now X_k denote the principal character mod k and χ_k a general non-principal character mod k so that for $(a, k) = 1$

$$\theta(x; a, k) = \frac{\theta(x, X_k)}{\phi(k)} + \frac{1}{\phi(k)} \sum_{\chi_k} \overline{\chi}_k(a) \theta(x, \chi_k)$$

with the implication that

$$H(x, k) = \sum_{\substack{0 < a \leq k \\ (a,k)=1}} E^2(x; a, k)$$

$$= \sum_{\substack{0 < a \leq k \\ (a,k)=1}} \left(\theta(x; a, k) - \frac{\theta(x, X_k)}{\phi(k)} \right)^2 + \frac{(x - \theta(x, X_k))^2}{\phi(k)}$$

$$= \frac{1}{\phi(k)} \sum_{\chi_k} |\theta(x, \chi_k)|^2 + O\left(\frac{x^2 e^{-A\sqrt{\log x}}}{\phi(k)} \right)$$

because $\theta(x, X_k)$ is the mean value of $\theta(x; a, k)$ for given k. Thence, having consequently expressed the moment

$$G(x, Q) = \sum_{k \leq Q} H(x, k) = \sum_{k \leq Q} \sum_{\substack{0 < a \leq k \\ (a,k)=1}} E^2(x; a, k) \tag{3.3}$$

as

$$\sum_{k \leq Q} \frac{1}{\phi(k)} \sum_{\chi_k} |\theta(x, \chi_k)|^2 + O\left(x^2 e^{-A\sqrt{\log x}}\right)$$

when $Q \leq x$, we proceed to consider the contribution to the first term in this due to those values of k for which χ_k is associated with a given primitive character χ_q^* mod q, where therefore $q|k$. Thus, as obviously

$$\theta(x, \chi_k) = \theta(x, \chi_q^*) + O\left(\sum_{p|k} \log p\right) = \theta(x, \chi_q^*) + O(\log 2k),$$

we deduce that

$$G(x, Q) = O\left(\sum_{q \leq Q} \sum_{\chi_q^*} |\theta(x, \chi_q^*)|^2 \sum_{\substack{k \leq Q \\ k \equiv 0 \bmod q}} \frac{1}{\phi(k)}\right)$$

$$+ O\left(\sum_{k \leq Q} \log^2 2k\right) + O\left(x^2 e^{-A\sqrt{\log x}}\right)$$

$$= O\left(\sum_{q \leq Q} \frac{1}{\phi(q)} \log \frac{2Q}{q} \sum_{\chi_q^*} |\theta(x, \chi_q^*)|^2\right) + O\left(x^2 e^{-A\sqrt{\log x}}\right)$$

$$= O\left(G_1(x, Q)\right) + O\left(x^2 e^{-A\sqrt{\log x}}\right), \quad \text{say.} \tag{3.4}$$

The proof is completed by using (3.2) in tandem with the estimate

$$\theta(x, \chi_q^*) = O\left(x \log^{-A} x\right),$$

from which in fact (2.3) above flows. Then, if we set $Q_1 = \log^{A+2} x$, we have[4]

$$G_1(x, Q) \leq \sum_{q \leq Q_1} + \sum_{Q_1 < q \leq Q} = G_1(x, Q_1) + G_2(x, Q), \quad \text{say,} \tag{3.5}$$

in which first

$$G_1(x, Q_1) = O\left(\frac{x^2 \log x}{\log^{2A+3} x} \sum_{q \leq Q_1} 1\right) = O\left(x^2 \log^{-A} x\right). \tag{3.6}$$

---

[4]The inequality sign is needed to take care of the case $Q < Q_1$, which in fact is of little significance.

Also, by (3.2) and partial summation, it is evident that

$$G_2(x, Q) = O\left(\frac{x^2 \log^2 x}{Q_1}\right) + O(Qx \log x)$$

$$= O\left(x^2 \log^{-A} x\right) + O(Qx \log x). \qquad (3.7)$$

We conclude from (3.4), (3.5), (3.6), and (3.7) the truth of

Theorem 1 (the Barban-Davenport-Halberstam theorem; Gallagher). *Let $G(x, Q)$ be defined through (2.2) and (3.3). Then, for $Q \leq x$ and any positive constant A, we have*

$$G(x, Q) = O(Qx \log x) + O\left(x^2 \log^{-A} x\right).$$

Although we associate this proposition with the names of Barban, Davenport, and Halberstam, we should stress that in its present form it is due to Gallagher and is stronger then its precursors obtained by the earlier writers. As here enunciated, it is certainly best possible in two of its aspects, while the earlier results only asserted that

$$G(x, Q) = O\left(x^2 \log^{-A} x\right),$$

for $Q \leq x \log^{-B} x$ where $B = B(A)$ in Barban's work and $B = A + 5$ in that of Davenport and Halberstam. Thus our theorem asserts one thing the others cannot, namely, that for $x \log^{-A} x < Q \leq x$ and any A, we have

$$G(x, Q) = O(Qx \log x), \qquad (3.8)$$

which as we shall soon see is an echo of two aspects of affairs and which is consistent with the expectation that (2.6) can be almost improved to

$$E(x; a, k) = O\left((x/\phi(k))^{1/2} \log^{1/2} x\right)$$

for most values of k bounded away from 1. Similarly, no betterment in the second term in the estimate for $G(x, Q)$ is possible at present because otherwise by considering $\sqrt{G(x, Q)}$ for smallish values of Q we could infer a better version of the classical estimate (2.3) that is constrained by our lack of knowledge of 'Siegel' zeros; this is a comment that has been attributed to Montgomery.

Apart from their basically different structures, one other distinction between the Bombieri-Vinogradov theorem and the Barban-Davenport-Halberstam theorem (shortened now to 'B.D.H. theorem' for convenience) is that the upper bound symbol before E only appears in (2.8). It is indeed tempting to think that this symbol could also be affixed to the enunciation of the latter because Montgomery [32], forming what he terms a maximal large sieve inequality, has shown that (3.1) may be generalized to

$$\sum_{k \leq Q} \sum_{0 < h \leq k} \max_{1 \leq y \leq x} \left| \sum_{n \leq y} a_n e^{2\pi i n h / k} \right|^2 = O\left((x + Q^2) \sum_{h \leq x} |a_n|^2 \right)$$

by appealing to deep work by Hunt on Fourier series. However, it is naïve to think that this aspiration can be easily fulfilled since the connecting link between (3.1) and Theorem 1 is insufficiently direct. Nevertheless, even before the appearance of Montgomery's paper, Uchiyama [38] had used an ingenious argument to get an estimate that implies that

$$G^\dagger(x, Q) = \sum_{k \leq Q} \sum_{\substack{0 < a \leq k \\ (a,k)=1}} \max_{1 \leq y \leq x} \left| E^2(y; a, k) \right| = O\left(Qx \log^3 x \right) \tag{3.9}$$

for $x \log^{-A} x < Q \leq x$ and that is a useful tool for some applications. Uchiyama's result, whose proof associates a combinational idea with minor adjustments to Theorem 1, is probably imperfect to the extent of a superfluous $\log^2 x$ in the estimate, as is rendered probable by the conclusion of the author (unpublished) to the effect that, if

$$E_1(x; a, k) = \sum_{\substack{p \leq x \\ p \equiv a \bmod k}} \left(1 - \frac{p}{x} \right) \log p - \frac{x}{2\phi(k)} \qquad ((a,k) = 1),$$

then

$$\sum_{k \leq Q} \sum_{\substack{0 < a \leq k \\ (a,k)=1}} \max_{1 \leq y \leq x} \left| E_1^2(y; a, k) \right| = O(Qx \log x)$$

for $x \log^{-A} x < Q \leq x$. Furthermore, unfolding the implications of this by means of a Tauberian argument, we were led to the relations

$$G^\dagger(x, Q) = \begin{cases} O(Qx \log x), & \text{if } x \log^{-1} x < Q \leq x, \\ O\left(Qx \log x (\log \log x)^2 \right), & \text{if } x \log^{-A} x < Q \leq x \log^{-1} x, \end{cases}$$

that were announced at our address to the I.C.M. in 1974 [12]. This improvement on (3.9), a proof of which may appear in due course, is a little disappointing because the refinement wrought by the technique would become nugatory as soon as we used methods, conditional or unconditional, for invading territory belonging to smaller values of Q. The elucidation of the status of $G^\dagger(x, Q)$ therefore remains a not unimportant aspect of the theory.

Theorem 1 and the earlier estimates, whether or not accompanied by variations in detail, are usually equally effective for the sort of applications we have in mind. But from the aspect of pure prime number theory the establishment and development of (3.8) for smaller values of Q is a desideratum. This is a matter to which we shall constantly return but, in the meanwhile, mention that in accord with Theorem 0 of our paper [17] equation (3.8) is seen to be valid on E.R.H. for $Q > x^{1/2} \log^3 x$ by substituting the use of (2.5) for (2.3) in our proof of Theorem 1 above.

4   Applications of the Barban-Davenport-Halberstam Theorem

We first illustrate the way in which Theorem 1 or its variants can be applied to problems in number theory by establishing a theorem of Bombieri-Vinogradov type for the set of almost primes s of the form $p_1 p_2$ affected with a weight

$$\psi(s) = \begin{cases} 2 \log p_1 \log p_2, & \text{if } s = p_1 p_2 \text{ (or } s = p_2 p_1) \text{ and } p_1 \neq p_2, \\ \log^2 p, & \text{if } s = p^2, \end{cases}$$

where p_1, p_2 do not exceed a large number $x^{1/2}$. Then, the analogues of $\theta(x)$, $\theta(x; a, k)$, and $E(x; a, k)$ in the prime number problem being

$$\Theta(x) = \sum_{s \leq x} \psi(s) = \sum_{p_1, p_2 \leq x^{1/2}} \log p_1 \log p_2 = \theta^2 \left(x^{1/2} \right) \sim x,$$

$$\Theta(x; a, k) = \sum_{\substack{s \leq x \\ s \equiv a \bmod k}} \psi(s),$$

and

$$E(x; a, k) = \Theta(x; a, k) - \frac{x}{\phi(k)},$$

the appropriate object of study in this instance is the sum

$$\Gamma(x, Q) = \sum_{\substack{k \leq Q \\ (a,k)=1}} \max_{\substack{0 < a \leq k}} |E(x; a, k)|.$$

First, letting $\bar{b}$ denote a solution of $bu \equiv 1 \bmod k$ when $(b,k) = 1$, and supposing that $(a,k) = 1$, we have

$$\Theta(x; a, k) = \sum_{\substack{p_1, p_2 \leq x^{1/2} \\ p_1 p_2 \equiv a \bmod k}} \log p_1 \log p_2$$

$$= \sum_{\substack{0 < b \leq k \\ (b,k)=1}} \sum_{\substack{p_1 \leq x^{1/2} \\ p_1 \equiv b \bmod k}} \log p_1 \sum_{\substack{p_2 \leq x^{1/2} \\ p_2 \equiv a\bar{b} \bmod k}} \log p_2$$

$$= \sum_{\substack{0 < b \leq k \\ (b,k)=1}} \theta\left(x^{1/2}; b, k\right) \theta\left(x^{1/2}; a\bar{b}, k\right)$$

$$= \sum_{\substack{0 < b \leq k \\ (b,k)=1}} \left(\frac{x^{1/2}}{\phi(k)} + E\left(x^{1/2}; b, k\right)\right) \left(\frac{x^{1/2}}{\phi(k)} + E\left(x^{1/2}; a\bar{b}, k\right)\right)$$

$$= \frac{x}{\phi(k)} + \frac{x^{1/2}}{\phi(k)} \sum_{\substack{0 < b \leq k \\ (b,k)=1}} \left(E\left(x^{1/2}; b, k\right) + E\left(x^{1/2} a\bar{b}, k\right)\right)$$

$$+ \sum_{\substack{0 < b \leq k \\ (b,k)=1}} E\left(x^{1/2}; b, k\right) E\left(x^{1/2}; a\bar{b}, k\right). \tag{4.1}$$

Since here, for $k \leq x^{1/2}$,

$$\sum_{\substack{0 < b \leq k \\ (b,k)=1}} E\left(x^{1/2}; b, k\right) = \sum_{\substack{p \leq x^{1/2} \\ p \nmid k}} \log p - x^{1/2}$$

$$= \theta\left(x^{1/2}\right) - x^{1/2} + O(\log k) = O\left(x^{1/2} e^{-A\sqrt{\log x}}\right)$$

with a similar estimate for the other sum in the second term on the last line of (4.1),

$$\Theta(x; a, k) - \frac{x}{\phi(k)} = O\left(\frac{x e^{-A\sqrt{\log x}}}{\phi(k)}\right) + \sum_{\substack{0 < b \leq k \\ (b,k)=1}} E\left(x^{1/2}; b, k\right) E\left(x^{1/2}; a\bar{b}; k\right)$$

so that

$$E(x; a, k) = O\left(\frac{x e^{-A\sqrt{\log x}}}{\phi(k)}\right) + O\left(\sum_{\substack{0 < c \leq k \\ (c,k)=1}} E^2\left(x^{1/2}; c, k\right)\right)$$

by the Cauchy-Schwarz inequality. Thus, through an argument that has recalled a well-known relation between covariances, Theorem 1 implies for $Q \leq x^{1/2}$ the estimate

$$\Gamma(x, q) = O\left(Qx^{1/2} \log x\right) + O\left(x \log^{-A} x\right),$$

which is of the type we sought. Alternatively, we may gain an almost equally useful estimate by replacing $E^{1/2}(x; a\bar{b}, k)$ in the last term of (4.1) by $O(x^{1/2} \log x / \phi(k))$ and then using the Cauchy-Schwarz inequality.

The latter reasoning serves as a model for the partial treatment of other problems in which two of the entities appearing are restricted to be prime numbers. Such a situation arises, for example, when we try to enlarge the scope of Lagrange's four square theorem by restricting two of the squares in the representation to be squares of primes. One avenue to this is to follow Greaves [11] by considering the set of positive numbers of the type $n - p_1^2 - p_2^2$ for $p_1, p_2 \leq (1/2)n$ that are congruent to 1 mod 4 and to try to use a sieve method to find a positive lower estimate for the cardinality of such numbers that are indivisible by primes congruent to 3 mod 4. To this end, as is familiar, good estimates for the number of p_1, p_2 satisfying the condition.

$$n - p_1^2 - p_2^2 \equiv 1 \bmod 4; \quad n - p_1^2 - p_2^2 \equiv 0 \bmod d; \quad p_1, p_2 \leq \frac{1}{2}n^{1/2}$$

will be needed either for individual square-free odd d or on average, this requirement being supplied through a variant of Theorem 1 by a modification in previous reasoning for values of d up to $n^{1/2} \log^{-A} n$. But, since it is only when d can be taken as large as $n^{1/2+\epsilon}$ that the $\frac{1}{2}$ residue sieve will show that a non-empty portion of the sequence has members with at most one prime divisor congruent to 3 mod 4, and hence none, the argument in this simple form fails and it was necessary for Greaves to combine it with other ideas in order to secure a satisfactory outcome.

Another approach, if one aims at the slightly more ambitious goal of finding an asymptotic formula for the number $\nu(n)$ of representations of n in the form $X^2 + Y^2 + p_1^2 + p_2^2$, is to use the equation

$$\nu(n) = \sum_{p_1^2 + p_2^2 < n} r(n - p_1^2 - p_2^2) + O(1),$$

where

$$r(\mu) = 4 \sum_{lm=\mu} \chi(\ell)$$

is the number of representations of a positive number μ as the sum of two squares and $\chi(\ell)$ is the non-principal character mod 4. Incorporating the second expression in the first, we obtain a double sum in which either ℓ or m does not exceed $n^{1/2}$ and which, by changes in the order of summation, consists of a sum over such values of ℓ and another over such values of m. The former, for example, can be taken to be

$$4 \sum_{\ell \leq n^{1/2}} \chi(\ell) \sum_{\substack{p_1^2 + p_2^2 < n \\ n - p_1^2 - p_2^2 \equiv 0 \bmod \ell}} 1$$

whose inner sum on average can be calculated adequately through our programmes for values of ℓ as large as $n^{1/2} \log^{-A} n$. The latter having similar attributes, there remains a narrow band of ℓ and m surrounding $n^{1/2}$ for which it is possible to extend in a lengthy way the author's operations in [12] (see §2). In such a manner the asymptotic formula for $\nu(n)$ was found by Shields [37] and independently and later by Plaksin [36].

These examples by no means exhaust the potentialities of the method. We need only point to the recent work of Daniel [5] on sums of squares of primes to confirm that theorems of B.D.H. type continue to play a useful rôle in the theory of numbers.

5 The Barban-Montgomery Theorem

The theme we now pursue is about developments of the B.D.H. theorem that pertain more to the structure of primes in arithmetical progressions than to applications elsewhere.

The first significant advance in this direction was made by Montgomery ([30]; see also his monograph [31]) when he converted the bound for $G(x, Q)$ into an asymptotic formula and incidentally showed that the treatment of Theorem 1 need not depend on the large sieve. In sketching the ideas behind his proof, we mention that he could first approximate closely to $H(x, k)$ in (3.3) by

$$\sum_{\substack{0 < a \leq k \\ (a,k)=1}} \theta^2(x; a, k) - \frac{x^2}{\phi(k)} \tag{5.1}$$

since $H(x, k)$ is almost a dispersion to which the usual rules of evaluation apply. Then, summing over k, he was left with the problem of assessing the sum

$$\sum_{k \leq Q} \sum_{\substack{0 < a \leq k \\ (a,k)=1}} \theta^2(x; a, k) \tag{5.2}$$

because of known asymptotic formulae for the sum

$$\sum_{k \leq Q} \frac{1}{\phi(k)}.$$

In continuation, he acted as though he expressed the inner sum in (5.2) for $k \leq x$ as

$$\sum_{\substack{0 < a \leq k \\ (a,k)=1}} \sum_{\substack{p,p' \leq x \\ p \equiv p' \equiv a \bmod k}} \log p \log p' = \sum_{\substack{p,p' \leq x \\ p - p \equiv 0 \bmod k \\ (pp',k)=1}} \log p \log p'$$

$$= \sum_{\substack{p \leq x \\ p \nmid k}} \log^2 p + 2 \sum_{\substack{p-p'=\ell k \\ p \leq x; (pp',k)=1}} \log p \log p'$$

$$= x \log x - x + O\left(x e^{-A\sqrt{\log x}}\right)$$

$$+ 2 \sum_{\substack{p-p'=\ell k \\ p \leq x}} \log p \log p' + O\left(\log^2 k\right)$$

$$= x \log x - x + O\left(x e^{-A\sqrt{\log x}}\right) + 2\theta_2(x, \ell k), \quad \text{say}, \qquad (5.3)$$

and, having summed over k to obtain (5.2), met the sum

$$\sum_{k \leq Q} \sum_{\ell \leq x/Q} \theta_2(x, \ell k). \qquad (5.4)$$

There being as yet no formula for the prime-twins counting function $\theta_2(x,m)$ available for insertion in this, Montgomery circumvented the apparent obstacle by using a theorem by Lavrik that asserts there is such a formula that is valid on average in a mean-square sense, namely, if $\Psi(h) = 0$ when h is odd but

$$\Psi(h) = 2 \prod_{p > 2}\left(1 - \frac{1}{(p-1)^2}\right) \prod_{\substack{p \mid h \\ p > 2}}\left(\frac{p-1}{p-2}\right)$$

when h is even and if

$$F(x, h) = \theta_2(x, h) - (x - h)\Psi(h),$$

then

$$\sum_{0 < h \leq x} F^2(x, h) = O(x^3 \log^{-A} x);$$

note here that we have departed slightly from Montgomery's notation and that there are a few minor misprints towards the end of his exposition on this topic in [31]. The sum (5.4) is therefore

$$\sum_{k \leq Q} k \sum_{\ell \leq x/k} \left(\frac{x}{k} - \ell\right) \Psi(\ell k) + \sum_{\substack{k \leq Q \\ \ell k \leq x}} F(x, \ell k),$$

the second term in which is

$$O\left(\sum_{m \leq x} d(m) \, |F(x, m)|\right) = O\left(\left(\sum_{m \leq x} d^2(m)\right)^{1/2} \left(\sum_{m \leq x} F^2(x, m)\right)^{1/2}\right)$$

$$= O\left(\left(x \log^3 x \cdot x^3 \log^{-2A-3} x\right)^{1/2}\right)$$

$$= O\left(x^2 \log^{-A} x\right). \tag{5.5}$$

Since the first term can be accurately calculated by routine methods, all the constituents in the broken-down form of $G(x, Q)$ have been evaluated when $Q \leq x$ and lead after appropriate cancellations to an asymptotic formula, which in the form originally found by Montgomery is

$$G(x, Q) = \begin{cases} Qx \log Q + O\left(Qx \log \frac{2x}{Q}\right) + O\left(x^2 \log^{-A} x\right) & \text{if } Q \leq x, \\ x^2 \log x + Bx^2 + O\left(x^2 \log^{-A} x\right) & \text{if } Q = x, \end{cases}$$

$$\tag{5.6}$$

but which he later slightly improved by the replacement of the first remainder term in the first line by $O(Qx)$ (see Croft [4]). Here, as almost always, we have curtailed attention to the most significant case where $Q \leq x$, although we should note that Montgomery makes no such restriction in his work.

In the designation of theorems stating formulae of the general type (5.6) Montgomery's name has been wont to be linked with that of Barban [2] because the latter had pre-empted the former to the extent of asserting the truth of (5.6) for the special case $Q = x$. Although Montgomery's contribution is undeniably the greater, this practice has at least the virtue of identifying unequivocally the theorems to which the nomenclature appertains. We are therefore ready to state the basic result in the second subject area as

Theorem 2 (Barban-Montgomery theorem). *Let $G(x, Q)$ be defined through (2.2) and (3.3). Then $G(x, Q)$ satisfies the asymptotic formulae in (5.6) for any positive constant A.*

This should be contrasted with Theorem 1 before we assay further developments. Having chosen any large A to establish a point of reference, we easily see that the formulae of these theorems are equivalent for $Q \leq x \log^{-A-1} x$ because each is then equivalent to $G(x, Q) = O\left(x^2 \log^{-A} x\right)$; on the other hand, for $Q > x \log^{-A-1+\epsilon} x$, the formula of Theorem 2 converts the formally dominant term $O(Qx \log x)$ in Theorem 1 into $Qx \log Q + O(Qx)$ and indicates an average value for the dispersion of $E(x; a, k)$ taken over values of a coprime to k, this being the origin of one of our later discussions about the distribution of $E(x; a, k)$.

As mentioned in §2, a constant thread throughout our work is our need to lower the values of Q—and hence indirectly of k—for which our results shed useful information on $E(x; a, k)$. To our comment that on E.R.H. the range of validity of (3.8) can be stretched to $Q > x^{1/2} \log^3 x$ we therefore add that on the same supposition we may replace the last remainder term in Theorem 2 by $O\left(x^{7/4+\epsilon}\right)$, thus obtaining an informative asymptotic formula for $Q > x^{3/4+\epsilon}$. This conclusion, of which Montgomery was most certainly aware, is deduced by ousting Lavrik's theorem in favour of a conditional version that is obtained along the lines of Hardy and Littlewood's work on Partitio Numerorum.

A new and easier way of dealing with theorems of the Barban-Montgomery type ('B.M. type' for brevity) was brought in by the author [17] in 1975 even though the analysis subsumed a prior acquaintance with Theorem 1. To expose the ideas involved, we should mention that the case $Q \leq x \log^{-A-1} x$ was first dismissed in the spirit of the last but one paragraph, whereupon, after setting

$$Q_1 = x \log^{-A-1} x, \quad Q_1 < Q \leq x,$$

and

$$G(x; Q_1, Q_2) = \sum_{Q_1 < k \leq Q_2} \sum_{\substack{0 < a \leq k \\ (a,k)=1}} E^2(x; a, k),$$

we worked with the equations

$$G(x, Q_2) = G(x; Q_1, Q_2) + O(x \log^{-A} x)$$

and

$$G(x; Q_1, Q_2) = G(x; Q_1, x) - G(x; Q_2, x).$$

Then, in analysing the sum $G(x; Q, x)$ for Q equal to Q_1 or Q_2, we followed the previous treatment of $G(x, Q)$ up to (5.3) and, summing over the revised range $Q < k \le x$, arrived at the sum

$$\sum_{\substack{p-p'=\ell k \\ p \le x; k > Q}} \log p \log p', \tag{5.7}$$

the condition $k > Q$ in which permits one to dispense with the use of Lavrik's theorem. Indeed, since now the dummy variable ℓ is less than $x/Q \le \log^{A+1} x$, a change in the order of summation permits one to use effectively the prime number theorem for arithmetical progressions for the small common difference ℓ in the revised formulation of (5.7) as

$$\sum_{\ell < x/Q} \sum_{\substack{p \equiv p' \bmod \ell \\ p \le x; p-p' > \ell Q}} \log p \log p_1 = \sum_{\ell < x/Q} \sum_{0 < b \le \ell} \sum_{\substack{p' < x - \ell Q \\ p' \equiv b \bmod \ell}} \log p' \sum_{\substack{\ell Q + p' < p \le x \\ p \equiv b \bmod \ell}} \log p.$$

After this and some consequential calculations there emerges a satisfactory expression for (5.7), which when combined with other items in the analysis recoups Theorem 2 in the slightly sharper form

$$G(x, Q) = Qx \log Q + B_1 Qx + O\left(Q^{5/4} x^{3/4}\right) + O\left(x^2 \log^{-A} x\right)$$

for $Q \le x$.

As well as its easy application to Theorem 2 and its wide-ranging generalizations that we later mention, this method has a number of other advantages over Montgomery's. Not only does it yield a more accurate form of the basic theorem but it also empowers us on the basis of E.R.H. to increase the range of usefulness of the theorem beyond the limit of $Q = x^{3/4+\epsilon}$ imposed by the earlier method. In fact, having already achieved the sharper $Q = x^{4/7+\epsilon}$ in the original paper, we went on to gain the yet better $Q = x^{1/2+\epsilon}$ in the second paper [13] by augmenting the technique through the use of primes in small intervals and a Tauberian argument. On the other hand, as we shall see in the next Section, the advantage does not lie wholly on one side because where applicable the Montgomery method can often yield more information than ours when suitably interpreted.

In continuing the story to the present day, we should remark that some clearly felt that the Hardy-Littlewood circle method was the most fitting instrument for the study of Theorem 2 along lines parallel to ours— a view in which we do not necessarily acquiesce for a reason we shortly give. Prompted by this outlook, Friedlander and Goldston [7] improved our bound of utility on E.R.H. from $x^{1/2+\epsilon}$ to $x^{1/2} \log^{5+\epsilon} x$ although they were partially forestalled by Vaughan, who somewhat similarly had obtained the better $Q = x^{1/2} \log^{3/2+\epsilon} x$ at about the time of our discoveries in [13] but who then laid the matter aside without proceeding to publication. Recently, however, in collaboration with Goldston, Vaughan has returned to the subject and published a proof of his bound with an account of associated topics [38].

We also have revisited earlier work and discovered that we were just a bit too clever in our application to our work in [14] of the properties of primes in small intervals. Had we used less sophisticated bounds in one instance, we would not only have materially simplified the treatment but would have obtained a lower bound for Q commensurate with that of Goldston and Vaughan's. An account of this revision will be given if a suitable opportunity arises.

6   Other Second Moments Containing $E(x; a, k)$

The information on the remainder $E(x; a, k)$ provided by Theorem 2 and its conditional versions is constrained amongst other things by the very wide domain of summation in $G(x, Q)$ and by the unchanging nature of the limit x for p as a and k vary. Therefore, with the object of obtaining partial confirmation of what the theorem presages, we seek to establish analogues of it in which either x is replaced by a number $x_{a,k}$ or the numbers a, k are subject to extra curtailment.

At one extreme, confining k to a single value in the summation, we wish to examine the sum

$$H(x,k) = \sum_{\substack{0 < a \leq k \\ (a,k)=1}} E^2(x; a, k),$$

for which on E.R.H. we had the Turán-Montgomery bound

$$H(x,k) = O(x \log^4 x) \qquad (k \leq x)$$

already stated in (2.5). This was developed by us [19] into

$$\sum_{\substack{0 < a \leq k \\ (a,k)=1}} \max_{1 \leq u \leq x} E^2(u; a, k) = O(x \log^4 x) \qquad (k \leq x) \qquad (6.1)$$

by an altogether more intricate argument that appealed to Selberg's work on the distribution of the zeros of Dirichlet's L-functions. Having validity for a range of k much wider than for most results of the type to which they belong, these bounds shed light on natural conjectures about $E(x; a, k)$

and are especially useful auxiliary tools in the pursuit of results such as the B.M. theorem in [14] or one to be mentioned. Yet they are probably wanting because by comparison with our bounds for $G(x, Q)$ they seem to contain a superfluous factor $\log^3 x$. We must therefore have recourse to other ideas to investigate whether $H(x, k)$ can be likened to the expected $x \log k$ as k veers away from 1, particularly when E.R.H. is not assumed.

Here Montgomery's ideas score over those used by others in connection with Theorem 2. Improved by minor modifications on which Croft [4] reported, his analysis of $H(x, k)$ gave an equation slightly more precise than

$$H(x, k) = x \log k + O\left(x\sigma_{-1/2}(k)\right) + O\left(\frac{x^2}{\phi(k) \log^{2A} x}\right) + O\left(\sum_{\ell \leq x/k} |F(x, \ell k)|\right)$$

$$= x \log k + O\left(x\sigma_{-1/2}(k)\right) + O\left(\frac{x^2}{\phi(k) \log^{2A} x}\right) + O(H_1(x, k)), \quad \text{say,}$$

the genesis of which can be discerned from the synopsis at the beginning of §5. Since

$$\sum_{Q < k \leq 2Q} H_1(x, k) = O\left(x \log^{-3A} x\right)$$

by (5.5), we infer as in our paper [14] that, if $x \log^{-A} x < Q \leq \frac{1}{2}x$, then

$$H(x, k) = x \log k + O\left(x\sigma_{-1/2}(k)\right) + O\left(x \log^{-A} x\right)$$

$$= x \log k + O\left(x\sigma_{-1/2}(k)\right)$$

for $Q < k \leq 2Q$ save for at most $O(Q \log^{-A} Q)$ exceptional values of k. Here the remainder term may be replaced by $O(x \log^{\epsilon} k)$ and actually more ambitiously by an explicit term added to a remainder term of lower order of magnitude; the latter refinement was confirmed and implemented by Friedlander and Goldston [7]. Thus, in summary, we can loosely say that almost always $H(x, k) \sim x \log k$ for values of k closer to x than $x \log^{-A} x$.

A deficiency in this sort of conclusion is that it lacks an absolute character in regard to the moduli k since it does not assert that those k for which $H(x, k) \sim x \log k$ can remain the same when x changes or, in other words, that the exceptional set of k can be chosen to be essentially independent of x. However, in a paper [16] we return to later, we remedied this omission by proving that almost all numbers k have the property that

$$H(x, k) = x \log k + O\left(x \log^{1/2} k\right) \tag{6.2}$$

for all x satisfying $k \leq x \leq k \log^A k$ and, indeed, on E.R.H. for all x satisfying $k \leq x \leq k^{5/4-\epsilon}$.

Significant progress was made when Friedlander and Goldston [7], viewing the matter from a different angle, obtained an estimate for *all* k in a given range by showing that

$$H(x, k) > \left(\frac{1}{2} - \epsilon\right) x \log k \qquad (6.3)$$

when

$$x \log^{-A} x < k \leq x. \qquad (6.4)$$

The source of their estimation lay in the properties of a surrogate prime number function $\Lambda_R(n)$, whose definition as later interpreted by us in [26] is tantamount to

$$\Lambda_R(n) = V(R) \sum_{\substack{d|n \\ d \leq R}} \lambda_d,$$

where

$$V(R) = \sum_{d \leq R} \frac{\mu^2(d)}{\phi(d)} = \log R + O(1),$$

and where the real numbers λ_d are chosen for square-free numbers $d \leq R$ as in Selberg's sieve method so that the quadratic form

$$\sum_{d_1, d_2 \leq R} \frac{\lambda_{d_1} \lambda_{d_2}}{[d_1, d_2]}$$

has a conditional minimum subject to the condition $\lambda_1 = 1$. Being likely from its form to mimic $\Lambda(n)$, the function $\Lambda_R(n)$ is used in the formation of sums

$$\psi_R(x; a, k) = \sum_{\substack{n \leq x \\ n \equiv a \bmod k}} \Lambda_R(n)$$

that can be indirectly compared with the sums $\theta(x; a, k)$ through the inequality

$$\theta^2(x; a, k) \geq 2\theta(x; a, k)\psi_R(x; a, k) - \psi_R^2(x; a, k).$$

Accordingly, as the first term in (5.1) is not less that

$$2 \sum_{0 < a \leq k} \theta(x; a, k)\psi_R(x; a, k) - \sum_{0 < a \leq k} \psi_R^2(x; a, k),$$

we meet two substitute sums that are easier to handle than their progenitor. Indeed, using the Bombieri-Vinogrodov theorem in combination with simply derived extensions of familiar properties of the numbers λ_d, we arrive at satisfactory estimates provided that R is not too large, whence (6.3) appears after all terms implicit in the total procedure are taken into account. Also, on adding the force of E.R.H. to this argument, these authors obtained the inequality

$$H(x,k) > \left(2 - \frac{3}{2\alpha} - \epsilon\right) x \log k \qquad (k = x^\alpha) \qquad (6.5)$$

that provides conditional information whenever $x^{3/4+\epsilon} < k \le x$.

The barrier of the type $k > x \log^{-A} x$ in the applicability of (6.3) was caused by the indirect appeal to (2.3) in the proof. However, by the use of an additional argument that nullified any deleterious effect from Siegel zeros, we showed that this impediment could be relaxed to $k > xe^{-A\sqrt{\log x}}$ and hinted that a further improvement might well be possible [26]. We also on this occasion sharpened the conditional (6.5) to

$$H(x,k) > \left(\frac{3}{2} - \frac{1}{\alpha} - \epsilon\right) x \log k \qquad (k = x^\alpha)$$

by using our estimate (6.1), as did Friedlander and Goldston simultaneously and independently by an identical argument [8].

Going back to theorems that are more recognisable as being of B.M. type, we should report first that our extension of Montgomery's method in [16] also gave the formulae

$$\sum_{k \le Q} \max_{1 \le y \le x} \sum_{\substack{0 < a \le k \\ (a,k)=1}} E^2(y; a, k) - Qx \log Q$$

$$= \begin{cases} O(Qx) + O\left(x^2 \log^{-A} x\right), & \text{unconditionally,} \\ O(Qx) + O\left(x^{9/5+\epsilon}\right), & \text{on E.R.H.,} \end{cases}$$

which provide a partial answer to the question raised in §3 about Uchiyama's bound (3.9). Also, restricting the domain of the variables of summation in a different way, we [25] have investigated the sum

$$\sum_{k \le Q} \sum_{\substack{0 < a \le \lambda k \\ (a,k)=1}} E^2(x; a, k) \qquad (\lambda \le 1)$$

and established an extension of the B.M. theorem by demonstrating that it equals

$$x\lambda \left(Q \log Q + b_1 Q\right) + O\left(\lambda Q^{5/4} x^{3/4}\right) + O\left(Q^2 \lambda \log x\right) + O\left(x^2 \log^{-A} x\right)$$

for $Q \leq x$ and any $\lambda \leq 1$. In its origin, the method used is more akin to our treatment of Theorem 2 than that of Montgomery's; however, in execution, it involves substantial new procedures.

Before we quit our survey of results in the genre of Theorem 1 and Theorem 2, we should call to mind the reason given after the statement of the former as to why they cannot currently be fully informative in ranges below $Q = x \log^{-A} x$. However, no like consideration debars the verification of useful lower bounds in a wider span of k, since awkward values of $E(x; a, k)$ can be eliminated from $G(x, Q)$ in such a quest; already, in fact, it is easily deduced from our improved version of Friedlander and Goldston's (6.1) that

$$G(x, Q) > \left(\frac{1}{2} - \epsilon\right) Q x \log Q \qquad \left(Q > x e^{-A\sqrt{\log x}}\right). \qquad (6.6)$$

Before the antecedent of this implication was available, Liu [29] had already proved that

$$G(x, Q) > \left(\frac{1}{4} - \epsilon\right) Q x \log Q \qquad \left(Q > x \exp\left(-\log^{3/5-\epsilon} x\right)\right), \qquad (6.7)$$

which is better than (6.6) in terms of range but worse in terms of the multiplier of $Q x \log Q$. Beyond this, Perelli [34] showed that there are certain positive constants A and η with the property that

$$G(x, Q) > A Q x \log Q \qquad \left(Q > x^{1-\eta}\right),$$

being the first to breach the limit $Q > x^{1-\epsilon}$ for problems of this sort.

It is of course desirable to replace the coefficient of $Q x \log Q$ in these bounds by $1 - \epsilon$ without weakening the range of k. This we largely achieved in [27] when we improved Liu's result above to

$$G(x, Q) > (1 - \epsilon) Q x \log Q \qquad \left(Q > x \exp\left(-\log^{3/5-\epsilon} x\right)\right), \qquad (6.8)$$

by associating our method of proving Theorem 2 with the use of the function $\Lambda_R(n)$. Yet, the application of $\Lambda_R(n)$ was more radical than before because a reappraisal of part of Selberg's sieve method was needed in order to accommodate the larger values of R involved. Also on the assumption that the Riemann zeta function has no zeros with real part exceeding $3/4$, we proved that

$$G(x, Q) > \left(2 - \frac{1}{\alpha} - \epsilon\right) Q x \log Q \qquad (6.9)$$

when $Q = x^\alpha$ and $1/2 < \alpha \leq 1$; this result, which is interesting if only because it involves a hypothesis weaker than E.R.H., depends on extra ideas including one related to the large sieve.

7   The Sum $H(x, k)$ again and the Distribution of $E(x; a, k)$

The reader will have noticed the apparent anomaly whereby the known forms of the prime number theorem mentioned in §2 are only valid for smaller values of the common difference k, whereas the results springing from the B.D.H. and B.M. theorems are usually only fully significant for the larger values of k. The most striking illustration of this contrast is perhaps furnished by the conditional theorems available on E.R.H., which are valid for $k \leq x^{1/2-\epsilon}$ in the former instance and essentially $k > x^{1/2+\epsilon}$ in the latter. Indeed, the only important overlap in the two types of ranges of k that has so far occurred has been in connection with the Turán-Montgomery estimate (2.5) and its generalization in (6.1), which as already stated in §5 are almost surely of imperfect sharpness. It is therefore desirable to elicit further theorems that shall illuminate the behaviour of $G(x, Q)$ or $H(x, k)$ for smaller values of Q or k.

Accordingly, emboldened to supply something to bridge this gap, we considered the integral

$$\int_2^x \frac{H(u, k)}{u^2} du$$

in 1975 [18] and showed on E.R.H. that it is $O(\log x \log 2k)$ for $k \leq x$, a result that was certainly consistent with our previously implicitly stated conjecture that $H(x, k) = O(x \log k)$ as soon as k tends to infinity; moreover, on assuming in addition that the zeros of every Dirichlet's L-function formed with a primitive character are simple, we sharpened one aspect of this estimate to

$$\lim_{k \to \infty} \lim_{x \to \infty} \frac{1}{\log x \log k} \int_2^x \frac{H(u, k)}{u^2} du = 1$$

and fortified our expectation of the equivalence

$$H(x, k) \sim x \log k \tag{7.1}$$

that had stemmed from (6.2).

Although the method used is a development of that given by Landau for the mean

$$\int_0^x \frac{(\psi(u) - u)^2}{u^2} du,$$

considerable refinements and modifications in the analysis are needed to achieve uniformity in k and, as for (6.1) above, it is necessary to use Selberg's theory of the distribution of the zeros of L-functions.

Cognate ideas serve to crystallize some notions we have loosely expressed from time to time about $E(x; a, k)$ and its distribution. What we have obtained so far is compatible with the likelihood that $E(x; a, k)$ is usually of size about $(x/\phi(k))^{1/2} \log^{1/2} k$, whatever be the stock of triplets (x, a, k) over which it is taken. Furthermore, since our belief in (7.1) would mean that

$$\frac{E(x; a, k)}{(x/\phi(k))^{1/2} \log^{1/2} k}$$

had a dispersion roughly equal to 1, we are further invited to guess that it has a distribution function of constant form in many underlying circumstances. To progress from these speculations we therefore identified one situation in which a substantial extension of the last method would yield information under the twin assumption of E.R.H. and the linear independence of the imaginary parts of the zeros of the Dedekind zeta-function taken over the (cyclotomic) field $\mathbb{Q}(\sqrt[k]{1})$. On the supposition that a and k were given co-prime integers, the random variable

$$T = \frac{E(e^t; a, k)}{e^{(1/2)t} \log^{1/2} k/\phi^{1/2}(k)} \qquad (t > 0)$$

was considered and shown to have a distribution function $F(y)$ that was independent of a for given k, it then being further demonstrated that

$$\lim_{k \to \infty} F(y) = \Phi(y)$$

where $\Phi(y)$ is the distribution function of the normal distribution $N(0, 1)$ [20].

Let now

$$\mu_r = \begin{cases} 1.3 \ldots (2\nu - 1), & \text{if } r = 2\nu \\ 0, & \text{if } r \text{ is odd,} \end{cases}$$

denote the rth moment of $N(0, 1)$. Then, as stated in [20], the suggested implication that

$$\lim_{k \to \infty} \lim_{v \to \infty} \frac{\phi^{1/2}(k)}{\log v \log^{(1/2)r} k} \int_1^v \frac{E^r(x; a, k)dx}{x^{(1/2)r+1}} = \mu_r$$

is confirmed on the same hypothesis as before by varying the procedure in a less elegant direction. Thus, in the manner of [20], we might be led to believe in the generalized B.M. theorem enunciated in the following

Conjecture. *For some appropriate function X of x such that $X > x \log^{-A} x$, we have, as $x \to \infty$,*

$$\sum_{k \le Q} \phi^{(1/2)r-1}(k) \sum_{\substack{0 < a \le k \\ (a,k)=1}} E^r(x; a, k) = (\mu_r + o(1))Q x^{(1/2)r} \log^{(1/2)r} Q$$

when $X < Q = o(x/\log x)$.

Couched in terms of a factor $\phi^{(1/2)r-1}(k)$ between the inner and outer summation symbols, the conjecture can no doubt be appropriately modified by the insertion of other factors such as $k^{(1/2)r-1}$ or $\phi^{(1/2)r-1/2}(k)$. Anyhow, whatever be the precise form of the conjecture we care to take, the cases $r = 1$ or 2 are trivial or correspond to the B.M. theorem, respectively, so that the real difficulties lying athwart our road begin when $r \geq 3$. For the next case $r = 3$, the author [22] resolved the situation when he proved that

$$\sum_{k \leq Q} \phi(k) \sum_{\substack{0 < a \leq k \\ (a,k)=1}} E^3(x;a,k) = o\left(Q^{3/2}X^{3/2}\log^{3/2}x\right) + O\left(x^3\log^{-A}x\right)$$

when $Q = o(x/\log x)$ by what turned out to be an elaborate and finely tuned analysis. Beyond this, nothing is yet known although conditional treatments may well have a chance of success.

8   Analogues of Barban-Davenport-Halberstam Type Theorems for Other Sequences

Not surprisingly the promulgation of the B.D.H. and B.M. theorems led to a search for individual sequences that answered to analogues of these results. Being in some ways the closest companions of the primes and having been much studied, the square-free numbers denoted here by s were probably the first to come under this sort of scrutiny, especially as there was already a known asymptotic formula of the type

$$S(x;a,k) = \sum_{\substack{0 < s \leq x \\ s \equiv a \bmod k}} 1 \sim f(a,k)x \tag{8.1}$$

valid in the lower reaches of k. For example, in terms of the notation

$$E(x;a,k) = S(x;a,k) - f(a,k)x \tag{8.2}$$

$$H(x,k) = \sum_{0 < a \leq k} E^2(x;a,k), \quad G(x,Q) = \sum_{k \leq Q} H(x,k) \tag{8.3}$$

that reflects our previous usage, Orr [33] and Warlimont [41] obtained bounds of the type

$$G(x,Q) = O(Qx)$$

in ranges of the type $x^{1-\alpha} < Q \leq x$, while the latter [42] later improved these to

$$G(x,Q) = O\left(Q^{3/2}x^{1/2}\right)$$

in a similar range. That this was best possible as an upper bound was then substantiated by Croft [4], who obtained the B.M. type asymptotic formula

$$G(x,Q) \sim B_2 Q^{3/2} x^{1/2} \tag{8.4}$$

for $x^{2/3+\epsilon} < Q \leq x$.

The author, however, felt that, instead of considering individual sequences, we would find it more rewarding in the long run to identify the properties a sequence must have in order that it should be the subject of a B.D.H. type theorem. In setting up this investigation we confined our attention to sequences of essentially positive density in the interests of simplicity, believing that any principles established would be applicable to some degree to other sequences. First, if now s denotes a general member of the sequence and if then the meaning of the previous notation in (8.2) and (8.3) is accordingly extended, it is clear there must be an asymptotic formula of type (8.1) since otherwise there could neither be a sensible way of defining $G(x,Q)$ nor any useful estimate for it however $f(a,k)$ were chosen within it. Then, mildly quantifying this condition as

$$S(x;a,k) = f(a,k)x + O\left(\Delta_k(x)\right) \tag{8.5}$$

where $\Delta_k(x) = O(x\log^{-A}x)$ and adding the not unnatural requirement that

$$f(a,k) = g(k,(a,k)) \tag{8.6}$$

(i.e. that $f(a,k)$ depends only on k and the h.c.f. of a and k, to obtain a situation not unlike that afforded by the prime numbers, we showed in [15] that

$$G(x,Q) = O(Qx) + O\left(x^2 \log^{-A} x\right). \tag{8.7}$$

Proved originally through the use of character sums and the large sieve in a manner reminiscent of our description of the proof of Theorem 1, this proposition later received a superior and more illuminating treatment by way of exponential sums under the impetus of our work on certain aspects of Waring's problem [21].

A few years ago we returned to the subject to investigate the circumstances in which (8.7) could be converted to a B.M. type result and proved that [23]

$$G(x,Q) = BQx + O\left(x^2 \log^{-A} x\right) \tag{8.8}$$

where it was further assumed that $g(k,k)$ in (8.6) was a non-zero multiple C of a multiplicative function $\psi(k)$. Of particular importance was the non-negative constant B, the value of which depended only on the form of $f(a,k)$—or, equivalently, C and $\psi(k)$—and was seen to be sometimes positive and sometimes zero. Whereas a genuine asymptotic formula answered to the former situation, a further study of the Dirichlet's series associated with $\psi(k)$ would in the opposite situation sometimes yield a formula similar to the one that was established for square-free numbers (for which our assumed conditions are in place). Nevertheless our work failed to provide a satisfactory criterion for determining the status of B.

A brief word on the condition (8.6) is appropriate before our account unfolds further. Although its presence was felt to be natural, it also seemed hard to dispense with in any of the treatments yet mentioned. This was bound up with a number of associated reasons, the most potent of which is that $H(x,k)$ is no longer generally a dispersion as in the case of the primes and that consequently the sum

$$\sum_{0<a\leq k} f(a,k)S(x;a,k) \tag{8.9}$$

occurring in the middle term of its expansion is a covariance that is not necessarily easy to estimate with an adequate remainder term. This difficulty, however, is abated when (8.6) is in place because then (8.9) can be expressed as a sum of dispersions taken over the divisors δ of k.

It was therefore not without some satisfaction that we shortly afterwards demonstrated [24] that all conditions subsidiary to the main (8.5) were in fact superfluous with the corollary that the existence of an asymptotic formula is strictly equivalent to that of a B.D.H. type theorem and even to a B.M. type theorem. The proofs involved were enhancements of the previous ones in which the large sieve method played additional rôles; as before, the B.D.H. type estimate (8.7) was first established as a stepping stone to the more precise (8.8), it therefore being important terminologically to keep both the descriptive titles of B.D.H. and B.M. even though the stronger result is often true in the same circumstances as the weaker.

Alongside the picture painted in [23] and [24], there is the important one portrayed almost simultaneously by Vaughan in two papers [39] and [40]. In the second of these, slightly generalizing the underlying circumstances by attaching weights to the members of the sequence but imposing the condition (8.6) that we removed in [24], he employed the circle method to obtain (8.8) with very accurate remainder terms that depended on a function $\Delta(x)$ appearing in lieu of $\Delta_k(x)$ in the analogue of (8.5). His method and insight also led, amongst other things, to a full understanding of the essence of the constant B in (8.8) that was lacking in our work.

This aspect of his findings we must now briefly describe because, apart from its obvious interest, it lies behind some further developments we shall adumbrate.

Let the formula (8.5) for $k = 1$ be expressed as

$$S(x) = S(x; 0, 1) = Cx + O\left(\Delta(x)\right)$$

and suppose we attempt formally to calculate $S(x)$ through its representation as

$$\int_0^1 |F(\theta)|^2 \, d\theta \qquad (8.10)$$

where

$$F(\theta) = \sum_{s \leq x} e^{2\pi i \theta s}.$$

To this assignment the circle method of Hardy and Littlewood is partially applicable because (8.5) with (8.6) make possible an accurate rendering of $F(h/k)$ when $(h, k) = 1$ and k is small and hence of $F(\theta)$ when θ is the vicinity of h/k. Regarding such values of h and k as determining the so called major arcs, one can then obtain a contribution to $S(x)$ that is asymptotically equivalent to $\mathfrak{G}x$ because the singular series $\mathfrak{G}$ formally derived is convergent. The remaining portion of $S(x)$ being non-negative because the integrand in (8.10) is non-negative, one can express Vaughan's conclusion in the form[5]

$$B = C - \mathfrak{G},$$

namely, that the main term in a formula of type (8.8) is derived from the minor arcs when the circle method is directed at (8.10). Thus, in the case of the square-free numbers for which Croft's formula (8.4) expresses the situation, the minor arcs make no principal contribution (of order x). For clarity, we should add that this has been a picturesque description, the essential point being the connection between C and the unequivocal meaning of $\mathfrak{G}$. Finally, as shown in the paper at which we now arrive, this interpretation of B remains in place when stipulation (8.6) is removed as in [24].

Our last paper [28] to date stems from ideas similar to those expressed by Professor Montgomery (see [4]) when he stated that Croft's formula (8.4) for square-free numbers could be foreseen from the expected truth of the asymptotic formula

$$S(x) = S(x; 0, 1) = \frac{6x}{\pi^2} + O\left(x^{1/4 + \epsilon}\right)$$

---

[5]In our work we obtained the same formula but failed to recognize its connection with a singular series.

for their sum function. We investigate what can be deduced about $H(x,k)$ and $G(x,Q)$ for large k and Q from our knowledge of an estimate $E(x;a,k) = O(\Delta_k(x))$ for small values of k. Initially our enquiry might have been whether $G(x,Q)$ for large Q is roughly the same as the sum

$$\sum_{k \leq Q} k\Delta_k^2(x). \tag{8.11}$$

Yet, we would be over ambitious in making such an assertion without qualification because, as we have already implied, the formally dominant term in $G(x,Q)$ depends only on the shape of the function $f(a,k)$. In fact, since the addition of an appropriately thick rogue sequence of zero density to the given sequence would increase the size of $\Delta_k(x)$ without altering $f(a,k)$ and hence the main constituent in $G(x,Q)$, we must reduce our speculations to the possible existence of one sided inequalities between $G(x,Q)$ and (8.11). And we are reinforced in this view by (8.7) that stated that $G(x,Q) = O(Qx)$ for Q close to x even when $\Delta_k(x)$ is merely assured to be not larger than $O\left(x \log^{-A} x\right)$.

We have said enough to indicate that it would be illuminating to study this problem in the case where there is still usually no explicitly stated restriction on the form of $f(a,k)$ and where

$$\Delta_k(x) = O\left((x/k)^\alpha\right) \tag{8.12}$$

for

$$k \leq x^{1/2} \tag{8.13}$$

and $0 < \alpha < 1/2$. In these circumstances, affected by Vaughan's observations on the constant B, we combine in [28] the procedures of [24] with a slightly novel use of the circle method to come close to our conjecture by showing that

$$G(x,Q) = O\left(Q^{2-2\alpha} x^{2\alpha} \log^2 2x/Q\right) \tag{8.14}$$

for largish values of Q up to x; thus $B = 0$ in particular. Much else can also be seen to be true but here we must confine ourselves to the observation that

$$H(x,k) = O\left(k^{1-2\alpha} x^{2\alpha} (x/k)^\epsilon\right)$$

when (8.6) also holds and k is a prime number close to x.

Many consequential matters arise, some of which are taken up in the later part of the paper. There is, for instance, the question as to whether (8.14) is best possible since it would appear to contain a superfluous factor $\log^2 2x/Q$; here we show that this can expunged if we suppose the hypothesis (8.12) to hold in a range of k rather longer than (8.13). Similarly, we would like to know when (8.14) can be converted into an asymptotic formula with a main term of type $B_2 Q^{2-2\alpha} x^{2\alpha}$. Vaughan, indeed, showed in [40] that there is such a formula when the function $f(a,k) = g(k, (a,k))$ is of certain type, although it must be said it is neither clear what sequences conform to such a requirement nor how they can be characterised in terms of the size of $E(x; a, k)$ for small k. This aspect of the theory was therefore followed up in two directions. The first was to consider other circumstances in which an asymptotic formula held, the second being then to construct a sequence for which any such formula failed.

A scrutiny of what we have said with the help of [28] makes it manifest that there is at least a loose connection between the form of $f(a,k)$ and the size of $\Delta_k(x)$. Just what it constitutes is one of several obvious aspects of our work that require further elucidation.

We have been deliberately selective in our provision of references for work cited in the text, confining ourselves in the main to those that relate to the main theme of the survey.

9   Postscript

The value of this report would be diminished if we did not mention some developments that have taken place between the time it was written and its final preparation for the press. These relate to the last two paragraphs of §6 in which lower bounds for $G(x, Q)$ were discussed.

Owing to the history of the subject workers have tended to regard $E(x; a, k)$ and the associated difference

$$E^*(x; a, k) = \theta(x; a, k) - \frac{1}{\phi(k)} \sum_{\substack{p \le x \\ p \nmid k}} \log p$$

as being on a par and have therefore sometimes failed to appreciate the distinction between them. This is that, being a true dispersion, the sum

$$H^*(x, k) = \sum_{\substack{0 < a \le k \\ (a,k)=1}} E^{*2}(x; a, k)$$

bounds $G(x, k)$ from below in accordance with a well known theorem in elementary probability theory. Thus, in regard to problems involving lower bounds, it is the sum

$$G^*(x, Q) = \sum_{k \le Q} H^*(x, k)$$

that is of primary importance even though of course the other sum $G(x, Q)$ still retains its interest.

As a rescrutiny of Liu's paper reveals, it was his failure to remember this point that led him to deduce his lower bound for $G(x, Q)$ in the form (6.7), while actually he could have inferred it for the wider range

$$Q > x^{1-0(1/\log\log x)} \tag{9.1}$$

because of his earlier result for $G^*(x, Q)$. Consequently, both the inequalities

$$G(x, Q), \; G^*(x, Q) > \left(\frac{1}{4} - \epsilon\right) Qx \log Q \tag{9.2}$$

for the range (9.1) should be essentially ascribed to Liu. On the other hand, our inequality (6.8) was derived from the comparable result for $G^*(x, Q)$ by using the minimal property of the latter, which may therefore be added to the left side of the statement.

Recently we developed the method in [27] to show that the inequality (6.9)—previously proved on the strength of a quasi-Riemann hypothesis— was actually true unconditionally; this contains the desired multiplier of $Qx \log Q$ when $\alpha \to 1$ and provides bound of requisite size down to about $Q = x^{\frac{1}{2}}$. Since, however, the method did not extend to $G^*(x, Q)$, this left us confronted by the task of improving the unconditional bounds for this sum implied by our discussion. In response, by a radical reappraisal of Liu's method, we first proved that

$$G^*(x, Q) > \{A(\delta) - \epsilon\}Qx \log Q \qquad (Q = x^{1-\delta}),$$

where $A(\delta)$ is a decreasing function of δ tending to $5\pi^2/6$ as $\delta \to 0$, thus improving (9.2) for a longer range in which the multiplier $1/4$ is increased to a number not far short of the desired 1. Then later, leaning on ideas taken from Heath-Brown's new version of the circle method, we found that $A(\delta)$ could be replaced by a similar function whose limiting value as $\delta \to 0$ was actually 1. There is thus a satisfactory theory for the lower bound for $G^*(x, Q)$ when Q is of the form $x^{1-\delta}$ for small δ.

References

[1] M. B. Barban, *Analogues of the divisor problem of Titchmarsh*, Vestnik Leningrad. Univ. Ser. Mat. Meh. Astronom **18** (1963), 5–13.

[2] ———, *On the average error in the generalized prime number theorem*, Doklady Akademiya Nauk UZ SSR (1964), 5–7.

[3] E. Bombieri, *On the large sieve*, Mathematika **12** (1965), 201–225.

[4] M. J. Croft, *Square-free numbers in arithmetic progressions*, Proc. London Math. Soc. (3) **30** (1975), 143–159.

[5] S. Daniel, *On the sum of a square and a square of a prime*, Math. Proc. Cambridge Philos. Soc. **131** (2001), 1–22.

[6] H. Davenport and H. Halberstam, *Primes in arithmetic progressions*, Michigan Math. J. **13** (1966), 485–489.

[7] J. B. Friedlander and D. A. Goldston, *Variance of distribution of primes in residue classes*, Quart. J. Math. Oxford Ser. (2) **47** (1996), 313–336.

[8] ――――, *Note on a variance in the distribution of primes*, Number theory in progress, Vol. 2 (Zakopane-Kościelisko, 1997), de Gruyter, Berlin, 1999, pp. 841–848.

[9] P. X. Gallagher, *The large sieve*, Mathematika **14** (1967), 14–20.

[10] D. A. Goldston and R. C. Vaughan, *On the Montgomery-Hooley asymptotic formula*, Sieve methods, exponential sums, and their applications in number theory (Cardiff, 1995), Cambridge Univ. Press, Cambridge, 1997, pp. 117–142.

[11] G. Greaves, *On the representation of a number in the form $x^2 + y^2 + p^2 + q^2$ where p, q are odd primes*, Acta Arith. **29** (1976), 257–274.

[12] C. Hooley, *The distribution of sequences in arithmetic progressions*, Proceedings of the International Congress of Mathematicians (Vancouver, B.C., 1974), Vol. 1, Canad. Math. Congress, Montreal, Que., 1975, pp. 357–364.

[13] ――――, *On the Barban-Davenport-Halberstam theorem: I*, J. Reine Angew. Math. **274/275** (1975), 206–223.

[14] ――――, *On the Barban-Davenport-Halberstam theorem: II*, J. London Math. Soc. (2) **9** (1975), 625–636.

[15] ――――, *On the Barban-Davenport-Halberstam theorem: III*, J. London Math. Soc. (2) **10** (1975), 249–256.

[16] ――――, *On the Barban-Davenport-Halberstam theorem: IV*, J. London Math. Soc. (2) **11** (1975), 399–407.

[17] _____, *Applications of sieve methods to the theory of numbers*, Cambridge University Press, Cambridge, 1976, Cambridge Tracts in Mathematics, No. 70.

[18] _____, *On the Barban-Davenport-Halberstam theorem: V*, Proc. London Math. Soc. (3) **33** (1976), 535–548.

[19] _____, *On the Barban-Davenport-Halberstam theorem: VI*, J. London Math. Soc. (2) **13** (1976), 57–64.

[20] _____, *On the Barban-Davenport-Halberstam theorem: VII*, J. London Math. Soc. (2) **16** (1977), 1–8.

[21] _____, *On a new approach to various problems of Waring's type*, Recent progress in analytic number theory, Vol. 1 (Durham, 1979), Academic Press, London, 1981, pp. 127–191.

[22] _____, *On the Barban-Davenport-Halberstam theorem: VIII*, J. Reine Angew. Math. **499** (1998), 1–46.

[23] _____, *On the Barban-Davenport-Halberstam theorem: IX*, Acta Arith. **83** (1998), 17–30.

[24] _____, *On the Barban-Davenport-Halberstam theorem: X*, Hardy-Ramanujan J. **21** (1998), 9 pp.

[25] _____, *On the Barban-Davenport-Halberstam theorem: XI*, Acta Arith. **91** (1999), 1–41.

[26] _____, *On the Barban-Davenport-Halberstam theorem: XII*, Number theory in progress, Vol. 2 (Zakopane-Kościelisko, 1997), de Gruyter, Berlin, 1999, pp. 893–910.

[27] _____, *On the Barban-Davenport-Halberstam theorem: XIII*, Acta Arith. **94** (2000), 53–86.

[28] _____, *On the Barban-Davenport-Halberstam theorem: XIV*, Acta Arith. **101** (2002), 247–292.

[29] H. Q. Liu, *Lower bounds for sums of Barban-Davenport-Halberstam type (supplement)*, Manuscripta Math. **87** (1995), 159–166.

[30] H. L. Montgomery, *Primes in arithmetic progressions*, Michigan Math. J. **17** (1970), 33–39.

[31] _____, *Topics in multiplicative number theory*, Springer-Verlag, Berlin, 1971, Lecture Notes in Mathematics, Vol. 227.

[32] _____, *Maximal variants of the large sieve*, J. Fac. Sci. Univ. Tokyo Sect. IA Math. **28** (1981), 805–812 (1982).

[33] R. C. Orr, *Remainder estimates for squarefree integers in arithmetic progression*, J. Number Theory **3** (1971), 474–497.

[34] A. Perelli, *The L^1 norm of certain exponential sums in number theory: a survey*, Rend. Sem. Mat. Univ. Politec. Torino **53** (1995), 405–418, Number theory, II (Rome, 1995).

[35] V. A. Plaksin, *Representation of numbers by the sum of four squares of integers, two of which are prime numbers*, Dokl. Akad. Nauk SSSR **257** (1981), 1064–1066.

[36] P. Shields, Ph.D. thesis, Cardiff, 1978.

[37] E. C. Titchmarsh, *A divisor problem*, Rendiconti del Circolo Matematico di Palermo **54** (1930), 414–429.

[38] S. Uchiyama, *The maximal large sieve*, Seminar on Modern Methods in Number Theory (Inst. Statist. Math., Tokyo, 1971), Paper No. 37, Inst. Statist. Math., Tokyo, 1971, p. 5.

[39] R. C. Vaughan, *On a variance associated with the distribution of general sequences in arithmetic progressions. I,*, R. Soc. Lond. Philos. Trans. Ser. A Math. Phys. Eng. Sci. **356** (1998), 781–791.

[40] _____, *On a variance associated with the distribution of general sequences in arithmetic progressions. II*, R. Soc. Lond. Philos. Trans. Ser. A Math. Phys. Eng. Sci. **356** (1998), 793–809.

[41] R. Warlimont, *On squarefree numbers in arthmetic progressions*, Monatsh. Math. **73** (1969), 433–448.

[42] _____, *Über die kleinsten quadratfreien Zahlen in arithmetischen Progressionen mit primen Differenzen*, J. Reine Angew. Math. **253** (1972), 19–23.

Elementary Evaluation of Certain Convolution Sums Involving Divisor Functions

James G. Huard,[1] Zhiming M. Ou,[2] Blair K. Spearman,[3] and Kenneth S. Williams[4]

1 Introduction

Let $\sigma_m(n)$ denote the sum of the m th powers of the positive divisors of the positive integer n. We set $\tau(n) = \sigma_0(n)$ and $\sigma(n) = \sigma_1(n)$. If l is not a positive integer we set $\sigma_m(l) = 0$. In Section 2 we prove an elementary arithmetic identity (Theorem 1), which generalizes a classical formula of Liouville given in [21]. In Section 3 we use this identity to evaluate in an elementary manner thirty-seven convolution sums involving the function σ_m treated by Lahiri in [14] using more advanced techniques. Some of these convolution sums had been considered earlier by Glaisher [9], [10], [11], MacMahon [24], [25, pp. 303–341] and Ramanujan [30], [31, pp. 136–162 and commentary pp. 365–368]. MacMahon used his formulae to deduce theorems about partitions. In Sections 4 and 5 we prove in an elementary manner some extensions of convolution formulae proved by Melfi in [26] using the theory of modular forms. In Section 6 we use Theorem 1 to derive in an elementary manner some formulae for the number of representations of positive integers as sums of triangular numbers proved by Ono, Robins and Wahl in [27] using more sophisticated methods. In Section 7 we give one illustrative example of how Theorem 1 can be used to determine the number of representations of a positive integer by a quaternary form. Finally in Section 8 we consider some further convolution sums of the type considered in Section 4.

[1] Research supported by a Canisius College Faculty Fellowship.

[2] Research supported by the China Scholarship Council.

[3] Research supported by a Natural Sciences and Engineering Research Council of Canada grant.

[4] Research supported by Natural Sciences and Engineering Research Council of Canada grant A-7233.

2   Generalization of Liouville's Formula

We prove

Theorem 1. *Let $f : \mathbb{Z}^4 \to \mathbb{C}$ be such that*

$$f(a, b, x, y) - f(x, y, a, b) = f(-a, -b, x, y) - f(x, y, -a, -b) \qquad (2.1)$$

for all integers a, b, x and y. Then

$$\sum_{ax+by=n} (f(a, b, x, -y) - f(a, -b, x, y) + f(a, a - b, x + y, y)$$

$$-f(a, a + b, y - x, y) + f(b - a, b, x, x + y) - f(a + b, b, x, x - y))$$

$$= \sum_{d \mid n} \sum_{x < d} (f(0, n/d, x, d) + f(n/d, 0, d, x) + f(n/d, n/d, d - x, -x)$$

$$-f(x, x - d, n/d, n/d) - f(x, d, 0, n/d) - f(d, x, n/d, 0)), \qquad (2.2)$$

where the sum on the left hand side of (2.2) is over all positive integers a, b, x, y satisfying $ax + by = n$, the inner sum on the right hand side is over all positive integers x satisfying $x < d$, and the outer sum on the right hand side is over all positive integers d dividing n.

Proof. We set

$$g(a, b, x, y) = f(a, b, x, y) - f(x, y, a, b)$$

so that

$$g(a, -b, x, y) = g(-a, b, x, y)$$

and

$$g(a, b, x, y) = -g(x, y, a, b).$$

Then

$$\sum_{ax+by=n} (f(a, b, x, -y) - f(a, -b, x, y) + f(a, a - b, x + y, y)$$

$$-f(a, a + b, y - x, y) + f(b - a, b, x, x + y) - f(a + b, b, x, x - y))$$

$$= \sum_{ax+by=n} (f(a, b, x, -y) - f(x, -y, a, b) + f(a, a - b, x + y, y)$$

$$-f(y, x + y, a - b, a) + f(a - b, a, y, x + y) - f(x + y, y, a, a - b))$$

$$= \sum_{ax+by=n} (g(a, a - b, x + y, y) + g(a - b, a, y, x + y) + g(a, b, x, -y))$$

and

$$
\sum_{d\,|\,n}\sum_{x<d}(f(0,n/d,x,d) + f(n/d,0,d,x) + f(n/d,n/d,d-x,-x)
$$

$$
-f(x,x-d,n/d,n/d) - f(x,d,0,n/d) - f(d,x,n/d,0))
$$

$$
= \sum_{d\,|\,n}\sum_{t<d}(f(0,n/d,t,d) + f(n/d,0,d,t) + f(n/d,n/d,d-t,-t)
$$

$$
-f(d-t,-t,n/d,n/d) - f(t,d,0,n/d) - f(d,t,n/d,0))
$$

$$
= \sum_{d\,|\,n}\sum_{t<d}(g(n/d,0,d,t) + g(0,n/d,t,d) + g(n/d,n/d,d-t,-t))
$$

so we must prove that

$$
\sum_{ax+by=n}(g(a,a-b,x+y,y) + g(a-b,a,y,x+y) + g(a,b,x,-y))
$$

$$
= \sum_{d\,|\,n}\sum_{t<d}(g(n/d,0,d,t) + g(0,n/d,t,d) + g(n/d,n/d,d-t,-t)).
$$

First we consider the terms with $a = b$ in the left hand sum. We have

$$
\sum_{\substack{ax+by=n\\a=b}}(g(a,a-b,x+y,y) + g(a-b,a,y,x+y) + g(a,b,x,-y))
$$

$$
= \sum_{a(x+y)=n}(g(a,0,x+y,y) + g(0,a,y,x+y) + g(a,a,x,-y))
$$

$$
= \sum_{d\,|\,n}\sum_{t<d}(g(n/d,0,d,t) + g(0,n/d,t,d) + g(n/d,n/d,d-t,-t)).
$$

Secondly we consider the terms with $a < b$. We have

$$
\sum_{\substack{ax+by=n\\a<b}}(g(a,a-b,x+y,y) + g(a-b,a,y,x+y) + g(a,b,x,-y))
$$

$$
= \sum_{\substack{a(x+y)+(b-a)y=n\\a<b}}(g(a,a-b,x+y,y) + g(a-b,a,y,x+y))
$$

$$
+ \sum_{\substack{ax+by=n\\a<b}}g(a,b,x,-y)
$$

$$
= \sum_{\substack{ax+by=n\\x>y}}(g(a,-b,x,y) + g(-b,a,y,x)) + \sum_{\substack{ax+by=n\\x<y}}g(x,y,a,-b)
$$

$$= \sum_{\substack{ax+by=n \\ x>y}} g(a,-b,x,y) + \sum_{\substack{ax+by=n \\ x>y}} g(b,-a,y,x)$$

$$- \sum_{\substack{ax+by=n \\ x<y}} g(a,-b,x,y)$$

$$= - \sum_{\substack{ax+by=n \\ x>y}} g(x,y,a,-b) + \sum_{\substack{ax+by=n \\ x<y}} g(a,-b,x,y)$$

$$- \sum_{\substack{ax+by=n \\ x<y}} g(a,-b,x,y)$$

$$= - \sum_{\substack{ax+by=n \\ x>y}} g(x,y,a,-b).$$

Thirdly we consider the terms with $a > b$. We have

$$\sum_{\substack{ax+by=n \\ a>b}} (g(a,a-b,x+y,y) + g(a-b,a,y,x+y) + g(a,b,x,-y))$$

$$= S_1 + S_2,$$

where

$$S_1 = \sum_{\substack{ax+by=n \\ a>b}} (g(a,a-b,x+y,y) + g(a-b,a,y,x+y))$$

$$= \sum_{\substack{ax+b(x+y)=n}} (g(a+b,a,x+y,y) + g(a,a+b,y,x+y))$$

$$= \sum_{\substack{ax+by=n \\ y>x}} (g(a+b,a,y,y-x) + g(a,a+b,y-x,y))$$

$$= \sum_{\substack{ax+by=n \\ a>b}} (g(x+y,y,a,a-b) + g(y,x+y,a-b,a))$$

$$= -S_1,$$

so that $S_1 = 0$, and

$$S_2 = \sum_{\substack{ax+by=n \\ a>b}} g(a,b,x,-y)$$

$$= \sum_{\substack{ax+by=n \\ x>y}} g(x,y,a,-b).$$

This completes the proof of the theorem. □

We remark that if f satisfies

$$f(a, -b, x, y) = f(-a, b, x, y), \quad f(a, b, x, -y) = f(a, b, -x, y), \quad (2.3)$$

for all integers a, b, x and y then f satisfies (2.1).

Finally we observe that if we choose $f(a, b, x, y) = F(a - b, x - y)$ with $F(x, y) = F(-x, y) = F(x, -y)$ then (2.2) becomes Liouville's identity [21, p. 284] (see also [8, p. 331])

$$\sum_{ax+by=n} (F(a - b, x + y) - F(a + b, x - y))$$

$$= \sum_{d|n} (d - 1)(F(0, d) - F(d, 0)) + 2 \sum_{d|n} \sum_{e<n/d} (F(d, e) - F(e, d)). \quad (2.4)$$

We also note that the choice $f(a, b, x, y) = h(a, b) + h(-a, -b)$ with $h(b, b - a) = h(a, b)$ gives Skoruppa's combinatorial identity [33, p. 69]

$$\sum_{ax+by=n} (h(a, b) - h(a, -b)) = \sum_{d|n} \left(\frac{n}{d} h(d, 0) - \sum_{j=0}^{d-1} h(d, j) \right).$$

Thus our identity is an extension of those of Liouville and Skoruppa. Conversely if one starts with $f(a, b)$ satisfying $f(a, b) = f(-a, -b)$ and defines

$$h(a, b) = f(a, b) + f(b, b - a) + f(b - a, -a)$$

so that $h(b, b - a) = h(a, b)$ then Skoruppa's identity is the special case of Theorem 1 with $f(a, b, x, y) = f(a, b)$.

For convenience in the rest of the paper, we set

$$\begin{aligned} E &= E(a, b, x, y) \\ &= f(a, b, x, -y) - f(a, -b, x, y) + f(a, a - b, x + y, y) \\ &\quad - f(a, a + b, y - x, y) + f(b - a, b, x, x + y) - f(a + b, b, x, x - y), \end{aligned}$$
$$(2.5)$$

and, for k a positive integer and a an integer, we set

$$F_k(a) = \begin{cases} 1, & \text{if } k \mid a, \\ 0, & \text{if } k \nmid a. \end{cases} \quad (2.6)$$

3   Application of Theorem 1 to Lahiri's Identities

Let e, f, g, h be integers such that

$$e \geq 0, \ f \geq 0, \ g \geq 1, \ h \geq 1, \ g \equiv h \equiv 1 \ (\text{mod } 2), \tag{3.1}$$

and define the sum $S(e \ f \ g \ h)$ by

$$S(e \ f \ g \ h) := \sum_{m=1}^{n-1} m^e (n-m)^f \sigma_g(m) \sigma_h(n-m). \tag{3.2}$$

The change of variable $m \longrightarrow n - m$ shows that

$$S(e \ f \ g \ h) = S(f \ e \ h \ g). \tag{3.3}$$

Thus we may suppose further that

$$g < h \text{ or } g = h, \ e \geq f. \tag{3.4}$$

For nonnegative integers r, s, t, u we set

$$[r \ s \ t \ u] := \sum_{ax+by=n} a^r b^s x^t y^u. \tag{3.5}$$

As the eight permutations (in cycle notation)

$$(a \ b \ x \ y)^i (a \ x)^j \quad (i = 0, 1, 2, 3; \ j = 0, 1)$$

leave $ax + by$ invariant, we have

$$
\begin{aligned}
[r \ s \ t \ u] &= [r \ u \ t \ s] &= [s \ r \ u \ t] &= [s \ t \ u \ r] \\
&= [t \ u \ r \ s] &= [t \ s \ r \ u] &= [u \ r \ s \ t] &= [u \ t \ s \ r].
\end{aligned} \tag{3.6}
$$

Moreover

$$S(e \ f \ g \ h) = \sum_{m=1}^{n-1} m^e (n-m)^f \sum_{ax=m} x^g \sum_{by=n-m} y^h = \sum_{ax+by=n} a^e b^f x^{e+g} y^{f+h}$$

so that

$$S(e \ f \ g \ h) = [e \ f \ e+g \ f+h] \tag{3.7}$$

and

$$[e \ f \ g \ h] = S(e \ f \ g-e \ h-f), \text{ if } g \geq e, \ h \geq f. \tag{3.8}$$

w	e	f	g	h	w	e	f	g	h	w	e	f	g	h
2	0	0	1	1	10	0	1	1	7	12	0	2	3	5
4	0	0	1	3	10	0	1	3	5	12	0	3	1	5
4	1	0	1	1	10	0	2	1	5	12	0	4	1	3
6	0	0	1	5	10	0	3	1	3	12	1	0	1	9
6	0	0	3	3	10	1	0	1	7	12	1	0	3	7
6	0	1	1	3	10	1	0	3	5	12	1	0	5	5
6	1	0	1	3	10	1	1	1	5	12	1	1	1	7
6	1	1	1	1	10	1	1	3	3	12	1	1	3	5
6	2	0	1	1	10	1	2	1	3	12	1	2	1	5
8	0	0	1	7	10	2	0	1	5	12	1	3	1	3
8	0	0	3	5	10	2	0	3	3	12	2	0	1	7
8	0	1	1	5	10	2	1	1	3	12	2	0	3	5
8	0	2	1	3	10	2	2	1	1	12	2	1	1	5
8	1	0	1	5	10	3	0	1	3	12	2	1	3	3
8	1	0	3	3	10	3	1	1	1	12	2	2	1	3
8	1	1	1	3	10	4	0	1	1	12	3	0	1	5
8	2	0	1	3	12	0	0	1	11	12	3	0	3	3
8	2	1	1	1	12	0	0	3	9	12	3	1	1	3
8	3	0	1	1	12	0	0	5	7	12	3	2	1	1
10	0	0	1	9	12	0	1	1	9	12	4	0	1	3
10	0	0	3	7	12	0	1	3	7	12	4	1	1	1
10	0	0	5	5	12	0	2	1	7	12	5	0	1	1

Table 1. Values of e, f, g, h, corresponding to weights $w \leq 12$

We define the weight of the sum $S(e\ f\ g\ h)$ to be the positive even integer

$$w = w(S) = 2e + 2f + g + h. \tag{3.9}$$

The integers e, f, g, h satisfying (3.1) and (3.4) for which the sum $S(e\ f\ g\ h)$ has weight less than or equal to 12 are given in Table 1. Lahiri [14], making use of ideas of Ramanujan [30], [31, pp. 136–162], has evaluated the single sum of weight 2, the two sums of weight 4, the six sums of weight 6, the ten sums of weight 8, and the first three of the twenty-eight sums of weight 12. We show that the values of these twenty-two sums follow from Theorem 1.

Weight = 2. We take $f(a,b,x,y) = xy$ in Theorem 1. Then $E = 2xy$ and the left hand side of (2.2) is

$$\sum_{ax+by=n} 2xy = 2[0\ 0\ 1\ 1] = 2S(0\ 0\ 1\ 1).$$

The right hand side of (2.2) is

$$\sum_{d\mid n}\sum_{x<d}\left(x^2 + dx - \frac{n^2}{d^2}\right) = \frac{5}{6}\sigma_3(n) + \left(\frac{1}{6} - n\right)\sigma(n).$$

Then Theorem 1 gives

$$\sum_{m=1}^{n-1}\sigma(m)\sigma(n-m) = \frac{1}{12}\left(5\sigma_3(n) + (1-6n)\sigma(n)\right). \qquad (3.10)$$

Formula (3.10) is equivalent to formula (3.1) in Lahiri [14] and originally appeared in a letter from Besge to Liouville [4]. Dickson erroneously attributed (3.10) to Lebesgue in [8, p. 338]. Rankin attributed it to Besgue in [32, p. 115]. Lützen [23, p. 81] asserts that Besge/Besgue is a pseudonym for Liouville. The formula (3.10) also appears in the work of Glaisher [9], [10], [11], Lehmer [17, p. 106], [18, p. 678], Ramanujan [30, Table IV], [31, p. 146] and Skoruppa [33].

Weight = 4. We begin by noting that

$$\sum_{m=1}^{n-1} m\sigma(m)\sigma(n-m) = \sum_{m=1}^{n-1} (n-m)\sigma(n-m)\sigma(m)$$

so that

$$\sum_{m=1}^{n-1} m\sigma(m)\sigma(n-m) = \frac{n}{2}\sum_{m=1}^{n-1} \sigma(m)\sigma(n-m).$$

Hence, by (3.10), we have

$$\sum_{m=1}^{n-1} m\sigma(m)\sigma(n-m) = \frac{n}{24}\left(5\sigma_3(n) + (1-6n)\sigma(n)\right), \qquad (3.11)$$

which is formula (3.2) of Lahiri [14]. This formula is due to Glaisher [11]. (Note that the multiplier 12 on the left hand side of Glaisher's formula should be replaced by 24.)

Taking $f(a, b, x, y) = xy^3 + x^3 y$, we find that

$$E = 8xy^3 + 8x^3 y$$

so that the left hand side of (2.2) is

$$\sum_{ax+by=n} (8xy^3 + 8x^3 y) = 8[0\ 0\ 1\ 3] + 8[0\ 0\ 3\ 1] = 16[0\ 0\ 1\ 3] = 16S(0\ 0\ 1\ 3).$$

Evaluating the right hand side of (2.2), we obtain

$$\sum_{m=1}^{n-1} \sigma(m)\sigma_3(n-m) = \frac{1}{240}(21\sigma_5(n) + (10-30n)\sigma_3(n) - \sigma(n)), \quad (3.12)$$

which is a result attributed to Glaisher by MacMahon [24, p. 101], [25, p. 329]. It also appears in Ramanujan [30, Table IV], [31, p. 146] and is also formula (5.1) of Lahiri [14].

Weight = 6. We begin with the sums $S(1\ 0\ 1\ 3)$ and $S(0\ 1\ 1\ 3)$. Clearly

$$
\begin{aligned}
S(1\ 0\ 1\ 3) + S(0\ 1\ 1\ 3) &= \sum_{m=1}^{n-1} m\sigma(m)\sigma_3(n-m) \\
&\quad + \sum_{m=1}^{n-1} (n-m)\sigma(m)\sigma_3(n-m) \\
&= n \sum_{m=1}^{n-1} \sigma(m)\sigma_3(n-m) \\
&= \frac{n}{240}(21\sigma_5(n) + (10-30n)\sigma_3(n) - \sigma(n)),
\end{aligned}
$$

by (3.12). Taking

$$f(a, b, x, y) = 3b^2 xy^3 + 5abx^3 y - 3abx^2 y^2 - 2b^2 x^3 y,$$

we find that

$$E = 12b^2 x^3 y - 6b^2 xy^3 - 4abx^4 + 12a^2 xy^3 - 8aby^4 + 6a^2 x^3 y$$

so that the left hand side of (2.2) is

$$12[0\ 2\ 3\ 1] - 6[0\ 2\ 1\ 3] - 4[1\ 1\ 4\ 0] + 12[2\ 0\ 1\ 3] - 8[1\ 1\ 0\ 4] + 6[2\ 0\ 3\ 1]$$

$$= 24[1\ 0\ 2\ 3] - 12[0\ 1\ 1\ 4] = 24S(1\ 0\ 1\ 3) - 12S(0\ 1\ 1\ 3).$$

The right hand side of (2.2) is found to be after some calculation

$$\frac{1}{10}((6n^2 - 5n)\sigma_3(n) - n\sigma(n)).$$

Solving for $S(1\ 0\ 1\ 3)$ and $S(0\ 1\ 1\ 3) = S(1\ 0\ 3\ 1)$ we obtain

$$\sum_{m=1}^{n-1} m\sigma(m)\sigma_3(n-m) = \frac{1}{240}(7n\sigma_5(n) - 6n^2\sigma_3(n) - n\sigma(n)) \qquad (3.13)$$

and

$$\sum_{m=1}^{n-1} m\sigma_3(m)\sigma(n-m) = \frac{1}{120}(7n\sigma_5(n) + (5n - 12n^2)\sigma_3(n)). \qquad (3.14)$$

Equations (3.13) and (3.14) are given in MacMahon [24, p. 103], [25, p. 331]. They are also formulae (5.4) and (5.3) of Lahiri [14].

   To evaluate the remaining four sums of weight 6, namely $S(1\ 1\ 1\ 1)$, $S(2\ 0\ 1\ 1)$, $S(0\ 0\ 3\ 3)$ and $S(0\ 0\ 1\ 5)$, we choose respectively

$$
\begin{aligned}
f(a,b,x,y) &= abx^3y - b^2xy^3, \\
f(a,b,x,y) &= b^2y^4 - b^2xy^3, \\
f(a,b,x,y) &= xy^5 + x^5y - 2x^3y^3, \\
f(a,b,x,y) &= xy^5 + x^5y - 20x^3y^3.
\end{aligned}
$$

The corresponding values of E are

$$
\begin{aligned}
E &= -4b^2x^3y - 2b^2xy^3 - 2abx^4 + 4a^2xy^3 + 2aby^4 - 6abx^2y^2 + 2a^2x^3y, \\
E &= 2b^2x^3y + 6b^2xy^3 - 2a^2xy^3, \\
E &= 36x^3y^3, \\
E &= -108x^5y - 108xy^5.
\end{aligned}
$$

The left hand sides of (2.2) are respectively

$$-4[0\ 2\ 3\ 1] - 2[0\ 2\ 1\ 3] - 2[1\ 1\ 4\ 0] + 4[2\ 0\ 1\ 3]$$
$$+ 2[1\ 1\ 0\ 4] - 6[1\ 1\ 2\ 2] + 2[2\ 0\ 3\ 1]$$
$$= -6[1\ 1\ 2\ 2] = -6S(1\ 1\ 1\ 1),$$

$$2[0\ 2\ 3\ 1] + 6[0\ 2\ 1\ 3] - 2[2\ 0\ 1\ 3] = 6[2\ 0\ 3\ 1] = 6S(2\ 0\ 1\ 1),$$

$$36[0\ 0\ 3\ 3] = 36S(0\ 0\ 3\ 3),$$

$$-108[0\ 0\ 5\ 1] - 108[0\ 0\ 1\ 5] = -216[0\ 0\ 1\ 5] = -216S(0\ 0\ 1\ 5).$$

Evaluating the right hand side of (2.2) for these choices of f, by Theorem 1 we obtain respectively

$$\sum_{m=1}^{n-1} m(n-m)\sigma(m)\sigma(n-m) = \frac{1}{12}(n^2\sigma_3(n) - n^3\sigma(n)), \qquad (3.15)$$

$$\sum_{m=1}^{n-1} m^2\sigma(m)\sigma(n-m) = \frac{1}{24}(3n^2\sigma_3(n) + (n^2 - 4n^3)\sigma(n)), \qquad (3.16)$$

$$\sum_{m=1}^{n-1} \sigma_3(m)\sigma_3(n-m) = \frac{1}{120}(\sigma_7(n) - \sigma_3(n)), \qquad (3.17)$$

$$\sum_{m=1}^{n-1} \sigma(m)\sigma_5(n-m) = \frac{1}{504}(20\sigma_7(n) + (21 - 42n)\sigma_5(n) + \sigma(n)).$$

$$(3.18)$$

Equations (3.15), (3.16), (3.17) and (3.18) are formulae (3.3), (3.4), (7.1) and (7.2) of Lahiri [14] respectively. We remark that formula (3.15) is due to Glaisher [11, p. 35]. Formula (3.16) follows from (3.11) and (3.15), and so is implicit in the work of Glaisher. Formula (3.17) is due to Glaisher [11, p. 35]. It also appears in the work of Ramanujan [30, Table IV], [31, p. 146]. Formula (3.18) is due to Ramanujan [30, Table IV], [31, p. 146]. It is also given in the work of MacMahon [24, p. 103], [25, p. 331]. The recurrence relations of van der Pol [28, eqns. (63), (65)] for $\sigma(n)$ and $\sigma_3(n)$ are simple consequences of (3.15) and (3.16).

Weight $= 8$. We begin with the sum $S(1\,0\,3\,3)$. We have

$$
\begin{aligned}
S(1\,0\,3\,3) &= \sum_{m=1}^{n-1} m\sigma_3(m)\sigma_3(n-m) = \sum_{m=1}^{n-1} (n-m)\sigma_3(n-m)\sigma_3(m) \\
&= nS(0\,0\,3\,3) - S(1\,0\,3\,3)
\end{aligned}
$$

so that

$$S(1\,0\,3\,3) = \frac{n}{2}S(0\,0\,3\,3).$$

Appealing to (3.17) we obtain

$$\sum_{m=1}^{n-1} m\sigma_3(m)\sigma_3(n-m) = \frac{n}{240}(\sigma_7(n) - \sigma_3(n)). \qquad (3.19)$$

Equation (3.19) is formula (7.5) of Lahiri [14].

Next we evaluate $S(2\,1\,1\,1) = S(1\,2\,1\,1)$ and $S(3\,0\,1\,1)$. We have

$$
\begin{aligned}
S(1\,2\,1\,1) - S(3\,0\,1\,1) &= \sum_{m=1}^{n-1} (m(n-m)^2 - m^3)\sigma(m)\sigma(n-m) \\
&= n^2 \sum_{m=1}^{n-1} m\sigma(m)\sigma(n-m) \\
&\quad -2n \sum_{m=1}^{n-1} m^2\sigma(m)\sigma(n-m) \\
&= n^2 S(1\,0\,1\,1) - 2n S(2\,0\,1\,1).
\end{aligned}
$$

Appealing to (3.11) and (3.16), we obtain

$$
S(3\,0\,1\,1) - S(1\,2\,1\,1) = \frac{n^3}{24}(\sigma_3(n) + (1 - 2n)\sigma(n)).
$$

Now we choose $f(a,b,x,y) = b^4 x y^3$ so that

$$
E = 2b^4 x y^3 + 12a^2 b^2 x y^3 - 8ab^3 y^4 - 8a^3 b y^4 + 2a^4 x y^3 + 6b^4 x^3 y.
$$

In this case the left hand side of (2.2) is

$$
\begin{aligned}
2[0\,4\,1\,3] + 12[2\,2\,1\,3] &- 8[1\,3\,0\,4] - 8[3\,1\,0\,4] + 2[4\,0\,1\,3] \\
&\quad + 6[0\,4\,1\,3] \\
&= -6[3\,0\,4\,1] + 12[1\,2\,2\,3] \\
&= 12 S(1\,2\,1\,1) - 6 S(3\,0\,1\,1).
\end{aligned}
$$

Evaluating the right hand side of (2.2), we find by Theorem 1 that

$$
12 S(1\,2\,1\,1) - 6 S(3\,0\,1\,1) = \frac{n^3}{4}(n-1)\sigma(n).
$$

Solving these two linear equations for $S(1\,2\,1\,1)$ and $S(3\,0\,1\,1)$, we obtain

$$
\sum_{m=1}^{n-1} m(n-m)^2 \sigma(m)\sigma(n-m) = \frac{n^3}{24}(\sigma_3(n) - n\sigma(n)), \tag{3.20}
$$

$$
\sum_{m=1}^{n-1} m^3 \sigma(m)\sigma(n-m) = \frac{n^3}{24}(2\sigma_3(n) + (1 - 3n)\sigma(n)). \tag{3.21}
$$

Formula (3.20) is due to Glaisher [11, p. 36]. Formulae (3.20) and (3.21) are formulae (3.5) and (3.6) in Lahiri [14].

Next we treat $S(0\ 2\ 1\ 3) = S(2\ 0\ 3\ 1)$ and $S(2\ 0\ 1\ 3)$. First we note that

$$
\begin{aligned}
& S(2\ 0\ 3\ 1) - S(2\ 0\ 1\ 3) \\
&= S(0\ 2\ 1\ 3) - S(2\ 0\ 1\ 3) \\
&= \sum_{m=1}^{n-1} ((n-m)^2 - m^2)\sigma(m)\sigma_3(n-m) \\
&= n^2 \sum_{m=1}^{n-1} \sigma(m)\sigma_3(n-m) - 2n \sum_{m=1}^{n-1} m\sigma(m)\sigma_3(n-m) \\
&= n^2 S(0\ 0\ 1\ 3) - 2n S(1\ 0\ 1\ 3) \\
&= \frac{1}{240}(7n^2\sigma_5(n) + (10n^2 - 18n^3)\sigma_3(n) + n^2\sigma(n)),
\end{aligned}
$$

by (3.12) and (3.13). Secondly taking

$$
f(a,b,x,y) = aby^6 - 5a^2x^3y^3 + 4a^2x^5y,
$$

we find that

$$
\begin{aligned}
E = \ & 70a^2x^3y^3 - 2aby^6 - 22a^2x^5y - 10b^2x^5y + 30abx^4y^2 \\
& -30abx^2y^4 + 10a^2xy^5 + 30b^2x^3y^3 + 12b^2xy^5 + 2abx^6
\end{aligned}
$$

so that the left hand side of (2.2) is

$$
70[2\ 0\ 3\ 3] - 2[1\ 1\ 0\ 6] - 22[2\ 0\ 5\ 1] - 10[0\ 2\ 5\ 1] + 30[1\ 1\ 4\ 2]
$$

$$
-30[1\ 1\ 2\ 4] + 10[2\ 0\ 1\ 5] + 30[0\ 2\ 3\ 3] + 12[0\ 2\ 1\ 5] + 2[1\ 1\ 6\ 0]
$$

$$
= 100[2\ 0\ 3\ 3] - 10[2\ 0\ 5\ 1] = 100S(2\ 0\ 1\ 3) - 10S(2\ 0\ 3\ 1).
$$

Evaluating the right hand side of (2.2), Theorem 1 gives

$$
100S(2\ 0\ 1\ 3) - 10S(2\ 0\ 3\ 1) = \frac{5}{12}(2n^2\sigma_5(n) - n^2\sigma_3(n) - n^2\sigma(n)).
$$

Solving the two linear equations for $S(2\ 0\ 1\ 3)$ and $S(2\ 0\ 3\ 1)$ we obtain

$$
\sum_{m=1}^{n-1} m^2\sigma_3(m)\sigma(n-m) = \frac{1}{24}(n^2\sigma_5(n) + (n^2 - 2n^3)\sigma_3(n)) \qquad (3.22)
$$

$$
\sum_{m=1}^{n-1} m^2\sigma(m)\sigma_3(n-m) = \frac{1}{240}(3n^2\sigma_5(n) - 2n^3\sigma_3(n) - n^2\sigma(n)). \qquad (3.23)
$$

These are formulae (5.7) and (5.8) of Lahiri [14]. Now we turn to the determination of $S(0\ 1\ 1\ 5) = S(1\ 0\ 5\ 1)$ and $S(1\ 0\ 1\ 5)$. First we observe that

$$S(1\ 0\ 1\ 5) + S(1\ 0\ 5\ 1) = S(1\ 0\ 1\ 5) + S(0\ 1\ 1\ 5)$$
$$= nS(0\ 0\ 1\ 5)$$
$$= \frac{1}{504}(20n\sigma_7(n) + (21n - 42n^2)\sigma_5(n) + n\sigma(n)),$$

by (3.18). The choice

$$f(a,b,x,y) = -11aby^6 + 30abx^2y^4 + 20a^2x^3y^3 + 6a^2x^5y$$

in Theorem 1 yields

$$540S(1\ 0\ 1\ 5) - 180S(1\ 0\ 5\ 1) = \frac{15}{14}((6n^2 - 7n)\sigma_5(n) + n\sigma(n)).$$

Solving for $S(1\ 0\ 1\ 5)$ and $S(1\ 0\ 5\ 1)$ we obtain

$$\sum_{m=1}^{n-1} m\sigma(m)\sigma_5(n-m) = \frac{1}{504}(5n\sigma_7(n) - 6n^2\sigma_5(n) + n\sigma(n)), \quad (3.24)$$

$$\sum_{m=1}^{n-1} m\sigma_5(m)\sigma(n-m) = \frac{1}{168}(5n\sigma_7(n) + (7n - 12n^2)\sigma_5(n)). \quad (3.25)$$

Formulae (3.24) and (3.25) are formulae (7.7) and (7.6) of Lahiri [14].

Finally we treat $S(1\ 1\ 1\ 3)$, $S(0\ 0\ 3\ 5)$ and $S(0\ 0\ 1\ 7)$. We choose

$$f(a,b,x,y) = -aby^6 - 3abx^2y^4 + a^2x^3y^3 + 3a^2x^5y,$$
$$f(a,b,x,y) = xy^7 + x^7y - x^3y^5 - x^5y^3,$$
$$f(a,b,x,y) = 11xy^7 + 11x^7y - 56x^3y^5 - 56x^5y^3,$$

so that

$$E = 12a^2x^5y + 36abx^2y^4 + 62a^2x^3y^3 + 8aby^6 + 24a^2xy^5$$
$$- 24b^2x^5y - 62b^2x^3y^3 - 12b^2xy^5 - 8abx^6 + 54abx^4y^2,$$
$$E = 90x^3y^5 + 90x^5y^3,$$
$$E = -720xy^7 - 720x^7y,$$

respectively. The left hand sides of (2.2) are

$$12[2\ 0\ 5\ 1] + 36[1\ 1\ 2\ 4] + 62[2\ 0\ 3\ 3] + 8[1\ 1\ 0\ 6] + 24[2\ 0\ 1\ 5]$$
$$-24[0\ 2\ 5\ 1] - 62[0\ 2\ 3\ 3] - 12[0\ 2\ 1\ 5] - 8[1\ 1\ 6\ 0] + 54[1\ 1\ 4\ 2]$$
$$= 90[1\ 1\ 2\ 4] = 90S(1\ 1\ 1\ 3),$$

$$90[0\ 0\ 3\ 5] + 90[0\ 0\ 5\ 3] = 180[0\ 0\ 3\ 5] = 180S(0\ 0\ 3\ 5),$$
$$-720[0\ 0\ 1\ 7] - 720[0\ 0\ 7\ 1] = -1440[0\ 0\ 1\ 7] = -1440S(0\ 0\ 1\ 7),$$

respectively. The right hand sides are

$$\frac{3}{2}(n^2\sigma_5(n) - n^3\sigma_3(n)),$$

$$\frac{1}{28}(11\sigma_9(n) - 21\sigma_5(n) + 10\sigma_3(n)),$$

$$-3(11\sigma_9(n) + (20 - 30n)\sigma_7(n) - \sigma(n)),$$

respectively. Hence, by Theorem 1, we have

$$\sum_{m=1}^{n-1} m(n-m)\sigma(m)\sigma_3(n-m) = \frac{1}{60}(n^2\sigma_5(n) - n^3\sigma_3(n)), \tag{3.26}$$

$$\sum_{m=1}^{n-1} \sigma_3(m)\sigma_5(n-m) = \frac{1}{5040}(11\sigma_9(n) - 21\sigma_5(n) + 10\sigma_3(n)),$$
$$\tag{3.27}$$

$$\sum_{m=1}^{n-1} \sigma(m)\sigma_7(n-m) = \frac{1}{480}(11\sigma_9(n) + (20 - 30n)\sigma_7(n) - \sigma(n)).$$
$$\tag{3.28}$$

These are formulae (5.6), (9.1) and (9.2) of Lahiri [14]. Formula (3.27) is due to Glaisher [11, p. 35]. It also appears in the work of Ramanujan [30, Table IV], [31, p. 146]. Formula (3.28) is due to Ramanujan [30, Table IV], [31, p. 146].

Weight = 12. Of the twenty-eight sums $S(e\ f\ g\ h)$ of weight 12, we know of only three which can be evaluated using Theorem 1, these are $S(0\ 0\ 5\ 7)$, $S(0\ 0\ 3\ 9)$ and $S(0\ 0\ 1\ 11)$. Choosing

$$f(a,b,x,y) = 8xy^{11} + 8x^{11}y - 35x^3y^9 - 35x^9y^3 + 27x^5y^7 + 27x^7y^5,$$
$$f(a,b,x,y) = 2xy^{11} + 2x^{11}y - 11x^3y^9 - 11x^9y^3 + 9x^5y^7 + 9x^7y^5,$$
$$f(a,b,x,y) = 271xy^{11} + 271x^{11}y - 1540x^3y^9 - 1540x^9y^3 + 1584x^5y^7$$
$$+1584x^7y^5,$$

we find that

$$E = 2520x^5y^7 + 2520x^7y^5,$$
$$E = -180x^3y^9 - 180x^9y^3,$$
$$E = 7560xy^{11} + 7560x^{11}y,$$

respectively. The left hand sides of (2) are $5040S(0\ 0\ 5\ 7)$, $-360S(0\ 0\ 3\ 9)$ and $15120S(0\ 0\ 1\ 11)$ respectively. Evaluating the right hand sides of (2.2) and applying Theorem 1, we obtain

$$\sum_{m=1}^{n-1} \sigma_5(m)\sigma_7(n-m) = \frac{1}{10080}(\sigma_{13}(n) + 20\sigma_7(n) - 21\sigma_5(n)),$$

$$(3.29)$$

$$\sum_{m=1}^{n-1} \sigma_3(m)\sigma_9(n-m) = \frac{1}{2640}(\sigma_{13}(n) - 11\sigma_9(n) + 10\sigma_3(n)), \quad (3.30)$$

$$\sum_{m=1}^{n-1} \sigma(m)\sigma_{11}(n-m) = \frac{1}{65520}(691\sigma_{13}(n) + 2730(1-n)\sigma_{11}(n)$$

$$-691\sigma(n)). \tag{3.31}$$

These are formulae (13.1), (13.2), (13.4) of Lahiri [14]. Formulae (3.29) and (3.30) are due to Glaisher [11, p. 35]. They also appear in Ramanujan [30, Table IV], [31, p. 146]. Formula (3.31) is due to Ramanujan [30, Table IV], [31, p. 146].

Weight = 10.   Formulae (3.1)–(3.3), (5.1)–(5.4), (7.1)–(7.5), (9.1)–(9.4), (11.1), (11.3), (11.4) of Lahiri [15] show that the nineteen sums $S(e\ f\ g\ h)$ of weight 10 require the Ramanujan tau function $\tau(n)$ in their evaluation. Thus it is likely that, in order to be able to deduce their evaluations from Theorem 1, representations of Ramanujan's tau function in terms of divisor functions would be needed. We have not explored this.

In addition to the twenty-two identities (3.10)–(3.31) that we have just considered, Lahiri [14] gave fifteen additional identities which give the value of the sum

$$\sum_{m_1+\cdots+m_r=n} m_1{}^{a_1}\cdots m_r{}^{a_r}\sigma_{b_1}(m_1)\cdots\sigma_{b_r}(m_r) \quad (r \geq 3)$$

where the sum is over all positive integers $m_1,...,\ m_r$ satisfying $m_1 + \cdots + m_r = n$, for the values of $a_1,...,\ a_r,b_1,...,\ b_r$ given in Table 2 below. The authors have checked that all of these sums can be evaluated using the sums (3.10)–(3.31) and so can be evaluated in an elementary manner. All of our evaluations agreed with those of Lahiri [14] except for that of

$$\sum_{m_1+m_2+m_3+m_4=n} m_1\sigma(m_1)\sigma(m_2)\sigma(m_3)\sigma(m_4),$$

where a typo had crept into Lahiri's evaluation. In Lahiri's equation (7.10) the term $-2^3 3^2(2,5)$ and one of the two terms $3\cdot 5(1,4)$ should be deleted.

a_1	a_2	a_3	b_1	b_2	b_3	Lahiri's eqn
0	0	0	1	1	1	(5.2)
1	0	0	1	1	1	(5.5)
1	1	0	1	1	1	(5.9)
2	0	0	1	1	1	(5.10)
0	0	0	1	1	3	(7.3)
1	0	0	3	1	1	(7.8)
1	0	0	1	1	3	(7.9)
0	0	0	1	3	3	(9.3)
0	0	0	1	1	5	(9.4)
0	0	0	3	3	5	(13.3)

a_1	a_2	a_3	a_4	b_1	b_2	b_3	b_4	Lahiri's eqn
0	0	0	0	1	1	1	1	(7.4)
1	0	0	0	1	1	1	1	(7.10)
0	0	0	0	1	1	1	3	(9.5)

a_1	a_2	a_3	a_4	a_5	b_1	b_2	b_3	b_4	b_5	Lahiri's eqn
0	0	0	0	0	1	1	1	1	1	(9.6)
1	0	0	0	0	1	1	1	1	1	(9.7)

Table 2. Lahiri's identities for $r = 3, 4, 5$

The correct evaluation of the above sum is then

$$\frac{1}{13824}(5n\sigma_7(n) + (21n - 42n^2)\sigma_5(n) + (15n - 90n^2 + 108n^3)\sigma_3(n)$$

$$+(n - 18n^2 + 72n^3 - 72n^4)\sigma(n)).$$

We illustrate the derivation of these fifteen identities with just one example. The rest can be treated similarly. We evaluate the sum

$$\sum_{m_1+m_2+m_3=n} m_1 m_2 \sigma(m_1)\sigma(m_2)\sigma(m_3).$$

We have appealing to (3.15), (3.21) and (3.22)

$$288 \sum_{m_1+m_2+m_3=n} m_1 m_2 \sigma(m_1)\sigma(m_2)\sigma(m_3)$$

$$= 288 \sum_{m_3=1}^{n-2} \sigma(m_3) \sum_{m_1+m_2=n-m_3} m_1 m_2 \sigma(m_1) \sigma(m_2)$$

$$= 24 \sum_{m_3=1}^{n-2} \sigma(m_3)((n-m_3)^2 \sigma_3(n-m_3) - (n-m_3)^3 \sigma(n-m_3))$$

$$= 24 \sum_{m=1}^{n-2} (n-m)^2 \sigma(m) \sigma_3(n-m) - 24 \sum_{m=1}^{n-2} (n-m)^3 \sigma(m) \sigma(n-m)$$

$$= 24 \sum_{m=2}^{n-1} m^2 \sigma_3(m) \sigma(n-m) - 24 \sum_{m=2}^{n-1} m^3 \sigma(m) \sigma(n-m)$$

$$= 24 \sum_{m=1}^{n-1} m^2 \sigma_3(m) \sigma(n-m) - 24 \sum_{m=1}^{n-1} m^3 \sigma(m) \sigma(n-m)$$

$$= (n^2 \sigma_5(n) + (n^2 - 2n^3) \sigma_3(n)) - (2n^3 \sigma_3(n) + (n^3 - 3n^4) \sigma(n))$$

$$= n^2 \sigma_5(n) + (n^2 - 4n^3) \sigma_3(n) - (n^3 - 3n^4) \sigma(n).$$

We remark that Lahiri's equations (5.2), (7.4) and (9.6) are implicit in Glaisher [11, p. 33]. Lahiri's identity (5.2) is proved in Bambah and Chowla [1, eqn. (28)], see also Chowla [5], [6, p. 669].

This concludes our proof that all thirty-seven convolution formulae of Lahiri [14] are consequences of Theorem 1.

For further work on convolution sums, see Grosjean [12], [13] and Levitt [19].

4   Application of Theorem 1 to Melfi's Identities

In this section we consider sums of the type

$$\sum_{m<n/k} \sigma(m) \sigma(n - km), \qquad (4.1)$$

where k is a given positive integer. Recently Melfi [26] has treated these sums for $k=$ 2, 3, 4, 5, 9 under the restriction that $\gcd(n, k) = 1$ using the theory of modular forms. We evaluate these sums using Theorem 1 for $k=$ 2, 3, 4 and all positive integers n thereby extending Melfi's results in these cases. We begin by expressing the sum (4.1) in terms of the quantity

$$\sum_{\substack{ax+by=n \\ x \equiv -y \,(\mathrm{mod}\ k)}} ab \; + \sum_{\substack{ax+by=n \\ x \equiv y \,(\mathrm{mod}\ k)}} ab, \qquad (4.2)$$

which can be evaluated explicitly for $k = 1, 2, 3, 4$.

Lemma 1. *Let k be a positive integer. Then*

$$\sum_{m<n/k} \sigma(m)\sigma(n-km) = -\frac{1}{24}\sigma_3(n) + \frac{1}{24}\sigma(n) + \frac{1}{4}\sigma_3(n/k) - \frac{n}{4}\sigma(n/k)$$

$$+\frac{1}{4}\sum_{\substack{ax+by=n \\ x\equiv-y \,(\mathrm{mod}\ k)}} ab + \frac{1}{4}\sum_{\substack{ax+by=n \\ x\equiv y \,(\mathrm{mod}\ k)}} ab. \qquad (4.3)$$

Proof. The identity (4.3) follows from Theorem 1 by taking $f(a,b,x,y) = (2a^2 - b^2)F_k(x)$. With this choice, after a long calculation, we find that the left hand side of (2.2) is

$$2\sum_{\substack{ax+by=n \\ x\equiv-y \,(\mathrm{mod}\ k)}} ab + 2\sum_{\substack{ax+by=n \\ x\equiv y \,(\mathrm{mod}\ k)}} ab - 8\sum_{m<n/k} \sigma(m)\sigma(n-km)$$

and the right hand side is

$$\frac{1}{3}\sigma_3(n) - \frac{1}{3}\sigma(n) - 2\sigma_3(n/k) + 2n\sigma(n/k).$$

Lemma 1 now follows by Theorem 1. $\qquad\square$

When $k=1$ the sum (4.2) is $2\sum_{m=1}^{n-1}\sigma(m)\sigma(n-m)$ and Lemma 1 gives the identity (3.10).

When $k=2$ the sum (4.2) is

$$2\sum_{\substack{ax+by=n \\ x\equiv y \,(\mathrm{mod}\ 2)}} ab = 4\sum_{\substack{ax+by=n \\ 2\mid x, 2\mid y}} ab + 2\sum_{ax+by=n} ab - 2\sum_{\substack{ax+by=n \\ 2\mid x}} ab$$

$$- 2\sum_{\substack{ax+by=n \\ 2\mid y}} ab$$

$$= 4\sum_{m<n/2} \sigma(m)\sigma(n/2-m) + 2\sum_{m=1}^{n-1}\sigma(m)\sigma(n-m)$$

$$- 4\sum_{m<n/2} \sigma(m)\sigma(n-2m).$$

Then, from Lemma 1 and (3.10), we obtain

Theorem 2 (see Melfi [26], (8)).

$$\sum_{m<n/2} \sigma(m)\sigma(n-2m) = \frac{1}{24}(2\sigma_3(n) + (1-3n)\sigma(n)$$

$$+ 8\sigma_3(n/2) + (1-6n)\sigma(n/2)). \qquad (4.4)$$

Replacing n by $n+1$ in (4.4), we obtain the following companion formula

$$\sum_{m \leq n/2} \sigma(m)\sigma(n - (2m - 1)) = \frac{1}{24}(2\sigma_3(n + 1) - (2 + 3n)\sigma(n + 1)$$

$$+ 8\sigma_3((n + 1)/2) - (5 + 6n)\sigma((n + 1)/2)). \quad (4.5)$$

When $k = 3$ the sum (4.2) is

$$3 \sum_{\substack{ax+by=n \\ 3\,|\,x,\,3\,|\,y}} ab + \sum_{ax+by=n} ab - \sum_{\substack{ax+by=n \\ 3\,|\,x}} ab - \sum_{\substack{ax+by=n \\ 3\,|\,y}} ab$$

$$= 3 \sum_{ax+by=n/3} ab + \sum_{ax+by=n} ab - 2 \sum_{3ax+by=n} ab$$

$$= 3 \sum_{m<n/3} \sigma(m)\sigma(n/3 - m) + \sum_{m=1}^{n-1} \sigma(m)\sigma(n - m)$$

$$-2 \sum_{m<n/3} \sigma(m)\sigma(n - 3m).$$

Appealing to Lemma 1 (with $k = 3$) and (3.10), we obtain

Theorem 3 (see Melfi [26], (12)).

$$\sum_{m<n/3} \sigma(m)\sigma(n - 3m) = \frac{1}{24}(\sigma_3(n) + (1 - 2n)\sigma(n)$$

$$+ 9\sigma_3(n/3) + (1 - 6n)\sigma(n/3)).$$

When $k = 4$ the sum (4.2) is

$$2 \sum_{\substack{ax+by=n \\ 4\,|\,x,\,4\,|\,y}} ab + 2 \sum_{\substack{ax+by=n \\ 2\,\|\,x,\,2\,\|\,y}} ab + \sum_{\substack{ax+by=n \\ 2\nmid x,\,2\nmid y}} ab.$$

Next we set

$$F(n) := \sum_{ax+by=n} ab, \quad G(n) := \sum_{2ax+by=n} ab,$$

so that the above sum is

$$2F(n/4) + 2(F(n/2) - 2G(n/2) + F(n/4)) + (F(n) - 2G(n) + F(n/2))$$

$$= F(n) + 3F(n/2) + 4F(n/4) - 2G(n) - 4G(n/2).$$

Since

$$F(n) = \sum_{m=1}^{n-1} \sigma(m)\sigma(n-m), \quad G(n) = \sum_{m<n/2} \sigma(m)\sigma(n-2m),$$

appealing to (3.10), Theorem 2 and Lemma 1 (with $k=4$), we obtain

Theorem 4 (see Melfi [26], (11)).

$$\sum_{m<n/4} \sigma(m)\sigma(n-4m) = \frac{1}{48}(\sigma_3(n) + (2-3n)\sigma(n) + 3\sigma_3(n/2)$$

$$+ 16\sigma_3(n/4) + (2-12n)\sigma(n/4)).$$

The authors have not been able to use Theorem 1 to evaluate the sum (4.1) for $k \geq 5$.

As a consequence of Theorems 2 and 4 we obtain the following new evaluation.

Theorem 5.

$$\sum_{m<n/2} \sigma(2m)\sigma(n-2m) = \frac{1}{24}(5\sigma_3(n) + 21\sigma_3(n/2) - 16\sigma_3(n/4)$$

$$+ (1-6n)(\sigma(n) + 3\sigma(n/2) - 2\sigma(n/4))).$$

Proof. As $\sigma(2m) = 3\sigma(m) - 2\sigma(m/2)$ we have

$$\sum_{m<n/2} \sigma(2m)\sigma(n-2m)$$

$$= 3 \sum_{m<n/2} \sigma(m)\sigma(n-2m) - 2 \sum_{\substack{m<n/2 \\ 2\,|\,m}} \sigma(m/2)\sigma(n-2m)$$

$$= 3 \sum_{m<n/2} \sigma(m)\sigma(n-2m) - 2 \sum_{m<n/4} \sigma(m)\sigma(n-4m).$$

The theorem now follows from Theorems 2 and 4. $\qquad\square$

As a corollary of Theorem 5 we have the companion result where, instead of running through even integers $2m$ $(m < n/2)$, we run through odd integers $2m - 1$ $(m \leq n/2)$.

Corollary 1.

$$\sum_{m\leq n/2} \sigma(2m-1)\sigma(n-(2m-1))$$

$$= \frac{1}{24}(5\sigma_3(n) - 21\sigma_3(n/2) + 16\sigma_3(n/4)$$

$$+ (1-6n)(\sigma(n) - 3\sigma(n/2) + 2\sigma(n/4))).$$

Proof. Corollary 1 follows from Theorem 5 and (3.10).   □

Taking $n = 2M$ with M odd in Corollary 1, we obtain

$$\sum_{m=1}^{M} \sigma(2m-1)\sigma(2M-2m+1) = \frac{1}{24}(5\sigma_3(2M) - 21\sigma_3(M))$$
$$+(1-12M)(\sigma(2M)-3\sigma(M)),$$

that is

$$\sum_{m=1}^{M} \sigma(2m-1)\sigma(2M-2m+1) = \sigma_3(M), \qquad (4.6)$$

which is a result of Liouville [20, p. 146], see also [7, p. 287], [8, p. 329].

Our next application of Theorem 1 extends two more of Melfi's identities [26, eqns. (9), (10)] to all positive integers n.

Theorem 6.

$$\sum_{m<n/2} \sigma_3(m)\sigma(n-2m) = \frac{1}{240}(\sigma_5(n) - \sigma(n) + 20\sigma_5(n/2)$$
$$+(10-30n)\sigma_3(n/2)),$$

$$\sum_{m<n/2} \sigma(m)\sigma_3(n-2m) = \frac{1}{240}(5\sigma_5(n) + (10-15n)\sigma_3(n)$$
$$+16\sigma_5(n/2) - \sigma(n/2)).$$

Proof. We set

$$X := \sum_{m<n/2} \sigma_3(m)\sigma(n-2m), \quad Y := \sum_{m<n/2} \sigma(m)\sigma_3(n-2m).$$

Choosing $f(a,b,x,y) = a^4 F_2(x)$ in Theorem 1, the left hand side of (2.2) is

$$\sum_{ax+by=n} ((b-a)^4 - (b+a)^4)F_2(x) = \sum_{2ax+by=n} (-8a^3b - 8ab^3) = -8X - 8Y$$

and the right hand side of (2.2) is (after some calculation)

$$-\frac{1}{5}\sigma_5(n) + \left(\frac{n}{2}-\frac{1}{3}\right)\sigma_3(n) + \frac{1}{30}\sigma(n) - \frac{6}{5}\sigma_5(n/2)$$
$$+\left(n-\frac{1}{3}\right)\sigma_3(n/2) + \frac{1}{30}\sigma(n/2).$$

Next setting $f(a, b, x, y) = b^4 F_2(a)$ in Theorem 1, the left hand side of (2.2) is

$$\sum_{ax+by=n} ((a-b)^4 - (a+b)^4) F_2(a) = \sum_{2ax+by=n} (-64a^3 b - 16ab^3) = -64X - 16Y$$

and the right hand side is (after some calculation)

$$- \frac{3}{5}\sigma_5(n) + \left(n - \frac{2}{3}\right)\sigma_3(n) + \frac{4}{15}\sigma(n) - \frac{32}{5}\sigma_5(n/2)$$
$$+ \left(8n - \frac{8}{3}\right)\sigma_3(n/2) + \frac{1}{15}\sigma(n/2).$$

Solving the two linear equations for X and Y resulting from Theorem 1, we obtain the assertions of Theorem 6. $\qquad\square$

As a consequence of Theorem 6 we have the following identity.

Corollary 2.

$$\sum_{k=0}^{n-1} \sigma(2k+1)\sigma_3(n-k) = \frac{1}{240}(\sigma_5(2n+1) - \sigma(2n+1)). \qquad (4.7)$$

Proof. We have

$$\sum_{k=0}^{n-1} \sigma(2k+1)\sigma_3(n-k) = \sum_{m=1}^{n} \sigma(2n - 2m + 1)\sigma_3(m)$$
$$= \sum_{m<(2n+1)/2} \sigma((2n+1) - 2m)\sigma_3(m)$$
$$= \frac{1}{240}(\sigma_5(2n+1) - \sigma(2n+1)),$$

by Theorem 6. This completes the proof of Corollary 2. $\qquad\square$

Corollary 2 was explicitly stated but never proved by Ramanujan [30], [31, p. 146]. A result equivalent to (4.7) was first proved by Masser, see Berndt [2, p. 329] and Berndt and Evans [3, p. 136], with later proofs by Atkin (see Berndt [2, p. 329]) and Ramamani [29]. None of these proofs is elementary. The above proof is the first elementary proof of (4.7). Berndt [2, p. 329] has indicated that it would be interesting to have such a proof. As a further consequence of Theorem 6, we obtain the following identity of Liouville [20, p. 147].

Corollary 3. *Let M be an odd positive integer. Then*

$$\sum_{m=0}^{M-1} \sigma(2m+1)\sigma_3(2M-2m-1) = \sigma_5(M).$$

Proof. As

$$\sigma(2k) = 3\sigma(k) - 2\sigma(k/2), \quad \sigma_3(2k) = 9\sigma_3(k) - 8\sigma_3(k/2),$$

we have

$$\begin{aligned}
\sum_{m=1}^{M-1} \sigma(2m)\sigma_3(2M-2m) &= \sum_{m=1}^{M-1} (3\sigma(m) - 2\sigma(m/2))(9\sigma_3(M-m) \\
&\quad -8\sigma_3((M-m)/2)) \\
&= 27S_1 - 18S_2 - 24S_3 + 16S_4,
\end{aligned}$$

where

$$S_1 = \sum_{m=1}^{M-1} \sigma(m)\sigma_3(M-m),$$

$$S_2 = \sum_{\substack{m=1 \\ 2\,|\,m}}^{M-1} \sigma(m/2)\sigma_3(M-m),$$

$$S_3 = \sum_{\substack{m=1 \\ 2\,|\,M-m}}^{M-1} \sigma(m)\sigma_3((M-m)/2),$$

$$S_4 = \sum_{\substack{m=1 \\ 2\,|\,m \\ 2\,|\,M-m}}^{M-1} \sigma(m/2)\sigma_3((M-m)/2).$$

By (3.12) we have

$$S_1 = \frac{21}{240}\sigma_5(M) + \frac{(1-3M)}{24}\sigma_3(M) - \frac{1}{240}\sigma(M).$$

By Theorem 6, we obtain

$$S_2 = \sum_{m<M/2} \sigma(m)\sigma_3(M-2m) = \frac{1}{48}\sigma_5(M) + \frac{(2-3M)}{48}\sigma_3(M).$$

Further, by Corollary 2, we have

$$
\begin{aligned}
S_3 &= \sum_{\substack{m=1 \\ 2 \mid M-m}}^{M-1} \sigma(m)\sigma_3((M-m)/2) \\
&= \sum_{k=0}^{(M-3)/2} \sigma(2k+1)\sigma_3\left(\frac{M-1}{2} - k\right) \\
&= \frac{1}{240}\sigma_5(M) - \frac{1}{240}\sigma(M).
\end{aligned}
$$

Finally, as m and $N - m$ are of opposite parity, we have $S_4 = 0$. Putting these evaluations together, we obtain

$$
\sum_{m=1}^{M-1} \sigma(2m)\sigma_3(2M - 2m) = \frac{151}{80}\sigma_5(M) + \frac{(3 - 18M)}{8}\sigma_3(M) - \frac{1}{80}\sigma(M).
$$

Then, appealing to (3.12), we obtain

$$
\begin{aligned}
&\sum_{m=0}^{M-1} \sigma(2m+1)\sigma_3(2M - 2m - 1) \\
&= \sum_{m=1}^{2M-1} \sigma(m)\sigma_3(2M - m) - \sum_{m=1}^{M-1} \sigma(2m)\sigma_3(2M - 2m) \\
&= \left(\frac{7}{80}\sigma_5(2M) + \frac{(1 - 6M)}{24}\sigma_3(2M) - \frac{1}{240}\sigma(2M)\right) \\
&\quad - \left(\frac{151}{80}\sigma_5(M) + \frac{(3 - 18M)}{8}\sigma_3(M) - \frac{1}{80}\sigma(M)\right) \\
&= \sigma_5(M).
\end{aligned}
$$

This completes the proof of Corollary 3. $\qquad\qquad\qquad\square$

We conclude this section with the following result, which is analogous to Theorem 5. We make use of Theorem 3 and a result of Melfi [26, eqn. (14)].

Theorem 7. *If $n \equiv 0 \,(\mathrm{mod}\ 3)$ then*

$$
\sum_{m<n/3} \sigma(3m)\sigma(n - 3m) = \frac{1}{36}(7\sigma_3(n) + (3 - 18n)\sigma(n) + 8\sigma_3(n/3)).
$$

If $n \equiv 1 \pmod 3$ and there exists a prime $p \equiv 2 \pmod 3$ with $p \,\|\, n$, or if $n \equiv 2 \pmod 3$, then

$$\sum_{m<n/3} \sigma(3m)\sigma(n-3m) = \frac{1}{72}(11\sigma_3(n) + (3-18n)\sigma(n)).$$

Proof. As $\sigma(3m) = 4\sigma(m) - 3\sigma(m/3)$ we have

$$\sum_{m<n/3} \sigma(3m)\sigma(n-3m)$$

$$= 4 \sum_{m<n/3} \sigma(m)\sigma(n-3m) - 3 \sum_{\substack{m<n/3 \\ 3\,|\,m}} \sigma(m/3)\sigma(n-3m)$$

$$= 4 \sum_{m<n/3} \sigma(m)\sigma(n-3m) - 3 \sum_{m<n/9} \sigma(m)\sigma(n-9m)$$

$$= \frac{1}{6}(\sigma_3(n) + (1-2n)\sigma(n) + 9\sigma_3(n/3) + (1-6n)\sigma(n/3))$$

$$\quad -3 \sum_{m<n/9} \sigma(m)\sigma(n-9m),$$

by Theorem 3. We first consider the case $n \equiv 0 \pmod 3$, say $n = 3N$. In this case we have

$$\sum_{m<n/9} \sigma(m)\sigma(n-9m)$$

$$= \sum_{m<N/3} \sigma(m)\sigma(3N-9m)$$

$$= 4 \sum_{m<N/3} \sigma(m)\sigma(N-3m) - 3 \sum_{m<N/3} \sigma(m)\sigma(N/3-m)$$

$$= \frac{1}{6}(\sigma_3(N) + (1-2N)\sigma(N) + 9\sigma_3(N/3) + (1-6N)\sigma(N/3))$$

$$\quad -\frac{1}{4}(5\sigma_3(N/3) + (1-2N)\sigma(N/3))$$

$$= \frac{1}{12}(2\sigma_3(N) + (2-4N)\sigma(N) + 3\sigma_3(N/3) - (1+6N)\sigma(N/3))$$

$$= \frac{1}{36}(6\sigma_3(n/3) + (6-4n)\sigma(n/3) + 9\sigma_3(n/9) - (3+6n)\sigma(n/9)),$$

by (3.10) and Theorem 3. Hence

$$\sum_{m<n/3} \sigma(3m)\sigma(n-3m)$$

$$= \frac{1}{12}(2\sigma_3(n) + 12\sigma_3(n/3) - 9\sigma_3(n/9) + (2-4n)\sigma(n)$$
$$-(4+8n)\sigma(n/3) + (3+6n)\sigma(n/9)).$$

Since

$$\sigma_3(n) = 28\sigma_3(n/3) - 27\sigma_3(n/9), \quad \sigma(n) = 4\sigma(n/3) - 3\sigma(n/9),$$

for $n \equiv 0 \pmod 3$, we have

$$\sum_{m<n/3} \sigma(3m)\sigma(n-3m) = \frac{1}{36}(7\sigma_3(n) + 8\sigma_3(n/3) + (3-18n)\sigma(n)).$$

Finally, if $n \equiv 1 \pmod 3$ and there exists a prime $p \equiv 2 \pmod 3$ such that $p \parallel n$, or if $n \equiv 2 \pmod 3$, then, by Melfi [26, eqn. (14)], we have

$$\sum_{m<n/9} \sigma(m)\sigma(n-9m) = \frac{1}{216}(\sigma_3(n) + (9-6n)\sigma(n))$$

so that

$$\sum_{m<n/3} \sigma(3m)\sigma(n-3m) = \frac{1}{72}(11\sigma_3(n) + (3-18n)\sigma(n)).$$

This completes the proof of Theorem 7. $\qquad\square$

We are unable to give the value of $\sum_{m<n/3} \sigma(3m)\sigma(n-3m)$ for those $n \equiv 1 \pmod 3$ such that $p^2 \mid n$ for every prime $p \equiv 2 \pmod 3$ with $p \mid n$ since the value of $\sum_{m<n/9} \sigma(m)\sigma(n-9m)$ is not known for such n.

5 Sums $\sum_{m=1, m\equiv a \pmod{b}}^{n-1} \sigma(m)\sigma(n-m)$

Let a and b be integers satisfying $b \geq 1$ and $0 \leq a \leq b-1$. We set

$$S(a,b) = \sum_{\substack{m=1 \\ m\equiv a \pmod b}}^{n-1} \sigma(m)\sigma(n-m). \qquad (5.1)$$

Clearly, by (3.10), Theorem 5 and Corollary 1, we have

$$S(0,1) = \frac{1}{12}(5\sigma_3(n) + (1 - 6n)\sigma(n)), \tag{5.2}$$

$$
\begin{aligned}
S(0,2) &= \frac{1}{24}(5\sigma_3(n) + 21\sigma_3(n/2) - 16\sigma_3(n/4) + (1 - 6n)(\sigma(n) \\
&\quad + 3\sigma(n/2) - 2\sigma(n/4))), \tag{5.3}
\end{aligned}
$$

$$
\begin{aligned}
S(1,2) &= \frac{1}{24}(5\sigma_3(n) - 21\sigma_3(n/2) + 16\sigma_3(n/4) + (1 - 6n)(\sigma(n) \\
&\quad - 3\sigma(n/2) + 2\sigma(n/4))). \tag{5.4}
\end{aligned}
$$

Melfi [26, eqn. (7)] has shown that

$$S(1,3) = \frac{1}{9}\sigma_3(n), \quad \text{if } n \equiv 2 \pmod 3. \tag{5.5}$$

Appealing to Theorem 7, we obtain the following partial evaluation of $S(i,3)$ $(i = 0,1,2)$.

Theorem 8. *If* $n \equiv 0 \pmod 3$ *then*

$$
\begin{aligned}
S(0,3) &= \frac{1}{36}(7\sigma_3(n) + (3 - 18n)\sigma(n) + 8\sigma_3(n/3)), \\
S(1,3) &= \frac{1}{9}(\sigma_3(n) - \sigma_3(n/3)), \\
S(2,3) &= \frac{1}{9}(\sigma_3(n) - \sigma_3(n/3));
\end{aligned}
$$

if $n \equiv 1 \pmod 3$ *and there exists a prime* $p \equiv 2 \pmod 3$ *such that* $p \parallel n$ *then*

$$
\begin{aligned}
S(0,3) &= \frac{1}{72}(11\sigma_3(n) + (3 - 18n)\sigma(n)), \\
S(1,3) &= \frac{1}{72}(11\sigma_3(n) + (3 - 18n)\sigma(n)), \\
S(2,3) &= \frac{1}{9}\sigma_3(n);
\end{aligned}
$$

and if $n \equiv 2 \pmod 3$ *then*

$$
\begin{aligned}
S(0,3) &= \frac{1}{72}(11\sigma_3(n) + (3 - 18n)\sigma(n)), \\
S(1,3) &= \frac{1}{9}\sigma_3(n), \\
S(2,3) &= \frac{1}{72}(11\sigma_3(n) + (3 - 18n)\sigma(n)).
\end{aligned}
$$

Proof. First we note that

$$S(0,3) + S(1,3) + S(2,3) = S(0,1),$$

so that by (5.2) we have

$$S(0,3) + S(1,3) + S(2,3) = \frac{1}{12}(5\sigma_3(n) + (1 - 6n)\sigma(n)). \tag{5.6}$$

Secondly we note that the change of variable $m \longrightarrow n - m$ yields $S(i,3) = S(n-i,3)$ so that

$$\begin{cases} S(1,3) = S(2,3), & \text{if } n \equiv 0 \pmod 3, \\ S(0,3) = S(1,3), & \text{if } n \equiv 1 \pmod 3, \\ S(0,3) = S(2,3), & \text{if } n \equiv 2 \pmod 3. \end{cases} \tag{5.7}$$

Then

$$S(0,3) = \sum_{\substack{m=1 \\ m \equiv 0 \, (\text{mod } 3)}}^{n-1} \sigma(m)\sigma(n-m) = \sum_{m < n/3} \sigma(3m)\sigma(n - 3m),$$

and the evaluation of $S(0,3)$ follows from Theorem 7. The values of $S(1,3)$ and $S(2,3)$ then follow from (5.6) and (5.7). This completes the proof of Theorem 8. □

It would be interesting to determine $S(i,3)$ $(i = 0,1,2)$ for all $n \equiv 1$ (mod 3).

We conclude this section by giving some partial results for the sums $S(i,4)$ $(i = 0,1,2,3)$. We do not know of any evaluations of $S(a,b)$ for $b \geq 5$.

Theorem 9. *If* $n \equiv 0 \pmod 4$ *then*

$$S(1,4) = S(3,4) = \frac{1}{16}(\sigma_3(n) - \sigma_3(n/2))$$

and

$$S(0,4) + S(2,4) = \frac{1}{24}(7\sigma_3(n) + (2 - 12n)\sigma(n) + 3\sigma_3(n/2)).$$

If $n \equiv 1 \pmod 4$ *then*

$$S(0,4) = S(1,4), \quad S(2,4) = S(3,4)$$

and

$$S(0,4) + S(2,4) = \frac{1}{24}(5\sigma_3(n) + (1-6n)\sigma(n)).$$

If $n \equiv 2 \pmod 4$ then

$$S(0,4) = S(2,4) = \frac{1}{72}(11\sigma_3(n) + (3-18n)\sigma(n))$$

and

$$S(1,4) + S(3,4) = \frac{1}{9}\sigma_3(n).$$

If $n \equiv 3 \pmod 4$ then

$$S(0,4) = S(3,4), \quad S(1,4) = S(2,4)$$

and

$$S(0,4) + S(1,4) = \frac{1}{24}(5\sigma_3(n) + (1-6n)\sigma(n)).$$

Proof. First we note that by (5.1) and (5.2) we have

$$S(0,4) + S(1,4) + S(2,4) + S(3,4) = \frac{1}{12}(5\sigma_3(n) + (1-6n)\sigma(n)). \quad (5.8)$$

Secondly the change of variable $m \longrightarrow n - m$ yields $S(i,4) = S(n-i,4)$ so that

$$\begin{cases} S(1,4) = S(3,4), & \text{if } n \equiv 0 \pmod 4, \\ S(0,4) = S(1,4), \quad S(2,4) = S(3,4), & \text{if } n \equiv 1 \pmod 4, \\ S(0,4) = S(2,4), & \text{if } n \equiv 2 \pmod 4, \\ S(0,4) = S(3,4), \quad S(1,4) = S(2,4), & \text{if } n \equiv 3 \pmod 4. \end{cases} \quad (5.9)$$

The asserted results for $n \equiv 1 \pmod 4$ and $n \equiv 3 \pmod 4$ now follow from (5.8) and (5.9).

For $n \equiv 2 \pmod 4$ we have by (5.9) and (5.3)

$$\begin{aligned} S(0,4) = S(2,4) &= \frac{1}{2}(S(0,4) + S(2,4)) \\ &= \frac{1}{2}S(0,2) \\ &= \frac{1}{48}(5\sigma_3(n) + 21\sigma_3(n/2) + (1-6n)(\sigma(n) + 3\sigma(n/2))) \\ &= \frac{1}{72}(11\sigma_3(n) + (3-18n)\sigma(n)). \end{aligned}$$

The value of $S(1,4) + S(3,4)$ then follows from (5.8).

For $n \equiv 0 \pmod 4$ we have by (5.9) and (5.4)

$$
\begin{aligned}
S(1,4) = S(3,4) &= \frac{1}{2}(S(1,4) + S(3,4)) \\
&= \frac{1}{2}S(1,2) \\
&= \frac{1}{48}(5\sigma_3(n) - 21\sigma_3(n/2) + 16\sigma_3(n/4) \\
&\quad + (1 - 6n)(\sigma(n) - 3\sigma(n/2) + 2\sigma(n/4))) \\
&= \frac{1}{16}(\sigma_3(n) - \sigma_3(n/2)).
\end{aligned}
$$

The value of $S(0,4) + S(2,4)$ now follows from (5.8). This completes the proof of Theorem 9. $\qquad\square$

6 Application of Theorem 1 to Triangular Numbers

The triangular numbers are the nonnegative integers

$$
\frac{1}{2}m(m+1), \quad m = 0, 1, 2, \dots .
$$

For k a positive integer, we let $\delta_k(n)$ denote the number of representations of n as the sum of k triangular numbers. It is an easily proved classical result that

$$
\delta_2(n) = \sum_{a \mid 4n+1} \left(\frac{-4}{a}\right), \quad n = 0, 1, 2, \dots . \tag{6.1}
$$

We derive the corresponding formulae for $\delta_4(n), \delta_6(n)$ and $\delta_8(n)$ from (6.1) using Theorem 1.

Theorem 10 ([27], Theorem 3).

$$
\delta_4(n) = \sigma(2n + 1).
$$

Proof. First we choose $f(a, b, x, y) = F_4(b)$ in Theorem 1 with n replaced by $4n + 2$. We obtain

$$\sum_{ax+by=4n+2} (F_4(a-b) - F_4(a+b))$$

$$= \sum_{d \mid 4n+2} \sum_{x<d} \left(F_4\left(\frac{4n+2}{d}\right) + F_4(0) + F_4\left(\frac{4n+2}{d}\right) - F_4(x-d) \right.$$
$$\left. - F_4(d) - F_4(x) \right)$$

$$= \sum_{d \mid 4n+2} \sum_{x<d} (1 - F_4(x-d) - F_4(x))$$

$$= \sum_{d \mid 4n+2} \left\{ d - 1 - 2 \left[\frac{d}{4}\right] \right\}$$

$$= \sigma(4n+2) - \tau(4n+2) - 2 \sum_{d \mid 4n+2} \left[\frac{d}{4}\right]$$

$$= 3\sigma(2n+1) - 2\tau(2n+1) - 2 \sum_{d \mid 2n+1} \left[\frac{d}{4}\right] - 2 \sum_{d \mid 2n+1} \left[\frac{d}{2}\right]$$

$$= 2\sigma(2n+1) - \tau(2n+1) - 2 \sum_{d \mid 2n+1} \left[\frac{d}{4}\right].$$

Secondly we choose $f(a, b, x, y) = F_4(a)F_2(x)$ in Theorem 1 with n replaced by $4n + 2$. We obtain

$$\sum_{2ax+by=4n+2} (F_4(a-b) - F_4(a+b))$$

$$= \sum_{d \mid 4n+2} \sum_{x<d} \left(F_4(0)F_2(x) + F_4\left(\frac{4n+2}{d}\right) F_2(d) - F_4(x)F_2\left(\frac{4n+2}{d}\right) \right.$$
$$\left. + F_4\left(\frac{4n+2}{d}\right) F_2(d-x) - F_4(x)F_2(0) - F_4(d)F_2\left(\frac{4n+2}{d}\right) \right)$$

$$= \sum_{d \mid 4n+2} \sum_{x<d} \left(F_2(x) - F_4(x)F_2\left(\frac{4n+2}{d}\right) - F_4(x) \right)$$

$$= \sum_{d \mid 4n+2} \sum_{x<d/2} 1 - \sum_{d \mid 2n+1} \sum_{x<d/4} 1 - \sum_{d \mid 4n+2} \sum_{x<d/4} 1$$

$$= \sum_{d \mid 2n+1} (d-1) + \sum_{d \mid 2n+1} \frac{(d-1)}{2} - \sum_{d \mid 2n+1} \left[\frac{d}{4}\right] - \sum_{d \mid 2n+1} \frac{(d-1)}{2}$$

$$- \sum_{d \mid 2n+1} \left[\frac{d}{4}\right]$$

$$= \sigma(2n+1) - \tau(2n+1) - 2 \sum_{d \mid 2n+1} \left[\frac{d}{4} \right].$$

Subtracting these two results, we deduce that

$$\sum_{ax+by=4n+2} (F_4(a-b) - F_4(a+b)) - \sum_{2ax+by=4n+2} (F_4(a-b) - F_4(a+b))$$

$$= \sigma(2n+1),$$

that is

$$\sum_{\substack{ax+by=4n+2 \\ x \equiv 1 \,(\mathrm{mod}\ 2)}} (F_4(a-b) - F_4(a+b)) = \sigma(2n+1).$$

A simple consideration of the residues of a and b modulo 4 shows that

$$F_4(a-b) - F_4(a+b) = \left(\frac{-4}{ab} \right)$$

so that

$$\sum_{\substack{ax+by=4n+2 \\ x \equiv 1 \,(\mathrm{mod}\ 2)}} \left(\frac{-4}{ab} \right) = \sigma(2n+1).$$

As $\left(\dfrac{-4}{ab} \right) = 0$, for $a \equiv 0 \,(\mathrm{mod}\ 2)$, we have

$$\sum_{\substack{ax+by=4n+2 \\ ax \equiv 1 \,(\mathrm{mod}\ 2)}} \left(\frac{-4}{ab} \right) = \sigma(2n+1).$$

Now

$$\sum_{\substack{ax+by=4n+2 \\ ax \equiv 3 \,(\mathrm{mod}\ 4)}} \left(\frac{-4}{ab} \right) = \sum_{\substack{ax+by=4n+2 \\ ax \equiv 3 \,(\mathrm{mod}\ 4)}} \left(\frac{-4}{xb} \right) = - \sum_{\substack{ax+by=4n+2 \\ ax \equiv 3 \,(\mathrm{mod}\ 4)}} \left(\frac{-4}{ab} \right),$$

as $\left(\dfrac{-4}{x} \right) = - \left(\dfrac{-4}{a} \right)$, so that

$$\sum_{\substack{ax+by=4n+2 \\ ax \equiv 3 \,(\mathrm{mod}\ 4)}} \left(\frac{-4}{ab} \right) = 0.$$

Thus

$$\sum_{\substack{ax+by=4n+2 \\ ax\equiv 1\,(\mathrm{mod}\,4)}} \left(\frac{-4}{ab}\right) = \sigma(2n+1).$$

Finally we have

$$\begin{aligned}
\delta_4(n) &= \sum_{m=0}^{n} \delta_2(m)\delta_2(n-m) \\
&= \sum_{m=0}^{n} \sum_{a\,|\,4m+1} \left(\frac{-4}{a}\right) \sum_{b\,|\,4(n-m)+1} \left(\frac{-4}{b}\right) \\
&= \sum_{\substack{ax+by=4n+2 \\ ax\equiv 1\,(\mathrm{mod}\,4)}} \left(\frac{-4}{ab}\right) \\
&= \sigma(2n+1).
\end{aligned}$$

This is the asserted formula for $\delta_4(n)$.   □

Theorem 10 was known to Legendre [16].

Theorem 11 ([27], Theorem 4).

$$\delta_6(n) = -\frac{1}{8}\sum_{d\,|\,4n+3} \left(\frac{-4}{d}\right) d^2.$$

Proof. We have by (6.1) and Theorem 10

$$\begin{aligned}
\delta_6(n) &= \sum_{m=0}^{n} \delta_2(m)\delta_4(n-m) \\
&= \sum_{m=0}^{n} \sum_{a\,|\,4m+1} \left(\frac{-4}{a}\right) \sum_{b\,|\,2(n-m)+1} b \\
&= \sum_{ax+2by=4n+3} \left(\frac{-4}{a}\right) b,
\end{aligned}$$

as

$$\begin{aligned}
\sum_{\substack{ax+2by=4n+3 \\ b\equiv 0(\mathrm{mod}\,2)}} \left(\frac{-4}{a}\right) b &= 2 \sum_{ax+4by=4n+3} \left(\frac{-4}{a}\right) b \\
&= \sum_{ax+4by=4n+3} \left\{\left(\frac{-4}{a}\right) + \left(\frac{-4}{x}\right)\right\} b \\
&= 0
\end{aligned}$$

and similarly

$$\sum_{\substack{ax+2by=4n+3 \\ y\equiv 0 \,(\mathrm{mod}\ 2) \\ b\equiv 1 \,(\mathrm{mod}\ 2)}} \left(\frac{-4}{a}\right) b = 0.$$

Next, as

$$\left(\frac{-4}{a+2b}\right) = \left(\frac{-4}{a}\right), \text{ if } b \equiv 0 \ (\mathrm{mod}\ 2),$$

and

$$\left(\frac{-4}{a+2b}\right) = \left(\frac{-4}{a+2}\right) = -\left(\frac{-4}{a}\right), \text{ if } b \equiv 1 \ (\mathrm{mod}\ 2),$$

we have

$$\sum_{ax+2by=4n+3} \left\{ \left(\frac{-4}{a}\right) b + \left(\frac{-4}{a+2b}\right) b \right\}$$

$$= 2 \sum_{\substack{ax+2by=4n+3 \\ b\equiv 0 \,(\mathrm{mod}\ 2)}} \left(\frac{-4}{a}\right) b$$

$$= 0,$$

so that

$$\delta_6(n) = - \sum_{ax+2by=4n+3} \left(\frac{-4}{a+2b}\right) b.$$

Now choose

$$f(a,b,x,y) = \left(\frac{-4}{a}\right) b$$

in Theorem 1. Then

$$\begin{aligned} E &= \left(\frac{-4}{a}\right) b - \left(\frac{-4}{a}\right)(-b) + \left(\frac{-4}{a}\right)(a-b) - \left(\frac{-4}{a}\right)(a+b) \\ &\quad + \left(\frac{-4}{b-a}\right) b - \left(\frac{-4}{a+b}\right) b \\ &= \left\{ \left(\frac{-4}{b-a}\right) - \left(\frac{-4}{a+b}\right) \right\} b \end{aligned}$$

and

$$\sum_{ax+by=4n+3} \left\{ \left(\frac{-4}{b-a}\right) - \left(\frac{-4}{a+b}\right) \right\} b$$

$$= \sum_{\substack{ax+by=4n+3 \\ a+1\equiv b\equiv 0 \,(\mathrm{mod}\ 2)}} \left\{ \left(\frac{-4}{b-a}\right) - \left(\frac{-4}{a+b}\right) \right\} b$$

$$+ \sum_{\substack{ax+by=4n+3 \\ a\equiv b+1\equiv 0 \,(\mathrm{mod}\ 2)}} \left\{ \left(\frac{-4}{b-a}\right) - \left(\frac{-4}{a+b}\right) \right\} b$$

$$= 2 \sum_{ax+2by=4n+3} \left\{ \left(\frac{-4}{2b-a}\right) - \left(\frac{-4}{2b+a}\right) \right\} b$$

$$+ \sum_{2ax+by=4n+3} \left\{ \left(\frac{-4}{b-2a}\right) - \left(\frac{-4}{b+2a}\right) \right\} b$$

$$= -4 \sum_{ax+2by=4n+3} \left(\frac{-4}{a+2b}\right) b,$$

as

$$\left(\frac{-4}{2b-a}\right)\left(\frac{-4}{2b+a}\right) = \left(\frac{-4}{4b^2-a^2}\right) = \left(\frac{-4}{-a^2}\right) = \left(\frac{-4}{3}\right) = -1,$$

for a odd, and

$$\left(\frac{-4}{b-2a}\right)\left(\frac{-4}{b+2a}\right) = \left(\frac{-4}{b^2-4a^2}\right) = \left(\frac{-4}{b^2}\right) = 1,$$

for b odd. We have shown that

$$\sum_{ax+by=4n+3} E = 4\delta_6(n).$$

Then, by Theorem 1, we have

$$4\delta_6(n) = \sum_{d\,|\,4n+3} \sum_{x<d} \left\{ \left(\frac{-4}{(4n+3)/d}\right)\frac{4n+3}{d} - \left(\frac{-4}{x}\right)x - \left(\frac{-4}{d}\right)x \right\}$$

$$= A_1 - A_2 - A_3,$$

where

$$
\begin{aligned}
A_1 &= \sum_{d \,|\, 4n+3} \sum_{x < (4n+3)/d} \left(\frac{-4}{d}\right) d \\
&= \sum_{d \,|\, 4n+3} \left(\frac{-4}{d}\right) d \left\{ \frac{4n+3}{d} - 1 \right\} \\
&= (4n+3) \sum_{d \,|\, 4n+3} \left(\frac{-4}{d}\right) - \sum_{d \,|\, 4n+3} \left(\frac{-4}{d}\right) d \\
&= - \sum_{d \,|\, 4n+3} \left(\frac{-4}{d}\right) d; \\
A_2 &= \sum_{d \,|\, 4n+3} \sum_{x < d} \left(\frac{-4}{x}\right) x \\
&= - \sum_{d \,|\, 4n+3} \left(\frac{-4}{d}\right) \frac{d-1}{2} \\
&= -\frac{1}{2} \sum_{d \,|\, 4n+3} \left(\frac{-4}{d}\right) d; \\
A_3 &= \sum_{d \,|\, 4n+3} \left(\frac{-4}{d}\right) \sum_{x < d} x \\
&= \sum_{d \,|\, 4n+3} \left(\frac{-4}{d}\right) \frac{(d-1)d}{2} \\
&= \frac{1}{2} \sum_{d \,|\, 4n+3} \left(\frac{-4}{d}\right) d^2 - \frac{1}{2} \sum_{d \,|\, 4n+3} \left(\frac{-4}{d}\right) d,
\end{aligned}
$$

so that

$$
4\delta_6(n) = -\frac{1}{2} \sum_{d \,|\, 4n+3} \left(\frac{-4}{d}\right) d^2,
$$

which gives the asserted formula. □

Theorem 12 ([27], Theorem 5).

$$
\delta_8(n) = \sigma_3(n+1) - \sigma_3((n+1)/2).
$$

Proof. We have appealing to Theorem 10 and Corollary 1

$$
\begin{aligned}
\delta_8(n) &= \sum_{m=0}^{n} \delta_4(m)\delta_4(n-m) \\
&= \sum_{m=0}^{n} \sigma(2m+1)\sigma(2n-2m+1) \\
&= \sum_{m=1}^{n+1} \sigma(2m-1)\sigma((2n+2)-(2m-1)) \\
&= \frac{1}{24}(5\sigma_3(2n+2) - 21\sigma_3(n+1) + 16\sigma_3((n+1)/2) \\
&\quad -(11+12n)(\sigma(2n+2) - 3\sigma(n+1) + 2\sigma((n+1)/2))) \\
&= \sigma_3(n+1) - \sigma_3((n+1)/2),
\end{aligned}
$$

as

$$
\sigma_3(2n+2) = 9\sigma_3(n+1) - 8\sigma_3((n+1)/2)
$$

and

$$
\sigma(2n+2) = 3\sigma(n+1) - 2\sigma((n+1)/2).
$$

This completes the proof of Theorem 12.   □

7 Application of Theorem 1 to the Representations of a Positive Integer by Certain Quaternary Forms

Theorem 1 can be used to determine in an elementary manner the classical formulae for the number of representations of a positive integer n by certain quaternary forms such as $x^2 + xy + y^2 + u^2 + uv + v^2$, $x^2 + y^2 + u^2 + v^2$, $x^2 + 2y^2 + u^2 + 2v^2$, etc. We just give one example to illustrate the ideas.

Theorem 13. *The number of representations of a positive integer n by the quaternary form $x^2 + xy + y^2 + u^2 + uv + v^2$ is $12\sigma(n) - 36\sigma(n/3)$.*

Proof. Let k be a nonnegative integer. We set

$$
r(k) = \mathrm{card}\{(x,y) \in \mathbb{Z}^2 \mid k = x^2 + xy + y^2\}
$$

and

$$
R(k) = \mathrm{card}\{(x,y,u,v) \in \mathbb{Z}^4 \mid k = x^2 + xy + y^2 + u^2 + uv + v^2\}.
$$

Let n be a positive integer. Clearly

$$
R(n) = \sum_{k=0}^{n} r(k)r(n-k).
$$

so that as $r(0) = 1$ we have

$$R(n) - 2r(n) = \sum_{k=1}^{n-1} r(k)r(n-k).\tag{7.1}$$

It is a classical result that

$$r(n) = 6\sum_{d|n}\left(\frac{-3}{d}\right) = 6\tau_{1,3}(n) - 6\tau_{2,3}(n),\tag{7.2}$$

where

$$\tau_{i,3}(n) = \sum_{\substack{d|n \\ d\equiv i\,(\text{mod }3)}} 1, \quad i = 0,1,2.$$

Since $\tau_{0,3}(n) = \tau(n/3)$ and $\tau_{0,3}(n) + \tau_{1,3}(n) + \tau_{2,3}(n) = \tau(n)$ we have

$$\begin{aligned}
\tau_{0,3}(n) &= \tau(n/3),\\
\tau_{1,3}(n) &= \frac{1}{2}\tau(n) - \frac{1}{2}\tau(n/3) + \frac{1}{12}r(n),\\
\tau_{2,3}(n) &= \frac{1}{2}\tau(n) - \frac{1}{2}\tau(n/3) - \frac{1}{12}r(n).
\end{aligned}\tag{7.3}$$

From (7.1) and (7.2) we obtain

$$R(n) - 2r(n) = 36\sum_{k=1}^{n-1}\sum_{a|k}\left(\frac{-3}{a}\right)\sum_{b|n-k}\left(\frac{-3}{b}\right) = 36\sum_{ax+by=n}\left(\frac{-3}{ab}\right)$$

so that

$$\sum_{ax+by=n}\left(\frac{-3}{ab}\right) = \frac{1}{36}R(n) - \frac{1}{18}r(n).\tag{7.4}$$

We now choose

$$f(a,b,x,y) = \left(\frac{-3}{ab}\right).$$

Clearly this choice of f satisfies (2.1) so we may apply Theorem 1. We obtain after a little simplification

$$\sum_{ax+by=n}\left\{2\left(\frac{-3}{ab}\right) + 2\left(\frac{-3}{a(a-b)}\right) - 2\left(\frac{-3}{a(a+b)}\right)\right\}$$
$$= \sum_{d|n\,x<d}\sum\left\{\left(\frac{-3}{(n/d)^2}\right) - \left(\frac{-3}{x(x-d)}\right) - 2\left(\frac{-3}{dx}\right)\right\}.\tag{7.5}$$

A simple examination of the possible residues of a and b modulo 3 shows that

$$\left(\frac{-3}{a(a-b)}\right) - \left(\frac{-3}{a(a+b)}\right) = \left(\frac{-3}{ab}\right)$$

so that the left hand side of (7.5) is

$$4 \sum_{ax+by=n} \left(\frac{-3}{ab}\right) = \frac{1}{9}R(n) - \frac{2}{9}r(n)$$

by (7.4). Next we determine the sums on the right hand side of (7.5). First

$$
\begin{aligned}
\sum_{d\,|\,n}\sum_{x<d} \left(\frac{-3}{(n/d)^2}\right) &= \sum_{d\,|\,n}\sum_{x<n/d} \left(\frac{-3}{d^2}\right) \\
&= \sum_{d\,|\,n,\,3\nmid d}\sum_{x<n/d} 1 \\
&= \sum_{d\,|\,n,\,3\nmid d} \left(\frac{n}{d} - 1\right) \\
&= (\sigma(n) - \tau(n)) - (\sigma(n/3) - \tau(n/3)).
\end{aligned}
$$

Secondly

$$
\begin{aligned}
-\sum_{d\,|\,n}\sum_{x<d} \left(\frac{-3}{x(x-d)}\right) \\
= \sum_{\substack{d\,|\,n\\3\nmid x}}\sum_{x<d} \left(\frac{-3}{x(d-x)}\right) \\
= \sum_{\substack{d\,|\,n\\3\nmid x}}\sum_{x<d} \left(\frac{-3}{xd-1}\right) \\
= \sum_{d\,|\,n} \left(\frac{-3}{d-1}\right) \sum_{\substack{x<d\\x\equiv 1\,(\mathrm{mod}\,3)}} 1 + \sum_{d\,|\,n} \left(\frac{-3}{2d-1}\right) \sum_{\substack{x<d\\x\equiv 2\,(\mathrm{mod}\,3)}} 1 \\
= \sum_{d\,|\,n} \left(\frac{-3}{d-1}\right) \left[\frac{d+1}{3}\right] + \sum_{d\,|\,n} \left(\frac{-3}{2d-1}\right) \left[\frac{d}{3}\right] \\
= \frac{1}{3}\sigma(n) - 3\sigma(n/3) - \frac{1}{18}r(n),
\end{aligned}
$$

by (7.2). Thirdly

$$-2\sum_{d\,|\,n}\sum_{x<d}\left(\frac{-3}{dx}\right) = -2\sum_{d\,|\,n}\left(\frac{-3}{d}\right)\left(\left[\frac{d+1}{3}\right]-\left[\frac{d}{3}\right]\right)$$

$$= 2\tau_{2,3}(n)$$

$$= \tau(n)-\tau(n/3)-\frac{1}{6}r(n),$$

by (7.3). Thus the right hand side of (7.5) is

$$\frac{4}{3}\sigma(n)-4\sigma(n/3)-\frac{2}{9}r(n).$$

Hence

$$\frac{1}{9}R(n)-\frac{2}{9}r(n)=\frac{4}{3}\sigma(n)-4\sigma(n/3)-\frac{2}{9}r(n)$$

so that

$$R(n)=12\sigma(n)-36\sigma(n/3)$$

as asserted.   □

Theorem 13 can be found for example in [22].

8 Further Convolution Sums

We conclude this paper by considering the sums

$$R = \sum_{m<n/3}\sigma(n-3m)\sigma_3(m),$$

$$S = \sum_{m<n/3}\sigma(m)\sigma_3(n-3m),$$

$$A = \sum_{m<n/2}\sigma(m)\sigma_5(n-2m),$$

$$B = \sum_{m<n/2}\sigma_3(m)\sigma_3(n-2m),$$

$$C = \sum_{m<n/2}\sigma_5(m)\sigma(n-2m).$$

Although we cannot evaluate any of R, S, A, B, C individually, Theorem 1 enables us to determine certain linear combinations of them.

Theorem 14.

$$3R + S = \frac{1}{240}(3\sigma_5(n) + (10 - 10n)\sigma_3(n) - 3\sigma(n)$$
$$+ 81\sigma_5(n/3) + (30 - 90n)\sigma_3(n/3) - \sigma(n/3)).$$

Proof. We set

$$G(x, y) = \begin{cases} 12, & \text{if } x \equiv y \equiv 0 \ (\text{mod } 3), \\ 9, & \text{if } x \equiv y \not\equiv 0 \ (\text{mod } 3), \\ 1, & \text{if } x \not\equiv y \ (\text{mod } 3), \end{cases}$$

and choose

$$f(a, b, x, y) = a^3 b \, G(x, y).$$

Clearly (2.3) is satisfied and a straightforward calculation shows that

$$E = \begin{cases} -24a^3b - 72a^3b, & \text{if } x \equiv y \equiv 0 \ (\text{mod } 3), \\ -18a^3b - 6ab^3, & \text{if } x \equiv 0 \ (\text{mod } 3), y \not\equiv 0 \ (\text{mod } 3), \\ -18a^3b - 54ab^3, & \text{if } x \not\equiv 0 \ (\text{mod } 3), y \equiv 0 \ (\text{mod } 3), \\ 6a^3b - 6ab^3, & \text{if } x \not\equiv 0 \ (\text{mod } 3), y \not\equiv 0 \ (\text{mod } 3). \end{cases}$$

Firstly we have

$$\sum_{\substack{ax+by=n \\ x \equiv y \equiv 0 \, (\text{mod } 3)}} (-24a^3b - 72ab^3) = \sum_{ax+by=n/3} (-24a^3b - 72ab^3)$$

$$= -96 \sum_{ax+by=n/3} ab^3$$

$$= -96 \sum_{m<n/3} \sigma(m)\sigma_3(n/3 - m).$$

Secondly

$$\sum_{\substack{ax+by=n \\ x \equiv 0 \, (\text{mod } 3) \\ y \not\equiv 0 \, (\text{mod } 3)}} (-18a^3b - 6ab^3)$$

$$= \sum_{3ax+by=n} (-18a^3b - 6ab^3) - \sum_{ax+by=n/3} (-18a^3b - 6ab^3)$$

$$= -18R - 6S + 24 \sum_{m<n/3} \sigma(m)\sigma_3(n/3 - m).$$

Thirdly

$$\sum_{\substack{ax+by=n \\ x\not\equiv0\,(\mathrm{mod}\,3) \\ y\equiv0\,(\mathrm{mod}\,3)}} (-18a^3b - 54ab^3)$$

$$= \sum_{\substack{ax+by=n \\ x\equiv0\,(\mathrm{mod}\,3) \\ y\not\equiv0\,(\mathrm{mod}\,3)}} (-18ab^3 - 54a^3b)$$

$$= -54R - 18S + 72 \sum_{m<n/3} \sigma(m)\sigma_3(n/3 - m)$$

as above. Fourthly the change of variables $(a, b, x, y) \longrightarrow (b, a, y, x)$ shows that

$$\sum_{\substack{ax+by=n \\ x,y\not\equiv0\,(\mathrm{mod}\,3)}} (6a^3b - 6ab^3) = \sum_{\substack{ax+by=n \\ x,y\not\equiv0\,(\mathrm{mod}\,3)}} (6ab^3 - 6a^3b)$$

so that

$$\sum_{\substack{ax+by=n \\ x,y\not\equiv0\,(\mathrm{mod}\,3)}} (6a^3b - 6ab^3) = 0.$$

Set

$$T := \sum_{m<n/3} \sigma(m)\sigma_3(n/3 - m).$$

Hence the left hand side of (2.2) is

$$(-96T - 18R - 6S) + (24T - 54R - 18S) + 72T = -72R - 24S.$$

The right hand side of (2.2) is (after a long straightforward calculation)

$$-\frac{1}{10}(3\sigma_5(n) + (10 - 10n)\sigma_3(n) - 3\sigma(n) + 81\sigma_5(n/3)$$

$$+(30 - 90n)\sigma_3(n/3) - \sigma(n/3)).$$

Theorem 14 follows by equating both sides of (2.2). $\qquad\square$

Theorem 15.

$$3A + 8B = \frac{1}{840}(28\sigma_7(n) + (105 - 105n)\sigma_5(n) - 28\sigma_3(n) + 128\sigma_7(n/2)$$
$$-28\sigma_3(n/2) + 5\sigma(n/2)),$$

$$2B + 3C = \frac{1}{840}(2\sigma_7(n) - 7\sigma_3(n) + 5\sigma(n) + 112\sigma_7(n/2)$$
$$+(105 - 210n)\sigma_5(n/2) - 7\sigma_3(n/2)).$$

Proof. The choice

$$f(a, b, x, y) = (-ab^5 + 10a^3b^3 - 12a^4b^2)F_2(x)$$

in Theorem 1 yields $36A + 96B$ and the choice

$$f(a, b, x, y) = (ab^5 - 10a^3b^3 + 12a^4b^2 - 36a^6)F_2(x)$$

yields $24B + 36C$. The details are left to the reader.   □

9   Conclusion

It is very likely that there are other choices of $f(a, b, x, y)$ for which Theorem 1 will yield new arithmetic identities. Moreover Theorem 1 itself may be capable of being generalized. It would also be interesting to know if the sums R, S, A, B, C of Section 8 can be determined individually, perhaps in terms of Ramanujan's tau function, and also whether $\delta_{16}(n)$ can be evaluated using Theorem 1.

The authors would like to thank Elizabeth S. Morcos who did some numerical calculations for them in connection with this research. They would also like to thank an unknown referee who drew their attention to the work of MacMahon [24], [25].

References

[1]   R. P. Bambah and S. Chowla, *The residue of Ramanujan's function* $\tau(n)$ *to the modulus* 2^8, J. London Math. Soc. **22** (1947), 140–147.

[2]   B. C. Berndt, *Ramanujan's Notebooks, Part II*, Springer-Verlag, New York, 1989.

[3]   B. C. Berndt and R. J. Evans, *Chapter 15 of Ramanujan's Second Notebook: Part 2, Modular forms*, Acta Arith. **47** (1986), 123–142.

[4]   M. Besge, *Extrait d'une lettre de M. Besge à M. Liouville*, J. Math. Pures Appl. **7** (1862), 256.

[5]   S. Chowla, *Note on a certain arithmetical sum*, Proc. Nat. Inst. Sci. India **13** (1947), 233.

[6]   ———, *Collected Papers, Vol. II (1936-1961)*, Centre de Recherches Math., Montréal, Canada, 1999.

[7]   L. E. Dickson, *History of the Theory of Numbers, Vol. I*, Chelsea Publ. Co., New York, 1952.

[8] _____, *History of the Theory of Numbers, Vol. II*, Chelsea Publ. Co., New York, 1952.

[9] J. W. L. Glaisher, *On the square of the series in which the coefficients are the sums of the divisors of the exponents*, Mess. Math. **14** (1884), 156–163.

[10] _____, *On certain sums of products of quantities depending upon the divisors of a number*, Mess. Math. **15** (1885), 1–20.

[11] _____, *Expressions for the first five powers of the series in which the coefficients are the sums of the divisors of the exponents*, Mess. Math. **15** (1885), 33–36.

[12] C. C. Grosjean, *An infinite set of recurrence formulae for the divisor sums, Part I*, Bull. Soc. Math. Belgique **29** (1977), 3–49.

[13] _____, *An infinite set of recurrence formulae for the divisor sums, Part II*, Bull. Soc. Math. Belgique **29** (1977), 95–138.

[14] D. B. Lahiri, *On Ramanujan's function $\tau(n)$ and the divisor function $\sigma(n)$, I*, Bull. Calcutta Math. Soc. **38** (1946), 193–206.

[15] _____, *On Ramanujan's function $\tau(n)$ and the divisor function σ_k, II*, Bull. Calcutta Math. Soc. **39** (1947), 33–52.

[16] A. M. Legendre, *Traité des fonctions elliptiques et des intégrales Euleriennes, Vol. III*, Huzard-Courcier, Paris, 1828.

[17] D. H. Lehmer, *Some functions of Ramanujan*, Math. Student **27** (1959), 105–116.

[18] _____, *Selected Papers, Vol. II*, Charles Babbage Research Centre, St. Pierre, Manitoba, 1981.

[19] J. Levitt, *On a problem of Ramanujan*, M.Phil. thesis, University of Nottingham, 1978.

[20] J. Liouville, *Sur quelques formules générales qui peuvent être utiles dans la théorie des nombres*, J. Math. Pures Appl. **3** (1858), 143–152.

[21] _____, *Sur quelques formules générales qui peuvent être utiles dans la théorie des nombres (fifth article)*, J. Math. Pures Appl. **3** (1858), 273–288.

[22] G. A. Lomadze, *Representation of numbers by sums of the quadratic forms $x_1^2 + x_1 x_2 + x_2^2$ (in Russian)*, Acta Arith. **54** (1989), 9–36.

[23] J. Lützen, *Joseph Liouville 1809-1882: Master of Pure and Applied Mathematics*, Springer-Verlag, Berlin/Heidelberg, 1998.

[24] P. A. MacMahon, *Divisors of numbers and their continuations in the theory of partitions*, Proc. London Math. Soc. (2) **19** (1920), 75–113.

[25] _____, *Collected Papers, Vol. II, Number Theory, Invariants, and Applications*, MIT Press, Cambridge, MA, 1986.

[26] G. Melfi, *On some modular identities*, Number Theory (K. Györy, A. Pethö, and V. Sós, eds.), de Gruyter, Berlin, 1998, pp. 371–382.

[27] K. Ono, S. Robins, and P. T. Wahl, *On the representation of integers as sums of triangular numbers*, Aequationes Math. **50** (1995), 73–94.

[28] B. van der Pol, *On a non-linear partial differential equation satisfied by the logarithm of the Jacobian theta-functions, with arithmetical applications. II*, Indag. Math. **13** (1951), 272–284.

[29] V. Ramamani, *On some identities conjectured by Srinivasa Ramanujan found in his lithographed notes connected with partition theory and elliptic modular functions*, Ph.D. thesis, University of Mysore, 1970.

[30] S. Ramanujan, *On certain arithmetical functions*, Trans. Cambridge Philos. Soc. **22** (1916), 159–184.

[31] _____, *Collected Papers*, AMS Chelsea Publishing, Providence, RI, 2000.

[32] R. A. Rankin, *Elementary proofs of relations between Eisenstein series*, Proc. Roy. Soc. Edinburgh **76A** (1976), 107–117.

[33] N.-P. Skoruppa, *A quick combinatorial proof of Eisenstein series identities*, J. Number Theory **43** (1993), 68–73.

Integer Points, Exponential Sums and the Riemann Zeta Function

M. N. Huxley

1 Lattice Points

One of the problems at the origin of mathematics is to define and calculate the area of a plane region D. If parts of the boundary of D are curved, then this problem leads to the integral calculus. Archimedes' construction used two polygons, one inside D, the other containing D. After Descartes' coordinate system, it is natural to consider a zigzag polygon whose vertices are of the form $x = m\delta$, $y = n\delta$, where δ is the size of the smallest square of the printed grid on the graph paper. To find the area of the inner and outer polygons, one counts squares of side δ. This gives upper and lower bounds for the area. A better estimate is given by counting the square bounded by the four points $x = m\delta$, $(m + 1)\delta$, $y = n\delta$, $(n + 1)\delta$ if and only if the point $(m\delta, n\delta)$ lies in D, so the integer lattice point (m, n) lies in $\delta^{-1}D$.

This method links geometry, analysis and number theory. Voronoi saw that an error estimate can be found by approximating the curve by a polygon whose gradients are rational numbers a/q of small height (defined by $h(a/q) = \log \max(|a|, q)$ when $q \geq 1$, $(a, q) = 1$). Voronoi [33] and Sierpiński [31] used these approximating polygons for the divisor and circle problems, respectively, and van der Corput extended the method to a general smooth convex region in his thesis [5]. The error has a contribution from each gradient a/q, which depends on the continued fraction for a/q.

Pfeiffer considered lattice point problems by Fourier methods in the nineteenth century (see van der Corput [5] and Landau [28], [29]). Voronoi [34] gave formal Fourier expansions for the divisor and circle problems. The general convex region had to wait for Kendall [24], followed by Hlawka [14] for a general convex body in higher dimensions. They evaluated the terms of the series to a good approximation; the approximate values correspond to integer vectors normal to the curve or surface, with a small correction from the rest of the curve. These expansions are useful for mean square and omega results, as we heard in Tsang's lecture at this conference.

Van der Corput's less symmetric treatment [6], [7], [8], [9] was more successful. The discrepancy (number of lattice points minus the area of $\delta^{-1}D$) is calculated piecewise on arcs with local coordinates $y = g(x)$. The discrepancy contribution is $\sum \rho(g(m))$, where

$$\rho(t) = [t] - t + \frac{1}{2} \simeq \sum_{h \neq 0} \frac{e(ht)}{2\pi i h}$$

is the usual row-of-teeth function. This leads to the van der Corput sums $\sum e(F(m))$, which he estimated by an iterative method that was purely analytic, with no further input from number theory. There are parameters M, T with

$$m \asymp M, \quad |F^{(r)}(x)| \asymp T/M^r.$$

(The order of magnitude notation used in this paper is '$f = O(g)$' or '$f \ll g$' for '$g > 0$, $|f|/g$ is bounded', '$\asymp g$' for '$f > 0$, $g > 0$, f/g, g/f are both bounded'.)

Step A, Poisson summation, replaces $F(x)$ by its complementary function $G(y)$, for which $F'(x)$, $G'(y)$ are inverse functions, and replaces the size parameter M by T/M. Step B, Weyl differencing, is a mean-to-max step, estimating the maximum of one sum in terms of the mean square of another sum. It introduces a new integer summand $h \ll H$, replaces $F(x)$ by $F(x + h) - F(x) \simeq hF'(x)$, and replaces the parameter T by hT/M. These steps are used in succession to reduce the parameters M and T until the resulting sums are short enough to estimate trivially. Each B step introduces extra integer variables. The multidimensional van der Corput iteration (see Graham and Kolesnik [12]) also uses A and B steps with respect to these new variables. However the errors from each step must be summed trivially over the other variables. This limits the power of the method.

2 Spacing Problems

Although van der Corput's mean-to-max step is purely analytic, other mean-to-max lemmas are useful in analytic number theory. They often lead to a spacing problem, to show that the sums whose mean square is estimated are sufficiently distinct. The sums are labelled by an integer parameter, n say, and we require different values of n to give sums which are non-overlapping or different in some sense. Often, as in Burgess' estimates for character sums [4], some overlapping does occur, and we must count the number of pairs of integers m, n for which the sums overlap. In van der Corput's Step A there is trouble only if $h = m - n$ is zero, giving trivial terms in the upper bound. A useful idea for spacing problems is to show that if the sums labelled by m and n overlap, then there are two integers a and b, constructed from m and n, which satisfy an equation $b = h(a)$ approximately, so that the integer point (a, b) is close to the curve $y = h(x)$. The spacing problem is translated into counting the integer points (a, b) in a certain plane region bounded above and below by arcs of the curves $y = h(x) \pm \delta$, where δ (which can be zero) measures the 'closeness'. If the region for (a, b) is bounded, then there is a nice geometric idea: given $R \geq d + 1$ integer points in d-dimensional space, then either they lie on a $(d-1)$-dimensional hyperplane, or their convex hull has volume at least $(R - d)/d!$.

Estimating the gap between square-free numbers is rapidly reduced to a spacing problem of points-close-to-a-curve type. If no number in $N \leq n \leq N + H$ is square-free, then many of them must be divisible by the square of some large prime. We need a non-trivial bound for the number of solutions of $N \leq p^2 q \leq N + H$, where p runs through large primes. The integer point (p, q) lies close to the curve $x^2 y = N$.

The expected number of lattice points (m, n) with m in an interval of length M, $|n - f(m)| \leq \delta$, is $2\delta M$ (with a possible endpoint correction). For δ large the Fourier method using the series for $\rho(t)$, cut off at $H \asymp 1/\delta$, gives $2\delta M$ plus an error term. The error term increases with H, so there is an optimal value $\delta = \delta_0$ at which both terms are equal. For $\delta < \delta_0$ it is better to count solutions of the weaker inequality $|n - f(m)| \leq \delta_0$. This is partly explained by the possibility of solutions with $\delta = 0$. The curve $y = \sqrt{x}$, $M \leq m \leq 2M$, is suitable for van der Corput's method, but goes through about $(\sqrt{2} - 1)\sqrt{M}$ integer points, so we cannot expect an estimate that tends to zero as $\delta \to 0$.

There is an elementary method that gives upper bounds, not asymptotic formulae, but works better when δ is small. It is built around a determinant formed from the coordinates of r points on a curve $y = f(x)$, which is equal to the Vandermonde determinant of the n values of x, multiplied by $f^{(r-1)}(\xi)/(r-1)!$. For $r \geq 2$ the derivative is small, but the determinant is approximately an integer when the n points are approximately integer points. In the simplest case $n = 3$, the determinant formed from three integer points P, Q and R is twice the area of the triangle PQR. We call P, Q, R a major triplet if the determinant is zero, so PQR is a straight line, and a minor triplet otherwise. A major arc is a region of the curve where all the integer points close to it lie on the same straight line. Major arcs cannot exist if δ is very small. A minor arc is a region where no three consecutive integer points close to the curve lie on a straight line. The Vandermonde determinant is bounded below, so there cannot be three integer points very close together on a minor arc. Major arcs have rational gradients of small height, and they can be enumerated (Branton and Sargos [3]).

Swinnerton-Dyer's refinement [32] (for the case $\delta = 0$, but the method can be made to work for $\delta \neq 0$, as in Huxley [17]), uses a 4×4 determinant and ingenious divisibility arguments to show that the 3×3 determinants are usually greater than one. The argument is a mean-to-max, considering the mean fourth power of the number of points close to a curve in a short interval.

Bombieri and Pila [2] use high order determinants and intersection theory to give good estimates for the number of points on a curve ($\delta = 0$). The upper bounds are now ineffective and non-uniform, unless the curve is algebraic. The method also works for δ extremely close to 0.

Points-close-to-curve problems often have $m \asymp M$, $n \asymp N$, where N is larger than M. There is an iteration analogous to that of van der Corput, with $T = MN$. Step A, interchanging m and n, replaces $f(x)$ by its inverse function, and replaces M by T/M. Step B, differencing, in its simplest form, introduces a new integer variable $h \leq H$, replaces $f(x)$ by $f(x+h) - f(x) \simeq hf'(x)$, and T by hT/M. In general Step B has r integers $h_1, \ldots, h_r$. We select the integer points (m, n) for which the r points $(m + h_i, n_i)$ are also integer points close to the curve for some $n_1, \ldots, n_r$, and use the $(r + 1) \times (r + 1)$ determinant to construct an r-th differenced function $f[x; h_1, \ldots, h_r]$. A related method (see Filaseta and Trifonov [10]) is to write down the $(r + 1) \times (r + 1)$ determinant of $r + 1$ consecutive integer points, and use it to define major and minor arcs analogously. The major arcs correspond to polynomials in $Q[x]$ of degree at most $r - 1$.

More complicated spacing problems give points close to a curve $y = f(x)$ where one or both of x and y is a rational number of small height, not necessarily an integer (Huxley [15], [18], [20]). The determinant method can be applied, and the major arcs correspond to rational functions in the latter two papers. Konyagin [25] has an interesting alternative approach in which the approximation on a major arc is found by Dirichlet's pigeon-hole principle, not constructed explicitly. No Swinnerton-Dyer refinement has been accomplished in these problems.

3 The Bombieri-Iwaniec-Mozzochi Method

This method was introduced by Bombieri and Iwaniec [1] for the van der Corput sum $\sum e(F(m))$, and adapted by Iwaniec and Mozzochi [23] for the lattice point sum $\sum \rho(g(m))$ by way of the double exponential sum $\sum \sum e(hg(m))$ over h and m, and by Heath-Brown and Huxley [13] for the double sum

$$\sum \sum e(F(m + h) - F(m - h)), \qquad (3.1)$$

which is related to the short interval mean square of a van der Corput sum. The method is most easily motivated for the lattice point case, with a domain D bounded by a smooth curve $y = g(x)$. The size parameters are H, the cut-off in the Fourier series for $\rho(t)$, M, the range for x, and T, with the dimensions of area, with $y \ll T/M$. They correspond to the T and M in the discussion above of the van der Corput exponential sum. The curve is divided up into arcs, the Farey arcs, by the vertices of a Voronoi polygon, labelled by the rational gradient a/q of the polygon side. The discrepancy contribution of each Farey arc is expressed as an exponential sum as in van der Corput, which is transformed by Poisson summation in both h and m to get new integer variables k and ℓ. If the Farey arcs correspond to intervals in x of length $\asymp N$, then the lengths K and L of the ranges for k and ℓ are proportional to the denominator q. When a/q has small height, the major arc case, then we get an estimate at once. Other Farey arcs are minor arcs. The denominator q has normal order R, where

$$NR^2 \asymp M^3/T.$$

Larger denominators occur for Farey arcs close to major arcs of small height. In this notation the ranges for k and ℓ have sizes

$$K \asymp Nq/R^2, \quad L \asymp Hq/R^2. \tag{3.2}$$

The estimate for major arcs is so good that the major arcs can be taken longer than N. Their contribution becomes larger, but then there are no Farey arcs with q very large. We can also arrange that $q \ll R^2/H$ on major arcs, $q \gg R$ on minor arcs. If the major arcs are so long that there are no minor arcs ($R \asymp H$), then we recover the simplest van der Corput bound for the discrepancy, $O(T^{1/3+\epsilon})$.

The savings come from a mean-to-max argument on the minor arcs, using the large sieve for the exponential sum bilinear form

$$E = \sum\sum a(k)b(j)e\left(\mathbf{x}^{(k)} \cdot \mathbf{y}^{(j)}\right), \tag{3.3}$$

which comes from the fourth power of the transformed exponential sum over k and ℓ, so there is a formal analogy with the Swinnerton-Dyer refinement for integer points close to curves. There are ingenious technical devices to remove smaller order terms, and to make the ranges for k and ℓ the same for all Farey arcs with q in a range $Q \leq q \leq 2Q$, 'but that's not important right now'. The multi-index $\mathbf{k} = (k_1, \ell_1, k_2, \ell_2)$ consists of the summands k and ℓ from two factors. The vectors $\mathbf{x}$ and $\mathbf{y}$ have four entries, from the four leading terms in the exponent, with

$$\mathbf{x}^{(\mathbf{k})} = \left(\ell_1 k_1 - \ell_2 k_2, \ell_1\sqrt{k_1} - \ell_2\sqrt{k_2}, \ell_1 - \ell_2, \ell_1/\sqrt{k_1} - \ell_2/\sqrt{k_2}\right).$$

The index j runs through the minor arcs, and the $\mathbf{y}$ vectors are constructed from the approximation to $g(x)$ on the minor arc. The $\mathbf{x}$ vectors in Heath-Brown and Huxley [13] are similar but more complicated. For the single exponential sum the $\mathbf{x}$ vectors are

$$\left(\sum k_i^2, \sum k_i^{3/2}, \sum k_i, \sum k_i^{1/2} \right). \tag{3.4}$$

The $\mathbf{y}$ vectors are essentially the same in all three cases, with $F'(x)$ replacing $g(x)$. The Large Sieve inequality (Huxley [16, Lemma 8.4.1]) says

$$|E|^2 = O\left(\frac{ABKL^4NR^2V}{Q^3} \right) = O\left(\frac{ABH^4N^2Q^2V}{R^8} \right),$$

where $V \geq 1$ is a free parameter, and A is the sum of $|a(\mathbf{k})a(\mathbf{k}')|$ taken over a 'neighbourhood of the diagonal', all pairs of vectors $\mathbf{x}^{(\mathbf{k})}$, $\mathbf{x}^{(\mathbf{k}')}$ whose difference lies in a box of size determined by the ranges for the components of the $\mathbf{y}$ vectors. The factor B is analogous, but the size of the box depends on V as well as on the ranges for the $\mathbf{x}$ vectors.

The coefficients $a(\mathbf{k})$ are bounded, so we estimate A by counting pairs of vectors in a neighbourhood of the diagonal. The trivial solutions have $\mathbf{k}'$ obtained from $\mathbf{k}$ by permuting the entries, and the First Spacing Problem is to show that most solutions of the inequalities are trivial. This is what Heath-Brown calls a paucity problem. Theorem 13.2.4 of Huxley [16] gives

$$A = O\left(KL^3 + K^2L^2 \log K + \eta K^3 L^2 \log K \right), \tag{3.5}$$

where $\eta \asymp R^2/HN$, so $\eta K \asymp Q/H$, a satisfactory result. For the double sum (3.1) Lemma 13.3.1 on the next page of Huxley [16] gives

$$A = O\left(KL^4 + K^2L^2 \log K + \eta K^3 L^2 \log K \right), \tag{3.6}$$

where the first term is bigger than we want. For the single exponential sum we can vary the number of summands $i = 1, \ldots, r$ in (3.4). We indicate this as $A(r)$. For reasonable Q we have

$$A(3) = O(K^3 \log K),$$
$$A(4) = O(K^4 + \eta K^5 \log^3 K),$$
$$A(5) = O(\eta K^7);$$

see Huxley [16, Lemma 17.1.1] and Watt [35]. Here $\eta \asymp R^2/N^2$, so $\eta K \asymp Q/N$. We conjecture that $A(5) = O(K^5)$ and $A(6) = O(K^6)$ for relevant values of η, implying better bounds for the exponential sum.

4   The Second Spacing Problem

The **y** vectors are indexed by minor arcs of the boundary curve of the domain D. The difference of two **y** vectors is close to the diagonal when the two Farey arcs can be superposed (to the required accuracy) by an affine map

$$(x'y') = \begin{pmatrix} a & b \\ c & d \end{pmatrix} \begin{pmatrix} x \\ y \end{pmatrix} + \begin{pmatrix} e \\ f \end{pmatrix} \qquad (4.1)$$

with a, b, c, d, e, f integers, $ad - bc = 1$. If we fix the 'magic matrix' $\begin{pmatrix} a & b \\ c & d \end{pmatrix}$, then such coincidences (which may persist for a block of consecutive Farey arcs) correspond to integer points close to some curve in a dual space, called the 'resonance curve' in Huxley [16]. There is a choice of magnification in defining the resonance curve, given by a matrix $\begin{pmatrix} A & B \\ C & D \end{pmatrix}$ acting on the gradients a/q, with A, B, C, D integers, $AD - BC = 1$. A beautiful result is that the resonance curve is magnified by an affine map

$$(x' \ \ y') = (x \ \ y) \begin{pmatrix} A & B \\ C & D \end{pmatrix} + (\xi \ \ \eta),$$

where ξ and η are very close to integers. Professor Coates approves this appearance of functoriality in analytic number theory. This is quite hard to prove; approximation theory wants to be linear, but the Poisson summation makes everything non-linear.

The estimates in the Second Spacing Problem are more complicated. There are $O(M/N)$ minor arcs, of which $O(MR^2/NQ^2)$ have q in a range $Q \leq q \leq 2Q$. The coincidence of two four-dimensional **y** vectors is measured by four numbers

$$\Delta_1 \asymp \frac{R^4}{HNQ^2V}, \qquad \Delta_2 \asymp \frac{R^2}{HN}, \qquad \Delta_3 \asymp \frac{R^2}{HQ}, \qquad \Delta_4 \asymp \frac{Q}{H},$$

and the probabilistic estimate for the number of consecutive pairs is

$$B = O\left(\Delta_1 \Delta_2 \Delta_3 \Delta_4 \frac{M^2 Q^2}{N^2 R^2}\right).$$

The parameter Δ_1 bounds the entries of the magic matrix in (4.1) by $c = O(\Delta_1 Q^2)$. Taking V larger makes the resonance curves longer. The parameter Δ_2 measures the distortion in area under the affine map (4.1), and Δ_3 and Δ_4 measure the shift in position.

The Second Spacing Problem is not a paucity problem: for good choices of the parameters N, R and V, the non-trivial (off-diagonal) coincidences dominate. Without using resonance curves, Bombieri and Iwaniec [1] got

$$B = O\left(\Delta_1(\Delta_1 + \Delta_2)\frac{M^2 Q^2}{N^2 R^2}\right) \qquad (4.2)$$

for the special case when $g(x) = C/x$ with C constant. Huxley and Watt [22] generalised (4.2). It would be useful to replace the factor $\Delta_1 + \Delta_2$ by Δ_2.

The resonance curves are constructed by differential equations. They have endpoints where the gradient is zero and minus infinity, and usually a cusp in between. In Huxley [16] we chose the parameters so that the resonance curves had bounded length, and the number of integer points close to a resonance curve was also bounded. For further progress we assume an hypothesis about the number of integer points close to a smooth plane curve.

Hypothesis $H(\kappa, \lambda)$. *When $F(x)$ is a bounded function, three times continuously differentiable on an interval containing the unit interval $[1, 2]$, with $F''(x) \neq 0$, $F^{(3)}(x) \neq 0$, when M and N are real parameters with*

$$M \geq 2, \qquad \sqrt{M} \leq N \leq M^2,$$

and $f(x) = NF(x/M)$, and when δ is real with $0 \leq \delta \leq 1/2$, then R, the number of integer points (m, n) with

$$|n - f(m)| \leq \delta, \qquad M \leq m \leq 2M,$$

satisfies

$$R = O\left(\delta M + (MN)^{\kappa}(\log MN)^{\lambda}\right).$$

The implied constant depends on $f(x)$ only by way of the upper and lower bounds for the derivatives of $F(x)$.

Using Hypothesis $H(\kappa, \lambda)$ we get, in the dominant case, with suitable choices of N, R, and V,

$$B = O\left(\frac{M^2 R^4 (\log N)^{\lambda}}{H^2 N^4 V_0^{2/3} V}\right),$$

which corresponds to a saving by $V_0^{2/3}$ over (3.6). Here

$$V_0 = \left(\frac{H}{R}\right)^{6\kappa/(9\kappa - 1)}. \tag{4.3}$$

If $\alpha = (\log M)/(\log T)$ is near $1/2$, then we take $V = V_0$ (plan A). If $\alpha \leq (67\kappa - 6)/(156\kappa - 14)$, then we choose V so that all magic matrices are upper triangular (plan B). If $\alpha \geq (89\kappa - 8)/(156\kappa - 14)$, then we choose V so that all magic matrices are lower triangular (plan C).

The hypothesis $H(\kappa, \lambda)$ cannot be used near the cusp or the endpoints. There are additional contributions from cusps of resonance curves, and the regions near the endpoints where the gradient of the resonance curve is very large or very small.

The main contributions to the Iwaniec-Mozzochi sum $\sum\sum e(hg(m))$ and to the exponential sum with a difference (3.1) are

$$O\left(\frac{MR^3}{NQ^2}\sqrt{\frac{|E|N}{M}}\log^2 N\right). \tag{4.4}$$

The main contribution to the single exponential sum $\sum e(F(m))$ is

$$O\left(\frac{MR^3}{NQ^{3/2}}\left(\frac{|E|N}{M}\right)^{1/r}\log N\right),$$

where r is the number of summands in (3.4).

The full force of $H(\kappa, \lambda)$ is known only for the van der Corput values $\kappa = 1/3$, $\lambda = 0$. The results of Huxley [17] give the bound corresponding to $H(3/10, 1/10)$ for shorter ranges of N. By subdividing the major arcs according to whether the continued fraction for the gradient a/q has a large partial quotient early on, we get the bounds which would follow from $H(3/10, 57/140)$ for the exponential sum (Huxley [21]), and the lattice point problem (Huxley [19]). The new exponents in the classical problems are:

$$\text{Circle problem:} \quad \frac{131}{208} = 0.6298\ldots \quad < \quad \frac{46}{73} = 0.6301\ldots$$

$$\text{Divisor problem:} \quad \frac{131}{416} = 0.3149\ldots \quad < \quad \frac{23}{73} = 0.3151\ldots$$

$$\text{Lindelöf problem:} \quad \frac{32}{205} = 0.156098\ldots < \quad \frac{89}{570} = 0.156140\ldots$$

The Bombieri-Iwaniec method is essentially restricted to sums involving a function of only one variable, because simultaneous approximation to several independent partial derivatives is not accurate enough. The only method is the van der Corput iteration in several variables (see Krätzel [26]), arranged, if possible, to end with an appeal to the Bombieri-Iwaniec-Mozzochi method (Krätzel and Nowak [27]). Other applications of the large sieve for exponential sums are found in Fouvry and Iwaniec [11] and Sargos [30].

5 Mean Values of Exponential Sums

We sketch the proof of new results on the short interval mean square of exponential sums and the mean square of the Riemann zeta function. The

Second Spacing Problem is the same as for the Iwaniec-Mozzochi sum $\sum\sum e(hg(m))$, with $g(x) = 2TF'(x/M)/M$, so the same choices of N, R and V are made as in Huxley [19]. The details of the argument sketched in section 3 are as in Huxley [16, Chapter 9]. There is an extra requirement (9.1.7):

$$H^3 \ll N^2 R, \tag{5.1}$$

and the bound in (3.6) for the First Spacing Problem is bigger than (3.5) by a factor

$$1 + O\left(\frac{L^2}{K}\right) = 1 + O\left(\frac{H^2 Q}{NR^2}\right) \tag{5.2}$$

for $Q \le q \le 2Q$. This factor is greater than 1 at the top end of the range for H. The factor Q in (5.2) causes the powers of Q in (4.4) to cancel, so that all ranges for Q contribute equally, and there is an extra logarithm factor in the upper bounds of Proposition 1 in the terms in (5.7), (5.9) and (5.11), and the bounds for H in (5.6), (5.8) and (5.10) have a term from (5.1). Otherwise Proposition 1 would have the same conclusions as the bounds for the Iwaniec-Mozzochi sum in Proposition 1 of Huxley [19].

Proposition 1. *Suppose that $H(\kappa, \lambda)$ holds for some κ, λ with $1/4 \le \kappa \le 1/3$, $\lambda \ge 0$. Let $F(x)$ be a real function four times continuously differentiable for $1 \le x \le 2$, and let $g(x)$, $G(x)$ be bounded functions of bounded variation on $1 \le x \le 2$. Let $C_2, \ldots, C_7$ be real numbers ≥ 1. Suppose that*

$$\left|F^{(r)}(x)\right| \le C_r \tag{5.3}$$

for $r = 2, 3, 4$,

$$\left|F^{(r)}(x)\right| \ge 1/C_r \tag{5.4}$$

for $r = 2, 3$. In some ranges we require extra conditions, either (5.4) for $r = 4$, or

$$\left|F''(x)F^{(4)}(x) - 3F^{(3)}(x)^2\right| \ge 1/C_5 \tag{5.5}$$

Let H, M and T be large parameters, and let S denote the sum

$$S = \sum_{h=H}^{2H-1} g\left(\frac{h}{H}\right) \sum_{m=M+2H-1}^{2M-2H} G\left(\frac{m}{M}\right) e\left(TF\left(\frac{m+h}{M}\right) - TF\left(\frac{m-h}{M}\right)\right).$$

Then we have the following bounds, in which B_1 and B_2 are positive constants constructed from $C_2, \ldots, C_7$, κ, λ and from the functions $g(x)$ and $G(x)$.
(A) *In the range*

$$C_6^{-1}T^{67\kappa-6}(\log T)^{(45\kappa-4)\lambda} \le M^{156\kappa-14} \le C_6 T^{89\kappa-8}(\log T)^{-(45\kappa-4)\lambda},$$

$$H \leq B_1 M T^{-\frac{119\kappa-11}{384\kappa-36}} (\log T)^{\frac{(9\kappa-1)\lambda}{128\kappa-12}}, \tag{5.6}$$

we have

$$S \leq B_2 H \left(\frac{H}{M}\right)^{\frac{6\kappa-1}{100\kappa-10}} T^{\frac{67\kappa-7}{200\kappa-20}} (\log T)^{\frac{9}{4}+\frac{(39\kappa-4)\lambda}{200\kappa-20}}$$

$$+ B_2 H \left(\frac{H}{M}\right)^{\frac{126\kappa-13}{200\kappa-20}} T^{\frac{207\kappa-21}{400\kappa-40}} (\log T)^{\frac{13}{4}+\frac{(69\kappa-7)\lambda}{400\kappa-40}}, \tag{5.7}$$

subject to the extra conditions that (5.4) *holds for* $r = 4$ *when*

$$M^{156\kappa-14} \leq C_6 T^{69\kappa-6} (\log T)^{-3(9\kappa-1)\lambda},$$

that (5.5) *holds when*

$$M^{156\kappa-14} \geq C_6^{-1} T^{87\kappa-8} (\log T)^{3(9\kappa-1)\lambda},$$

that

$$H \geq C_7^{-1} T^4 (\log T)^{3\lambda}/M^9 \quad for \quad M \leq C_6^{-1} T^{7/16} (\log T)^{\lambda/32\kappa},$$

and that

$$H \geq C_7^{-1} M^{11} (\log T)^{3\lambda}/T^6 \quad for \quad M \geq C_6 T^{9/16} (\log T)^{-\lambda/32\kappa}.$$

(B) *If* (5.4) *holds for* $r = 4$, *then in the ranges*

$$C_6^{-1} T^{1/3} \leq M \leq C_6 T^{7/16} (\log T)^{3\lambda/16},$$

$$H \leq \min\left(B_1 M^{\frac{35\kappa-3}{53\kappa-5}} T^{-\frac{26\kappa-2}{159\kappa-15}} (\log T)^{\frac{(9\kappa-1)\lambda}{53\kappa-5}}, \right.$$

$$\left. B_1 M^{\frac{9}{7}} T^{-\frac{3}{7}}, \quad C_7 T^4 M^{-9} (\log T)^{3\lambda} \right), \tag{5.8}$$

we have

$$S \leq B_2 H^{\frac{87\kappa-9}{80\kappa-8}} M^{\frac{15\kappa-1}{80\kappa-8}} T^{\frac{9\kappa-1}{40\kappa-4}} (\log T)^{\frac{9}{4}+\frac{(9\kappa-1)\lambda}{80\kappa-8}}$$

$$+ B_2 H^{\frac{267\kappa-27}{160\kappa-16}} M^{-\frac{45\kappa-5}{160\kappa-16}} T^{\frac{29\kappa-3}{80\kappa-8}} (\log T)^{\frac{13}{4}+\frac{(9\kappa-1)\lambda}{160\kappa-16}}$$

$$+ B_2 H^{\frac{107\kappa-13}{12(9\kappa-1)}} M^{-\frac{7\kappa-1}{4(9\kappa-1)}} T^{\frac{43\kappa-5}{12(9\kappa-1)}} (\log T)^{\frac{9+\lambda}{4}}$$

$$+ B_2 H^{\frac{331\kappa-39}{24(9\kappa-1)}} M^{-\frac{59\kappa-7}{8(9\kappa-1)}} T^{\frac{131\kappa-15}{24(9\kappa-1)}} (\log T)^{\frac{13+\lambda}{4}}. \tag{5.9}$$

(C) *If* (5.5) *holds, then in the ranges*

$$C_6^{-1}T^{9/16}(\log T)^{-5\lambda/16} \le M \le C_6 T^{2/3},$$

$$H \le \min\left(B_1 M^{\frac{71\kappa-7}{53\kappa-5}} T^{-\frac{80\kappa-8}{159\kappa-15}} (\log T)^{\frac{(9\kappa-1)\lambda}{53\kappa-5}}, \right.$$

$$\left. B_1 M^{\frac{5}{7}} T^{-\frac{1}{7}}, C_6 M^{11} T^{-6} (\log T)^{3\lambda} \right), \tag{5.10}$$

we have

$$S \le B_2 H^{\frac{87\kappa-9}{80\kappa-8}} M^{-\frac{29\kappa-3}{80\kappa-8}} T^{\frac{1}{2}} (\log T)^{\frac{9}{4}+\frac{(9\kappa-1)\lambda}{80\kappa-8}}$$

$$+ B_2 H^{\frac{267\kappa-27}{160\kappa-16}} M^{-\frac{169\kappa-17}{160\kappa-16}} T^{\frac{3}{4}} (\log T)^{\frac{13}{4}+\frac{(9\kappa-1)\lambda}{160\kappa-16}}$$

$$+ B_2 H^{\frac{107\kappa-13}{12(9\kappa-1)}} M^{\frac{23\kappa-1}{12(9\kappa-1)}} T^{\frac{7\kappa-1}{4(9\kappa-1)}} (\log T)^{\frac{9+\lambda}{4}}$$

$$+ B_2 H^{\frac{331\kappa-39}{24(9\kappa-1)}} M^{-\frac{53\kappa-9}{24(9\kappa-1)}} T^{\frac{23\kappa-3}{8(9\kappa-1)}} (\log T)^{\frac{13+\lambda}{4}}. \tag{5.11}$$

After some tidying steps, described in Huxley [16, Chapter 19], the estimation of the short interval mean square of an exponential sum can be reduced to the estimation of the sums in Proposition 1. A 'mathematics made difficult' interpretation of the A step in van der Corput's method is as a mean-to-max from the value at a fixed T to the mean square over a short interval of T, which is then expressed in terms of sums like those in Proposition 1.

Proposition 2. *Suppose that $H(\kappa, \lambda)$ holds with some κ in $1/4 \le \kappa \le 1/3$ and some $\lambda \ge 0$. Let the functions $F(x)$ and $G(x)$ satisfy the conditions of Proposition 1, with* (5.3) *and* (5.4) *also holding for $r = 1$. Let $S(t)$ be the exponential sum*

$$S(t) = \sum_{M}^{2M-1} G\left(\frac{m}{M}\right) e\left(tF\left(\frac{m}{M}\right)\right),$$

and let Δ be any real number with

$$\Delta \gg T^{\frac{207\kappa-21}{652\kappa-66}} (\log T)^{\frac{130(10\kappa-1)+(69\kappa-7)\lambda}{652\kappa-66}}.$$

Then for

$$T^{545\kappa-57}(\log T)^{\theta} \le M^{1304\kappa-32} \le T^{759\kappa-75}(\log T)^{-\theta} \tag{5.12}$$

where

$$\theta = 26(34\kappa - 1) + (399\kappa - 41)\lambda,$$

we have the mean square bound

$$\int_{T-\Delta}^{T+\Delta} |S(t)|^2 dt \ll \Delta M.$$

Proof of Proposition 2. From the discussion in Huxley [16, Chapter 19] we must sum Proposition 1 over values of H, with H doubling at each step, and obtain a total bound $O(M)$. The main contribution to the upper bound from (4.4) can be written as

$$O\left(\frac{M}{V_0^{1/6}}\left(\left(\frac{H^2}{NR}\right)^{\frac{1}{2}}(\log T)^{\frac{9+\lambda}{4}} + \left(\frac{H^2}{NR}\right)^{\frac{3}{4}}(\log T)^{\frac{13+\lambda}{4}}\right)\right),$$

where V_0 is given by (4.3). If H is so large that this expression is $O(M)$, then $H^2 \gg NR$, and the even-numbered terms in (5.7), (5.9) and (5.11) dominate. When we substitute for V_0, then the structural condition (5.1) becomes

$$R^{19\kappa-3} \ll N^{19\kappa-3}(\log T)^{(39\kappa+3\lambda)(9\kappa-1)},$$

a consequence of the structural condition $R \leq N$. In case (A) we take the largest value of H for which the sum is $O(M)$, and then $\Delta \asymp M/H$ in (5.11). In cases (B) and (C) we take the same upper bound for H, and (5.12) is the condition for the terms in (5.8) or (5.10) to be $O(M)$. The values of H in cases (B) and (C) satisfy (5.7) and (5.8) respectively.   □

Following the arguments of Heath-Brown and Huxley [13], we obtain the asymptotic formula for the mean square of the zeta function. At the maximum value of H, the sums in Proposition 1 are $O(M \log M)$, so the exponents are related to those in Proposition 2.

Proposition 3. *Suppose that $H(\kappa, \lambda)$ holds with some κ in $1/4 \leq \kappa \leq 1/3$ and some $\lambda \geq 0$. Then for T large,*

$$\int_0^T \left|\zeta\left(\frac{1}{2} + it\right)\right|^2 dt = T\log\frac{T}{2\pi} + (2\gamma - 1)T + E(T),$$

with

$$E(T) \ll T^{\frac{207\kappa-21}{652\kappa-66}}(\log T)^{\frac{2204\kappa-212+(69\kappa-7)\lambda}{652\kappa-66}}.$$

With the van der Corput values $\kappa = 1/3$, $\lambda = 0$, Propositions 1, 2 and 3 become the results given in Huxley [16, section 19.1] (with the logarithm factor corrected; the extra logarithm from the ranges of Q was overlooked in Huxley [16]). As in Huxley [21], [19], using the bounds by Swinnerton-Dyer's method from Huxley [18], and a further subdivision if the continued fraction for a/q has some partial quotient $\gg \log M$, gives the result that would follow from $H(3/10, 57/140)$.

Theorem. *The results of Propositions 1, 2 and 3 with $\kappa = 3/10$, $\lambda = 57/140$ are true unconditionally.*

We note that with $\kappa = 3/10$ the main exponent in Propositions 2 and 3, $(207\kappa - 21)/(652\kappa - 66)$, is

$$\frac{137}{432} = 0.31713\cdots < 0.31718\cdots = \frac{72}{227},$$

the latter being the exponent in Huxley [16] from $\kappa = 1/3$.

References

[1]   E. Bombieri and H. Iwaniec, *On the order of $\zeta(1/2 + it)$*, Ann. Scuola Norm. Sup. Pisa Cl. Sci. (4) **13** (1986), 449–472.

[2]   E. Bombieri and J. Pila, *The number of integral points on arcs and ovals*, Duke Math. J. **59** (1989), 337–357.

[3]   M. Branton and P. Sargos, *Points entiers au voisinage d'une courbe plane à très faible courbure*, Bull. Sci. Math. **118** (1994), 15–28.

[4]   D. A. Burgess, *On character sums and L-series. II*, Proc. London Math. Soc. **13** (1963), 524–536.

[5]   J. G. van der Corput, *Over Roosterpunten in het Platte Vlak*, Noordhof, Groningen, 1919.

[6]   ———, *Über Gitterpunkte in der Ebene*, Math. Ann. **81** (1920), 1–20.

[7]   ———, *Zahlentheoretische Abschätzungen*, Math. Ann. **84** (1921), 53–79.

[8]   ———, *Verschärfung der Abschätzung beim Teilerproblem*, Math. Ann. **87** (1922), 39–65.

[9]   ———, *Neue zahlentheoretische Abschätzungen*, Math. Ann. **89** (1923), 215–254.

[10] M. Filaseta and O. Trifonov, *The distribution of fractional parts with applications to gap results in number theory*, Proc. London Math. Soc. **73** (1996), 241–278.

[11] E. Fouvry and H. Iwaniec, *Exponential sums for monomials*, J. Number Theory **33** (1989), 311–333.

[12] S. W. Graham and G. Kolesnik, *Van der Corput's method for exponential sums*, London Math. Soc. Lecture Notes 126, Cambridge University Press, 1991.

[13] D.R. Heath-Brown and M. N. Huxley, *Exponential sums with a difference*, Proc. London Math. Soc. **61** (1990), 227–250.

[14] E. Hlawka, *Integrale auf konvexen Körpern*, Monatsh. Math. **54** (1950), 1–36, 81–99.

[15] M. N. Huxley, *The rational points close to a curve*, Ann. Scuola Norm. Sup. Pisa Cl. Sci. (4) **21** (1994), 357–375.

[16] ———, *Area, lattice points and exponential sums*, London Math. Soc. Monographs 13, Oxford University Press, 1996.

[17] ———, *The integer points close to a curve III*, Number Theory in Progress, de Gruyter, Berlin, 1999, pp. 911–940.

[18] ———, *The rational points close to a curve II*, Acta Arith. **93** (2000), 201–219.

[19] ———, *Exponential sums and lattice points III*, Proc. London Math. Society, to appear.

[20] ———, *The rational points close to a curve III*, to appear.

[21] ———, *Exponential sums and the Riemann zeta function V*, to appear.

[22] M. N. Huxley and N. Watt, *Exponential sums and the Riemann zeta function*, Proc. London Math. Soc. **57** (1988), 1–24.

[23] H. Iwaniec and C. J. Mozzochi, *On the divisor and circle problems*, J. Number Theory **29** (1988), 60–93.

[24] D. G. Kendall, *On the number of lattice points inside a random oval*, Quart. J. Math. (Oxford) **19** (1948), 1–26.

[25] S. Konyagin, *Estimates of the least prime factor of a binomial coefficient*, Mathematika **45** (1999), 41–55.

[26] E. Krätzel, *Lattice points*, Deutscher Verlag Wiss., Berlin, 1988.

[27] E. Krätzel and W. G. Nowak, *Lattice points in large convex bodies II*, Acta Arith **62** (1992), 285–295.

[28] E. Landau, *Die Bedeutung der Pfeiffer'schen Methode für die analytische Zahlentheorie*, Sitzungsberichte Akad. Wiss. Wien, math.-naturwiss. Kl. **121** (1912), 2195–2322.

[29] _____, *Die Bedeutungslosigkeit der Pfeiffer'schen Methode für die analytische Zahlentheorie*, Monatsh. Math. Phys. **34** (1926), 1–36.

[30] P. Sargos, *Points entiers au voisinage d'une courbe, sommes trigonométriques courtes et paires d'exposants*, Proc. London Math. Soc. **70** (1995), 285–312.

[31] W. Sierpiński, *Sur un problème du calcul des fonctions asymptotiques*, Prace Mat.-Fiz. **17** (1906), 77–118.

[32] H. P. F. Swinnerton-Dyer, *The number of lattice points on a convex curve*, J. Number Theory **6** (1974), 128–135.

[33] G. Voronoi, *Sur un problème du calcul des fonctions asymptotiques*, J. Reine Angew. Math. **126** (1903), 241–282.

[34] _____, *Sur une fonction transcendente et ses applications à la sommation de quelques séries*, Ann. École Norm. Sup. **21** (1904), 207–267, 459–533.

[35] N. Watt, *A problem on semicubical powers*, Acta Arith. **52** (1989), 119–140.

Euler Products and Abstract Trace Formulas

Georg Illies

1 Introduction

In [6], [8], [9], and [10] Deninger showed how to obtain the analytic continuation (along with other conjectured properties) of motivic L-functions from certain "cohomological data" delivered by a conjectural infinite dimensional cohomology theory. We briefly sketch his idea: Let $\hat{L}(z) = \prod_{\mathfrak{p} \in \hat{S}} L_{\mathfrak{p}}(z)$ be a motivic L-function; for the precise definition of this most general class of arithmetic L-functions, which in particular contains the Dedekind zeta functions of number fields, see [7], but we will formulate our theorems for abstract Euler products (2.1). The data are infinite dimensional "(Frobenius) operators"

$$\theta \colon H^i \to H^i, \quad i = 0, 1, \ldots, n,$$
$$\theta \colon H^0_{\mathfrak{p}} \to H^0_{\mathfrak{p}}, \quad \mathfrak{p} \in \hat{S}$$

with either **C**-vector spaces or Hilbert spaces H^i and $H^0_{\mathfrak{p}}$ such that:

- There is a countable (Hilbert space) basis of eigenvectors, with

$$\#\{\text{eigenvalues} < r\} = o(r^m), \quad r \to \infty.$$

- The eigenvalues for $H^0_{\mathfrak{p}}$ are the poles[1] of $L_{\mathfrak{p}}(z)$.

- There is a functional calculus for θ and certain test functions $\Phi(w)$ such that the *"Lefschetz trace formula"*

$$\operatorname{Tr}_g(\Phi(\theta)| \oplus H^i) = \sum_{\mathfrak{p} \in \hat{S}} \operatorname{Tr}_g(\Phi(\theta)|H^0_{\mathfrak{p}}) \qquad (1.1)$$

with the alternating trace $\operatorname{Tr}_g := \sum (-1)^{i+1} \operatorname{Tr}_{|H^i}$ is valid.

Taking

$$\Phi(w) = (z - w)^{-s}$$

---

[1] $L_{\mathfrak{p}}^{-1}(z)$ is entire of order 1 in the case of a motivic L-function.

and formally applying the operation $\exp(-\frac{d}{ds}\dots|_{s=0})$ to the Lefschetz trace formula yields

$$\underbrace{\det_\infty^g(z - \theta| \oplus H^i)}_{\hat{L}(z)} = \prod_{\mathfrak{p}\in\hat{S}} \underbrace{\det_\infty^g(z - \theta|H_\mathfrak{p}^0)}_{L_\mathfrak{p}(z)} \qquad (1.2)$$

with the alternating zeta-regularized determinant

$$\det_\infty^g(z-\theta|\oplus H^i) := \exp\left(\left(-\frac{d}{ds}\sum_{i=1}^n(-1)^{i+1}\sum_{\rho \text{ eigenv. of } \theta|_{H^i}}(z-\rho)^{-s}\right)\Bigg|_{s=0}\right),$$

assuming that the Dirichlet series in the exponent has a holomorphic continuation. By the general theory of zeta-regularized products, as developed in [15], one can strictly justify the above formal calculation and show that the local factors on the right-hand side of (1.2) are the functions $L_\mathfrak{p}(z)$ and that the alternating zeta-regularized determinant on the left-hand side (i.e., $\hat{L}(z)$) is a meromorhic function of finite order (i.e., a quotient of two entire functions of finite order), whose poles and zeroes are the eigenvalues of θ on the H^i.[2]

One would like to know how realizations of the "Lefschetz trace formula" (1.1) might look like. The first question is for which test functions $\Phi(w)$ can one expect such a trace formula to be satisfied; an answer is given by Theorem 1 below, which improves [5], Theorem 1.7, and [11], Theorem 4.1. The *"abstract trace formulas"* of Theorem 1 are a kind of *"explicit formulas"* for Euler products for a class of test functions that is smaller than the classes for the Weil-type explicit formulas of [19], [1], and [16], but *does not require a functional equation* for $\hat{L}(z)$, in contrast to these results. In other words the "Lefschetz trace formula" is equivalent to $\hat{L}(z)$ being a meromorphic function of finite order.

A concrete conceptual construction of Deninger's cohomological data would indeed be extremely interesting, but is not known, even in the case of the Riemann zeta function $\zeta(s)$. The second part of the paper is motivated by this fact and by an article of Goldfeld [13], who, in the case of $\zeta(s)$, tries to imitate the proof of the Selberg trace formula for a compact Riemann surface (compare [18], [14]): Goldfeld constructs (we simplify here a little bit) a compact topological group X and a map

$$\Phi \mapsto \mathcal{A}(\Phi)$$

---

[2]This approach is certainly analogous to the proof of the Weil conjectures using determinants of the Frobenius operator in the l-adic cohomology (compare [4], [17]).

for certain test functions with an operator $\mathcal{A}(\Phi)$ acting on $L^2(X)$ and comes close to a proof that[3]

$$\operatorname{Tr}\mathcal{A}(\Phi) = \text{ right-hand side of explicit formula of [19], [1].} \qquad \text{(A1)}$$

He leaves open the question whether the eigenvalues of $\mathcal{A}(\Phi)$ (for a simultaneous base of eigenfunctions) are exactly the values $\Phi(\rho)$ with the zeroes ρ of $\zeta(s)$, which would give a "spectral interpretation" of the zeroes of $\zeta(s)$. An obvious necessary condition for this is

$$\mathcal{A}(\Phi_1\Phi_2) = \mathcal{A}(\Phi_1)\mathcal{A}(\Phi_2). \qquad \text{(A2)}$$

Theorem 2 ($n = 1$) and Theorem 3 (arbitrary n) below state the perhaps surprising fact that, assuming (A1), condition (A2) is essentially also sufficient. This means that a "synthetic" approach, such as that of [13], for a spectral interpretation of "abstract trace formulas" and "explicit formulas" is perhaps not hopeless.

Concluding Remark. The only known partial construction of Deninger's cohomology[4] is that for zeta functions of varieties over finite fields [8]; this is an elementary construction based on the well known l-adic cohomology. To find a spectral interpretation for the "abstract trace formulas", say in the case of Dedekind zeta functions, would give the very next piece of the whole picture; Theorems 2 and 3 are perhaps useful for this task. It should be pointed out that this may well be much less difficult than a spectral interpretation of the Weil-type explicit formulas as attempted, for example, in [2] (though it does not yield much information about the Riemann hypothesis).

2 The Theorems

In this article we consider an absolutely convergent abstract Euler product

$$\hat{L}(z) = \prod_{\mathfrak{p}\in\hat{S}} L_{\mathfrak{p}}(z), \quad \Re(z) > \sigma_0 \qquad \text{(2.1)}$$

---

[3]$\mathcal{A}(\Phi)$ is given by an integral kernel $k_\Phi(x,y)$, and Goldfeld proves that $\int_X k_\Phi(x,x)dx$ is equal to the right-hand side of the explicit formula. But A. Deitmar pointed out to me that $\operatorname{Tr}\mathcal{A}(\Phi)$, in contrast to Goldfeld's expectation, cannot equal this diagonal integral, so the construction fails already because of this reason.

[4]An interesting construction for a non-arithmetic situation is A. Deitmar's "reduced tangential cohomology" [3] for a large class of Selberg zeta functions; this shows several features of Deninger's cohomology.

with a set of "places" $\hat{S} = S \uplus S_\infty$, the disjoint union of a countable set S ("finite places") and a finite set S_∞ ("infinite places"), and Euler factors

$$L_\mathfrak{p}(z) = (1 - \beta_\mathfrak{p} e^{-\alpha_\mathfrak{p} z})^{\pm 1}, \qquad\qquad \mathfrak{p} \in S,$$
$$L_\mathfrak{p}(z) = e^{P_\mathfrak{p}(z)} \Gamma^{\pm 1}(\alpha_\mathfrak{p}(z - \beta_\mathfrak{p})), \qquad \mathfrak{p} \in S_\infty,$$

with $\beta_\mathfrak{p} \in \mathbf{C}$, $\alpha_\mathfrak{p} > 0$, and a polynomial $P_\mathfrak{p}(z)$. The absolute convergence of the Euler product certainly implies $|\beta_\mathfrak{p}| < e^{\alpha_\mathfrak{p} \sigma_0}$ for $\mathfrak{p} \in S_0$, and $\Re(\beta_\mathfrak{p}) < \sigma_0$ for $\mathfrak{p} \in S_\infty$. The Euler factors of motivic L-functions are products of our simple Euler factors, but our method also works for more complicated Euler factors such as higher Gamma functions, and thus applies to Selberg zeta functions. However, for simplicity we restrict ourselves to these special Euler products.

We recall some basic facts about *Hardy spaces*. For $a \in \mathbf{R}$ and $0 < p < \infty$ the Hardy space $\mathcal{H}^p(\Re(w) < a)$ consists of all functions $\Phi(w)$ which are holomorphic for $\Re(s) < a$ and satisfy

$$\sup_{\sigma < a} ||\Phi_\sigma||_p < \infty$$

with $\Phi_\sigma(t) := \Phi(\sigma + it)$ and the p-norm $|| \cdot ||_p$. According to [12], Cor. 2 of Thm. 11.3, for every function $\Phi \in \mathcal{H}^p(\Re(w) < a)$ one has $\Phi(w) \to 0$ for $|w| \to \infty$ uniformly in every half plane $\Re(w) \le a'$ with $a' < a$. The central property of these functions is given by [12], Thm. 11.8: If $1 \le p < \infty$, then for all $\Phi \in \mathcal{H}^p(\Re(w) < a)$ and $a' < a$

$$\frac{1}{2\pi i} \int_{a'-i\infty}^{a'+i\infty} \frac{\Phi(w)}{w - z} dw = \begin{cases} \Phi(z) & \Re(z) < a', \\ 0 & \Re(z) > a', \end{cases} \qquad (2.2)$$

with absolute convergence of the integral. If $\int_{a'-i\infty}^{a'+i\infty} |\Phi(w)dw| < \infty$ then we also have (see [5], (2.19))

$$\int_{a'-i\infty}^{a'+i\infty} \Phi(w)dw = 0. \qquad (2.3)$$

Theorem 1. *Suppose that for some* $1 \le p < \infty$ *and* $\sigma_0 < a' < a$ *the function* $\Phi(w)$ *satisfies the following conditions.*

(α) $\Phi \in \mathcal{H}^p(\Re(w) < a)$,

(β) $\int_{-\infty}^{\infty} |\Phi(a' + it)| \log(|t| + 2)dt < \infty$.

Then:

(a) $\text{Tr}_{\mathfrak{p}}(\Phi) := \sum_{\rho \in \mathbf{C}} \text{ord}_{L_{\mathfrak{p}},\rho} \cdot \Phi(\rho)$ *is absolutely convergent for all* $\mathfrak{p} \in \hat{S}$.

(b) $\sum_{\mathfrak{p} \in \hat{S}} \text{Tr}_{\mathfrak{p}}(\Phi)$ *is absolutely convergent.*

(c) *If* $\hat{L}(z)$ *is meromorphic of finite order and if*

$$\text{Tr}(\Phi) := \sum_{\rho \in \mathbf{C}} \text{ord}_{\hat{L},\rho} \cdot \Phi(\rho)$$

converges absolutely, then the following abstract trace formula *is valid:*

$$\text{Tr}(\Phi) = \sum_{\mathfrak{p} \in \hat{S}} \text{Tr}_{\mathfrak{p}}(\Phi). \tag{2.4}$$

Observe that the poles and zeroes of $L_{\mathfrak{p}}(z)$, as well as those of $\hat{L}(z)$, all lie in the half plane $\Re(z) \leq \sigma_0$, and that $\Phi(z)$ is defined in this half plane.

Remark. Theorem 1 improves Theorem 1.7 of [5] (for Hecke L-series) and Theorem 4.1 of [11] (for motivic L-functions) insofar as it weakens the conditions imposed on the test function, but also because we do not need a functional equation for the abstract Euler product in our proof. So (2.4) should be regarded as a type of explicit formula for Euler products without functional equation for a smaller class of test functions. Its relation to the Weil-type explicit formulas of [19] and [1] for Hecke L-series and of [16] for abstract Euler products with functional equation, which express $\text{Tr}(\Phi)$ in terms of the Fourier transform φ of Φ, is as follows: The "typical" function $\Phi(w)$ satisfying conditions (α) and (β) of the theorem is the Fourier transform of a function $\varphi : \mathbf{R}^+ \to \mathbf{C}$ with $\varphi(t) = t^\alpha$, $t \to 0$, for some $\alpha \gg 0$. Then $\text{Tr}_{\mathfrak{p}}(\Phi)$ can be expressed in terms of φ, for $\mathfrak{p} \in S$ by applying the Poisson summation formula, and for $\mathfrak{p} \in S_\infty$ by simply inserting the Fourier integral of φ instead of Φ. This yields the expressions given in [19], [1] and [16] for these special test functions.

For a Hilbert space H we denote by $\text{L}_{\text{nt}}(H)$ the space of normal trace class operators on H; it is a subspace of $\text{L}_{\text{nc}}(H)$, the space of normal compact operators. By the spectral theorem, for every normal compact operator A there is an orthonormal basis of eigenvectors, the set of eigenvalues is bounded, and every nonzero eigenvalue λ has finite multiplicity. We also introduce the functions

$$\Phi_{a,s}(w) := (a - w)^{-s} \quad \text{for } \Re(w) < \Re(a),$$

with the principal branch of the logarithm.

Theorem 2. *Suppose $\mathcal{T}$ is an $\mathbf{R}$-algebra of test functions $\Phi(w)$ satisfying the following conditions:*

- *For all $\Phi \in \mathcal{T}$ there is an $\varepsilon > 0$ such that $\Phi(w)$ is holomorphic for $\Re(w) < \sigma_0 + \varepsilon$.*

- *For all $\Phi \in \mathcal{T}$ there is a $\delta > 0$ such that*

$$\Phi(w) = O(w^{-1-\delta}), \quad |w| \to \infty.$$

- *There are $n_0 \in \mathbf{N}^{\geq 2}$ and $a_0 \in \mathbf{R}$ with $\Re(a_0) > \sigma_0$ such that*

$$\Phi_{a_0,n} \in \mathcal{T} \quad \text{for } n \geq n_0.$$

Suppose H is a Hilbert space and $\mathcal{A}$ an $\mathbf{R}$-linear map

$$\mathcal{A} \colon \mathcal{T} \to \mathrm{L_{nt}}(H)$$
$$\Phi \mapsto \mathcal{A}(\Phi)$$

satisfying the following axioms:

$$\mathrm{Tr}\,\mathcal{A}(\Phi) = \sum_{\mathfrak{p} \in \hat{S}} \sum_{\rho \in \mathbf{C}} \mathrm{ord}_{L_{\mathfrak{p}},\rho} \cdot \Phi(\rho), \qquad (\text{A1})$$

$$\mathcal{A}(\Phi_1 \Phi_2) = \mathcal{A}(\Phi_1)\mathcal{A}(\Phi_2). \qquad (\text{A2})$$

Then $\hat{L}(z)$ is entire of finite order $\leq n_0$. There exists a uniquely defined orthogonal decomposition

$$H = \tilde{H} \oplus H_0$$

and a uniquely defined (unbounded) normal operator θ in $\tilde{H}$ with a basis consisting of eigenfunctions and discrete eigenvalue spectrum satisfying

$$\mathrm{mult}(\vartheta, \theta) = \mathrm{ord}_{\hat{L},\vartheta} \quad \text{for all } \vartheta \in \mathbf{C}$$

and such that

$$\mathcal{A}(\Phi) = \Phi(\theta) \oplus 0_{|H_0} \quad \text{for all } \Phi \in \mathcal{T},$$

where $\Phi(\theta)$ is the operator on $\tilde{H}$ defined by the functional calculus for normal operators and $0_{|H_0}$ is the zero-operator on H_0.

This means that we have a spectral interpretation of the zeroes of $\hat{L}(z)$ as the eigenvalues of the operator θ. Observe that all $\Phi \in \mathcal{T}$ satisfy conditions (α) and (β) of Theorem 1, so the two series on the right-hand side of (A1) converge absolutely (while in general the double sum does not).

Remarks.

(1) The condition $\Phi_{a_0,n} \in \mathcal{T}$ ensures that there is some minimal subset of interesting functions in $\mathcal{T}$. It is probably possible to give alternative versions of the theorem with other such minimal sets of functions.

(2) We concentrate here on spectral interpretations of our abstract trace formulas, but it is easily possible to prove an analogous theorem for Weil-type explicit formulas with larger classes of test functions, as in [19], [1], [16].

(3) Because of its pole at $s = 1$ the (completed) Riemann zeta function is not explicitly covered by Theorem 2, but the theorem remains valid with an additional Euler factor that is a rational function.

The following theorem, whose proof is an easy modification of that of Theorem 2, states what remains valid when there exist "higher cohomology groups" and when eigenvalues can cancel. After (in a non-constructive way) getting rid of a trivial component in which the eigenvalues always completely cancel, one also gets the representation of $\mathcal{A}$ as the functional calculus of an operator.

Theorem 3. *Suppose $\mathcal{T}$ is an $\mathbf{R}$-algebra of test functions $\Phi(w)$ satisfying the three conditions of Theorem 2, H^i, $i = 0, 1, \ldots, n$, are Hilbert spaces, and $\mathcal{A}^i$ are $\mathbf{R}$-linear maps*

$$\mathcal{A}^i \colon \mathcal{T} \to \mathrm{L}_{\mathrm{nt}}(H^i)$$
$$\Phi \mapsto \mathcal{A}^i(\Phi)$$

satisfying the axioms

$$\mathrm{Tr}_g(\oplus \mathcal{A}^i(\Phi)) = \sum_{\mathfrak{p} \in \hat{S}} \sum_{\rho \in \mathbf{C}} \mathrm{ord}_{L_{\mathfrak{p}},\rho} \cdot \Phi(\rho), \qquad (\mathrm{A1})$$

$$\mathcal{A}^i(\Phi_1 \Phi_2) = \mathcal{A}^i(\Phi_1) \mathcal{A}^i(\Phi_2). \qquad (\mathrm{A2})$$

Then $\hat{L}(z)$ is meromorphic of finite order. There is an orthogonal decomposition

$$H^i = \tilde{H}^i \oplus H_0^i,$$
$$\tilde{\mathcal{A}}^i \colon \mathcal{T} \to \mathrm{L}_{\mathrm{nt}}(\tilde{H}^i),$$
$$\mathcal{A}_0^i \colon \mathcal{T} \to \mathrm{L}_{\mathrm{nt}}(H_0^i),$$
$$\mathcal{A}^i = \tilde{\mathcal{A}}^i \oplus \mathcal{A}_0^i$$

with

$$\sum_{i=0}^{n} (-1)^{i+1} \mathrm{mult}(\lambda, \mathcal{A}_0^i(\Phi)) = 0$$

for all $\lambda \in \mathbf{C}^*$, $\Phi \in \mathcal{T}$, and such that

$$\tilde{A}^i(\Phi) = \Phi(\theta^i) \quad \text{for all } \Phi \in \mathcal{T}$$

with (unbounded) normal operators θ^i in $\tilde{H}^i$ whose spectra are discrete and satisfy

$$\sum_{i=0}^{n} (-1)^{i+1} \operatorname{mult}(\vartheta, \theta^i) = \operatorname{ord}_{\hat{L},\vartheta} \quad \text{for all } \vartheta \in \mathbf{C}.$$

3   Proof of Theorem 1

The proof is a modification of that in [5] and uses the partial fraction expansions of the logarithmic derivatives of $L_{\mathfrak{p}}(z)$ and $\hat{L}(z)$ which are related by

$$\frac{\hat{L}'}{\hat{L}}(z) = \sum_{\mathfrak{p} \in \hat{S}} \frac{L_{\mathfrak{p}}'}{L_{\mathfrak{p}}}(z) \tag{3.1}$$

and absolutely convergent for $\Re(z) > \sigma_0$. We first describe these expansions. By differentiating the logarithm of the canonical Weierstrass product for $L_{\mathfrak{p}}(z)$ one gets the partial fraction expansion

$$\frac{L_{\mathfrak{p}}'}{L_{\mathfrak{p}}}(z) = \operatorname{ord}_{L_{\mathfrak{p}},0} \cdot \frac{1}{z} + \sum_{\rho \in \mathbf{C}^*} \operatorname{ord}_{L_{\mathfrak{p}},\rho} \cdot \left(\frac{1}{z-\rho} + \frac{1}{\rho} \right), \tag{3.2}$$

which satisfies the estimate

$$\left| \operatorname{ord}_{L_{\mathfrak{p}},0} \cdot \frac{1}{z} \right| + \sum_{\rho \in \mathbf{C}^*} \left| \operatorname{ord}_{L_{\mathfrak{p}},\rho} \cdot \left(\frac{1}{z-\rho} + \frac{1}{\rho} \right) \right| = O\left(\log(|\Im(z-\sigma_0)| + 2)\right) \tag{3.3}$$

uniformly in every half plane $\Re(z) > \sigma_0 + \varepsilon$, $\varepsilon > 0$. This standard estimate follows easily from

$$\sum_{|\rho| < 2|z|} \left| \operatorname{ord}_{L_{\mathfrak{p}},\rho} \cdot \frac{1}{z-\rho} \right| = O\left(\log(|\Im(z-\sigma_0)| + 2)\right), \tag{3.4}$$

by applying the Taylor series expansion of $(1 - \frac{z}{\rho})^{-1}$ to those ρ with $|\rho| \geq 2|z|$, and using the estimate

$$\sum_{x_1 < |\rho_k| < x_2} |\operatorname{ord}_{L_{\mathfrak{p}},\rho} \cdot \rho_k^{-n}| \leq \begin{cases} c'(\log x_2 - \log x_1 + 1) & \text{for } n = 1, \\ c'(x_1^{-(n-1)} - x_2^{-(n-1)} + 1) & \text{for } n > 1, \end{cases}$$

for given $0 < x_1 < x_2$ with a suitable $c' > 0$. The latter estimate, as well as (3.4), follow from a trivial comparison with an integral.

If $\hat{L}(z)$ is meromorphic of finite order, i.e., a quotient of two entire functions of finite order, then we also have a partial fraction expansion for $\frac{\hat{L}'}{\hat{L}}(z)$: If g is the *genus* of $\hat{L}(z)$, i.e., the minimal element $g \in \mathbf{N}_0$ such that

$$\sum_{\rho \in \mathbf{C}^*} |\operatorname{ord}_{\hat{L},\rho} \cdot \rho^{-g-1}| < \infty,$$

then, for some polynomial $P_0(z)$, one has the absolutely convergent canonical Weierstrass product

$$\hat{L}(z) = e^{P_0(z)} z^{\operatorname{ord}_{\hat{L},0}} \prod_{\rho \in \mathbf{C}^*} \left\{ \left(1 - \frac{z}{\rho}\right) \exp\left(\sum_{k=1}^{g} \frac{1}{k}\left(\frac{z}{\rho}\right)^k\right) \right\}^{\operatorname{ord}_{\hat{L},\rho}}.$$

Thus,

$$\frac{\hat{L}'}{\hat{L}}(z) = P_0'(z) + \operatorname{ord}_{\hat{L},0} \cdot \frac{1}{z} + \sum_{\rho \in \mathbf{C}^*} \operatorname{ord}_{\hat{L},\rho} \cdot \left(\frac{1}{z-\rho} + \sum_{k=1}^{g} \frac{z^{k-1}}{\rho^k}\right) \quad (3.5)$$

and we have the following trivial estimate, which can be derived in the same manner as (3.3):

$$|P_0'(z)| + |\operatorname{ord}_{\hat{L},0} \cdot \frac{1}{z}| + \sum_{\rho \in \mathbf{C}^*} \left|\operatorname{ord}_{\hat{L},\rho} \cdot \left(\frac{1}{z-\rho} + \sum_{k=1}^{g} \frac{z^{k-1}}{\rho^k}\right)\right| = O((z-\sigma_0)^l)$$
$$(3.6)$$

for some $l > 0$ (the exact value of l is irrelevant here), in every half plane $\Re(z) > \sigma_0 + \varepsilon, \varepsilon > 0$.

Now, using (3.2), dominated convergence, (3.3) and (β), and applying (2.2) and (2.3), one gets

$$\frac{1}{2\pi i} \int_{a'+i\infty}^{a'+i\infty} \Phi(w) \frac{L_{\mathfrak{p}}'}{L_{\mathfrak{p}}}(w)dw = \sum_{\rho \in \mathbf{C}} \operatorname{ord}_{L_{\mathfrak{p}},\rho} \cdot \Phi(\rho), \quad (3.7)$$

with absolute convergence of the integral and the sum. This implies (a).

Because of (3.1),

$$\sum_{\mathfrak{p} \in \hat{S}} \left|\frac{L_{\mathfrak{p}}'}{L_{\mathfrak{p}}}(w)\right| = \underbrace{\sum_{\mathfrak{p} \in S} \left|\frac{L_{\mathfrak{p}}'}{L_{\mathfrak{p}}}(w)\right|}_{=O(1)} + \underbrace{\sum_{\mathfrak{p} \in S_\infty} \left|\frac{L_{\mathfrak{p}}'}{L_{\mathfrak{p}}}(w)\right|}_{=O(\log(|\Im(w-\sigma_0)|+2))} \quad (3.8)$$

(the estimate of S is clear from the definition of $L_\mathfrak{p}(z)$, and that of S_∞ follows from (3.2) and (3.3)), by dominated convergence we then get

$$\frac{1}{2\pi i} \int_{a'+i\infty}^{a'+i\infty} \Phi(w)\frac{\hat{L}'}{\hat{L}}(w)dw = \sum_{\mathfrak{p}\in\hat{S}} \left(\frac{1}{2\pi i} \int_{a'+i\infty}^{a'+i\infty} \Phi(w)\frac{L_\mathfrak{p}'}{L_\mathfrak{p}}(w)dw \right), \quad (3.9)$$

with absolute convergence of the integrals and the sum. This and (3.7) imply (b).

To prove (c) we fix some b with $\Re(b) > a$ and introduce

$$\Phi_s(w) := (b-w)^{-s}\Phi(w), \quad \Re(w) < a, \ \Re(s) \geq 0.$$

Obviously, $\Phi_s(w) \in \mathcal{H}^p(\Re(w) < a)$, and for given $n \in \mathbf{N}_0$ and sufficiently large $\Re(s)$ the function $w^n\Phi_s(w)$ is also in $\mathcal{H}^p(\Re(w) < a)$ and satisfies

$$\int_{a'-i\infty}^{a'+i\infty} |w^n\Phi_s(w)dw| < \infty.$$

Thus, if $\Re(s)$ is large enough, we get using (3.5), dominated convergence and (3.6), and applying (2.2) and (2.3):

$$\frac{1}{2\pi i} \int_{a'+i\infty}^{a'+i\infty} \Phi_s(w)\frac{\hat{L}'}{L}(w)dw = \sum_{\rho\in\mathbf{C}} \operatorname{ord}_{\hat{L},\rho} \cdot \Phi_s(\rho), \quad (3.10)$$

where the integral and the sum are absolutely convergent. By (a) and (b) and the assumption in (c) all three series in

$$\sum_{\rho\in\mathbf{C}} \operatorname{ord}_{\hat{L},\rho} \cdot \Phi_s(\rho) = \sum_{\mathfrak{p}\in\hat{S}}\sum_{\rho\in\mathbf{C}} \operatorname{ord}_{L_\mathfrak{p},\rho} \cdot \Phi_s(\rho)$$

converge absolutely for $\Re(s) \geq 0$, and by (3.7), (3.9) and (3.10) we already know that the equation holds when $\Re(s) \gg 0$. But both sides are holomorphical in $\Re(s) > 0$ and continuous in $\Re(s) \geq 0$. The identity theorem and an evaluation at $s = 0$ then yield (2.4).

4   Proof of Theorem 2

To simplify some of the formulas, we introduce the notation

$$R_{\hat{L}}(\Phi) := \sum_{\mathfrak{p}\in\hat{S}} R_{L_\mathfrak{p}}(\Phi), \quad R_{L_\mathfrak{p}}(\Phi) := \sum_{\rho\in\mathbf{C}} \operatorname{ord}_{L_\mathfrak{p},\rho} \cdot \Phi(\rho)$$

for the right-hand side of (2.4). Also we use $\Phi(\theta)$ to denote the operator on H given by $\Phi(\theta) \oplus 0_{|H_0}$. We proceed in several steps.

(1) The operators $\mathcal{A}(\Phi)$, $\Phi \in \mathcal{T}$, are a family of commuting normal trace class operators; thus there is an orthonormal base $(\varphi_i)_{i=1,2,\ldots}$ of simultaneous eigenvectors.

(2) Before constructing the operator θ we consider the following general situation:

(F) There exists $n_1 \in \mathbf{N}$ and elements $A_{n_1}, A_{n_1+1}, A_{n_1+2}, \ldots$ of $\mathrm{L}_{\mathrm{nt}}(H)$ satisfying

$$A_{n+m} = A_n \cdot A_m \quad \text{for all } n, m \geq n_1$$

and such that the vectors φ_i from (1) are simultaneously eigenvectors for all A_n.

In this situation there are uniquely defined λ_i, $i \in \mathbf{N}$, with $\lambda_i \to 0$ for $i \to \infty$, such that

$$A_n = A^n \quad \text{for all } n \geq n_1,$$

where the operator $A \in \mathrm{L}_{\mathrm{nc}}(H)$ is defined by

$$A\varphi_i := \lambda_i \varphi_i, \quad i \in \mathbf{N}.$$

This follows from the elementary fact that for any $n_1 \in \mathbf{N}$ and any sequence $\alpha_{n_1}, \alpha_{n_1+1}, \ldots$ of complex numbers satisfying

$$(\alpha_n)^m = (\alpha_m)^n \quad \text{for all } n \geq n_1, m \geq n_1 \tag{4.1}$$

there exists a unique $\alpha \in \mathbf{C}$ such that

$$\alpha_n = \alpha^n \quad \text{for all } n \geq n_1.$$

This can easily be seen as follows: Assuming, without loss of generality, that $\alpha_{n_1} \neq 0$, the only choice for α is $\alpha := \frac{\alpha_{n_1+1}}{\alpha_{n_1}}$, and then $\alpha^{n_1} = \alpha_{n_1}$ and $\alpha^{n_1+1} = \alpha_{n_1+1}$ are immediate from (4.1), and one has

$$\left(\frac{\alpha_n}{\alpha^n}\right)^{n_1} = \left(\frac{\alpha_{n_1}}{\alpha^{n_1}}\right)^n = 1 = \left(\frac{\alpha_{n_1+1}}{\alpha^{n_1+1}}\right)^n = \left(\frac{\alpha_n}{\alpha^n}\right)^{n_1+1}$$

and thus $\frac{\alpha_n}{\alpha^n} = 1$.

(3) The construction of $\tilde{H}$, H_0 and of θ is now easy. Choose

$$A_n := \mathcal{A}(\Phi_{a_0,n}) \text{ for } n \geq n_0,$$

so that we are in situation (F). We get the operator $A \in \mathrm{L}_{\mathrm{nc}}(H)$ from the discussion in (2) with the eigenvectors φ_i and the eigenvalues λ_i. We define $\tilde{H}$ to be the span of those φ_i with $\lambda_i \neq 0$, and H_0 to be its orthogonal complement in H, and we define the operator θ on $\tilde{H}$ by

$$\theta \varphi_i = \vartheta_i \varphi_i,$$

where the complex number ϑ_i is defined by the equation

$$\lambda_i = \frac{1}{a_0 - \vartheta_i}.$$

In particular, we already have $\mathcal{A}(\Phi_{a_0,n}) = \Phi_{a_0,n}(\theta)$ for all $n \geq n_0$.

(4) We need another general construction: Take $A \in L_{nc}(H)$ with eigenvalues λ_i, $i \in \mathbf{N}$, and assume $A^{n_1} \in L_{nt}(H)$ so that

$$\mathrm{Tr}\, A^n = \sum_{i=1}^{\infty} \lambda_i^n, \quad n \geq n_1$$

converges absolutely. Then we define

$$g(A, n_1, z) := \sum_{n=n_1}^{\infty} \mathrm{Tr}(A^n) \cdot \binom{n-1}{n_1 - 1} z^{n-n_1} \tag{4.2}$$

$$= (-1)^{n_1} \sum_{\lambda_i \neq 0} (z - \lambda_i^{-1})^{-n_1}. \tag{4.3}$$

(4.3) follows easily from the Taylor series expansion of $(z - \lambda_i^{-1})^{-n_1}$; in particular, the radius of convergence of the series (4.2) is

$$r_A = \min_{\lambda_i \neq 0} |\lambda_i^{-1}|.$$

(5) We now return to the situation in (3) and prove that $\Re(\vartheta_i) \leq \sigma_0$ and that $\hat{L}(z)$ is entire of finite order $\leq n_0$ with zeroes exactly the ϑ_i with the right multiplicities. For this it is enough to show that

$$\left(\frac{d}{dz}\right)^{n_0} [\log \hat{L}(z)] = (-1)^{n_0 - 1}(n_0 - 1)! \sum (z - \vartheta_i)^{-n_0}, \tag{4.4}$$

as the right hand side is obviously the n_0-th logarithmic derivative of the absolutely convergent Weierstrass product

$$\Delta(z) = z^{\mathrm{mult}(\vartheta_i = 0)} \prod_{\vartheta_i \neq 0} (1 - \frac{z}{\vartheta_i}) \exp \left(\sum_{k=1}^{n_0 - 1} \frac{1}{k} \left(\frac{z}{\vartheta_i}\right)^k \right),$$

which is entire and of finite order $\leq n_0$, according to the Hadamard-Weierstrass theory. But the right-hand side of (4.4) equals $(-1)^{n_0 - 1}(n_0 -$

$1)!g(A, n_0, a_0 - z)$ because of (4.3), and we will now show that for $|z - a_0|$ small,

$$g(A, n_0, a_0 - z) = \sum_{\mathfrak{p} \in \hat{S}} R_{L_{\mathfrak{p}}}(\Phi_{z, n_0})$$

$$= \sum_{\mathfrak{p} \in \hat{S}} \frac{(-1)^{n_0 - 1}}{(n_0 - 1)!} (\frac{d}{dz})^{n_0} [\log L_{\mathfrak{p}}(z)]$$

(with absolute convergence of the sums), which implies (4.4). The second equality is clear by (3.2), and to prove the first equality we use the Taylor series expansion of $((a_0 - w) - (a_0 - z))^{-n_0}$ in powers of $(a_0 - z)$, which yields

$$\Phi_{z, n_0}(w) = \sum_{n=n_0}^{\infty} \Phi_{a_0, n}(w) \binom{n-1}{n_0 - 1} (a_0 - z)^{n - n_0},$$

with absolute convergence for $|z - a_0| < |w - a_0|$. Thus, taking z such that $|z - a_0| < \max(r_A, \Re(a_0) - \sigma_0)$, where r_A is the convergence radius of (4.2) and $0 < \varepsilon' < \varepsilon$ is such that $\varepsilon' + |z - a_0| < \Re(a_0) - \sigma_0$, we obtain, on applying (4.2), (A1), (3.7), (3.8) and dominated convergence,

$$g(A, n_0, a_0 - z) = \sum_{n=n_0}^{\infty} R_{\hat{L}}(\Phi_{a_0, n}) \binom{n-1}{n_0 - 1} (a_0 - z)^{n - n_0}$$

$$= \sum_{\mathfrak{p} \in \hat{S}} \frac{1}{2\pi i} \int_{\sigma_0 + \varepsilon' - i\infty}^{\sigma_0 + \varepsilon' + i\infty} \frac{L'_{\mathfrak{p}}}{L_{\mathfrak{p}}}(w) \Phi_{z, n_0}(w) dw.$$

By (3.7) this is what we need.

(6) Two operators $A, B \in L_{nc}(H)$ are called *0-equivalent* if for all $\lambda \neq 0$ we have $\text{mult}(\lambda, A) = \text{mult}(\lambda, B)$. We now take a function $\Phi \in \mathcal{T}$ which in addition satisfies

$$\Phi(w) = O(|w|^{-n_0}) \quad \text{for } \Re(w) < \sigma_0 + \varepsilon \tag{4.5}$$

and show that then $\mathcal{A}(\Phi)$ and $\Phi(\theta)$ are 0-equivalent. The proof is easy: From Theorem 1(c) we obtain

$$R_{\hat{L}}(\Phi^n) = \sum \Phi^n(\vartheta_i) = \text{Tr } \Phi^n(\theta), \quad n \in \mathbf{N},$$

as (4.5) and $\Phi_{a_0,n_0}(\theta) \in L_{nt}(H)$ ensure the absolute convergence of the sum. Hence, by (4.2), (A1) and (A2), we have

$$
\begin{aligned}
g(\mathcal{A}(\Phi), 1, z) &= \sum_{n=1}^{\infty} \operatorname{Tr} \mathcal{A}(\Phi^n) z^{n-1} \\
&= \sum_{n=1}^{\infty} R_{\hat{L}}(\Phi^n) z^{n-1} \\
&= \sum_{n=1}^{\infty} \operatorname{Tr} \Phi^n(\theta) z^{n-1} \\
&= g(\Phi(\theta), 1, z),
\end{aligned}
$$

and the claim follows from (4.3).

(7) In the situation of (6) we now show that, in fact,

$$\mathcal{A}(\Phi) = \Phi(\theta).$$

For all $i \in \mathbf{N}$ we have an $a_i \in \mathbf{C}$ such that $\mathcal{A}(\Phi)\varphi_i = a_i\varphi_i$; we need to show that $a_i = \Phi(\vartheta_i)$ (we set $\vartheta_i = \infty$ if $\lambda_i = 0$ and $\Phi(\infty) = 0$). In the proof the linearity comes into play. For $\alpha \in [0,1]$, consider the operator

$$\Phi_\alpha(w) := (1-\alpha)\Phi(w) + \alpha\Phi_{a_0,n_0}(w).$$

Then

$$\mathcal{A}(\Phi_\alpha) = (1-\alpha)\mathcal{A}(\Phi) + \alpha\Phi_{a_0,n_0}(\theta),$$

which we already know to be 0-equivalent to

$$\Phi_\alpha(\theta) = (1-\alpha)\Phi(\theta) + \alpha\Phi_{a_0,n_0}(\theta).$$

This means that if we fix i then for all α there exists $j_\alpha \in \mathbf{N}$ such that

$$(1-\alpha)a_i + \alpha\Phi_{a_0,n_0}(\vartheta_i) = (1-\alpha)\Phi(\vartheta_{j_\alpha}) + \alpha\Phi_{a_0,n_0}(\vartheta_{j_\alpha}). \qquad (4.6)$$

The set $[0,1]$ is not countable, but for each of the countably many values of ϑ_{j_α} there is only one value of α that satisfies (4.6), except in the case when $a_i = \Phi(\vartheta_{j_\alpha})$ and $\Phi_{a_0,n_0}(\vartheta_i) = \Phi_{a_0,n_0}(\vartheta_{j_\alpha})$. This means $\vartheta_i = \vartheta_{j_\alpha}$, i.e., $a_i = \Phi(\vartheta_i)$.

(8) Now take an arbitrary operator $\Phi \in \mathcal{T}$. Then there is $n_1 \in \mathbf{N}$ such that the functions $\Phi^n(z)$, $n \geq n_1$, satisfy all conditions needed for steps (6) and (7). This means that for all $n \geq n_1$

$$(\mathcal{A}(\Phi))^n = \mathcal{A}(\Phi^n) = \Phi^n(\theta) = (\Phi(\theta))^n.$$

But by the discussion in (2) this implies

$$\mathcal{A}(\Phi) = \Phi(\theta).$$

(9) Finally the uniqueness of the construction is a trivial consequence of the fact that the injective function Φ_{a_0,n_0} already determines the eigenvalues of the operator θ.

References

[1]   K. Barner, *On A. Weil's explicit formula*, J. Reine Angew. Math. **323** (1981), 139–152.

[2]   A. Connes, *Trace formula in noncommutative geometry and the zeroes of the Riemann zeta function*, Selecta Math., New Ser. **5** (1999), 29–106.

[3]   A. Deitmar, *Selberg zeta functions for groups of higher rank*, preprint.

[4]   P. Deligne, *La conjecture de Weil I*, Publ. Math. IHES **43** (1974), 273–307.

[5]   C. Deninger, *Lefschetz trace formulas and explicit formulas in analytic number theory*, J. Reine Angew. Math. **441** (1993), 1–15.

[6]   ———, *Evidence for a cohomological approach to analytic number theory*, Proc. of the First Europ. Cong. of Math., Birkhäuser, 1994, pp. 491–510.

[7]   ———, *L-functions of mixed motives*, Motives, Proc. Symp. Pure Math., vol. 55/1, AMS, Providence, 1994, pp. 517–525.

[8]   ———, *Motivic L-functions and regularized determinants*, Motives, Proc. Symp. Pure Math., vol. 55/1, AMS, Providence, 1994, pp. 707–743.

[9]   ———, *Motivic L-functions and regularized determinants II*, F. Catanese (Hrsg.) Proc. Arithmetic Geometry (Cortona), 1994.

[10] ———, *Some analogies between number theory and dynamical systems on foliated spaces*, Documenta Mathematica, extra vol., Plenary Talks, vol. I, ICM, 1998, pp. 23–46.

[11] C. Deninger and M. Schröter, *A distribution theoretic proof of Guinand's functional equation for Cramér's V-Function and generalizations*, J. London Math. Soc. **52** (1995), 48–60.

[12] P. L. Duren, *Theory of H^p spaces*, Academic Press, New York, 1970.

[13] D. Goldfeld, *Explicit formulae as trace formulae*, Number theory, trace formulas and discrete groups (K. E. Aubert, E. Bombieri, and D. Goldfeld, eds.), Academic Press, 1989, pp. 281–288.

[14] D. A. Hejhal, *The Selberg trace formula for PSL(2,R), Volume I*, Lecture Notes in Mathematics, vol. 548, Springer, Berlin, 1976.

[15] G. Illies, *Regularized products and determinants*, Commun. Math. Phys. **220** (2001), 69–94.

[16] J. Jorgenson, S. Lang, and D. Goldfeld, *Explicit formulas*, Lecture Notes in Mathematics, vol. 1593, Springer, Berlin, 1994.

[17] N. Katz, *Review of l-adic cohomology*, Motives, Proc. Symp. in Pure Math., vol. 55/1, AMS, Providence, 1994, pp. 21–30.

[18] A. Selberg, *Harmonic analysis and discontinuous groups in weakly symmetric Riemannian spaces with applications to Dirichlet series*, J. Indian Math. Soc. **20** (1956), 47–87.

[19] A. Weil, *Sur les "formules explicites" de la théorie des nombres premier*, Comm. Séminaire Math. Université de Lund (1952), 252–265.

On a Binary Diophantine Inequality Involving Prime Powers

A. Kumchev and M. B. S. Laporta

1 Introduction

Given $c > 1$, let $H(c)$ denote the least number s such that, for every fixed $\varepsilon > 0$ and for all real $N \geq N_0(\varepsilon)$, the inequality

$$\left| p_1^c + \cdots + p_s^c - N \right| < \varepsilon \tag{1.1}$$

has solutions in prime numbers $p_1, \ldots, p_s$. In 1952 Piatetski-Shapiro [12] established the existence of $H(c)$ for every (non-integer) $c > 1$ and proved that

$$\limsup_{c \to +\infty} \frac{H(c)}{c \log c} \leq 4 .$$

For c close to 1, the celebrated three primes theorem by Vinogradov [15] suggests that one should expect $H(c) \leq 3$. In [12] Piatetski-Shapiro showed that the upper bound $H(c) \leq 5$ holds if $1 < c < 3/2$. In 1992, Tolev [14] established the bound $H(c) \leq 3$ for $1 < c < 15/14$. (Tolev also showed that one can let ε be a function of N which tends to zero as N tends to infinity.) Subsequently, several authors sharpened Tolev's result, by improving on the range for c (see [2], [7], [8]). The most recent improvement is due to the first author [7], who used Harman's sieve [3], [4] to show that Tolev's theorem holds for $1 < c < 61/55$.

Our purpose in this paper is to improve a recent result [9] of the second author regarding the solvability of inequality (1.1) for $s = 2$. The result states that, for *almost all* $y \in [N, 2N)$ (i.e., the Lebesgue measure of the set of the exceptions is $o(N)$), the following inequality has solutions in primes p_1, p_2:

$$\left| p_1^c + p_2^c - y \right| < \varepsilon, \tag{1.2}$$

where $\varepsilon = N^{1-15/(14c)} \log^C N$ with $1 < c < 15/14$ and some positive constant C. (In fact, a careful examination of the proof of [9, Theorem 2] reveals that one can obtain the same conclusion with $9/8$ in place of $15/14$ after a little bit of extra work.) Here we use the method of [7] in order to show that Laporta's result holds for $1 < c < 6/5$ if one is content with an

307

upper bound for the measure of the exceptional set which is slightly weaker than the one in [9]. More precisely, we prove:

Theorem 1. *Let c be fixed with* $1 < c < 6/5$ *and let* $\delta > 0$ *be a fixed number that is sufficiently small in terms of c. Also, let N be a sufficiently large real number, let* $\varepsilon \geq N^{1-6/(5c)+\delta}$, *and let g be a function satisfying*

$$\lim_{N \to \infty} g(N) = \infty.$$

There exists a measurable set $\mathcal{H}$ *having Lebesgue measure*

$$|\mathcal{H}| \ll g(N)^2 N (\log N)^{-1/2}$$

such that for each $y \in [N, 2N] \setminus \mathcal{H}$, *the number* $R(y)$ *of solutions of* (1.2) *satisfies*

$$R(y) \gg \frac{\varepsilon N^{2/c-1}}{\log^2 N}, \tag{1.3}$$

where the implied constants depend at most on c and δ.

In [7, Theorem 2] it was also shown that one should expect $H(c) \leq 3$, at least when $1 < c < 3/2$. We apply a similar probabilistic argument to prove the following result for the binary case.

Theorem 2. *For almost all (in the sense of Lebesgue measure)* $c \in (1,2)$, *one can find an* N_0 *such that if* $N \geq N_0$ *there is a set* $\mathcal{H}$ *having Lebesgue measure* $|\mathcal{H}| \ll N (\log N)^{-1/2}$ *and such that, for each* $y \in [N, 2N] \setminus \mathcal{H}$ *and each* $\varepsilon \geq N^{1-2/c} (\log N)^{10}$, *the number* $R(y, c)$ *of solutions of* (1.2) *satisfies* (1.3).

The Piatetski-Shapiro inequality (1.1) may be considered as a variant of the Waring-Goldbach problem on representations of natural numbers as sums of kth powers of primes, with a bounded number of summands. It is worth recalling that, for $k = 2$, every sufficiently large natural number (subject to some natural congruence conditions) is the sum of at most five prime squares; see [5]. Moreover, the problem can be solved with a smaller number of summands by showing that three prime squares are sufficient to represent almost all sufficiently large natural numbers. Our Theorem 2 provides a result which, at least for values c that are close to 2, is better than what one would expect, given the current state of knowledge of the Goldbach-Waring problem for squares. This suggests extending the investigation of (1.1) to the case when c is close to any positive integer. The authors hope to present an account of the developments arising from this remark in a future paper.

Notation. Throughout this paper, the letters p, q, r, with or without subscripts, always denote primes; d, k, ℓ, m, n denote integers. We choose $X = \frac{1}{2}N^{1/c}$ and $\eta = \delta^2$, where δ is the number from the statement of Theorem 1. We write $m \sim M$ if m runs through the interval $[M, 2M)$. Also, we will use a kernel K having the following properties:

1. There is a constant $C(\eta) > 0$, depending only on η, such that
$$|K(x)| \leq C(\eta) \min\left(\varepsilon, |x|^{-1}|\varepsilon x|^{-10/\eta}\right).$$

2. Both K and its Fourier transform
$$\widehat{K}(y) = \int_{\mathbb{R}} K(x)e(-xy)\,dx$$
are non-negative, where henceforth we set $e(x) = e^{2\pi i x}$.

3. If χ is the characteristic function of the interval $(-\varepsilon, \varepsilon)$, then, for all real x, one has
$$\tfrac{1}{3}\chi(4x) \leq \widehat{K}(x) \leq \chi(x).$$

One can construct such a function by dilating the kernel described in [1, Lemma 1]. Finally, throughout the paper, we say that an assertion holds for 'almost all' $y \in [N, 2N)$, if given a sufficiently large real N one can find a measurable set $\mathcal{H}$ with Lebesgue measure $|\mathcal{H}| \ll g(N)^2 N(\log N)^{-1/2}$ and such that the assertion is true for all $y \in [N, 2N) \setminus \mathcal{H}$.

2 The Sieve Method

Writing
$$P(z) = \prod_{p<z} p,$$
we set, for any sequence of integers $\mathcal{E}$ with weights $w(n)$, $n \in \mathcal{E}$,
$$S(\mathcal{E}, z) = \sum_{\substack{n \in \mathcal{E} \\ (n, P(z))=1}} w(n),$$
and we denote by $\mathcal{E}_d$ the subsequence of elements $n \in \mathcal{E}$ with $n \equiv 0$ (mod d). For every $y \in [N, 2N)$, we define $\mathcal{A} = \mathcal{A}(y)$ to be the sequence of integers $n \in [X, 2X)$ with weights
$$w(n) = w(n, y) = \sum_{p \sim X} \widehat{K}(p^c + n^c - y).$$

Since
$$R(y) \geq \sum_{p_1, p_2 \sim X} \widehat{K}(p_1^c + p_2^c - y) = S(\mathcal{A}, (2X)^{1/2}),$$

in order to prove Theorem 1, it suffices to show that the estimate

$$S(\mathcal{A}, (2X)^{1/2}) \gg \frac{\varepsilon X^{2-c}}{\log^2 X} \qquad (2.1)$$

holds for almost all $y \in [N, 2N)$.

In sieve theory, a common technique for obtaining sharp lower bounds is to combine upper bounds with (weaker) known lower bounds using combinatorial identities such as *Buchstab's identity*

$$S(\mathcal{E}, z_1) = S(\mathcal{E}, z_2) - \sum_{z_2 \leq p < z_1} S(\mathcal{E}_p, p). \qquad (2.2)$$

Our proof of (2.1) uses a version of this idea developed by Harman [3], [4]. Let $\mathcal{B}$ be the set of integers in $[X, 2X]$. The arithmetic information is provided in the form of asymptotic formulas

$$\sum_{m \sim M} a(m) S(\mathcal{A}_m, z(m)) = \lambda \sum_{m \sim M} a(m) S(\mathcal{B}_m, z(m)) + \text{ error terms, } (2.3)$$

where λ is suitably chosen. We expect the error terms here to be always 'small', but of course, we can prove this only for certain values of M and $z(m)$. We apply (2.2) repeatedly to express the left-hand side of (2.1) as a linear combination of sifting functions. This decomposition of $S(\mathcal{A}, (2X)^{1/2})$ is done so that we have asymptotic formulas for most of the sifting functions that arise and suitable upper and lower bounds for the few sifting functions for which we have no asymptotic formula. Combining the estimates for all the terms in the decomposition, we obtain the lower bound.

In the remainder of this section we set up this decomposition. We set $A = X^{8/75}$, $B = X^{1/5}$, $C = X^{11/41}$, $D = X^{1/3}$, and $F = X^{11/25}$. Applying (2.2), we find

$$S(\mathcal{A}, (2X)^{1/2}) = S(\mathcal{A}, A) - \sum_{A \leq p < B} S(\mathcal{A}_p, p)$$

$$- \sum_{B \leq p \leq C} S(\mathcal{A}_p, p) - \sum_{C < p < D} S(\mathcal{A}_p, p)$$

$$- \sum_{D \leq p \leq F} S(\mathcal{A}_p, p) - \sum_{F < p < \sqrt{2X}} S(\mathcal{A}_p, p)$$

$$= S_1 - S_2 - S_3 - S_4 - S_5 - S_6, \quad \text{say.} \qquad (2.4)$$

We decompose S_2 and S_4 further. Another application of (2.2) gives

$$S_2 = \sum_{A \leq p < B} S(\mathcal{A}_p, A) - \sum_{\substack{A \leq q < p < B \\ pq \leq C}} S(\mathcal{A}_{pq}, q)$$

$$- \sum_{\substack{A \leq q < p < B \\ C < pq < D}} S(\mathcal{A}_{pq}, q) - \sum_{\substack{A \leq q < p < B \\ pq \geq D}} S(\mathcal{A}_{pq}, q)$$

$$= S_7 - S_8 - S_9 - S_{10}, \quad \text{say.}$$

Similarly, we obtain

$$S_4 = \sum_{C < p < D} S(\mathcal{A}_p, A) - \sum_{\substack{C < p < D \\ A \leq q < B}} S(\mathcal{A}_{pq}, q)$$

$$- \sum_{\substack{C < p < D \\ B \leq q \leq C}} S(\mathcal{A}_{pq}, q) - \sum_{C < q < p < D} S(\mathcal{A}_{pq}, q)$$

$$= S_{11} - S_{12} - S_{13} - S_{14}, \quad \text{say.}$$

Consider now S_9. Let $S_9^\dagger$ and $S_9^\ddagger$ be the subsums of S_9 in which $pq^2 \leq F$ and $pq^2 > F$, respectively. Using (2.2) two more times, we decompose $S_9^\dagger$ further as

$$S_9^\dagger = \sum_{p,q:(\dagger)} S(\mathcal{A}_{pq}, A) - \sum_{\substack{p,q:(\dagger) \\ A \leq r < q}} S(\mathcal{A}_{pqr}, A)$$

$$+ \sum_{\substack{p,q:(\dagger) \\ A \leq s < r < q}} S(\mathcal{A}_{pqrs}, s)$$

$$= S_{15} - S_{16} + S_{17}, \quad \text{say;}$$

here r and s are primes and ($\dagger$) stands for the summation conditions in $S_9^\dagger$. Finally, we deal with the sum S_6, which counts almost primes having two prime factors,

$$S_6 = \sum_{\substack{pq \sim X \\ F < p \leq q}} w(pq).$$

It turns out to be more convenient to switch the sifting process from the product pq to the prime variable appearing in the definition of $w(n)$. In order to do so, we write S_6 as $S(\mathcal{A}^*, (2X)^{1/2})$, where $\mathcal{A}^* = \mathcal{A}^*(y)$ is the set of integers in $[X, 2X)$ with weights

$$w^*(n) = \sum_{\substack{pq \sim X \\ F < p \leq q}} \widehat{K}(n^c + (pq)^c - y).$$

Let S_i^* denote a sum analogous to S_i, with $\mathcal{A}$ replaced by $\mathcal{A}^*$. We decompose $S(\mathcal{A}^*, (2X)^{1/2})$ via (2.4) and then deal with S_2^* exactly as we did with S_2, but we do not decompose S_4^* and S_6^* further. Combining all these decompositions, we obtain the identity

$$
\begin{aligned}
S(\mathcal{A}, (2X)^{1/2}) = {} & S_1 - S_3 - S_5 - S_7 + S_8 + S_9^{\ddagger} + S_{10} \\
& - S_{11} + S_{12} + S_{13} + S_{14} + S_{15} - S_{16} \\
& + S_{17} - S_1^* + S_3^* + S_4^* + S_5^* + S_6^* + S_7^* \\
& - S_8^* - S_9^{\ddagger *} - S_{10}^* - S_{15}^* + S_{16}^* - S_{17}^*.
\end{aligned}
\qquad (2.5)
$$

In Section 4, we will establish asymptotic formulas of the form (2.3) for almost all $y \in [N, 2N)$ (cf. Lemmas 7 and 8). Those results will allow us to evaluate the terms on the right of (2.5). The most difficult part of the proofs of the asymptotic formulas is the estimation of certain double exponential sums; these are dealt with in the next section. In Section 5, we combine (2.5) with Lemmas 7 and 8, and after some numerical work, we are able to show that the lower bound resulting from the above procedure is non-trivial.

3   Exponential Sums

In this section we gather some estimates for exponential sums of the form

$$
U(x) = \sum_{m \sim M} \sum_{\ell \sim L} a(m) b(\ell) e(x(m\ell)^c),
\qquad (3.1)
$$

where $a(m)$, $b(\ell)$ are complex numbers of modulus ≤ 1 and $ML \asymp X$.

Lemma 1. *Let $1 < c < 6/5$ and let $X^{1-c-\eta} \leq |x| \leq X^{6/5-c+\eta}$. Let $U(x)$ be defined by (3.1). We then have*

$$
|U(x)| \ll X^{9/10+\eta},
\qquad (3.2)
$$

whenever

$$
X^{1/5} \ll L \ll X^{11/41},
\qquad (3.3)
$$

or

$$
X^{1/3} \ll L \ll X^{11/25}.
\qquad (3.4)
$$

Proof. Inequalities (3.4) and the restriction $c < 6/5$ are the assumptions under which [13, Theorem 9] provides the upper bound (3.2). The proof of

(3.3) is easier and, in fact, repeats that of [7, Lemma 6]. By the Weyl–van der Corput inequality,

$$|U|^2 \ll \frac{X^2}{Q} + \frac{X}{Q} \sum_{q \leq Q} \sum_{\ell \sim L} \left| \sum_{m \sim M} e(f(m)) \right|,$$

where $f(m) = x((\ell + q)^c - \ell^c)m^c$ and $Q \leq L$ is a parameter at our disposal. We choose $Q = X^{1/5 - 2\eta}$. Then, by using an exponent pair (κ, λ) and the hypotheses about x, we get

$$|U|^2 \ll X^{9/5 + 2\eta} + X^{1 + \lambda + 2\kappa/5} L^{1 - \lambda}.$$

Thus, the lemma follows provided that

$$L \ll X^{1 - (2\kappa + 1)/5(1 - \lambda)}.$$

Using $(\kappa, \lambda) = BABABA^3 B(0, 1) = (\frac{13}{49}, \frac{57}{98})$, we can now infer (3.2) from (3.3). $\qquad\square$

Lemma 2. *Let $1 < c < 6/5$ and let $X^{1-c-\eta} \leq |x| \leq X^{6/5-c+\eta}$. Also, let $U(x)$ be defined by (3.1) with $b(\ell) = 1$ for all ℓ, $\ell \sim L$. We then have*

$$|U(x)| \ll X^{9/10 + \eta}, \tag{3.5}$$

whenever

$$L \gg X^{1/2}. \tag{3.6}$$

Proof. For $L \geq X^{3/5}$, the desired bound follows by applying the exponent pair $(\frac{1}{6}, \frac{2}{3})$ to the summation over ℓ and then summing the resulting estimate over m. We now treat the case $X^{1/2} \leq L \leq X^{3/5}$. We first apply Cauchy's inequality and Weyl's lemma with $Q = X^{1/5}$ to obtain

$$|U|^2 \ll \frac{X}{Q} \sum_{q \leq Q} \sum_{m \sim M} \sum_{\ell} e(x((\ell + q)^c - \ell^c)m^c) + X^{9/5}, \tag{3.7}$$

where ℓ runs through a subinterval of $[L, 2L)$. Denote the sum over (m, ℓ) by $U_1(q)$. Applying the truncated Poisson formula and partial summation to the variables ℓ and m successively we find

$$|U_1(q)| \ll XF^{-1} \left| \sum_{\mu, \nu} e(f(\mu, \nu)) \right| + E,$$

where for $F = |x|qX^c L^{-1}$, $\mu \cong FM^{-1}$, $\nu \cong FL^{-1}$,

$$E = X^\eta \left(XF^{-1/2} + M + F^{1/2} \right),$$

and $f(\mu, \nu)$ is a C^∞ function satisfying

$$\frac{\partial^{i+j}}{\partial \mu^i \partial \nu^j} f(\mu, \nu) = B_{i,j} (xq)^{1/(2-2c)} \nu^{1/2-j} \mu^{c/(2c-2)-i} \left(1 + O\left(\frac{q}{L}\right)\right)$$
$$\asymp F \mu^{-i} \nu^{-j},$$

where $B_{i,j}$ is a constant depending on i and j. Substituting the estimate for $U_1(q)$ in (3.7), we get

$$|U|^2 \ll \frac{X^2}{Q} \sum_{q \leq Q} F^{-1} \left| \sum_{\mu, \nu} e(f(\mu, \nu)) \right| + X^{9/5}. \tag{3.8}$$

Estimating the sum over (μ, ν) in (3.8) via Kolesnik's AB-theorem [10, Lemma 9], we complete the proof of the lemma. $\qquad\square$

Lemma 3. *Let $1 < c < 3/2$ and let $X^{3/2-2c-2\eta} \leq |x| \leq X^{1-c-\eta}$. Also let $U(x)$ be defined by (3.1). Then,*

$$|U(x)| \ll X^{1-\eta/3}, \tag{3.9}$$

whenever

$$X^\eta \ll L \ll X^{1/2}. \tag{3.10}$$

Furthermore, if $b(\ell) = 1$ for all ℓ, $\ell \sim L$, (3.9) holds for

$$L \gg X^\eta. \tag{3.11}$$

Proof. We first consider (3.9) under the assumption (3.10). By Cauchy's inequality and a Weyl shift,

$$|U(x)|^2 \ll \frac{X^2}{Q} + \frac{X}{Q} \sum_{q \leq Q} \sum_{\ell \sim L} \left| \sum_{m \sim M} e(f(m)) \right|,$$

where $f(m) = x((\ell+q)^c - \ell^c)m^c$ and $Q \leq L$ is a parameter at our disposal. We now estimate the sum over m via the Kuz'min–Landau inequality and obtain

$$|U(x)|^2 \ll X^2 Q^{-1} + |x|^{-1} X^{2-c} Q^{-1} L \log X. \tag{3.12}$$

This is justified, provided that $|f'(m)| \leq 1/2$. To complete the proof of (3.9) we choose

$$Q = \begin{cases} X^{2\eta/3} & \text{if } |x| \geq X^{1-c-2\eta}, \\ \min\left(L, |x|^{-1} X^{1-c-\eta}\right) & \text{if } |x| < X^{1-c-2\eta}. \end{cases}$$

Note that in both cases

$$|f'(m)| \asymp |x|qX^{c-1} \ll X^{-\eta/3} = o(1),$$

and the expression

$$X^{2-2\eta/3} + X^{1/2+c+3\eta}$$

is an upper bound for the right side of (3.12).

We now consider the case when all the coefficients $b(\ell)$ are equal to one. For $L \leq X^{1-\eta}$ we can apply the first part of the lemma (with M and L interchanged if $L \geq X^{1/2}$). If $L \geq X^{1-\eta}$, we can prove (3.9) by applying the exponent pair $(\frac{1}{2}, \frac{1}{2})$ to the summation over ℓ. This gives

$$|U(x)| \ll M \left((|x|X^{c-1}M)^{1/2}L^{1/2} + (|x|X^{c-1}M)^{-1} \right)$$
$$\ll X^{1/2+\eta} + X^{c-1/2+2\eta}.$$

This completes the proof of (3.9). $\qquad\qquad\qquad\qquad\qquad\qquad$ □

4 Mean Square Estimates

We start this section by introducing some further notation. We will use the following exponential sums and integrals

$$S(x) = \sum_{p \sim X} e(xp^c), \qquad S_1(x) = \sum_{\substack{pq \sim X \\ X^{0.44} < p \leq q}} e(x(pq)^c),$$

$$I_0(x) = \int_X^{2X} \frac{e(xt^c)}{\log t}dt, \qquad I_1(x) = \sum_{X^{0.44} < p < \sqrt{2X}} \frac{1}{p} \int_X^{2X} \frac{e(xt^c)}{\log(t/p)}dt.$$

Also, we set $\tau = X^{1-c-\eta}$ and define

$$W_j(n,y) = \int_{-\tau}^{\tau} I_j(x)K(x)e((n^c - y)x)dx \qquad (j = 0,1).$$

Finally, X^σ denotes a function of the form $e^{a(\log X)^{1/4}}$ with some unspecified constant $a > 0$; in particular, we may write $X^{-\sigma}(\log X)^A \ll X^{-\sigma}$ instead of

$$e^{-a(\log X)^{1/4}}(\log X)^A \ll e^{-\frac{a}{2}(\log X)^{1/4}}.$$

Lemma 4. *Let* $1 < c < 6/5$. *Then*

$$\int_N^{2N} \left| \sum_{m \sim M} \sum_{\ell \sim L} a(m)b(\ell)\big(w(m\ell,y) - W_0(m\ell,y)\big) \right|^2 dy \ll \varepsilon^2 X^{4-c-\sigma}, \quad (4.1)$$

whenever L satisfies the inequalities (3.3) *or* (3.4). *Moreover, the same estimate holds when $b(\ell) = 1$ for all ℓ and L satisfies* (3.6).

Proof. Let $U(x)$ be given by (3.1). By the Fourier inversion formula,

$$\sum_{m\sim M}\sum_{\ell\sim L} a(m)b(\ell)w(m\ell, y) = \int_{\mathbb{R}} S(x)U(x)K(x)e(-yx)dx.$$

Denote the last integral by $D(X, y)$. It suffices to establish the estimate

$$\int_N^{2N} \left| D(X,y) - \int_{-\tau}^{\tau} I_0(x)U(x)K(x)e(-yx)dx \right|^2 dy \ll \varepsilon^2 X^{4-c-\sigma}. \quad (4.2)$$

We partition the line into three subsets

$$E_1 = (-\tau, \tau), \quad E_2 = \{x\colon \tau \le |x| \le H\}, \quad E_3 = \mathbb{R}\setminus(E_1\cup E_2),$$

where $H = X^{6/5-c+\eta}$, and write

$$D_i(X, y) = \int_{E_i} S(x)U(x)K(x)e(-yx)dx \qquad (i = 1, 2, 3).$$

In view of the rapid decay of K, we readily have

$$\int_N^{2N} |D_3(X,y)|^2 dy \ll 1. \quad (4.3)$$

We now show that

$$\int_N^{2N} \left| D_1(X,y) - \int_{E_1} I_0(x)U(x)K(x)e(-yx)dx \right|^2 dy \ll \varepsilon^2 X^{4-c-\sigma}, \quad (4.4)$$

$$\int_N^{2N} |D_2(X,y)|^2 dy \ll \varepsilon^2 X^{4-c-\eta}. \quad (4.5)$$

The first step towards both of these estimates is an application of Plancherel's theorem. For any function $F \in L^1(\mathbb{R}) \cap L^2(\mathbb{R})$, one has

$$\int_N^{2N} \left| \int_{\mathbb{R}} F(x)e(-yx)dx \right|^2 dy$$

$$\le \int_{\mathbb{R}} \left| \int_{\mathbb{R}} F(x)e(-yx)dx \right|^2 dy = \int_{\mathbb{R}} |F(x)|^2 dx. \quad (4.6)$$

Let $\mathbf{1}_{\mathfrak{B}}$ denote the indicator function of a measurable set $\mathfrak{B}$. Applying (4.6) to the function $F = (S - I_0)UK\mathbf{1}_{E_1}$, we find that the left side of (4.4) is

$$\leq \int_{E_1} |(S(x) - I_0(x))U(x)K(x)|^2 \, dx$$

$$\ll \varepsilon^2 \sup_{x \in E_1} |S(x) - I_0(x)|^2 \int_{E_1} |U(x)|^2 dx.$$

Thus, (4.4) follows from the estimates

$$\sup_{x \in E_1} |S(x) - I_0(x)| \ll X^{1-\sigma}, \quad \int_{E_1} |U(x)|^2 dx \ll X^{2-c}(\log X)^4. \quad (4.7)$$

The latter can be proven similarly to the bounds in [14, Lemma 7]. To prove the first inequality in (4.7) we use partial summation and the asymptotic relation

$$\sum_{X \leq p < Y} (\log p)e(xp^c) = \int_X^Y e(xt^c)dt + O(X^{1-\sigma}), \quad (4.8)$$

valid for $|x| \leq \tau$ and $X < Y \leq 2X$. This can be established similarly to [14, Lemma 14].

Finally, we prove (4.5). If $F = SUK\mathbf{1}_{E_2}$, we deduce from (4.6) that

$$\int_N^{2N} |D_2(X,y)|^2 dy \ll \varepsilon^2 \max_{x \in E_2} |U(x)|^2 \sum_{0 \leq h \leq H} \int_h^{h+1} |S(x)|^2 dx.$$

By [14, Lemma 7], $\int_h^{h+1} |S(x)|^2 dx \ll X$, so (4.5) follows from the upper bound

$$\max_{x \in E_2} |U(x)| \ll X^{9/10+\eta}$$

upon choosing η sufficiently small in terms of δ. Hence, we can complete the proof of the lemma by referring to (3.2), if L satisfies (3.3) or (3.4), or to (3.5), if L satisfies (3.6). $\qquad \square$

Lemma 5. *Let $1 < c < 6/5$. Then*

$$\int_N^{2N} \left| \sum_{m \sim M} \sum_{\ell \sim L} a(m)b(\ell)\big(w^*(m\ell, y) - W_1(m\ell, y)\big) \right|^2 dy \ll \varepsilon^2 X^{4-c-\sigma}, \quad (4.9)$$

whenever L satisfies the inequalities (3.3) or (3.4). Moreover, the same estimate holds when $b(\ell) = 1$ for all ℓ and L satisfies (3.6).

Proof. This is similar to the proof of Lemma 4. We need to prove an analogue of (4.2) with $S_1(x)$ in place of $S(x)$ and $I_1(x)$ in place of $I_0(x)$. The corresponding analogues of (4.3) and (4.5) can be established exactly as in the previous proof, but we need to approach the inequality

$$\int_N^{2N} \left| \int_{E_1} (S_1(x) - I_1(x))U(x)K(x)e(-xy)dx \right|^2 dy \ll \varepsilon^2 X^{4-c-\sigma}, \quad (4.10)$$

slightly differently. For, using (4.8) with X/p in place of X and xp^c in place of x, we now have the approximation

$$S_1(x) = I_1(x) + O(X^{1-\sigma}) \quad (4.11)$$

only for $|x| \leq \tau_1 := X^{3/2-2c-2\eta}$. Let

$$E_{1,1} = (-\tau_1, \tau_1) \quad \text{and} \quad E_{1,2} = E_1 \setminus E_{1,1}.$$

Using (4.11), we can show similarly to (4.4) that

$$\int_N^{2N} \left| \int_{E_{1,1}} (S_1(x) - I_1(x))U(x)K(x)e(-xy)dx \right|^2 dy \ll \varepsilon^2 X^{4-c-\sigma}.$$

To finish the proof we show that

$$\int_N^{2N} \left| \int_{E_{1,2}} (S_1(x) - I_1(x))U(x)K(x)e(-xy)dx \right|^2 dy \ll \varepsilon^2 X^{4-c-\eta/2}.$$

This inequality is similar to (4.5). Indeed, by virtue of the estimates (which can be established similarly to those in [14, Lemma 7])

$$\int_{E_1} |S_1(x)|^2 dx \ll X^{2-c}(\log X)^2 \quad \text{and} \quad \int_{E_1} |I_1(x)|^2 dx \ll X^{2-c}(\log X)^2,$$

one can adapt the proof of (4.5) so that it suffices to have the bound

$$\max_{x \in E_{1,2}} |U(x)| \ll X^{1-\eta/3},$$

which, under the given hypotheses, is provided by Lemma 3. $\qquad \square$

Lemma 6. *Let* $1 < c < 6/5$, $ML \asymp X$, *and suppose* L *satisfies one of the inequalities* (3.3) *or* (3.4). *Let* I, J *be integers and* $\mathfrak{I}_i$, $\mathfrak{J}_j$ *be intervals for* $1 \leq i \leq I$, $1 \leq j \leq J$. *Write*

$$a(m, \ell) = \sum_{\substack{rp_1 \cdots p_I = \ell \\ p_1 < p_2 < \cdots < p_I \\ p_i \in \mathfrak{I}_i}} c(\ell) \sum_{\substack{lq_1 \cdots q_J = m \\ q_1 < q_2 < \cdots < q_J \\ q_j \in \mathfrak{J}_j}} d(m)$$

with $|c(\ell)|$, $|d(m)| \le 1$ and $p_1, \ldots, p_I$ and $q_1, \ldots, q_J$ *satisfying* $O(1)$ *joint conditions of the form*

$$p_u \le q_v \quad or \quad q_v \le p_u,$$

or

$$\prod_{u \in \mathcal{U}} p_u \prod_{v \in \mathcal{V}} q_v \le Q \quad or \quad \prod_{u \in \mathcal{U}} p_u \ge \prod_{v \in \mathcal{V}} q_v$$

for given subsets $\mathcal{U} \subset \{1, \ldots, I\}$ *and* $\mathcal{V} \subset \{1, \ldots, J\}$, $Q \le X$, *or similar conditions. Then*

$$\int_N^{2N} \left| \sum_{m \sim M} \sum_{\ell \sim L} a(m, \ell)\big(w(m\ell, y) - W_0(m\ell, y)\big) \right|^2 dy \ll \varepsilon^2 X^{4-c-\sigma}.$$

Furthermore, the result still holds if we replace $w(n)$ *by* $w^*(n)$ *and* $W_0(n)$ *by* $W_1(n)$.

Proof. We combine Lemmas 4 and 5 with Perron's formula [4, (15)]. The details can be found in [4, Lemma 1] or in [7, Lemma 11].   □

$$\prod_{u \in \mathcal{U}} p_u \prod_{v \in \mathcal{V}} q_v \le Q \quad or \quad \prod_{u \in \mathcal{U}} p_u \ge \prod_{v \in \mathcal{V}} q_v$$

Let

$$I(x) = \int_X^{2X} e(xt^c)\, dt,$$

and define

$$J_0(X, y) = \int_{\mathbb{R}} I_0(x) I(x) K(x) e(-yx)\, dx,$$

$$J_1(X, y) = \int_{\mathbb{R}} I_1(x) I(x) K(x) e(-yx)\, dx.$$

We shall also use the *Buchstab function* $\omega(x)$ which is the continuous solution of the differential-difference equation

$$\begin{cases} \omega(x) = 1/x & \text{if } 1 < x \le 2, \\ (x\omega(x))' = \omega(x-1) & \text{if } x > 2. \end{cases}$$

It will enter the discussion through the asymptotic relation

$$\sum_{\substack{X \le n < Y \\ (n, \bar{P}(z)) = 1}} 1 = \frac{Y - X}{\log z}\, \omega\left(\frac{\log X}{\log z}\right) + O\left(X(\log X)^{-2}\right), \tag{4.12}$$

which is valid when $X < Y \le 2X$ and $z \ge X^a$ for some constant $a > 0$.

Lemma 7. *Let $1 < c < 6/5$ and $u \geq 1$, and suppose that, for some L satisfying (3.3) or (3.4), there exists a set $\mathcal{D} \subset \{1, \ldots, u\}$ with*

$$\prod_{j \in \mathcal{D}} p_j \asymp L.$$

Then

$$\int_N^{2N} \left| \sum_{p_1, \ldots, p_u} \left(S(\mathcal{A}_{p_1 \cdots p_u}, p_1) - \frac{J_0(X, y)}{X} S(\mathcal{B}_{p_1 \cdots p_u}, p_1) \right) \right|^2 dy$$

$$\ll \varepsilon^2 X^{4-c} (\log X)^{-4.5}. \quad (4.13)$$

Here the summation is over primes $p_1, \ldots, p_u \geq X^{8/75}$ satisfying $p_j > p_1$, together with $O(1)$ further conditions of the type

$$p_j \leq p_l \quad or \quad Q \leq \prod_{j \in \mathcal{F}} p_j \leq R$$

for some $\mathcal{F} \subset \{1, \ldots, u\}$ and $R \leq X$. Furthermore, the result still holds if we replace $\mathcal{A}$ by $\mathcal{A}^*$ and J_0 by J_1, and/or if, instead of L, X/L satisfies (3.3) or (3.4).

Proof. We have

$$\sum_{p_1, \ldots, p_u} S(\mathcal{A}_{p_1 \cdots p_u}, p_1) = \sum_{p_1, \ldots, p_u} \sum_{\substack{n \sim X/p_1 \cdots p_u \\ (n, P(p_1)) = 1}} w(np_1 \cdots p_u).$$

Upon writing the product $np_1 \cdots p_u$ as $m\ell$ where

$$\ell = \prod_{j \in \mathcal{D}} p_j \quad and \quad m = \left(\prod_{j \notin \mathcal{D}} p_j \right) \cdot n,$$

we can now use Lemma 6 to replace $\sum S(\mathcal{A}_{p_1 \cdots p_u}, p_1)$ in (4.13) by

$$\int_{-\tau}^{\tau} I_0(x) U(x) K(x) \, e(-yx) \, dx \quad (4.14)$$

where

$$U(x) = \sum_{p_1, \ldots, p_u} \sum_{\substack{n \sim X/p_1 \cdots p_u \\ (n, P(p_1)) = 1}} e(x(np_1 \cdots p_u)^c).$$

(Note that $|U(x)| \ll X(\log X)^{-1}$.) Then, using (4.6) and the bound (cf. [11, Lemma 5.1])

$$|I_0(x)| \ll \frac{1}{|x| X^{c-1} \log X}, \quad (4.15)$$

we obtain

$$\int_N^{2N} \left| \int_{|x| \geq \tau_1} I_0(x) U(x) K(x) e(-yx) dx \right|^2 dy \ll \frac{\varepsilon^2 X^{4-c}}{(\log X)^{4.5}},$$

where $\tau_1 = X^{-c}(\log X)^{1/2}$. Hence we may replace (4.14) by a similar integral with τ_1 in place of τ. Similarly, we may replace in (4.13) $J_0(X, y)$ by

$$\int_{-\tau_1}^{\tau_1} I_0(x) I(x) K(x) e(-yx) dx.$$

Therefore, it suffices to show that, for $|x| < \tau_1$, we have the approximation

$$U(x) = I(x) X^{-1} \sum_{p_1, \ldots, p_u} S(\mathcal{B}_{p_1 \cdots p_u}, p_1) + O\left(X(\log X)^{-3/2}\right);$$

this follows from (4.12) and partial summation. $\qquad \square$

Lemma 8. *Let $1 < c < 6/5$ and $M \leq X^{11/25}$. Suppose further that $a(m)$ are real numbers such that $a(m) \ll 1$ and $a(m) = 0$ unless all prime divisors of m are $\geq X^{8/75}$. Then*

$$\int_N^{2N} \left| \sum_{m \sim M} a(m)\left(S(\mathcal{A}_m, X^{8/75}) - \frac{J_0(X, y)}{X} S(\mathcal{B}_m, X^{8/75})\right) \right|^2 dy$$

$$\ll \varepsilon^2 X^{4-c}(\log X)^{-4.5}.$$

Furthermore, the result still holds if we replace $\mathcal{A}$ by $\mathcal{A}^*$ and J_0 by J_1.

Proof. The transition from $\sum a(m) S(\mathcal{A}_m, X^{8/75})$ to the integral (4.14), in which now

$$U(x) = \sum_{m \sim M} \sum_{\substack{n \sim X/m \\ (n, P(X^{8/75}))=1}} a(m) e(x(mn)^c),$$

is made by using the Eratosthenes–Legendre sieve as in [4, Lemma 2] or in [7, Lemma 13]. We then argue as in the proof of the previous lemma. $\qquad \square$

5   Proof of Theorem 1

5.1   The Lower Bound

We are now in position to complete the proof of Theorem 1 by combining (2.5) with Lemmas 7 and 8. Using these lemmas (and occasionally trivial estimates), we can evaluate the terms on the right side of (2.5) for almost all $y \in [N, 2N)$. For example, by Lemma 8, the measure of the set of values of y for which the inequality (see Section 2 for the definition of the quantities $A, B, \dots$)

$$\left| S_7 - \frac{J_0(X, y)}{X} \sum_{A \le p < B} S(\mathcal{B}_p, A) \right| < \frac{\varepsilon X^{2-c}}{g(N)(\log X)^2} \tag{5.1}$$

fails is $\ll g(N)^2 N (\log N)^{-1/2}$, i.e., (5.1) holds for almost all $y \in [N, 2N)$. We can apply similarly Lemma 8 to S_1, S_{11}, S_{15}, S_{16}, S_1^*, S_7^*, S_{15}^*, S_{16}^*; and Lemma 7 applies to S_3, S_5, S_8, S_{10}, S_{13}, S_3^*, S_5^*, S_8^*, S_{10}^*.

Let T_j be a sum similar to S_j or S_j^* in which the set $\mathcal{A}$ or $\mathcal{A}^*$ has been replaced by $\mathcal{B}$ (e.g., T_7 is the sum appearing on the left side of (5.1)), and write

$$\lambda_j = J_j(X, y)/X \qquad (j = 0, 1).$$

We have now shown that, for almost all $y \in [N, 2N)$,

$$\begin{aligned}
S(\mathcal{A}, (2X)^{1/2}) = {} & \lambda_0 \big(T_1 - T_3 - T_5 - T_7 + T_8 + T_{10} \\
& \qquad - T_{11} + T_{13} + T_{15} - T_{16} \big) \\
& - \lambda_1 \big(T_1 - T_3 - T_5 - T_7 + T_8 + T_{10} + T_{15} - T_{16} \big) \\
& + S_9^{\ddagger} + S_{12} + S_{14} + S_{17} + S_4^* + S_6^* - S_9^{\ddagger *} - S_{17}^* \\
& + O\left(\frac{\varepsilon X^{2-c}}{g(N)(\log X)^2} \right).
\end{aligned} \tag{5.2}$$

Next, we observe that, for almost all $y \in [N, 2N)$, we can obtain asymptotic formulas for S_{17} and S_{17}^*. Indeed, the summation conditions in these sums are such that either $pqr \ge X^{1/3}$ and pqr satisfies (3.4), or $rs \le X^{2/9}$ and rs satisfies (3.3). Thus, Lemma 7 can be used to evaluate S_{17} and S_{17}^*. Lemma 7 works also for the following subsums of S_{12} and S_{14}:

$$S_{12}' := \sum_{\substack{C < p < D \\ A \le q < B \\ D \le pq \le F}} S(\mathcal{A}_{pq}, q) \quad \text{and} \quad S_{14}' := \sum_{\substack{C < q < p < D \\ pq \ge X/D}} S(\mathcal{A}_{pq}, q).$$

Using the inequalities $S_{12} \ge S_{12}'$, $S_{14} \ge S_{14}'$, and the nonnegativity of $S_9^{\ddagger}$, S_4^* and S_6^*, we infer from (5.2) that, for almost all $y \in [N, 2N)$,

$$\begin{aligned}
S(\mathcal{A}, (2X)^{1/2}) \ge {} & \lambda_0 \big(T_1 - T_3 - T_5 - T_7 + T_8 + T_{10} - T_{11} \\
& \qquad + T_{12}' + T_{13} + T_{14}' + T_{15} - T_{16} + T_{17} \big) \\
& - \lambda_1 \big(T_1 - T_3 - T_5 - T_7 + T_8 + T_{10} + T_{15} - T_{16} + T_{17} \big) \\
& - S_9^{\ddagger *} + O\left(\frac{\varepsilon X^{2-c}}{g(N)(\log X)^2} \right).
\end{aligned}$$

Applying the decomposition for $S(\mathcal{A}, (2X)^{1/2})$ from Section 2 to the sifting function $S(\mathcal{B}, (2X)^{1/2})$, we see that the sum in the first set of parentheses equals

$$S(\mathcal{B}, (2X)^{1/2}) + T_6 - T_9^{\ddagger} - T_{12}'' - T_{14}'',$$

where

$$T_{12}'' := T_{12} - T_{12}' \quad \text{and} \quad T_{14}'' := T_{14} - T_{14}'.$$

Similarly, using the decomposition for $S(\mathcal{A}^*, (2X)^{1/2})$, we find that the sum in the second set of parentheses is

$$S(\mathcal{B}, (2X)^{1/2}) + T_4 + T_6 - T_9^{\ddagger}.$$

Finally, we note that $S_9^{\ddagger*}$ does not exceed

$$S_{18}^* := \sum_{\substack{A \leq q < p < B \\ C < pq < D \\ pq^2 > F}} S(\mathcal{A}_{pq}^*, A),$$

and evaluate the latter sum using Lemma 8. Therefore, for almost all $y \in [N, 2N)$,

$$
\begin{aligned}
S(\mathcal{A}, (2X)^{1/2}) \geq {} & \lambda_0 \big(S(\mathcal{B}, (2X)^{1/2}) + T_6 - T_9^{\ddagger} - T_{12}'' - T_{14}'' \big) \\
& - \lambda_1 \big(S(\mathcal{B}, (2X)^{1/2}) + T_4 + T_6 - T_9^{\ddagger} + T_{18} \big) \\
& + O\Big(\frac{\varepsilon X^{2-c}}{g(N)(\log X)^2} \Big).
\end{aligned}
\tag{5.3}
$$

In order to simplify the notation, we now drop the superscripts in T_{12}'', T_{14}'', and $T_9^{\ddagger}$. We observe that, for almost all $y \in [N, 2N)$,

$$J_1(X, y) = \log\big(\tfrac{14}{11}\big) \cdot J_0(X, y) + O\Big(\frac{\varepsilon X^{2-c}}{g(N)(\log X)} \Big).$$

This follows easily from the inequality

$$\big| I_1(x) - \log\big(\tfrac{14}{11}\big) I_0(x) \big| \ll X(\log X)^{-2}.$$

Using (4.12), the Prime Number Theorem and partial summation, we also obtain that

$$S(\mathcal{B}, (2X)^{1/2}) = \frac{X}{\log X} + O\big(X(\log X)^{-2} \big),$$

$$T_j = \frac{f_j X}{\log X} + O\big(X(\log X)^{-2} \big),$$

where the f_j's are constants, given by

$$f_4 = \int_2^{30/11} \omega(t)dt, \quad f_6 = \log\left(\tfrac{14}{11}\right),$$

$$f_{18} = \frac{75}{8} \iint_{\mathcal{D}_{18}} \omega\left(\frac{1-u-v}{8/75}\right) \frac{du\,dv}{uv},$$

$$f_j = \iint_{\mathcal{D}_j} \omega\left(\frac{1-u-v}{v}\right) \frac{du\,dv}{uv^2} \qquad (j = 9, 12, 14)$$

with

$$\mathcal{D}_9 = \mathcal{D}_{18} = \left\{(u,v): \tfrac{8}{75} < v < u < \tfrac{1}{5}, \tfrac{11}{41} < u+v < \tfrac{1}{3}, u+2v > \tfrac{11}{25}\right\},$$

$$\mathcal{D}_{12} = \left\{(u,v): \tfrac{8}{75} < v < \tfrac{1}{5}, \tfrac{11}{41} < u < \tfrac{1}{3}, u+v > \tfrac{11}{25}\right\},$$

$$\mathcal{D}_{14} = \left\{(u,v): \tfrac{11}{41} < v < u < \tfrac{1}{3}, u+v < \tfrac{14}{25}\right\}.$$

Substituting the above approximations into (5.3), we deduce, for almost all $y \in [N, 2N)$, that

$$S(\mathcal{A}, (2X)^{1/2}) \geq \frac{J_0(X,y)}{\log X}\left(1 - f_9 - f_{12} - f_{14}\right.$$

$$\left. - \log\left(\tfrac{14}{11}\right)\left(f_4 + f_6 - f_9 + f_{18}\right)\right) + O\left(\frac{\varepsilon X^{2-c}}{g(N)(\log X)^2}\right).$$

Hence, Theorem 1 will follow if we show that

$$J_0(X,y) \gg \varepsilon X^{2-c}(\log X)^{-1} \tag{5.4}$$

and

$$1 - f_9 - f_{12} - f_{14} - \log\left(\tfrac{14}{11}\right)\left(f_4 + f_6 - f_9 + f_{18}\right) > 0. \tag{5.5}$$

The proof of (5.4) is similar to that of [14, Lemma 6], and (5.5) will be established in Section 5.2. This completes the proof of Theorem 1.

5.2 Numerical Computations

We first observe that

$$\omega(t) = \begin{cases} 1/t & 1 \leq t \leq 2, \\ (1 + \log(t-1))/t & 2 \leq t \leq 3, \end{cases}$$

and $\omega(t) \leq 0.5644$ for $t \geq 3$; for the latter estimate see [6, Lemma 8]. Hence,

$$f_4 = \int_2^{30/11} \frac{1 + \log(t-1)}{t} dt \leq 0.3983.$$

Note that the numeric integration here and elsewhere in this section is safe as we use it only to calculate univariate integrals.

For $(u, v) \in \mathcal{D}_9$, we have $u + 4v < 1$, which implies that $(1 - u - v)/v > 3$. Thus,

$$f_9 \leq 0.5644 \iint_{\mathcal{D}_9} \frac{du\, dv}{uv^2}.$$

Since $\mathcal{D}_9$ can be represented as $\mathcal{D}_{9,1} \cup \mathcal{D}_{9,2}$, where

$$\mathcal{D}_{9,1} = \left\{ (u, v): \tfrac{11}{75} < u < \tfrac{1}{6}, \tfrac{11}{50} - u/2 < v < u \right\},$$
$$\mathcal{D}_{9,2} = \left\{ (u, v): \tfrac{1}{6} < u < \tfrac{1}{5}, \tfrac{11}{50} - u/2 < v < \tfrac{1}{3} - u \right\},$$

a direct computation of the last integral gives

$$f_9 \leq 0.5644 \times \left(\frac{50}{11} \log\left(\frac{5}{3}\right) - \frac{9}{11} - 3\log\left(\frac{3}{2}\right) \right) \leq 0.1622.$$

Similarly

$$f_{18} \leq 5.292 \left[\int_{11/75}^{1/6} \log\left(\frac{2u}{\frac{11}{25} - u}\right) \frac{du}{u} + \int_{1/6}^{1/5} \log\left(\frac{\frac{2}{3} - 2u}{\frac{11}{25} - u}\right) \frac{du}{u} \right] \leq 0.217.$$

Thus,

$$1 - f_9 - \log\left(\tfrac{14}{11}\right)\left(f_4 + f_6 - f_9 + f_{18}\right) > 0.6265. \tag{5.6}$$

We now consider f_{12}. We have $\mathcal{D}_{12} = \mathcal{D}_{12,1} \cup \mathcal{D}_{12,2}$, where

$$\mathcal{D}_{12,1} = \left\{ (u, v): \tfrac{11}{41} < u < \tfrac{1}{3}, \tfrac{11}{25} - u < v < \tfrac{1}{4}(1 - u) \right\},$$
$$\mathcal{D}_{12,2} = \left\{ (u, v): \tfrac{11}{41} < u < \tfrac{1}{3}, \tfrac{1}{4}(1 - u) < v < \tfrac{1}{5} \right\}.$$

For $(u, v) \in \mathcal{D}_{12,1}$, we have $u + 4v < 1$, so once again

$$\iint_{\mathcal{D}_{12,1}} \omega\left(\frac{1 - u - v}{v}\right) \frac{du\, dv}{uv^2} \leq 0.5644 \iint_{\mathcal{D}_{12,1}} \frac{du\, dv}{uv^2}$$

$$= 0.5644 \times \left(\frac{25}{11} \log 2 - 4\log\left(\frac{15}{11}\right) \right) \leq 0.189.$$

If $(u, v) \in \mathcal{D}_{12,2}$, we have

$$\omega\left(\frac{1 - u - v}{v}\right) = \frac{1 + \log\left((1 - u)/v - 2\right)}{(1 - u - v)/v} \leq \frac{1 + \log 2}{(1 - u - v)/v},$$

so we obtain

$$\iint_{\mathcal{D}_{12,2}} \omega \left(\frac{1-u-v}{v}\right) \frac{du\,dv}{uv^2} \leq 1.6932 \iint_{\mathcal{D}_{12,2}} \frac{du\,dv}{uv(1-u-v)}$$

$$= 1.6932 \int_{11/41}^{1/3} \frac{\log\left(3/(4-5u)\right)}{u(1-u)} du$$

$$\leq 0.0963.$$

Finally, we have

$$\mathcal{D}_{14} = \left\{(u,v)\colon \tfrac{7}{25} < u < \tfrac{299}{1025}, \tfrac{11}{41} < v < \tfrac{14}{25} - u\right\}$$
$$\cup \left\{(u,v)\colon \tfrac{11}{41} < u < \tfrac{7}{25}, \tfrac{11}{41} < v < u\right\},$$

and for $(u,v) \in \mathcal{D}_{14}$,

$$\omega\left(\frac{1-u-v}{v}\right) = \frac{v}{1-u-v}.$$

Hence,

$$f_{14} = \int_{7/25}^{299/1025} \frac{1}{u(1-u)} \log\left(\frac{(14-25u)(30-41u)}{121}\right) du$$

$$+ \int_{11/41}^{7/25} \frac{1}{u(1-u)} \log\left(\frac{u(30-41u)}{11(1-2u)}\right) du \leq 0.0041.$$

We now have that

$$f_{12} + f_{14} < 0.2894.$$

Combined with (5.6), this inequality shows that the left-hand side of (5.5) is $> \frac{1}{3}$.

Remark. Note that the limit of the method is determined by the condition $c < 6/5$ which arises in the application of [13, Theorem 9] in the proof of Lemma 1. In other words, the arithmetic information (2.3) 'fails' before the sieve machinery begins to yield trivial conclusions. This is not typical for the method—in most applications the sieve fails before the analytic part of the argument (see the comments in Harman [4, p. 256]). In fact, the authors are convinced that an improvement of the term $(XM_1^6 M_2^6)^{1/8}$ in [13, Theorem 9] will imply an immediate improvement on Theorem 1.

6 Proof of Theorem 2

For $1 < c < 2$, let $X = X(c) = \frac{1}{2}N^{1/c}$, $\varepsilon(c) = N^{1-2/c}(\log N)^{10}$. We define

$$R^*(y,c) = \sum_{p_1,p_2 \sim X} \widehat{K}(p_1^c + p_2^c - y) = \int_{\mathbb{R}} S(x,c)^2 K(x,c)e(-yx)dx,$$

$$J^*(y,c) = \int_{\mathbb{R}} I_0(x,c)^2 K(x,c)e(-yx)dx \gg \varepsilon(c)X^{2-c}(\log X)^{-2},$$

where $S(x,c)$, $K(x,c)$ and $I_0(x,c)$ are the functions $S(x)$, $K(x)$ and $I_0(x)$ from the previous sections. The theorem will follow if we prove that, for any fixed $\rho \in (0, \frac{1}{2})$,

$$\int_{1+\rho}^{2-\rho} \int_{N}^{2N} \left|R^*(y,c) - J^*(y,c)\right|^2 dy\, dc \ll N(\log N)^{14}. \tag{6.1}$$

We first note that we can show, analogously to (4.4), that the inequality

$$\int_{N}^{2N} \left| \int_{-\tau(c)}^{\tau(c)} \left(S(x,c)^2 - I_0(x,c)^2\right) K(x,c)e(-xy)dx \right|^2 dy \ll N$$

holds uniformly for $c \in (1 + \rho, 2 - \rho)$; here $\tau(c) = X^{1-c-\eta}$ with $\eta < \frac{1}{3}\rho$. Also, in view of (4.6) and (4.15), we have

$$\int_{N}^{2N} \left| \int_{|x|>\tau(c)} I_0(x,c)^2 K(x,c)e(-xy)dx \right|^2 dy \ll 1.$$

We now proceed to establish the estimate

$$\int_{1+\rho}^{2-\rho} \int_{N}^{2N} \left| \int_{|x|>\tau(c)} S(x,c)^2 K(x,c)e(-xy)dx \right|^2 dy\, dc \ll N(\log N)^{14}. \tag{6.2}$$

Splitting the interval $(1 + \rho, 2 - \rho)$ into $O(\log N)$ subintervals (a,b), where $b = a + (\log N)^{-1}$ and using (4.6) once again, we see that it suffices to prove that

$$\int_{a}^{b} \int_{|x|>\tau(c)} |S(x,c)|^4 K(x,c)^2 dx\, dc \ll N(\log N)^{13}, \tag{6.3}$$

whenever $1 + \rho < a < 2 - \rho$, $b = a + (\log N)^{-1}$ (so that $N^{1/a} \asymp N^{1/b}$). The left-hand side of (6.3) is

$$\ll \varepsilon(b) \int_{\tau(b)}^{\infty} |S(x,c_0)|^2 \left(\int_{a}^{b} |S(x,c)|^2 dc \right) K(x,c_0)dx,$$

where $c_0 \in (a, b)$. To estimate the inner integral we use the upper bound

$$\int_a^b |S(x, c)|^2 dc \ll N^{1/a} + N^{2/a-1}|x|^{-1}$$

which can be proven using standard techniques (cf. [14, Lemma 7]). Hence, the left-hand side of (6.3) is

$$\ll \varepsilon(b) \int_{\tau(b)}^{\infty} |S(x, c_0)|^2 \left(N^{1/a} + N^{2/a-1}|x|^{-1} \right) K(x, c_0) dx,$$

$$\ll \varepsilon(b) \left(N^{2/a-1} \varepsilon(c_0) \int_{\tau(b)}^{1} |S(x, c_0)|^2 |x|^{-1} dx \right.$$

$$\left. + N^{1/a} \int_0^{\infty} |S(x, c_0)|^2 K(x, c_0) dx \right).$$

Using the mean value estimate (see [14, Lemma 7])

$$\int_n^{n+1} |S(x, c)|^2 dx \ll N^{1/c},$$

we now infer that

$$\int_a^b \int_{|x|>\tau(c)} |S(x, c)|^4 K(x, c)^2 dx \, dc \ll \frac{\varepsilon^2(b) N^{3/b-1}}{\tau(b)} + \varepsilon(b) N^{2/b}.$$

By the choice of $\varepsilon(b)$, $\tau(b)$ and η, the last expression is $O(N(\log N)^{10})$ and, therefore, the proof of (6.1) is complete.

References

[1] J. Brüdern and A. Kumchev, *Diophantine approximation by cubes of primes and an almost prime II*, Illinois J. Math. **45** (2001), 309–321.

[2] Y.-C. Cai, *On a diophantine inequality involving prime numbers*, Acta Math. Sinica **39** (1996), 733–742, in Chinese.

[3] G. Harman, *On the distribution of αp modulo one*, J. London Math. Soc. **27** (1983), 9–18.

[4] ———, *On the distribution of αp modulo one. II*, Proc. London Math. Soc. **72** (1996), 241–260.

[5] L.-K. Hua, *Additive theory of prime numbers*, American Math. Soc., Providence, Rhode Island, 1965.

[6] C.-H. Jia, *On the Piatetski-Shapiro-Vinogradov theorem*, Acta Arith. **73** (1995), 1–28.

[7] A. Kumchev, *A diophantine inequality involving prime powers*, Acta Arith. **89** (1999), 311–330.

[8] A. Kumchev and T. Nedeva, *On an equation with prime numbers*, Acta Arith. **83** (1998), 117–126.

[9] M. B. S. Laporta, *On a binary diophantine inequality involving prime numbers*, Acta Math. Hungar. **83** (1999), 205–213.

[10] H.-Q. Liu, *On the number of abelian groups of a given order*, Acta Arith. **59** (1991), 261–277.

[11] H. L. Montgomery, *Ten lectures on the interface between analytic number theory and harmonic analysis*, CBMS Regional Conf. Ser. in Math., vol. 84, Amer. Math. Soc., Providence, RI, 1994.

[12] I. I. Piatetski-Shapiro, *On a variant of Waring-Goldbach's problem (in Russian)*, Mat. Sb., **30 72** (1952), 105–120.

[13] P. Sargos and J. Wu, *Multiple exponential sums with monomials and their applications in number theory*, Acta Math. Hungar. **87** (2000), 333–354.

[14] D. I. Tolev, *On a diophantine inequality involving prime numbers*, Acta Arith. **61** (1992), 289–306.

[15] I. M. Vinogradov, *Representation of an odd number as the sum of three primes (in Russian)*, Dokl. Akad. Nauk SSSR **15** (1937), 291–294.

Recent Developments in Automorphic Forms and Applications

Wen-Ching Winnie Li[1]

1 Introduction

In the past decade there has been tremendous progress in the area of automorphic forms. The crown jewel was Wiles' spectacular proof of Fermat's Last Theorem in 1994; this was a by-product of his proof of the Taniyama-Shimura conjecture for semi-stable elliptic curves defined over $\mathbb{Q}$. The remaining case of the Taniyama-Shimura conjecture was completely settled through the joint effort of Breuil, Conrad, Diamond, and Taylor. In another direction, the proof of the Local Langlands conjecture for $\mathrm{GL}(n)$ in all characteristics is now complete, with the finite characteristic case proved by Laumon, Rapoport and Stuhler and the characteristic zero case proved by Harris and Taylor and by Henniart. Finally, the proof of the Global Langlands conjecture for $\mathrm{GL}(n)$ over function fields is well underway. In addition to these developments, substantial progress has been made on many related subjects.

The purpose of this review article is two-fold. First, we describe the local and global Langlands conjectures on correspondences between representations of $\mathrm{GL}(n)$ and representations of Galois/Weil groups, and briefly summarize what is currently known about them. Second, we explain recent results giving explicit examples of matching automorphic and algebraic-geometric L-functions, with applications to conjectures in number theory, and results giving applications of automorphic forms in constructing good combinatorial objects.

The paper is organized as follows. In Sections 2 and 3 we introduce the basic local and global invariants, namely the L- and ε-factors attached to a representation of a Galois group, resp. a representation of $\mathrm{GL}(n)$. They are used in Sections 4 and 5 to describe the local and global Langlands conjecture for $\mathrm{GL}(n)$. We shall view the Taniyama-Shimura conjecture as part of

---

[1]The research of the author was supported in part by a grant from the National Science Foundation no. DMS-9970651. This work was done when the author was visiting the Institute for Advanced Study at Princeton, NJ, supported by the Ellentuck Fund, to which she expresses sincere thanks.

the global Langlands conjecture for GL(2) over $\mathbb{Q}$, and discuss its conjectural generalization to abelian varieties over $\mathbb{Q}$. In Section 6 we give explicit examples of the Langlands global correspondence over function fields; this is a joint work of the author with C.-L. Chai. We derive certain interesting consequences, among them the establishment of the Kloosterman sum conjecture over function fields and new examples of automorphic forms for GL(2) for which the Sato-Tate conjecture holds. These are discussed in Sections 7 and 8. Applications of automorphic forms to combinatorial constructions are presented in the last section. We review Ramanujan graphs constructed by Lubotzky, Phillips and Sarnak based on quaternion algebras and extend these to constructing Ramanujan 3-hypergraphs based on division algebras of degree 9 over a function field.

2 Invariants Attached to Representations of Galois Groups

One of the major goals of algebraic number theory is to study algebraic extensions of a given field, either a global field—a number field or a function field of one variable over a finite field—or a local field—the completion of a global field at a place. Typical examples of global fields are the field of rational numbers $\mathbb{Q}$ and the field of rational functions $\mathbb{F}_q(t)$; examples of local fields are the field of p-adic numbers $\mathbb{Q}_p$ and the field of Laurent series $\mathbb{F}_q((t))$. Let K be a global or local field. Denote by $\bar{K}$ the separable closure of K and by G_K the Galois group of $\bar{K}$ over K. To understand algebraic extensions of K amounts to understanding G_K, which is a profinite group. According to Tannaka duality, to know a profinite group is equivalent to knowing all of its finite-dimensional representations, not just as a set, but also with the tensor product. Therefore we study (linear) representations of G_K. To each representation we shall attach certain invariants, called L- and ε-factors, such that the family of invariants attached to the representation itself as well as its tensor product with certain representations will characterize the representation up to equivalence.

Assume that K is a global field. We describe representations of the Galois group G_K arising from geometry. Let X be a smooth variety defined over K. The action of G_K on X induces an action of G_K on the étale cohomology group $H^i_{et}(X_{/\bar{K}}, \overline{\mathbb{Q}_\ell})$. Let V be an irreducible subquotient of dimension n of this action. Hence we get a continuous irreducible ℓ-adic representation

$$\rho \colon G_K \longrightarrow \mathrm{Aut}(V).$$

It is known that ρ is unramified at all except finitely many places of K. Thus at a nonarchimedean place v where ρ is unramified, for each chosen decomposition group at v, the restriction of ρ to the inertia subgroup is trivial, and consequently ρ restricted to the Frobenius conjugacy class is single-valued. To take into account the different choices in G_K of a decomposition group at v, which are conjugate to each other, we obtain a conjugacy class $\rho(\mathrm{Frob}_v)$. Nonetheless the eigenvalues $\rho_{1,v}, \ldots, \rho_{n,v}$ of $\rho(\mathrm{Frob}_v)$ are well-defined. Put

$$L_v(s,\rho) = \prod_{1 \leq i \leq n} \frac{1}{1 - \rho_{i,v} Nv^{-s}},$$

which is a local invariant of ρ restricted to a decomposition group at v. Here Nv, the norm of v, is the cardinality of the residue field of K_v, the completion of K at v. At the places where the representation ρ ramifies, as well as the archimedean places if K is a number field, local invariants $L_v(s,\rho)$ can be defined in a more complicated way. Put together, one defines the L-function attached to ρ, which is a global invariant attached to ρ:

$$L(s,\rho) := \prod_v L_v(s,\rho) \approx \prod_{v \text{ good}} \prod_{1 \leq i \leq n} \frac{1}{1 - \rho_{i,v} Nv^{-s}},$$

Here $\approx$ means that we ignore finitely many local L-factors.

An example of what we described above is the case of an elliptic curve E defined over K. Given a prime ℓ, not equal to the characteristic of K if $\mathrm{char}(K) > 0$, the points on E with order dividing ℓ^n form an abelian group $E[\ell^n]$, which is isomorphic to $(\mathbb{Z}/\ell^n\mathbb{Z}) \times (\mathbb{Z}/\ell^n\mathbb{Z})$. The inverse limit of $E[\ell^n]$ as n approaches infinity is called a Tate module $T_\ell(E)$ of E, which is isomorphic to $\mathbb{Z}_\ell \times \mathbb{Z}_\ell$. In other words, it is a $\mathbb{Z}_\ell$-module of rank two. The Galois group G_K permutes the points in $E[\ell^n]$, and hence one gets a two-dimensional ℓ-adic representation of G_K. The associated L-function is the Hasse-Weil L-function of E:

$$L(s,E) \approx \prod_{v \text{ good}} \frac{1}{1 - a_v Nv^{-s} + Nv^{1-2s}},$$

where $1 + Nv - a_v$ is the number of rational points of the reduction of E at v over the residue field at v, which is independent of ℓ.

When ρ is nontrivial, $L(s,\rho)$ is expected to have a holomorphic continuation to the whole s-plane, bounded in each vertical strip of finite width, and satisfies a functional equation

$$L(s,\rho) = \varepsilon(s,\rho) L(1-s,\hat{\rho}),$$

where $\hat{\rho}$ denotes the contragradient of ρ. Here the ε-factor, $\varepsilon(s,\rho)$, is another global invariant attached to ρ. Deligne has shown in [13] that $\varepsilon(s,\rho)$

is a product of local ε-factors, each of which is a local invariant attached to the restriction of ρ to a decomposition group.

These analytic properties for $L(s, \rho)$ were proved by Grothendieck when K is a function field. When K is a number field, these are mostly unproved. The Hasse-Weil L-function attached to an elliptic curve defined over $\mathbb{Q}$ is one case where they are proved, as a consequence of the Taniyama-Shimura conjecture explained in Section 5.

When $K = \mathbb{Q}$, more properties of ρ are known:

(a) The restriction of ρ to a decomposition group is "potentially semi-stable" in the sense of Fontaine.

(b) The eigenvalues $\rho_{1,v}, \dots, \rho_{n,v}$ are algebraic for almost all v.

A conjecture of Fontaine and Mazur says that a continuous irreducible finite-dimensional ℓ-adic representation ρ of $G_{\mathbb{Q}}$ arises from geometry if and only if it is ramified at finitely many places and statement (a) above holds. Some progress on this conjecture has been made by Skinner, Taylor, and Wiles.

3   Invariants Attached to Automorphic Representations

For a global field K as described in Section 2, denote by $\mathbb{A}_K$ the ring of adeles of K. Let π be an automorphic irreducible representation of $\mathrm{GL}_n(\mathbb{A}_K)$. Then π is a restricted tensor product $\bigotimes'_v \pi_v$, where π_v is an admissible irreducible representation of $\mathrm{GL}_n(K_v)$ with K_v being the completion of K at v, which is unramified for all except finitely many places. The unramified local representations are determined by the action of the Hecke algebra $\mathcal{H}_v$, which is isomorphic to the algebra of symmetric polynomials in $\mathbb{C}[z_1, z_1^{-1}, \dots, z_n, z_n^{-1}]$. Therefore π_v is determined by n complex numbers

$$z_1(\pi_v), \dots, z_n(\pi_v).$$

Define

$$L_v(s, \pi) := L(s, \pi_v) = \prod_{1 \le i \le n} \frac{1}{1 - z_i(\pi_v) N v^{-s}},$$

which is a local invariant attached to π_v. At the places where π ramifies and at the archimedean places of K if K is a number field, one can also define local L-factors $L(s, \pi_v)$ in a more complicated way. The product of local L-factors is a global invariant attached to π,

$$L(s, \pi) := \prod_v L_v(s, \pi_v) \approx \prod_{\pi_v \text{ unramified}} \prod_{1 \le i \le n} \frac{1}{1 - z_i(\pi_v) N v^{-s}},$$

called an automorphic L-function of $\mathrm{GL}_n(\mathbb{A}_K)$.

Many familiar functions are examples of global automorphic L-functions. For instance, the Riemann zeta function $\zeta(s)$ and the Dirichlet L-function $L(s, \chi)$ are automorphic L-functions of $\mathrm{GL}_1(\mathbb{A}_{\mathbb{Q}})$, the Dedekind zeta function $\zeta_K(s)$ is an automorphic L-function for $\mathrm{GL}_1(\mathbb{A}_K)$ with K a number field, and the L-function attached to a newform f of weight k, level N, and character χ,

$$L(s, f) = \prod_{p|N} \frac{1}{1 - a_p p^{-s}} \prod_{p \nmid N} \frac{1}{1 - a_p p^{-s} + \chi(p) p^{k-1-2s}},$$

is an automorphic L-function for $\mathrm{GL}_2(\mathbb{A}_{\mathbb{Q}})$.

An automorphic L-function $L(s, \pi)$ is known to have meromorphic continuation to the whole s-plane, bounded at infinity in each vertical strip of finite width if K is a number field, while $L(s, \pi)$ is a rational function in q^{-s} if K is a function field with q elements in its field of constants, and $L(s, \pi)$ satisfies the functional equation

$$L(s, \pi) = \varepsilon(s, \pi) L(1 - s, \hat{\pi}).$$

Here $\hat{\pi}$ is the contragradient of π. Moreover, if π is a cuspidal representation, then $L(s, \pi)$ is holomorphic everywhere. In the case when K is a function field, $L(s, \pi)$ is in fact a polynomial in q^{-s}. When an L-function has the aforementioned analytic property, it is said to *behave nicely*. Therefore, automorphic L-functions behave nicely and the L-function attached to an ℓ-adic representation ρ of a Galois group G_K arising from geometry behaves nicely if K is a function field; when K is a number field, the analytic behavior of $L(s, \rho)$ is unknown for most ρ.

The $\varepsilon(s, \pi)$ occurring in the functional equation is another global invariant attached to π. It is a product of local ε-factors attached to π_v.

4 The Local Langlands Conjecture

Let K be a non-archimedean local field with residue field F. The residue field of $\bar{K}$ is an algebraic closure $\bar{F}$ of F. Since a Galois automorphism preserves integrality, it naturally induces an automorphism of $\bar{F}$ over F. Such a map from $\mathrm{Gal}(\bar{K}/K)$ to $\mathrm{Gal}(\bar{F}/F)$ is surjective with kernel equal to the inertia subgroup I_K of $\mathrm{Gal}(\bar{K}/K)$. Since $\mathrm{Gal}(\bar{K}/K)$ is compact, we replace it by a dense subgroup W_K, called the Weil group, consisting of the preimages in $\mathrm{Gal}(\bar{K}/K)$ of the dense subgroup generated by the Frobenius automorphism in $\mathrm{Gal}(\bar{F}/F)$. Endowed with the induced topology, the Weil group W_K not only captures the essence of $\mathrm{Gal}(\bar{K}/K)$, it also affords more continuous representations. It turns out to be the right group to work with.

Local Langlands Conjecture. *There is a natural bijection from the set of equivalence classes of degree n irreducible complex continuous representations of W_K to the set of equivalence classes of admissible irreducible supercuspidal representations of $\mathrm{GL}_n(K)$, so that the corresponding representations as well as their twists have the same L- and ε-factors.*

Remark. Henniart proved in [22] that the invariant L- and ε-factors as stated in the conjecture do determine the class of representations of $\mathrm{GL}_n(K)$.

Theorem. *The local Langlands conjecture for $\mathrm{GL}_n(K)$ holds for all local fields K.*

For $n = 1$, this is local class field theory. For $n = 2$, this was proved by Kutzko [28]. For $n = 3$, this is a result of Henniart [21]. Kutzko and Moy [29] proved the conjecture for the case where n is a prime. The general case with K of positive characteristic was proved by Laumon, Rapoport and Stuhler [32] in 1991. In 1998 the remaining case with K of zero characteristic was settled by M. Harris and R. Taylor [20]. Later Henniart [23] gave a much shorter proof.

5   The Global Langlands Conjecture

The situation for the case of a global field K is more complicated, and much less is known. To explain correspondences, we consider three kinds of representations:

(I) The set G_n of equivalence classes of degree n continuous irreducible complex representations of G_K.

(II) The set $G_{n,\ell}$ of equivalence classes of degree n continuous irreducible ℓ-adic representations of G_K arising from geometry.

(III) The set A_n of equivalence classes of cuspidal automorphic irreducible representations of $\mathrm{GL}_n(\mathbb{A}_K)$.

Note that the representations in G_n are those in $G_{n,\ell}$ whose underlying varieties consist of finitely many points.

Global Langlands Conjecture Over a Number Field K. *There is an injection from G_n into A_n which extends to a map from $G_{n,\ell}$ into A_n such* *that if a representation ρ of G_K corresponds to a representation $\pi = \otimes' \pi_v$ of $\mathrm{GL}_n(\mathbb{A}_K)$, then the corresponding representations as well as their twists have the same attached L- and ε-factors. Further, at each place v of K, the restriction of ρ to a decomposition group at v corresponds to π_v according to the local Langlands conjecture.*

An immediate consequence of the global correspondence is that the global L-functions attached to representations of G_K arising from geometry behave nicely.

When $n = 1$, this is global class field theory. We describe below what is known of the case $n = 2$ and $K = \mathbb{Q}$. A degree two representation ρ of $G_\mathbb{Q}$ is called *odd* if the action (via ρ) of complex conjugation has determinant -1. The cuspidal representations of $\mathrm{GL}_2(\mathbb{A}_\mathbb{Q})$ arise in two ways: either from classical holomorphic cuspidal newforms of integral weight $k \geq 1$ or from real analytic cuspidal new Maass wave forms. Not much is known about the latter, while the former has been studied extensively. Deligne and Serre [15] established an injection from the set of holomorphic cuspidal (normalized) newforms of weight 1 to the set $G_{2,\mathrm{odd}}$ of odd representations in G_2 satisfying the Langlands conjecture. For forms with higher weight, there is an injection from the set of holomorphic cuspidal newforms of weight $k \geq 2$ to the set $G_{2,\ell,\mathrm{odd}}$ of odd representations in $G_{2,\ell}$ satisfying Langlands' conjecture. This was proved by Eichler-Shimura for $k = 2$ and forms with rational coefficients, and by Deligne [12] in general for $k \geq 2$. For $n = 2$ and K a totally real field of degree r, Rogawski-Tunnell [44] together with a technical result from H. Reimann (see, for example, [42]) extended the result of Deligne-Serre to holomorphic forms of weight $(1, \ldots, 1)$, while Brylinski-Labesse [5] and Blasius-Rogawski [2] generalized Deligne's result to forms of weight $(k_1, \ldots, k_r)$, where $k_1, \ldots, k_r$ have the same parity modulo 2.

The progress made on the global Langlands conjecture had a far reaching impact on the advancement of number theory in the past decade. Here we elaborate a little. Recall first the result by Eichler and Shimura.

Theorem. *Let f be a weight 2 holomorphic cuspidal normalized newform with coefficients in $\mathbb{Q}$. Then $L(s,f) = L(s,E)$ for some elliptic curve E defined over $\mathbb{Q}$.*

In fact, E is a quotient A_f of the Jacobian of the modular curve $X_0(N)$ attached to f. Here N is the level of f. As explained in Section 2, to an elliptic curve E there is the associated ℓ-adic representation of $G_\mathbb{Q}$ arising from its action on the Tate module $T_\ell(E)$. This is the representation in $G_{2,\ell,\mathrm{odd}}$ to which f corresponds. The conjecture of Taniyama-Shimura says that the relation $L(s,f) = L(s,E)$ actually holds for all elliptic curves defined over $\mathbb{Q}$.

Taniyama-Shimura Conjecture. *Every elliptic curve E defined over $\mathbb{Q}$ is modular, that is, there is a holomorphic cuspidal normalized newform f of weight 2 such that $L(s,E) = L(s,f)$.*

Using the isogeny conjecture proved by Faltings [18], we can restate the above conjecture: every elliptic curve defined over $\mathbb{Q}$ is isogenous to some A_f.

Wiles [46] and Taylor-Wiles [45] proved the Taniyama-Shimura conjecture for semi-stable elliptic curves, which, combined with the work of Ribet [43] and others, establishes Fermat's Last Theorem. The remaining case of the conjecture was completely settled through the joint efforts of Breuil, Conrad, Diamond, and Taylor in 1999 [4].

If f is a holomorphic cuspidal newform of weight 2 for $\Gamma_1(N)$, that is, of level N and some nontrivial character, then so are its conjugates by elements in $G_{\mathbb{Q}}$. In fact, f has g distinct conjugates, where g is the degree over $\mathbb{Q}$ of the field of Fourier coefficients of f. The variety A_f attached to f in this case, being a quotient of the Jacobian of the modular curve $X_1(N)$, is a simple abelian variety over $\mathbb{Q}$ of dimension g. (It is also the variety attached to any conjugate of f.) Furthermore, the ring of endomorphisms of A_f contains an order of a field of degree g over $\mathbb{Q}$. An abelian variety with the properties described above is said to be of GL_2-type. A natural question along the same lines as the Taniyama-Shimura conjecture is to characterize the abelian varieties isogenous to A_f. The conjectural answer is stated below.

Generalized Taniyama-Shimura Conjecture. *Every simple abelian variety defined over $\mathbb{Q}$ which is of GL_2-type is modular, that is, isogenous to A_f for some holomorphic cuspidal newform f of weight 2.*

The global Langlands conjecture over a function field can be stated more precisely as follows.

Global Langlands Conjecture Over a Function Field K. *There is a bijection from the subset $G_{n,\ell}^f$ of $G_{n,\ell}$ consisting of representations whose determinants are of finite order to the subset A_n^f of A_n consisting of representations whose central characters are of finite order such that the corresponding representations as well as their twists have the same L- and ε-factors. Further, if a representation ρ of G_K corresponds to the representation $\pi = \otimes'_v \pi_v$ of $\mathrm{GL}_n(\mathbb{A}_K)$, then for each place v of K, the restriction* *of ρ to a decomposition group at v corresponds to π_v as described by the local Langlands conjecture.*

For $n = 1$ this is global class field theory. Drinfeld [16] proved the conjecture for the case $n = 2$, and Lafforgue's proof of the case of general n is well underway.

The Taniyama -Shimura conjecture over a function field K asserts that the Hasse-Weil L-function attached to an elliptic curve defined over K is an automorphic L-function of $GL_2(\mathbb{A}_K)$. This was proved by Deligne [13] using Grothendieck's result on the analytic behavior of the L-functions attached to the elliptic curve as well as its twists by characters, and the converse theorem for GL_2 proved by Jacquet and Langlands [24]. This is totally different from the approach in the case $K = \mathbb{Q}$.

6   Explicit Examples of Global Langlands Correspondence Over a Function Field

For arithmetic applications one sometimes wants more than the abstract existence of the Langlands global correspondence, namely, specific details for individual representations of interest. In this section we describe results in the function field case, which give completely explicit examples of the global correspondence between n-dimensional ℓ-adic representations of the Galois group and cuspidal representations of GL_n. The correspondence is expressed in terms of the attached L-functions. These results are joint work with C.-L. Chai (cf. [9], [10]).

We first illustrate through a familiar example the prototype of the L-functions which we shall be concerned with. Let K be a function field of one variable with field of constants F. Denote by F_v the residue field of the completion K_v of K at the place v. Its cardinality is Nv, the norm of v.

Let E be an elliptic curve defined over K. Supposing, for convenience, that the characteristic of K is not 2 or 3, we may assume that an affine part of E is given by the equation

$$y^2 = g(x),$$

where g is a polynomial over K of degree 3. Then

$$L(s, E) \approx \prod_{v \text{ good}} \frac{1}{1 - a_v Nv^{-s} + Nv^{1-2s}},$$

where

$$a_v = -\sum_{x \in F_v} \chi \circ N_{F_v/F}(g_v(x))$$

with χ the quadratic character of $F^\times$, $N_{F_v/F}$ the norm map from F_v to F, and g_v the polynomial over F_v obtained from the reduction of g mod v. Let η_v denote the quadratic idele class character of the rational function field $F_v(t)$ attached to the quadratic extension $y^2 = g_v(t)$. Then at a place w of $F_v(t)$ where η_v is unramified, its value at a uniformizer ϖ_w is given by

$$\eta_v(\varpi_w) = \chi \circ N_{F_v/F}\left(\prod_\alpha g_v(\alpha)\right).$$

Here α runs through all points of the projective line in an algebraic closure of F_v which constitute the closed point w. While each $g_v(\alpha)$ might lie outside F_v, after taking the product over all α, the resulting value does lie in F_v. Note that the L-function attached to η_v is the reciprocal of the factor at v occurring in $L(s, E)$; in other words,

$$L(s, \eta_v) = 1 - a_v N v^{-s} + N v^{1-2s}.$$

The L-functions we shall consider have the following features:

(1) Each is attached to a (global) rational function over K and either an additive character or a multiplicative character of F.

(2) Each has an Euler product with the local factor at a good place v being the reciprocal of a global L-function attached to an idele class character η_v of a rational function field.

(3) The character η_v in (2) is constructed using the reduction mod v of the global rational function in (1) and the additive/multiplicative character.

The details are as follows.

Let $f(x)$ be a rational function with coefficients in K. Only finitely many places of K occur as zeros or poles of the coefficients of f. These are the 'possibly bad' places, and the remaining places are called 'good' places of f. At a good place v, the coefficients of f are units in the completion of K at v. By passing to the residue field F_v, we get a rational function $f_v := f \mod v$ with coefficients in F_v.

Theorem A. *Let $f(x) \in K(x)$ be a non-constant rational function over K and ψ a nontrivial additive character of F.*

(A1) *At each good place v of f, there exists an idele class character η_{ψ, f_v} of $F_v(t)$ such that at the places w of $F_v(t)$ where η_{ψ, f_v} is unramified, its value at a uniformizer ϖ_w is given by the character sum*

$$\eta_{\psi, f_v}(\varpi_w) = \psi \circ \text{Tr}_{F_v/F}\left(\sum_\alpha f_v(\alpha)\right),$$

where α runs through all points of the projective line in an algebraic closure $\bar{F}$ of F which constitute the closed point w.

(A2) *Suppose that $f(x)$ is not of the form $h_1(x)^p - h_1(x) + h_2$ for any $h_1(x) \in K(x)$ and $h_2 \in K$. Then the L-function $L(s, \eta_{\psi, f_v})$ attached to η_{ψ, f_v} is a polynomial in Nv^{-s} of degree n equal to the degree of the conductor of η_{ψ, f_v} minus 2, and n is the same for almost all v.*

(A3) *With the same assumption as in (A2), there exists a compatible family of ℓ-adic representations of G_K, depending on ψ and f, such that its associated L-function $L(s, \psi, f)$ satisfies*

$$L(s, \psi, f) \approx \prod_{v \text{ good}} L(s, \eta_{\psi, f_v})^{-1}.$$

Further $L(s, \psi, f)$ and its twists by idele class characters of K and by all ℓ-adic representations of G_K behave nicely.

Theorem B. *Let $g(x) \in K(x)$ be a non-constant rational function over K and let χ be a nontrivial character of $F^{\times}$ of order $d > 1$.*

(B1) *At a good place v of g, there exists an idele class character η_{χ, g_v} of $F_v(t)$ such that at the places w of $F_v(t)$ where η_{χ, g_v} is unramified, its value is given by the character sum*

$$\eta_{\chi, g_v}(\varpi_w) = \chi \circ N_{F_v/F} \left(\prod_\alpha g_v(\alpha) \right),$$

where, as in Theorem A, α runs through all points in the projective line constituting the closed point w.

(B2) *Suppose that $g(x)$ is not of the form $h_1 h_2(x)^d$ for any $h_2(x) \in K(x)$ and any $h_1 \in K$. Then the L-function $L(s, \eta_{\chi, g_v})$ attached to η_{χ, g_v} is a polynomial in Nv^{-s} of degree n equal to the degree of the conductor of η_{χ, g_v} minus 2, and n is the same for almost all v.*

(B3) *Under the same assumption as in (B2), there exists a compatible family of ℓ-adic representations of G_K, depending on χ and g, such that the associated L-function $L(s, \chi, g)$ satisfies*

$$L(s, \chi, g) \approx \prod_{v \text{ good}} L(s, \eta_{\chi, g_v})^{-1}.$$

Further $L(s, \chi, g)$ and its twists by idele class characters of K and by all ℓ-adic representations of G_K behave nicely.

As a consequence of the global Langlands conjecture proved for $n \leq 2$ by Drinfeld [16] and the converse theorem for GL_n proved by Jacquet and Langlands [24] for $n = 2$, by Jacquet, Piatetski-Shapiro and Shalika [25] for $n = 3$, and by Cogdell and Piatetski-Shapiro [11] for $n = 4$, we conclude

Corollary. $L(s, \psi, f)$ *(resp.* $L(s, \chi, g))$ *is an automorphic L-function for* $\mathrm{GL}_n(\mathbb{A}_K)$ *if the common degree n as described in Theorem A (resp. B) is at most 4.*

It would follow from the global Langlands conjecture over function fields that all $L(s, \psi, f)$ and $L(s, \chi, g)$ are automorphic L-functions for $\mathrm{GL}_n(\mathbb{A}_K)$ if the common degree of local factors is n.

More concretely, we consider the following examples.

Example 1. Let $f(x) = x + b/x$ with any nonzero element b in K. We obtain an automorphic L-function $L(s, \psi, f)$ for $\mathrm{GL}_2(\mathbb{A}_K)$.

Example 2. Let $f_1(x) = x^2 + a/x$ and $f_2(x) = x + a/x^2$ for a nonzero element a in K. If char K is not 2, then $L(s, \psi, f_1)$ and $L(s, \psi, f_2)$ are automorphic L-functions for $\mathrm{GL}_3(\mathbb{A}_K)$.

Example 3. $L(s, \psi, f)$ is an automorphic L-function for $\mathrm{GL}_4(\mathbb{A}_K)$ if $f(x) = x^3 + a/x$ and char K is not equal to 3, or $f(x) = x^2 + a/x^2$ and char K is not equal to 2. Here a is any nonzero element in K as before.

Example 4. Suppose char K is odd. Let $g(x) = (x - 1)^2 + 4ax$ for $a \in K^\times$ and χ be a nontrivial character of $F^\times$. At a place v of K which is not a pole of a and where $a \not\equiv 0, 1 \pmod{v}$, there is an idele class character η_{χ, g_v} whose associated L-function is

$$L(s, \eta_{\chi, g_v}) = 1 + \lambda_\chi(F_v; a) N v^{-s} + N v^{1-2s},$$

where

$$\lambda_\chi(F_v; a) = \sum_{\substack{x, y \in \mathbb{F}_v \\ y^2 = g_v(x)}} \chi \circ \mathrm{N}_{F_v/F}(x).$$

One can show

Theorem C. *Given a nonzero a in K, there exists an automorphic form ϕ_a of* $\mathrm{GL}_2(\mathbb{A}_K)$ *such that*

$$L(s, \phi_a) \approx \prod_{v \ good} \frac{1}{1 + \lambda_\chi(F_v; a) N v^{-s} + N v^{1-2s}}.$$

Example 5. Suppose char K is odd. Denote by F' the quadratic extension of the field of constants F of K. Let ε be the quadratic character of $F^\times$,

ω be a regular character of $F'^{\times}$, and a be a nonzero element in K. At a place v of K which is not a pole of a and where $a \not\equiv 2, -2 \pmod{v}$, there is an idele class character $\eta_{a,\varepsilon,\omega,v}$ whose associated L-function is

$$L(s, \eta_{a,\varepsilon,\omega,v}) = 1 + \lambda_{\varepsilon,\omega}(F_v; a)Nv^{-s} + Nv^{1-2s},$$

with

$$\lambda_{\varepsilon,\omega}(F_v; a) = \sum_{u \in U(F_v)} \varepsilon \circ N_{F_v/F}(\text{Tr}_{F'_v/F_v}(u) + a_v)\omega \circ N_{F'_v/F'}(u).$$

Here a_v denotes $a \pmod{v}$, $U(F_v)$ is the norm 1 subgroup of F'_v over F_v, and $F'_v = F_v \otimes_F F'$, which is a quadratic extension of F_v (resp. $F_v \times F_v$) when $\deg v$ is odd (resp. even).

Similar to the previous example, we have

Theorem D. *Given a nonzero a in K, a quadratic character ε of $F^{\times}$ and a regular character ω of $F'^{\times}$, there exists an automorphic form $\phi_{a,\varepsilon,\omega}$ of $GL_2(\mathbb{A}_K)$ such that*

$$L(s, \phi_{a,\varepsilon,\omega}) \approx \prod_{v \text{ good}} \frac{1}{1 + \lambda_{\varepsilon,\omega}(F_v; a)Nv^{-s} + Nv^{1-2s}}.$$

7 Kloosterman Sum Conjectures

For a prime p denote by ψ_p the additive character of $\mathbb{Z}/p\mathbb{Z}$ given by $\psi_p(x) = \exp(2\pi i x/p)$. Given a finite field k of characteristic p, the composition $\psi_p \circ Tr_{k/(\mathbb{Z}/p\mathbb{Z})}$ is a nontrivial additive character ψ of k. A Kloosterman sum is defined for each $b \in k^{\times}$ as

$$\text{Kl}(k; b) = \sum_{x \in k^{\times}} \psi\left(x + \frac{b}{x}\right).$$

Example 1 in Section 6 deserves some elaboration. For a given $b \in K^{\times}$ and $f(x) = x + b/x$, a good place is a place v of K away from the zeros and poles of b. At such a place, $b \mod v$ is a nonzero element in the residue field F_v, and one finds that

$$L(s, \eta_{\psi,f_v}) = 1 + \text{Kl}(F_v; b)Nv^{-s} + Nv^{1-2s},$$

and we know that $L(s, \psi, f)$ is an automorphic L-function for $GL_2(\mathbb{A}_K)$. Stated in another way, we have proven [9]:

Theorem E (Kloosterman Sum Conjecture Over a Function Field).
*Let K be a function field with the field of constants F a finite field. Given
a nonzero element $b \in K$, there exists an automorphic form ϕ_b of $\mathrm{GL}_2(\mathbb{A}_K)$
which is an eigenfunction of the Hecke operator T_v with eigenvalue
$-\mathrm{Kl}(F_v; b)$ at each place v of K which is neither a zero nor a pole of b. In
other words,*

$$L(s, \phi_b) \approx \prod_{v \text{ good}} \frac{1}{1 + \mathrm{Kl}(F_v; b)Nv^{-s} + Nv^{1-2s}}.$$

It is interesting to compare this with its counterpart over the field $\mathbb{Q}$;
this is a question raised by Katz in [27].

Kloosterman Sum Question Over $\mathbb{Q}$. *Given a nonzero integer b, does
there exist a Maass wave form f which is an eigenfunction of the Hecke
operator T_p for all $p \nmid b$ with eigenvalue $-\mathrm{Kl}(\mathbb{Z}/p\mathbb{Z}; b)$? In other words,
does there exist a Maass wave form f such that*

$$L(s, f) \approx \prod_{p \nmid b} \frac{1}{1 + \mathrm{Kl}(\mathbb{Z}/p\mathbb{Z}; b)p^{-s} + p^{1-2s}}?$$

Using an idea of Sarnak, A. Booker [3] showed that if there existed a
Maass wave form f with the p-th Fourier coefficient given by $\pm \mathrm{Kl}((\mathbb{Z}/p\mathbb{Z}); 1)$
and of level N equal to a power of 2, then

$$N(\lambda + 3) > 2^{24},$$

where λ is the eigenvalue of the Laplacian at f. He also proved that it is
impossible to show the nonexistence of such a form numerically since there
always exists a Maass wave form for $SL_2(\mathbb{Z})$ whose normalized Fourier
coefficients at the prescribed finitely many places are as close to prescribed
values as we wish provided that the eigenvalue of the Laplacian is allowed
to grow unboundedly.

8   The Sato-Tate Conjecture

If a cuspidal automorphic representation π of $\mathrm{GL}_n(\mathbb{A}_K)$ corresponds to a
representation ρ of G_K, then they have the same L-functions. In particular,
the Ramanujan-Petersson conjecture holds for π, that is, when $L(s, \pi)$ is
written as an Euler product

$$L(s, \pi) \approx \prod_{v} \prod_{1 \le i \le n} \frac{1}{1 - z_i(\pi_v)Nv^{-s}},$$

we have

$$|z_1(\pi_v)| = \cdots = |z_n(\pi_v)|.$$

Hence we may ask how the angles distribute, or equivalently, how the normalized jth coefficients of

$$(1 - z_1(\pi_v)x)\ldots(1 - z_n(\pi_v)x) = 1 + a_{1,v}x + \cdots + a_{n,v}x^n$$

distribute as v varies. When $n = 2$, this is the Sato-Tate conjecture, which originally was stated for elliptic curves.

Sato-Tate Conjecture for Elliptic Curves. *If E is an elliptic curve defined over $\mathbb{Q}$ which does not have complex multiplications, then the family $\{a_p/\sqrt{p}\}$, where p runs through all primes not dividing the conductor N of E, is uniformly distributed with respect to the Sato-Tate measure*

$$\mu_{ST} = \frac{1}{\pi}\sqrt{1 - \frac{x^2}{4}}\,dx \qquad on \ [-2,2].$$

Here $a_p = 1 + p - N_p$, where N_p is the number of $\mathbb{Z}/p\mathbb{Z}$-rational points on the reduction of E mod p.

Since the Taniyama-Shimura conjecture has been established, we know that these a_p's are in fact eigenvalues of the Hecke operator at p acting on a cuspidal newform of weight 2 and level N. There is an obvious extension of the above conjecture to cuspidal newforms of level N, trivial character, any weight, and not of CM type.

There is good supporting numerical evidence for the conjecture, but it remains unproved for any elliptic curves over $\mathbb{Q}$ or holomorphic cuspidal automorphic forms for $GL_2(\mathbb{A}_\mathbb{Q})$.

The situation over a function field K is quite different. Indeed, the Sato-Tate conjecture holds for an elliptic curve defined over K with nonconstant j-invariant, as proved by Yoshida [47] in the 1970s. The Ramanujan-Petersson conjecture for $GL_2(\mathbb{A}_K)$ was proved by Drinfeld [17]. While the status of the global Langlands conjecture over K is unclear at the moment, the work of Lafforgue [31], [30] showed that the Ramanujan-Petersson conjecture always holds for cuspidal automorphic forms for $GL_n(\mathbb{A}_K)$.

In the joint papers [9], [10] we obtain more examples of cuspidal automorphic forms for $GL_2(\mathbb{A}_K)$ for which Sato-Tate conjecture holds.

Theorem F. *If a in Theorem C is not in the field of constants F of K, then the Sato-Tate conjecture holds for ϕ_a in Theorem C.*

Theorem G. *If a in Theorem D is not in the field of constants F of K, then the Sato-Tate conjecture holds for $\phi_{a,\varepsilon,\omega}$ in Theorem D.*

Theorem H. *If b in Theorem D is not in the field of constants F of K, then the Sato-Tate conjecture holds for ϕ_b in Theorem E.*

9 Applications

In this section we present applications of automorphic forms to constructing good combinatorial objects. Let X be a k-regular connected undirected graph on n vertices. Denote by A_X the adjacency matrix of X. Its rows and columns are parametrized by the vertices of X, and the ijth entry of A_X records the number of edges from vertex i to vertex j. The matrix A_X is best viewed as an operator which sends a function f defined on the vertices of X to another function Af whose value at the vertex x is given by

$$A_X f(x) = \sum_{y \text{ is a neighbor of } x} f(y).$$

Since A is a symmetric matrix, its eigenvalues are real. By the maximal modulus principle, one finds them lying between k and $-k$. So we can order them in non-increasing order

$$k = \lambda_1 > \lambda_2 \geq \cdots \geq \lambda_n \geq -k.$$

Here the largest eigenvalue is achieved once since the graph is k-regular and connected; the smallest eigenvalue $-k$ is achieved if and only if the graph X is bipartite. Denote by $\lambda^+(X)$ the largest eigenvalue $< k$ and by $\lambda^-(X)$ the smallest eigenvalue $> -k$. We are interested in the behavior of $\lambda^\pm(X)$ as X varies.

Let $\{X_j\}$ be a family of k-regular graphs with the size of X_j approaching infinity as j tends to infinity. The growth of $\lambda^+(X_j)$ is well-known:

Theorem (Alon-Boppana). $\liminf_{j\to\infty} \lambda^+(X_j) \geq 2\sqrt{k-1}$.

The behavior of $\lambda^-(X_j)$ is not always similar to that of $\lambda^+(X_j)$. For example, if $\{X_j\}$ is taken to be the family of line graphs of k-regular graphs, then the X_j's are $(2k-2)$-regular and all $\lambda^-(X_j) \geq -2$. On the other hand, I proved recently [36] that under an extra hypothesis one obtains a symmetrical result.

Proposition 1. *Assume the extra condition that the length of odd cycles (when they exist) in X_j tends to infinity as j tends to infinity. Then*

$$\limsup_{j\to\infty} \lambda^-(X_j) \leq -2\sqrt{k-1}.$$

This conclusion can also be derived from McKay's result [40] under the assumption that for each fixed length d, the number of cycles in X_j with length d divided by the size of X_j approaches zero as j tends to infinity.

The above analysis leads to the following definition.

Definition. A k-regular graph X is Ramanujan if

$$|\lambda^{\pm}(X)| \leq 2\sqrt{k-1}.$$

In other words, a Ramanujan graph has nontrivial eigenvalues that are small in absolute value. Such a graph has a large magnifying constant and large girth. Hence it is a good communication network. The reader is referred to [34] for a comprehensive review of Ramanujan graphs.

The universal cover of a k-regular graph is the k-regular infinite tree. Its spectrum is $[-2\sqrt{k-1}, 2\sqrt{k-1}]$. Hence a Ramanujan graph has its nontrivial eigenvalues stay within the limits of the spectrum of its universal cover, while the nontrivial spectrum of a large regular graph in general has the tendency to "test the limit".

When $k = q+1$ for some prime power q, the infinite k-regular tree is the Bruhat-Tits building attached to $\mathrm{PGL}_2(K)/\mathrm{PGL}_2(\mathcal{O})$, where K is a local field with q elements in its residue field and $\mathcal{O}$ is the ring of integers in K. For example, $K = \mathbb{Q}_p$ and $\mathcal{O} = \mathbb{Z}_p$, or $K = \mathbb{F}_q((t))$ and $\mathcal{O} = \mathbb{F}_q[[t]]$. In this case, vertices are equivalence classes of rank 2 lattices over $\mathcal{O}$; two classes $[L_1]$ and $[L_2]$ are adjacent if and only if one can choose a representative from each class, say, L_1 and L_2, such that L_2 is a sublattice of L_1 with index q.

A systematic way to construct an infinite family of $(q+1)$-regular Ramanujan graphs is to use the theory of automorphic forms on quaternion groups which we describe below.

Fix a prime p. The graphs to be constructed will have valency $k = p+1$. Let H be a definite quaternion algebra defined over $\mathbb{Q}$ unramified at p. Denote by D the multiplicative group of H divided by its center, viewed as an algebraic group. Our graph X has its vertices the double coset space $D(\mathbb{Q})\backslash D(\mathbb{A}_{\mathbb{Q}})/D(\mathbb{R})\prod_{\ell} D(\mathcal{O}_\ell)$, where ℓ runs through all primes. By strong approximation theory, we can choose coset representatives locally at p, that is, X can also be seen as the double coset space $D(\mathbb{Z}[\frac{1}{p}])\backslash D(\mathbb{Q}_p)/D(\mathbb{Z}_p)$, which is the same as $D(\mathbb{Z}[\frac{1}{p}])\backslash \mathrm{PGL}_2(\mathbb{Q}_p)/\mathrm{PGL}_2(\mathbb{Z}_p)$ since H is unramified at p.

The third expression gives the graph structure of X, namely, as a quotient of the infinite $(p+1)$-regular tree $\mathrm{PGL}_2(\mathbb{Q}_p)/\mathrm{PGL}_2(\mathbb{Z}_p)$ by the group $D(\mathbb{Z}[\frac{1}{p}])$ so that X is $(p+1)$-regular. The first expression of X allows us to view functions on vertices of X as automorphic forms on the quaternion group $D(\mathbb{A}_\mathbb{Q})$ left invariant by the global rational points of D, and right invariant by the real points and the product of the standard maximal compact subgroup at each nonarchimedean place ℓ. The space C of such automorphic forms is known to contain the constant functions, and, if X is bipartite, the alternating constant functions taking opposite value on adjacent vertices. The orthogonal complement C' of such functions in C can be viewed as certain cusp forms of weight 2 for $GL_2(\mathbb{A}_\mathbb{Q})$ via Jacquet-Langlands correspondence, as proved in [19]. The restriction of the adjacency matrix A_X to C' is nothing but the Hecke operator T_p, and the nontrivial eigenvalues of A_X are therefore eigenvalues of T_p on certain cusp forms of weight 2, which are known to satisfy the Ramanujan-Petersson conjecture, that is, have absolute value majorized by $2\sqrt{p} = 2\sqrt{k-1}$. This shows that X is a Ramanujan graph.

We can replace $D(\mathbb{Z}[\frac{1}{p}])$ by congruence subgroups to get an infinite family of $(p+1)$-regular Ramanujan graphs. These were first constructed by Lubotzky-Phillips-Sarnak [38], and independently by Margulis [39]. The parallel construction with $\mathbb{Q}$ replaced by a function field K was done by Morgenstern [41]. There the Ramanujan-Petersson conjecture was proved by Drinfeld [17]. For a totally real field K, the Ramanujan-Petersson conjecture holds for certain holomorphic forms for $GL_2(\mathbb{A}_K)$, such as those from the work of [44], [5] and [2], as discussed in Section 5, and from the work of Carayol [6] coupled with the results of Deligne in [14]. Using these, Jordan and Livné [26] extended the construction of Ranamujan graphs to totally real fields.

There are other number-theoretic constructions, which give rise to finitely many Ramanujan graphs for each fixed k. They turn out to be quotients of the Ramanujan graphs constructed over function fields [35]. For example, the Cayley graph on the additive group $\mathbb{F}_{q^2}$ with generator set N_1, the set of elements in $\mathbb{F}_{q^2}$ whose norm to $\mathbb{F}$ is 1, is a $(q+1)$-regular Ramanujan graph, called a norm graph. See [33] for more details. The nontrivial eigenvalues of a norm graph are $-\operatorname{Kl}(\mathbb{F}_q; b)$. Theorem E describes an explicit automorphic form for GL_2 over a function field which is an eigenfunction of Hecke operators at good places with eigenvalues equal to eigenvalues of norm graphs, given by Kloosterman sums.

Another example is the Cayley graph on the cosets $GL_2(\mathbb{F}_q)/M$, where M is an embedded image of $\mathbb{F}_{q^2}^\times$ in $GL_2(\mathbb{F}_q)$, with generator set the M-cosets of a double M coset S, constructed by Terras. It is also a $(q+1)$-regular Ramanujan graph, called a Terras graph [1], [8]. The quantities $-\lambda_\chi(\mathbb{F}_q; a)$ occurring in Section 6 are part of the eigenvalues of Terras graphs. Theorems C and D describe automorphic forms for GL_2 over a function field which are eigenfunctions of Hecke operators at good places with eigenvalues equal to eigenvalues of Terras graphs. Theorems F, G, and H may be regarded as describing the distribution of eigenvalues of Terras graphs and norm graphs, respectively.

We end this section by considering a generalization of Ramanujan graphs to 3-hypergraphs.

Again let K be a nonarchimedean local field with q elements in its residue field. Denote by $\mathcal{O}$ the ring of integers, and by ϖ a uniformizer of $\mathcal{O}$. The structure of the Bruhat-Tits building $\mathcal{B}_{3,K}$ for $\mathrm{PGL}_3(K)/\mathrm{PGL}_3(\mathcal{O})$ is as follows. Its vertices are equivalence classes of lattices of rank 3 over $\mathcal{O}$. There is an edge from vertex $[L_1]$ to vertex $[L_2]$ if these two classes can be represented by lattices L_1 and L_2 respectively such that L_2 is a sublattice of L_1 with index q. Thus lattices L_1, L_2, L_3 represent three vertices $[L_1]$, $[L_2]$, and $[L_3]$ such that there is a loop $[L_1] \to [L_2] \to [L_3] \to [L_1]$ of length 3 if and only if $L_1 \supset L_2 \supset L_3 \supset \varpi L_1$ and each lattice is a sublattice of its predecessor with index q. If this happens, we say that $[L_1], [L_2], [L_3]$ form a face. This makes the Bruhat-Tits building a two-dimensional simplicial complex, which is topologically contractible. Each vertex has $q^2 + q + 1$ out-neighbors and $q^2 + q + 1$ in-neighbors.

There are two Hecke operators T and S acting on functions f defined on vertices of the building as follows:

(a) $(Tf)(x) = \sum_{x \to y} f(y)$,

(b) $(Sf)(x) = \sum_{y \to x} f(y)$

for all vertices x on $\mathcal{B}_{3,K}$. The spectrum of T is known to be the set

$$\Omega_T := \{q(\alpha + \beta + \gamma) \mid \alpha, \beta, \gamma \in \mathbb{C}, |\alpha| = |\beta| = |\gamma| = 1, \alpha\beta\gamma = 1\}.$$

The region Ω_T is invariant by multiplication by a cube root of unity ζ_3, that is, rotation by $2\pi/3$. Its boundary, consisting of the points $q(2e^{i\theta} + e^{-2i\theta})$ for $0 \le \theta \le 2\pi$, is smooth except for three singularities at $3q$, $3\zeta_3 q$, and $3\bar{\zeta}_3 q$, respectively, with a hypocycloid connecting any two of the three singular points. With respect to the usual Hermitian inner product $<,>$ on the space of ℓ^2-functions on vertices of $\mathcal{B}_{3,K}$, S and T are transposes of each other. Hence the spectrum of S, being the complex conjugation of the spectrum of T, is also Ω_T. Further, $ST = TS$. The reader is referred to the article [7] by Cartwright and Młotkowski for details.

A finite (q^2+q+1)-regular oriented 3-hypergraph is a finite 2-dimensional simplicial complex for which the Bruhat-Tits building $\mathcal{B}_{3,K}$ is its universal cover; each 0-dimensional simplex is called a vertex and each 2-dimensional triangular face with oriented 1-dimensional boundary is called an oriented 3-hyperedge. Its 1-dimensional skeleton is a directed graph such that each vertex has $q^2 + q + 1$ out-neighbors and $q^2 + q + 1$ in-neighbors. Let $\{X_j\}$

be a family of such hypergraphs with $|X_j| \to \infty$ as $j \to \infty$. Denote by $T_j := T_{X_j}$ the operator acting on functions on vertices of X_j defined by (a).

Parallel to the Alon-Boppana theorem for graphs is the following unconditional lower bound for hypergraphs.

Theorem I. *On each X_j there is a real valued function f_j, perpendicular to the constant functions and of norm one, such that*

$$\liminf_{j \to \infty} \langle T_j f_j, f_j \rangle \geq 3q.$$

Similar to the conditional upper bound in Proposition 1 for graphs, we have the following conditional result which gives examples of families of hypergraphs whose spectra combined are at least dense in the region Ω_T.

Theorem J. *Let $\Gamma_1 \supset \Gamma_2 \supset \cdots$ be a tower of discrete subgroups of $\mathrm{PGL}_3(K)$ such that*
(1) $X_j := \Gamma_j \backslash \mathrm{PGL}_3(K) / \mathrm{PGL}_3(\mathcal{O})$ is a finite, $(q^2 + q + 1)$-regular 3-hypergraph;
(2) $\lim_{j \to \infty} \Gamma_j = \{identity\}$.
Then, given any $\lambda \in \Omega_T$, there exists a function f_j on X_j with norm 1 such that the norm of $T_j f_j - \lambda f_j$ tends to 0 as j approaches ∞.

Consequently, for every $\lambda \in \Omega_T$, there exist eigenvalues λ_j of T_j such that λ_j converges to λ as $j \to \infty$.

The above discussion leads us to the following definition of a Ramanujan 3-hypergraph, analogous to a Ramanujan graph.

Definition. A finite $(q^2 + q + 1)$-regular oriented 3-hypergraph X is *Ramanujan* if all eigenvalues of T_X other than $q^2 + q + 1, (q^2 + q + 1)\zeta_3, (q^2 + q + 1)\bar{\zeta}_3$ fall in Ω_T.

In a joint work with Yu [37], we construct, for each fixed q, an infinite family of $(q^2 + q + 1)$-regular oriented Ramanujan 3-hypergraphs.

Theorem K. *Let K be a rational function field $F(t)$ over a finite field F of q elements. Let v be a place of K of degree 1. Let $\mathcal{D}$ be a division algebra of degree 9 over K which is unramified at v but ramified at a place w of degree 1. Denote by D the multiplicative group of $\mathcal{D}$ divided by its center. Then for a congruence subgroup $\mathcal{K}$ of $\prod_{v' \neq v, w} D(\mathcal{O}_{v'})$, the double coset space*

$$X_{\mathcal{K}} = D(K) \backslash D(\mathbb{A}_K) / D(K_w) D(\mathcal{O}_v) \mathcal{K},$$

which can also be represented as local double coset space

$$\Gamma_{\mathcal{K}} \backslash \mathrm{PGL}_3(K_v) / \mathrm{PGL}_3(\mathcal{O}_v),$$

is a Ramanujan $(q^2 + q + 1)$-regular oriented 3-hypergraph. Here $\Gamma_{\mathcal{K}}$ is the intersection of $D(K)$ with $\mathcal{K}$.

The proof is parallel to our description of the explicit construction of Ramanujan graphs. The key ingredients are approximation theory, the correspondence from automorphic representations of division group $D(\mathbb{A}_K)$ to automorphic representations of $\mathrm{PGL}_3(\mathbb{A}_K)$ as described in [32], and the Ramanujan conjecture for GL_3 proved by Lafforgue [30].

References

[1] J. Angel, N. Celniker, S. Poulos, A. Terras, C. Trimble, and E. Velasquez, *Special functions on finite upper half planes*, Hypergeometric functions on domains of positivity, Jack polynomials, and applications (Tampa, FL, 1991), Contemp. Math., vol. 138, Amer. Math. Soc., Providence, RI, 1992, pp. 1–26.

[2] D. Blasius and J. Rogawski, *Motives for Hilbert modular forms*, Invent. Math. **114** (1993), 55–87.

[3] A. Booker, *A test for identifying Fourier coefficients of automorphic forms and application to Kloosterman sums*, Experimental Math. **9** (2000), 571–581.

[4] C. Breuil, B. Conrad, F. Diamond, and R. Taylor, *On the modularity of elliptic curves over* $\mathbb{Q}$, J. Amer. Math. Soc. **14** (2001), 843–939.

[5] J.-L. Brylinski and J.-P. Labesse, *Cohomologie d'intersection et fonctions L de certaines variétés de Shimura*, Ann Scient. Éc. Norm. Sup. **17** (1984), 361–412.

[6] H. Carayol, *Sur les représentations ℓ-adiques associées aux formes modulaires de Hilbert*, Ann. Sci. École Norm. Sup. **19** (1986), 409–468.

[7] D. Cartwright and W. Młotkowski, *Harmonic analysis for groups acting on triangle buildings*, J. Austral. Math. Soc. Ser. A **56** (1994), 345–383.

[8] N. Celniker, S. Poulos, A. Terras, C. Trimble, and E. Velasquez, *Is there life on finite upper half planes?*, A tribute to Emil Grosswald: number theory and related analysis, Contemp. Math., vol. 143, Amer. Math. Soc., Providence, RI, 1993, pp. 65–88.

[9] C.-L. Chai and W.-C. W. Li, *Character sums, automorphic forms, equidistribution, and Ramanujan graphs. Part I. The Kloosterman sum conjecture over function fields*, Math Forum, to appear.

[10] ———, *Character sums, automorphic forms, equidistribution, and Ramanujan graphs. Part II. Eigenvalues of Terras graphs*, preprint, 2000.

[11] J. W. Cogdell and I. I. Piatetski-Shapiro, *A converse theorem for GL_4*, Math. Research Letters **3** (1996), 67–76.

[12] P. Deligne, *Formes modulaires et représentations ℓ-adiques*, Séminaire Bourbaki, vol. 1968/69, Lecture Notes in Math., vol. 179, Springer, Berlin, 1971.

[13] ———, *Les constantes des équations fonctionelles des fonctions L*, Lecture Notes in Math., vol. 349, Springer, 1973, pp. 501–595.

[14] ———, *La conjecture de Weil II*, IHES Publ. Math. **52** (1980), 137–252.

[15] P. Deligne and J.-P. Serre, *Formes modulaires de poids 1*, Ann. Sci. École Norm. Sup. **7** (1974), 507–530.

[16] V. G. Drinfel'd, *Proof of the global Langlands conjecture for $GL(2)$ over a function field*, Functional Anal. Appl. **11** (1977), 223–225.

[17] ———, *The proof of Petersson's conjecture for $GL(2)$ over a global field of characteristic p*, Functional Anal. Appl. **22** (1988), 28–43.

[18] G. Faltings, *Endlichkeitssätze für abelsche Varietäten über Zahlkörpern*, Invent. Math. **73** (1983), 349–366.

[19] S. Gelbart and H. Jacquet, *Forms of $GL(2)$ from the analytic point of view*, Automorphic forms, representations and L-functions (Corvallis, Ore., 1977), Part 1, Proc. Sympos. Pure Math., vol. XXXIII, Amer. Math. Soc., 1979, pp. 213–251.

[20] M. Harris and R. Taylor, *On the geometry and cohomology of some simple Shimura varieties*, preliminary version, 1998.

[21] G. Henniart, *La conjecture de Langlands locale pour $GL(3)$*, Mémoire Soc. Math. France, France, 1984.

[22] ———, *Caractérisation de la correspondance de Langlands locals par les facteurs ε de paires*, Invent. Math. **113** (1993), 339–350.

[23] ———, *Une preuve simple des conjectures de Langlands pour GL_n sur un corps p-adique*, Invent. Math. **139** (2000), 439–455.

[24] H. Jacquet and R. P. Langlands, *Automorphic Forms on $GL(2)$*, Springer-Verlag, Berlin-Heidelberg-New York, 1970.

[25] H. Jacquet, I. I Piatetski-Shapiro, and J. Shalika, *Automorphic forms on GL_3, I & II*, Ann. of Math. **109** (1979), 169–258.

[26] B. Jordan and R. Livné, *The Ramanujan property for regular cubical complexes*, Duke Math. J. **105** (2000), 85–103.

[27] N. Katz, *Gauss Sums, Kloosterman Sums, and Monodromy Groups*, Princeton Univ. Press, Princeton, 1988.

[28] P. Kutzko, *The Langlands conjecture for GL_2 of a local field*, Ann. of Math. **112** (1980), 381–412.

[29] P. Kutzko and A. Moy, *On the local Langlands conjecture in prime dimension*, Ann. of Math. **121** (1985), 495–517.

[30] L. Lafforgue, *Chtoucas de Drinfeld et conjecture de Ramanujan-Petersson*, Astérisque **243** (1997).

[31] _____, *La correspondance de Langlands sur les corps de fonctions*, preprint, 2000.

[32] G. Laumon, M. Rapoport, and U. Stuhler, *$\mathcal{D}$-elliptic sheaves and the Langlands correspondence*, Invent. Math. **113** (1993), 217–338.

[33] W.-C. W. Li, *Character sums and abelian Ramanujan graphs*, J. Number Theory **41** (1992), 199–217.

[34] _____, *A survey of Ramanujan graphs*, Arithmetic, geometry and coding theory (Luminy, 1993), de Gruyter, Berlin, 1996, pp. 127–143.

[35] _____, *Eigenvalues of Ramanujan graphs*, Emerging applications of number theory (Minneapolis, MN, 1996), IMA Vol. Math. Appl., vol. 109, Springer, New York, 1999, pp. 387–403.

[36] _____, *On negative eigenvalues of regular graphs*, C. R. Acad. Sci. Paris Sér. I Math. **333** (2001), 907–912.

[37] W.-C. W. Li and R.-K. Yu, *Ramanujan 3-graphs*, preprint, 2000.

[38] A. Lubotzky, R. Phillips, and P. Sarnak, *Ramanujan graphs*, Combinatorica **8** (1988), 261–277.

[39] G. Margulis, *Explicit group theoretic constructions of combinatorial schemes and their application to the design of expanders and concentrators*, J. Prob. Info. Trans. (1988), 39–46.

[40] B. D. McKay, *The expected eigenvalue distribution of a large regular graph*, Linear Algebra Appl. **40** (1981), 203–216.

[41] M. Morgenstern, *Existence and explicit constructions of q + 1 regular Ramanujan graphs for every prime power q*, J. Comb. Theory Ser. B **62** (1994), 44–62.

[42] H. Reimann, *The semi-simple zeta function of quaternionic Shimura varieties*, Springer-Verlag, Berlin-Heidelberg-New York, 1997.

[43] K. Ribet, *On modular representations of* $Gal(\overline{\mathbb{Q}}/\mathbb{Q})$ *arising from modular forms*, Invent. Math. **100** (1990), 431–476.

[44] J. Rogawski and J. Tunnell, *On Artin L-functions associated to Hilbert modular forms of weight one*, Invent. Math. **74** (1983), 1–42.

[45] R. Taylor and A. Wiles, *Ring-theoretic properties of certain Hecke algebras*, Ann. of Math. **141** (1995), 553–572.

[46] A. Wiles, *Modular elliptic curves and Fermat's last theorem*, Ann. of Math. **141** (1995), 443–551.

[47] H. Yoshida, *On an analogue of the Sato-Tate conjecture*, Invent. Math. **19** (1973), 261–277.

Convergence of Corresponding Continued Fractions

Lisa Lorentzen

1 Introduction

There exist several methods to expand a given function $f(z)$ into a continued fraction

$$b_0 + K(a_n/b_n) = b_0 + \cfrac{a_1}{b_1} + \cfrac{a_2}{b_2} + \ldots = b_0 + \cfrac{a_1}{b_1 + \cfrac{a_2}{b_2 + \ldots}}, \qquad (1.1)$$

where the elements a_n and b_n are polynomials or other (mostly entire) functions of z. What we hope is that (1.1) *converges* to f; that is, that its sequence of *approximants*

$$f_n = b_0 + \cfrac{a_1}{b_1} + \cfrac{a_2}{b_2} + \cdots + \cfrac{a_n}{b_n}; \qquad n = 0, 1, 2, \ldots \qquad (1.2)$$

converges to f in some domain $D \subseteq \mathbb{C}$. We regard continued fraction expansions as alternatives to Laurent series expansions $L =: \mathcal{L}(f)$ of f at $z = 0$ or asymptotic series L of f. What we want is that

- $b_0 + K(a_n/b_n)$ converges in a larger domain than L, uniformly on compact subsets, to f;

- $b_0 + K(a_n/b_n)$ converges faster than L in the convergence disk of L;

- $b_0 + K(a_n/b_n)$ lends itself to easy ways of convergence acceleration;

- $b_0 + K(a_n/b_n)$ lends itself to easy ways of meromorphic continuation;

- the computation of f_n is stable.

And, luckily enough, this is what we get in a number of useful cases.

In this paper we shall mainly concentrate on the problem of when $b_0 + K(a_n/b_n)$ converges to f. There exist a number of theorems giving sufficient conditions for a continued fraction to converge, but very few of these use the additional information that $b_0 + K(a_n/b_n)$ is a continued fraction expansion of some given function.

In Section 2 we describe what it means that $b_0 + K(a_n/b_n)$ is an expansion of f. In Section 3 we give some basic results on the chordal metric and extend some earlier results. In Section 4 we present an alternative definition of convergence which extends the classical concept mentioned above. In Section 5 we give some general results on convergence of $b_0 + K(a_n/b_n)$, and in Section 6 we handle the simple but useful case when $b_0 + K(a_n/b_n)$ is limit periodic.

If the elements a_n and b_n are polynomials, then the approximants $f_n(z)$ are rational functions. Under additional conditions they are *Padé approximants* to f, and for these we can prove stronger results, as shown in Section 7. As an example, in Section 8 we treat *regular C-fractions* $b_0 + K(a_n z/1)$, where all a_n are complex numbers. A particularly beautiful case occurs when $a_n > 0$ for all n. These *Stieltjes fractions* are described in Section 9. Finally, in Section 10 we return to the regular C-fractions for additional results. Throughout this paper, $\widehat{\mathbb{C}} := \mathbb{C} \cup \{\infty\}$ denotes the Riemann sphere, and a *domain* is always an open, connected set in $\mathbb{C}$.

2 Correspondence

We shall here clarify what we mean when we say that (1.1) is a continued fraction expansion of f. Actually, what we really say is that $b_0 + K(a_n/b_n)$ corresponds to a Laurent series at $z = 0$. So let Λ denote the family of formal Laurent series $\sum_{n=n_0}^{\infty} c_n z^n$, $n_0 \in \mathbb{Z}$. In this family, we let $\text{ord}(\sum_{k=n_0}^{\infty} c_k z^k) := n_0$ when $c_{n_0} \neq 0$, with the limit form $\text{ord}(\ell_0) = \infty$ for $\ell_0 \equiv 0$. We let further $\mathcal{M}$ denote the family of functions that are meromorphic at $z = 0$. Then $\mathcal{L}(f) \in \Lambda$ for every $f \in \mathcal{M}$. For simplicity we also define $\text{ord}(\mathcal{L}(f)) := -\infty$ for the function $f(z) \equiv \infty$. We say that a sequence $\{g_n\}$ of functions *corresponds* (at $z = 0$) to some $L \in \Lambda$ if $g_n \in \mathcal{M}$ for sufficiently large n, and

$$\nu_n := \text{ord}(\mathcal{L}(g_n) - L) \to \infty \quad \text{as } n \to \infty. \tag{2.1}$$

ν_n is called the *order of correspondence*, and we write $\{g_n\} \sim L$. Evidently, if $\{g_n\} \sim L$ for some $L \in \Lambda$, then L is unique. We say that $b_0 + K(a_n/b_n)$ corresponds to $L \in \Lambda$ if its approximants satisfy $\{f_n\} \sim L$, and we write $b_0 + K(a_n/b_n) \sim L$.

We say that $b_0 + K(a_n/b_n)$ is a *corresponding continued fraction* if there exists an $L \in \Lambda$ such that $b_0 + K(a_n/b_n) \sim L$. According to Jones and Thron we have:

Theorem 2.1 ([14, Thm. 5.11, p. 151]). *A given sequence $\{g_n\}$ from $\mathcal{M}$ corresponds to some $L \in \Lambda$ if and only if $\lim_{n \to \infty} \text{ord}(\mathcal{L}(g_n) - \mathcal{L}(g_{n+1})) = \infty$.*

Hence, $b_0 + K(a_n/b_n)$ is a corresponding continued fraction if and only if $f_n \in \mathcal{M}$ for sufficiently large n, and

$$\text{ord}(\mathcal{L}(f_n) - \mathcal{L}(f_{n+1})) \to \infty \quad \text{as } n \to \infty. \tag{2.2}$$

If further $f \in \mathcal{M}$ and $b_0 + K(a_n/b_n) \sim L = \mathcal{L}(f)$, we say that $b_0 + K(a_n/b_n)$ is a *continued fraction expansion* of f (at $z = 0$), and we write $b_0 + K(a_n/b_n) \sim f$. We also do so if $D \subset \mathbb{C}$ is a domain with 0 on its boundary, f is meromorphic in D, and L is an asymptotic expansion of f in D when $b_0 + K(a_n/b_n) \sim L$.

One may consider correspondence at points in $\widehat{\mathbb{C}}$ other than the point $z = 0$, but in this paper we shall always mean correspondence at the origin. For more information on correspondence and other basic properties of continued fractions we refer to [14], [18], [25].

3 The Chordal Metric and Normal Families

For continued fractions it is natural to use the chordal metric

$$d(z_1, z_2) = \frac{2|z_1 - z_2|}{\sqrt{1 + |z_1|^2}\sqrt{1 + |z_2|^2}} \quad \text{for } z_1, \ z_2 \in \widehat{\mathbb{C}} \tag{3.1}$$

on the Riemann sphere, where $d(z_1, z_2)$ takes the obvious limit forms if z_1 or z_2 is infinite. The advantage of this metric is that $\{z_n\}$ converges to $z \in \widehat{\mathbb{C}}$ (in the usual sense) if and only if $d(z_n, z) \to 0$, regardless of whether z is finite or infinite. That is, convergence to ∞ is no different from convergence to any other number, when expressed in this metric. This allows us to treat meromorphic functions in the same way as we treat analytic functions. In particular, a family $\mathcal{F}$ of meromorphic functions on a domain D is *normal* if every sequence $\{f_n\}$ from $\mathcal{F}$ has a subsequence which converges spherically uniformly on compact subsets of D. Here *spherically uniform convergence* means uniform convergence with respect to the chordal metric. The following two standard results hold; see, for instance, [27, p. 72–75].

Theorem 3.1. *Let $\{f_n\}$ be a sequence of meromorphic functions on a domain D which converges spherically uniformly on compact subsets of D to f. Then f is either meromorphic in D or identically equal to ∞.*

Theorem 3.2. *Let $\mathcal{F}$ be a family of meromorphic functions on a domain D. If either*

(i) *there exist three distinct points* p_1, p_2, $p_3 \in \widehat{\mathbb{C}}$ *such that* $f(z) \neq p_j$ *for all* $f \in \mathcal{F}$, $z \in D$ *and* $j = 1, 2, 3$; *or*

(ii) *for every compact subset* $E \subset D$ *there exists a constant* $M > 0$ *such that the spherical derivative satisfies*

$$f^{\#}(z) := \frac{2|f'(z)|}{1 + |f(z)|^2} \leq M \quad \text{for all } f \in \mathcal{F} \text{ and } z \in E,$$

then $\mathcal{F}$ *is normal in* D.

The importance of Theorem 3.2 for our purpose lies in the following result, which generalizes results by Jones and Thron [13] and Baker and Graves-Morris [1].

Theorem 3.3. *Let* $\{f_n(z)\}$ *be a sequence of meromorphic functions on a domain* D *containing the origin. If* $\{f_n\} \sim L$ *and the family* $\mathcal{F} := \{f_n : n \geq n_0\}$ *is normal in* D *for some* $n_0 \in \mathbb{N}$, *then* $\{f_n\}$ *converges spherically uniformly on compact subsets of* D *to a function* f *with* $\mathcal{L}(f) = L$ *that is meromorphic in* D.

Proof. Since $\mathcal{F}$ is normal, there is a subsequence $\{f_{n_k}\}$ of $\{f_n\}$ which converges spherically uniformly on compact subsets of D to some function f. It suffices to prove that $\mathcal{L}(f) = L$, since this ensures that f is unique and $f \not\equiv \infty$.

Suppose first that $f(0) \neq \infty$. Then there exists a $\delta > 0$ such that f is analytic in the closed disk $E_\delta := \{z : |z| \leq \delta\} \subseteq D$. By the uniformity of the spherical convergence of $\{f_{n_k}\}$ to f on E_δ, it follows that f_{n_k} is analytic in E_δ for sufficiently large k. By Weierstrass' theorem it follows that the derivatives $f_{n_k}^{(m)}$ satisfy $f_{n_k}^{(m)}(0) \to f^{(m)}(0)$ as $k \to \infty$. This can only happen if $\mathcal{L}(f) = L$.

Next, suppose that $f(0) = \infty$. Then $g_{n_k} := 1/f_{n_k}$ converges spherically uniformly on compact subsets of D to $g := 1/f$. In particular, $g(0) = 0$, and thus it follows from the arguments above that $\mathcal{L}(g) = 1/L$; i.e., we have $\mathcal{L}(f) = L$. $\qquad\qquad\square$

4   General Convergence

We suppose first that the elements a_n and b_n of $b_0 + K(a_n/b_n)$ are complex numbers. We also require that $a_n \neq 0$ for all n, so that the continued fraction is non-terminating. In 1942 Paydon and Wall [24] described $b_0 + K(a_n/b_n)$ by means of the linear fractional transformations

$$s_0(w) := b_0 + w, \qquad s_n(w) := a_n/(b_n + w) \quad \text{for } n = 1, 2, 3, \ldots. \quad (4.1)$$

(Note that s_n is non-singular since $a_n \neq 0$.) They formed the compositions

$$S_n(w) := s_0 \circ s_1 \circ s_2 \circ \cdots \circ s_n(w) = b_0 + \frac{a_1}{b_1} + \frac{a_2}{b_2} + \cdots + \frac{a_n}{b_n + w}, \quad (4.2)$$

which again are (non-singular) linear fractional transformations. Then the approximants of $b_0 + K(a_n/b_n)$ can be written as $f_n = S_n(0)$. In [8] it was shown that for the periodic continued fraction

$$\frac{2}{1} + \frac{1}{1} - \frac{1}{1} + \frac{2}{1} + \frac{1}{1} - \frac{1}{1} + \frac{2}{1} + \frac{1}{1} - \frac{1}{1} + \dots \quad (4.3)$$

$S_{3n}(w) \to 1/2$ for every $w \neq 0$, $S_{3n+1}(w) \to 1/2$ for every $w \neq \infty$, and $S_{3n+2}(w) \to 1/2$ for every $w \neq -1$. Since $S_{3n}(0) = 0$ for all n, this shows that (4.3) diverges, although it has a very strong convergence behavior. To catch this kind of convergence, a new concept of convergence was introduced. Since we allow convergence to ∞, it was natural to base the concept on the chordal metric (3.1).

Definition 4.1 ([8]). *We say that $b_0 + K(a_n/b_n)$, with elements a_n, b_n from $\mathbb{C}$ and $a_n \neq 0$ for all n, converges generally to f if there exist two sequences $\{u_n\}$ and $\{v_n\}$ of numbers from $\widehat{\mathbb{C}}$ such that*

$$\liminf_{n\to\infty} d(u_n, v_n) > 0 \quad and \quad \lim_{n\to\infty} S_n(u_n) = \lim_{n\to\infty} S_n(v_n) = f. \quad (4.4)$$

Looking back at (4.3), any sequences $\{w_n\}$ with $\liminf d(w_{3n+2}, -1) > 0$, $\liminf d(w_{3n+1}, \infty) > 0$ and $\liminf d(w_{3n}, 0) > 0$ can be used for $\{u_n\}$ and $\{v_n\}$ in (4.4) to prove that (4.3) converges generally to $1/2$. This is a typical situation:

Theorem 4.2 ([8]). *Let $b_0 + K(a_n/b_n)$, with elements a_n, b_n from $\mathbb{C}$ and $a_n \neq 0$ for all n, converge generally to f. Then there exists a sequence $\{\zeta_n\}$ of numbers from $\widehat{\mathbb{C}}$ such that $S_n(w_n) \to f$ for every sequence $\{w_n\}$ satisfying $\liminf_{n\to\infty} d(w_n, \zeta_n) > 0$.*

We say that $\{\zeta_n\}$ is an *exceptional sequence* for the generally convergent continued fraction $b_0 + K(a_n/b_n)$. The proof of this theorem is so short that I include it here:

Proof. Without loss of generality we assume that $b_0 + K(a_n/b_n)$ converges to $f \neq \infty$. We can do this, since

$$\frac{1}{b_0} + \frac{a_1}{b_1} + \frac{a_2}{b_2} + \frac{a_3}{b_3} + \dots = \{b_0 + K(a_n/b_n)\}^{-1} \sim \frac{1}{f}. \quad (4.5)$$

Let $\zeta_n := S_n^{-1}(\infty)$ for all n, let $\{u_n\}$ and $\{v_n\}$ be chosen such that (4.4) holds, and let $\{w_n\} \subset \widehat{\mathbb{C}}$ be such that $\liminf d(w_n, \zeta_n) > 0$. We consider indices n for which $w_n \neq u_n, v_n$. Then u_n, v_n, ζ_n and w_n are distinct points for every sufficiently large n. The invariance of the cross ratio under linear fractional transformations therefore gives that

$$\frac{d(S_n(u_n), S_n(w_n))}{d(S_n(u_n), S_n(v_n))} = \frac{d(u_n, w_n) \cdot d(\zeta_n, v_n)}{d(u_n, v_n) \cdot d(\zeta_n, w_n)}. \qquad (4.6)$$

Since the right hand side is bounded and $d(S_n(u_n), S_n(v_n)) \to 0$, we must have $d(S_n(u_n), S_n(w_n)) \to 0$. $\qquad \square$

(Theorem 4.2 can be proved without this trick with the cross ratio. In [17] this was done in a more general setting, not only for convergence, but also for functions $\{S_n\}$ with a limiting structure [11].)

Theorem 4.2 shows that the limit f of a generally convergent continued fraction is unique. Moreover, classical convergence to f implies general convergence to f since $S_n(0) = S_{n+1}(\infty)$. The converse is not true, since (4.3) converges generally to $\frac{1}{2}$. In fact, this new definition captures every continued fraction that ought to converge, but which diverges in the classical sense because its exceptional sequence $\{\zeta_n\}$ has a limit point at 0.

Given $g \in \widehat{\mathbb{C}}$, we say that $g_n := S_n^{-1}(g)$ is a *tail sequence* for $b_0 + K(a_n/b_n)$. If $b_0 + K(a_n/b_n)$ converges generally to f and $\{\zeta_n\}$ is an exceptional sequence for $b_0 + K(a_n/b_n)$, then it follows from Theorem 4.2 that $\lim d(g_n, \zeta_n) = 0$ for every $g \neq f$. That is, every such tail sequence is an exceptional sequence. On the other hand, if $g = f$ and we choose $w_n = g_n$, then $S_n(w_n) = g = f$ for all n, which shows that this tail sequence is a particularly good sequence. In fact, an easy way to accelerate the convergence of $b_0 + K(a_n/b_n)$ is to use approximants $S_n(w_n)$ satisfying $\lim d(w_n, S_n^{-1}(f)) = 0$. This is in particular so for limit periodic continued fractions. In some cases this can even give a meromorphic extension of the limit function; see [15].

Also, the stability of the computation is improved if we choose w_n appropriately. If $f \neq \infty$, we choose w_n such that the denominator of the linear fractional transformation

$$S_n(w_n) = b_0 + \frac{a_1}{b_1 +} \cdots \frac{a_n}{b_n + w_n} = \frac{A_n w_n + B_n}{C_n w_n + D_n} \qquad (4.7)$$

stays away from 0, and if $f \neq 0$, we make sure that its numerator stays away from 0. But how can we choose w_n if f is totally unknown? One idea is to compute

$$S_n(0), \qquad S_n(\infty) = S_{n-1}(0) \quad \text{and} \quad S_n(-b_n) = S_{n-2}(0) \qquad (4.8)$$

and use the average of the two values that lie closest to each other on the Riemann sphere. This works if $\{b_n\}$ is bounded away from 0 and ∞.

Otherwise one can replace $S_n(-b_n)$ by $S_n(1)$. Another idea is to make use of *value sets* $\{V_n\}_{n=-1}^{\infty}$; that is, the V_n are proper subsets of $\widehat{\mathbb{C}}$ with non-empty interiors such that $s_n(V_n) \subset V_{n-1}$ for all n. Then there is an exceptional sequence $\{\zeta_n\}$ with $\zeta_n \notin V_n$ for all n; see [18]. Hence, if the chordal diameter of the interior of V_n stays $\geq 2\delta > 0$ for all n, then, for each n, we can choose w_n as the center of a δ–disk in V_n.

Let us return to the situation where $b_0 + K(a_n/b_n)$ is a corresponding continued fraction. It is important that the variable w of S_n not be confused with the variable z of a_n and b_n. For instance, in (4.7) the coefficients of S_n are functions of z, so S_n is really a function of two complex variables. To keep the notation simple, will still write $S_n(w)$. The modifying factors w_n will normally also depend on z. It is natural to choose $\{w_n\}$ such that the functions $S_n(w_n)$ are meromorphic functions of z in the domain D of interest, and to ask for spherically uniform convergence of $S_n(w_n)$ on compact subsets of D.

5 Some General Results

Let f be a function that is analytic at $z = 0$, and let $b_0 + K(a_n/b_n) \sim f$. It is then very natural to assume that if $b_0 + K(a_n/b_n)$ converges in some domain $D \subset \mathbb{C}$, then its limit is $f(z)$ in D, at least if the approximants are analytic in D and the convergence is uniform on compact subsets of D. However, this is not always the case. For instance, the periodic continued fraction

$$\frac{z}{1-z} + \frac{z}{1-z} + \frac{z}{1-z} + \dots \tag{5.1}$$

corresponds to the entire function $f(z) = z$. However, since

$$S_n(w) = z\,\frac{[1 - (-z)^{n-1}]w + 1 - (-z)^n}{[1 - (-z)^n]w + 1 - (-z)^{n+1}}, \tag{5.2}$$

we find that $S_n(0)$ converges uniformly on compact subsets of $|z| > 1$ to the value $f(z) = -1$. This shows that one has to be a little careful. Correspondence and uniform convergence of analytic approximants is not sufficient to secure convergence to the "right value." However, if we also have $0 \in D$, then life is much easier. From the three theorems in Section 3 we obtain:

Theorem 5.1. *Let $D \subseteq \mathbb{C}$ be a domain with $0 \in D$, and let $\{g_n\}$ be a sequence of functions that are meromorphic in D, such that $\{g_n\} \sim L$ for some $L \in \Lambda$. Then the following statements hold.*

(i) *If $\{g_n\}$ omits three distinct values p_1, p_2, $p_3 \in \widehat{\mathbb{C}}$ for $z \in D$ and n sufficiently large, then $\{g_n\}$ converges spherically uniformly on every closed subset E of D.*

(ii) *$\{g_n\}$ converges spherically uniformly on a closed subset E of D if its spherical derivatives $g_n^\#$ are uniformly bounded in E for sufficiently large n.*

(iii) *If $\{g_n\}$ converges spherically uniformly on closed subsets of D, then its limit function g is meromorphic in D and $L = \mathcal{L}(g)$.*

Theorem 5.1 applies to corresponding continued fractions whose classical approximants $f_n = S_n(0)$ are meromorphic in some domain D containing the origin. By turning to approximants $g_n = S_n(w_n)$ of $b_0 + K(a_n/b_n)$, we have even better chances of proving convergence by means of Theorem 5.1. We only have to make sure that if $b_0 + K(a_n/b_n) \sim L$, then also $\{S_n(w_n)\} \sim L$, but that is not difficult:

Theorem 5.2. *Let $b_0 + K(a_n/b_n) \sim L \in \Lambda$ with $a_n, b_n \in \mathcal{M}$ and $a_n \not\equiv 0$ for all n, and let $\zeta_n := S_n^{-1}(\infty)$. Then $S_n(w_n) \sim L$ for every sequence $\{w_n\}$ which satisfies $w_n \in \mathcal{M}$ for sufficiently large n and $\limsup_{n\to\infty} \operatorname{ord}\{\mathcal{L}(1 - w_n/\zeta_n)\} < \infty$.*

Proof. With the notation (4.7) we find that if $C_n D_n \not\equiv 0$, then

$$S_n(w) = \frac{A_n w + B_n}{C_n w + D_n} = \frac{A_n}{C_n} - \frac{A_n D_n - B_n C_n}{C_n^2 w + C_n D_n} = S_n(\infty) - \frac{S_n(0) - S_n(\infty)}{1 - w/\zeta_n},$$

where $S_n(\infty) = S_{n-1}(0)$, and thus $\operatorname{ord}(\mathcal{L}(S_n(\infty)) - L) \to \infty$. That $C_n D_n \not\equiv 0$ for sufficiently large n follows since $f_n = B_n/D_n = A_{n+1}/C_{n+1}$ and $\{f_n\} \sim L$. Since $\operatorname{ord}(\mathcal{L}(S_n(0)) - \mathcal{L}(S_n(\infty))) \to \infty$ by (2.2) and $\operatorname{ord}(A/B) = \operatorname{ord}(A) - \operatorname{ord}(B)$, the result follows. $\square$

The relationship to Theorem 4.2 is clear: $\{w_n\}$ just has to stay away from the same kind of exceptional sequence. (In [9] we used this to define *general correspondence* in the obvious way, but that is another story.) To prove that the conditions in Theorem 5.1(i) or (ii) hold, we may again use value sets.

6 Limit Periodic Continued Fractions

Here we let again $a_n \neq 0$ and b_n in $b_0 + K(a_n/b_n)$ be complex numbers. If the limit

$$q := \lim_{n\to\infty} q_n, \quad \text{with } q_1 := a_1/b_1, \quad q_n := a_n/b_{n-1}b_n \quad (n \geq 2) \qquad (6.1)$$

exists, we say that $b_0 + K(a_n/b_n)$ is limit periodic with period 1. A large class of important functions have limit periodic continued fraction expansions, mostly with period 1 (the case we treat here) or period 2. For limit periodicity with periods ≥ 2 we refer to [7] and [18].

If the limit q belongs to the cut plane

$$Q := \{u \in \mathbb{C} : \; |\arg(1 + 4u)| < \pi\}, \tag{6.2}$$

then $b_0 + K(a_n/b_n)$ converges in the classical sense. Otherwise, if $q = \infty$ or if q is real with $q \leq -1/4$, the situation is more unclear. The continued fraction may converge or diverge, depending on how q_n approaches its limit; see [5], [10], [29]. To keep this survey simple (and not too long), we shall restrict to the case $q \in Q$. Then the following results follow from [25, Satz 2.4, p. 93], [7, Thm. 3.2B], [18, Thm. 28, p. 151] and the equivalence transformation $b_0 + K(a_n/b_n) \approx b_0 + K(q_n/1)$. (If $q_n = \infty$ for some n, this can only happen for finitely many indices, and thus $b_N + K(a_{N+n}/b_{N+n}) \approx b_N(1 + K(q_{N+n}/1))$ for N sufficiently large.)

Theorem 6.1. *Let $b_0 + K(a_n/b_n)$ with elements from $\mathbb{C}$ and $a_n \neq 0$ for all n satisfy (6.1) with $q \in Q$, and let $x := -\frac{1}{2}(1 - \sqrt{1 + 4q})$, where $Re(\sqrt{1 + 4q} > 0$. Then $b_0 + K(a_n/b_n)$ has the following properties.*

(i) *$b_0 + K(a_n/b_n)$ converges in the classical sense to some $f \in \widehat{\mathbb{C}}$.*

(ii) *$\lim_{n \to \infty} S_n^{-1}(f)/b_n = x$.*

(iii) *$\lim_{n \to \infty} S_n^{-1}(g)/b_n = -1 - x$ for every $g \in \widehat{\mathbb{C}}$ with $g \neq f$.*

In particular, $\{b_n(-1-x)\}$ is an exceptional sequence, and thus $S_n(w_n) \to f$ as long as $\liminf d(w_n, b_n(-1 - x)) > 0$. Moreover, the approximants $S_n(b_n x)$ converge faster to f than $S_n(0)$ if $\liminf |b_n x| > 0$:

Theorem 6.2. *Let $b_0 + K(a_n/b_n)$, f and x be as in Theorem 6.1. Then $b_0 + K(a_n/b_n)$ has the following properties.*

(i) *If $f \neq \infty$, then for every $r > |x|/|1 + x|$ there exists a $C > 0$ such that $|f - S_n(0)| \leq Cr^n$ for every n for which $S_n(0) \neq \infty$.*

(ii) *To every $r > |x|/|1 + x|$ there exists a $C > 0$ such that $d(f, S_n(0)) \leq Cr^n$ for all $n \geq 1$.*

(iii) *If $\liminf_{n \to \infty} |b_n x| > 0$ and $\liminf_{n \to \infty} d(w_n, b_n(-1 - x)) > 0$, then*

$$\frac{d(f, S_n(w_n))}{d(f, S_n(0))} = \mathcal{O}(d(f^{(n)}, w_n)) \quad \text{as } n \to \infty, \quad f^{(n)} := S_n^{-1}(f).$$

Proof. Part (i) was essentially proved by Thron and Wadeland [31] for the case when $b_n = 1$ for all n and thus $a_n = q_n \to q \in Q$. We shall instead prove (ii), from which (i) follows immediately.

(ii) Let $f \neq \infty$. By using (4.6) with $u_n = f^{(n)}$, $w_n = 0$ and $v_n = \infty$, we get

$$\frac{d(f, S_n(0))}{d(f, S_n(\infty))} = \frac{d(f, f_n)}{d(f, f_{n-1})} = \frac{d(f^{(n)}, 0) \cdot d(\zeta_n, \infty)}{d(f^{(n)}, \infty) \cdot d(\zeta_n, 0)}$$

$$= \left|\frac{f^{(n)}}{\zeta_n}\right| \to \left|\frac{x}{1+x}\right|, \tag{6.3}$$

which proves the result for this case. If $f = \infty$, then (4.5) converges to $\hat{f} = 1/f = 0$, and thus its approximants $\hat{f}_n = 1/f_{n-1}$ satisfy

$$\frac{d(\hat{f}, \hat{f}_n)}{d(\hat{f}, \hat{f}_{n-1})} \to \left|\frac{x}{1+x}\right|, \tag{6.4}$$

by (6.3). Since $d(z_1, z_2) = d(1/z_1, 1/z_2)$, the result follows for this case as well.

(iii) Again we first assume that $f \neq \infty$, so that, by (4.6),

$$\frac{d(f, S_n(w_n))}{d(f, S_n(0))} = \frac{d(f^{(n)}, w_n) \cdot d(\zeta_n, 0)}{d(f^{(n)}, 0) \cdot d(\zeta_n, w_n)}$$

$$\sim \frac{d(f^{(n)}, w_n) \cdot d(b_n(-1-x), 0)}{d(b_n x, 0) \cdot d(b_n(-1-x), w_n)} \quad \text{as } n \to \infty,$$

which proves the result for this case. If $f = \infty$, then the result follows by the argument used in the proof of part (ii). $\qquad\square$

This way of improving the rate of convergence was essentially proved by Gill [6] and Thron and Waadeland [30] for $K(a_n/1)$. Indeed, if we can approximate $f^{(n)}$ even better, then we can improve the rate of convergence even further; see [15]. If the elements of $b_0 + K(a_n/b_n)$ are functions of z such that $b_0 + K(a_n/b_n) \sim L$, then we can combine Theorem 6.1 with Theorem 5.1 to obtain:

Corollary 6.3. *Let $D \subseteq \mathbb{C}$ be a domain containing the origin, and let a_n and b_n be functions that are analytic in D such that $a_n(z) \not\equiv 0$ for all n, $b_0 + K(a_n/b_n) \sim L \in \Lambda$ and $\lim_{n \to \infty} a_n(z)/(b_{n-1}(z)b_n(z)) = q(z) \in Q$ (where Q is given by (6.2)), for all $z \in D$. Then $b_0 + K(a_n/b_n)$ converges spherically uniformly to a meromorphic function f on compact subsets of D, and $\mathcal{L}(f) = L$.*

7 Padé Approximants

So far, the best we have done (apart from Corollary 6.3) is Theorem 5.1 which essentially says that if $b_0 + K(a_n/b_n) \sim L$, and if the approximants $\{S_n(0)\}$ converge uniformly on compact subsets of a domain D containing the origin, then $S_n(0) \to f$ in D, with $\mathcal{L}(f) = L$. Moreover, we may replace $\{S_n(0)\}$ by $\{S_n(w_n)\}$ as long as $\{w_n\}$ stays asymptotically away from an exceptional sequence.

In this section we shall see that we can remove the condition $0 \in D$ if the correspondence is good enough and if $L = \mathcal{L}(f)$ for a function which is meromorphic in the whole complex plane $\mathbb{C}$. We shall also assume that the elements a_n and b_n of $b_0 + K(a_n/b_n)$ are polynomials in the complex variable z. Then its classical approximants are rational functions $f_n = S_n(0) = B_n/D_n$. We also require that $\operatorname{ord}(L) \geq 0$ and $D_n(0) = 1$ and that B_n, D_n are the canonical numerators and denominators of $b_0 + K(a_n/b_n)$. Then f_n is a *Padé approximant* to f (or L) if

$$\operatorname{ord}(LD_n - B_n) \geq M_n + N_n + 1, \tag{7.1}$$

where M_n and N_n are the degrees of B_n and D_n respectively. We write $f_n = [M_n/N_n]_f$. We also write $[M_n/N_n; w_n]_f$ for the modified Padé approximant $(B_n + B_{n-1}w_n)/(D_n + D_{n-1}w_n)$ with $[M_n/N_n; \infty]_f := B_{n-1}/D_{n-1}$. Clearly, $[M_n/N_n; w_n]_f$ depends on the *sequence* $\{[M_n/N_n]_f\}$. Examples of continued fractions which have Padé approximants are:

- regular C-fractions $b_0 + K(a_n z/1)$, where $b_0, a_n \in \mathbb{C}$ with $a_n \neq 0$ for all n;

- associated continued fractions

$$b_0 + \frac{a_1 z}{1 + d_1 z} + \frac{a_2 z^2}{1 + d_2 z} + \frac{a_3 z^2}{1 + d_3 z} + \frac{a_4 z^2}{1 + d_4 z} + \dots, \tag{7.2}$$

  where $a_n, d_n \in \mathbb{C}$ with $a_n \neq 0$ for all n, or, equivalently, J-fractions $b_0 + K(a_n/(d_n + z))$;

- P-fractions $b_0 + K(1/b_n(z))$, where each $b_n(z)$ is a polynomial in z^{-1} of degree ≥ 1.

The approximants of (5.1) are not Padé approximants, since $M_n = n$ and $N_n = n$, whereas $\operatorname{ord}(LD_n - B_n) = n + 1$. (In fact, (5.1) has 2-point Padé approximants, but that is another story.) If the approximants of $b_0 + K(a_n/b_n)$ are Padé approximants, then the correspondence is best possible. In [16] the following generalization of work by Beardon [2] and Chisholm [4] was proved:

Theorem 7.1 ([16]). *Let f be meromorphic in $\mathbb{C}$ with $f(0) \neq \infty$, let $E \subset \mathbb{C}$ be a compact set containing no poles of f, and let $[M_n/N_n]_f$ be a sequence of Padé approximants for f such that*

$$N_n \leq M_n \leq M_{n+1}, \quad N_n \to \infty, \quad \gamma_n := N_n + M_n - N_{n-1} - M_{n-1} \geq 0$$
$$(7.3)$$

for sufficiently large n. If there exists a sequence $\{\beta_n\}$ of numbers from $\widehat{\mathbb{C}}$ such that the poles of $[M_n/N_n; \beta_n z^{\gamma_n}]_f$ have no limit point in E, then $[M_n/N_n; \beta_n z^{\gamma_n}]_f$ converges uniformly to f in E.

Remarks.

(1) The condition $N_n \leq M_n$ in (7.3) can be replaced by the condition $\limsup(N_n - M_n) < \infty$. This follows since $h(z) := z^J f(z)$ has Padé approximants $[M_n + J/N_n]_h = z^J [M_n/N_n]_f$ for $J \in \mathbb{N}$. Moreover, $[M_n + J/N_n; w_n]_h = z^J [M_n/N_n; w_n]_f$.

(2) Since $g(z) := z^J / f(z)$ has Padé approximants $[N_n + J/M_n]_g = z^J / [M_n/N_n]_f$ for $J \in \mathbb{N}$ when $f(0) \neq 0$, it also follows that $[M_n/N_n; \beta_n z^{\gamma_n}]_f$ converges spherically uniformly in E to f if $f(0) \neq 0$, and that

$$M_n \to \infty, \quad \limsup(M_n - N_n) < \infty,$$
$$\gamma_n := M_n + N_n - M_{n-1} - N_{n-1} \geq 0,$$

and the zeros of $[M_n/N_n; \beta_n z^{\gamma_n}]_f$ have no limit point in E.

One way to apply Theorem 7.1 is to let E be a one-point set. If the approximants are finite and their (spherical) derivatives uniformly bounded in E (for sufficiently large n), then the convergence is clear. Value sets can also be helpful. Still, it would have been better if we did not have to worry about the location of the poles! Unfortunately, Wallin [32] has proved that there exists an *entire* function f with a normal Padé table, such that its diagonal Padé approximants $[n/n]_f$ diverge at every $z \in \mathbb{C} \setminus \{0\}$. What happens is that the poles of $[n/n]_f$ are dense in $\widehat{\mathbb{C}}$. One has the feeling that $[n/n; w_n]_f$ still converges to f under mild conditions on w_n, but that has not yet been proved.

It is also a problem that Theorem 7.1 works only for functions that are meromorphic in the *whole* complex plane. There is a similar theorem for functions meromorphic in a disk (see [16]), but the conditions are too strong to allow for continued fraction approximants of the type considered in this paper.

In [23] Nuttall proved that if f is meromorphic in $\mathbb{C}$, then $[n, n]_f$ converges to f in measure in $\mathbb{C}$ as $n \to \infty$. This was generalized by Pommerenke [26] to convergence in logarithmic capacity, and further by Lubinsky [22] to functions meromorphic in some domain D (or even more generally, to functions of the Gonçar–Walsh class). However, to keep the paper to a reasonable length, we shall not go into this here.

8 Regular C-Fractions

As an example, we apply Theorem 7.1 to regular C-fractions $b_0 + K(a_n z/1)$, where b_0 and a_n are complex numbers and all $a_n \neq 0$. Its approximants $f_n = B_n/D_n$ satisfy

$$M_{2n-1} = n, \quad M_{2n} \leq n, \quad N_{2n} = n \quad \text{and} \quad N_{2n+1} \leq n. \qquad (8.1)$$

To satisfy (7.3) we shall assume that $b_0 + K(a_n z/1)$ is *normal*; i.e., that (8.1) holds with equalities for sufficiently large n. Then $\gamma_n = 1$, and we get:

Theorem 8.1. *Let the function f be meromorphic in $\mathbb{C}$ and correspond to a normal regular C-fraction $b_0 + K(a_n z/1)$. Let further E be a compact subset of $\mathbb{C}$. Then the following hold.*

(i) *If f is analytic in E and there exists a sequence $\{\beta_n\}$ of numbers from $\widehat{\mathbb{C}}$ such that the poles of $S_n(\beta_n z)$ have no limit point in E, then $S_n(\beta_n z)$ converges uniformly to f in E.*

(ii) *If f omits zero in E and there exists a sequence $\{\beta_n\}$ of numbers from $\widehat{\mathbb{C}}$ such that the zeros of $S_n(\beta_n z)$ have no limit point in E, then $S_n(\beta_n z)$ converges spherically uniformly to f in E.*

(iii) *If $b_0 + K(a_n z/1)$ converges generally in E, and there exists a subsequence $\{S_{n_k}\}$ and a sequence $\{\beta_k\}$ of numbers from $\widehat{\mathbb{C}}$ such that the poles (or zeros) of $S_{n_k}(\beta_k z)$ have no limit point in E, then $b_0 + K(a_n z/1)$ converges generally to f in E.*

Not every meromorphic function f has a normal regular C-fraction expansion. However, Lubinsky [21] proved that for every function f that is analytic at the origin there exists a point u such that if we change the center of our expansions from $z = 0$ to the point $z = u$, then $f \sim b_0 + K(a_n(z-u)/1)$ is normal. Indeed, to every such function f, there is at most a countable number of values for u that has to be avoided.

9 Stieltjes Fractions

Stieljes fractions, or S-fractions, are regular C-fractions with $b_0 > 0$ and $a_n > 0$ for all n. This is a very nice class of continued fractions studied by

Stieltjes [28]. We say that

$$L(z) = \sum_{n=0}^{\infty} (-1)^n c_n z^n \qquad (9.1)$$

is a *Stieltjes series* if it corresponds to an S-fraction. All sorts of nice things are true in this situation. For instance, S-fractions are normal, and $L(z)$ is a Stieltjes series if and only if there exists a distribution function ψ on $[0, \infty)$ such that

$$c_n = \int_0^\infty t^n \, d\psi(t) \qquad \text{for } n = 0, 1, 2, \ldots. \qquad (9.2)$$

(ψ is a distribution function on a real interval I if ψ is a real-valued, non-decreasing function taking infinitely many values there.) Hence, correspondence to S–fractions characterizes the sequences $\{c_n\}$ for which the moment problem has a solution on $[0, \infty)$. The convergence properties of S-fractions are also extremely beautiful. With ψ given by (9.2), Stieltjes proved:

Theorem 9.1 ([28]). *Let $\{f_n\}$ be the (classical) approximants of the S-fraction $b_0 + K(a_n z/1)$, and let Ω denote the cut plane $\Omega := \{z \in \mathbb{C} : |\arg z| < \pi\}$. Then the following hold.*

(A) f_n is analytic in $\Omega \cup \{0\}$ for all n.

(B) $\{f_{2n-1}\}$ and $\{f_{2n}\}$ converge uniformly on compact subsets of Ω.

(C) $\{f_n\}$ converges in Ω if and only if $\{f_n(z_0)\}$ converges for some $z_0 \in \Omega$.

(D) $\{f_n\}$ converges in Ω if and only if $\sum b_n = \infty$, where

$$b_1 := \frac{1}{a_1}, \quad b_{2n} := \frac{a_1 a_3 \ldots a_{2n-1}}{a_2 a_4 \ldots a_{2n}}, \quad b_{2n+1} := \frac{a_2 a_4 \ldots a_{2n}}{a_1 a_3 \ldots a_{2n+1}}. \qquad (9.3)$$

(E) The moment problem for $\{c_n\}$ on $[0, \infty)$ is determined (i.e., $d\psi$ is essentially unique) if and only if $\{f_n\}$ converges.

(F) If $\{f_n\}$, and thus $b_0 + K(a_n z/1)$, converges, then the convergence is uniform on compact subsets of Ω, and its limit is

$$f(z) = \int_0^\infty \frac{d\psi(t)}{1 + zt}. \qquad (9.4)$$

Remarks.

(1) Part (A) shows that the poles of f_n are all located on the negative real axis. In fact, Stieltjes [28] proved that the denominators $\{D_n\}$ of $f_n = B_n/D_n$ form an orthogonal polynomial sequence for the distribution function ψ, and thus all the zeros of D_n are simple and interlace the zeros of D_{n-1}.

(2) The limits of $\{f_{2n-1}\}$ and $\{f_{2n}\}$ both have the forms (9.4), but for essentially different distribution functions ψ if $\{f_n\}$ diverges. If $\{f_n\}$ diverges, then the continued fraction also diverges in the general sense, so S-fractions converge generally if and only if they converge in the classical sense. (This is a consequence of the Stern-Stolz theorem; see [18, Thm. 1, p. 94].)

(3) The expression (9.4) allows us to approximate ψ by using approximants of $b_0 + K(a_n z/1)$; see [25, p. 188].

(4) If we expand $(1 + zt)^{-1}$ into a geometric series and formally interchange the summation and integration in (9.4) (without justifying the validity of these operations), we obtain

$$\int_0^\infty \frac{d\psi(t)}{1+zt} \sim \int_0^\infty \sum_{n=0}^\infty (-zt)^n \, d\psi(t) \sim \sum_{n=0}^\infty (-1)^n \int_0^\infty t^n \, d\psi(t) = L(z)$$

which suggests that $\mathcal{L}(f) = L$. Moreover, since the denominator D_n of the rational function $f_n = B_n/D_n$ only has real and simple zeros $\zeta_{n,k} < 0$, f_n has a partial fraction decomposition which can be written as $\int_0^\infty d\psi_n(t)/(1+zt)$ with an appropriate step function $\psi_n(t)$ having steps at the points $-1/\zeta_{n,k}$. This suggests that $\psi_n \to \psi$ weakly, and thus that $f_n \to f$. Theorem 9.1 shows that these conclusions, derived here informally, are indeed correct.

A particularly useful result is due to Carleman:

Theorem 9.2 ([3]). *Let (9.1) be a Stieltjes series. If $\sum c_n^{-1/2n} = \infty$, then the corresponding S-fraction converges in Ω.*

To have conditions that are stated in terms of the numbers c_n, instead of the S-fraction itself, can be very useful indeed. Moreover, the condition on $\{c_n\}$ is a rather mild one: for instance, it is satisfied whenever the series (9.1) has a positive radius of convergence.

Theorem 9.3. *Let f be a function that is analytic in $|z| < R$ for some $0 < R \le \infty$. If f has an S-fraction expansion $c_0 + K(a_n z/1)$, then the following hold.*

(A) f has an analytic extension

$$f^*(z) := \int_0^{1/R} \frac{d\psi(t)}{1 + zt} \quad \text{for } z \in \Omega_R := \Omega \cup \{z \in \mathbb{C} : |z| < R\}.$$

(B) $c_0 + K(a_n z/1)$ converges uniformly on compact subsets of Ω_R to f^*.

(C) f is an entire function if and only if $a_n \to 0$.

If the Stieltjes series (9.1) converges only at $z = 0$, it is still possible that its corresponding S-fraction converges in Ω. For instance, this is the case if Carleman's criterion holds; in this case L is an asymptotic expansion of the limit function.

Results similar to Theorems 9.1–9.3 also hold for the associated continued fractions (7.2) with $a_n < 0$ and $d_n \in \mathbb{R}$ or the equivalent real J-fraction $b_0 + K(a_n/(d_n + z))$ with $a_n < 0$ and $d_n \in \mathbb{R}$.

10   More on Regular C-fractions

Inspired by Carleman's theorem for S-fractions, it is tempting to try to find convergence criteria for regular C-fractions based on the coefficients of their corresponding series. This is no easy task, but Lubinsky has come up with some very nice results in this direction. The first result even guarantees that the given entire function f has a normal regular C-fraction expansion. Indeed, it guarantees the stronger result that the Padé table of f is normal, but we restrict the result to C-fractions.

Theorem 10.1 ([19]). *Let $L(z) = \sum_{n=0}^{\infty} c_n z^n$ satisfy $c_n \neq 0$ for all n and $|c_{n-1} c_{n+1}/c_n^2| \leq \rho_0^2$ for all $n \in \mathbb{N}$, where $\rho_0 = 0.4559\ldots$ is the positive root of the equation $2 \sum_{n=1}^{\infty} \rho^{n^2} = 1$. Then $L \sim c_0 + K(a_n z/1)$, where $a_n \to 0$, and $c_0 + K(a_n z/1)$ is normal and converges uniformly on compact subsets of $\mathbb{C}$ to an entire function f where $\mathcal{L}(f) = L$.*

Remarks.

(1) The condition $|c_{n-1} c_{n+1}/c_n^2| \leq \rho_0^2$ shows that $|c_{n+1}/c_n| = \mathcal{O}(\rho_0^{2n})$; i.e., it is a very strong condition which not only implies that f is entire, but also that f has order zero.

(2) Lubinsky also proved that $c_0 + K(a_n z/1)$ converges separately in Theorem 10.1. That is, $D_n(z) \to 1$ and $B_n(z) \to f(z)$ as $n \to \infty$, where $f_n = B_n/D_n$.

Theorem 10.2 ([20]). *Let $L(z) = \sum_{n=0}^{\infty} c_n z^n$, with $c_n \neq 0$ for all sufficiently large n, be such that the limit $q := \lim_{n \to \infty} c_{j-1} c_{j+1}/c_j^2$ exists with*

$|q| < 1$. *If* $L \sim c_0 + K(a_n z/1)$, *then* $a_n \to 0$ *and the regular C-fraction converges uniformly on compact subsets of* $\mathbb{C}$ *to an entire function* f *where* $\mathcal{L}(f) = L$.

Again the condition implies that $c_{n+1}/c_n \to 0$ at a geometric rate, but this time we have $|c_{n+1}/c_n| = o(\rho^n)$ for every $\rho > |q|$.

References

[1]   G.A. Baker, Jr. and P.R. Graves-Morris, *The convergence of Padé approximants*, J. Math. Anal. Appl. **87** (1982), 382 – 394.

[2]   A.F. Beardon, *On the convergence of Padé approximants*, J. Math. Anal. Appl. **21** (1968), 344 – 346.

[3]   T. Carleman, *Les fonctions quasi analytiques*, Gauthier – Villars, 1926.

[4]   J.S.R. Chisholm, *Approximation by sequences of Padé approximants in regions of meromorphy*, J. Math. Phys. **7** (1966), 39 – 44.

[5]   J. Gill, *Infinite compositions of Möbius transformations*, Trans. Amer. Math. Soc. **176** (1973), 479 – 487.

[6]   ———, *The use of attractive fixed points in accelerating the convergence of limit-periodic continued fractions*, Proc. Amer. Math. Soc. **47** (1975), 119 – 126.

[7]   L. Jacobsen, *Convergence of limit k–periodic continued fractions $k(a_n/b_n)$, and of subsequences of their tails*, Proc. London Math. Soc. (3) **51** (1985), 563 – 576.

[8]   ———, *General convergence for continued fractions*, Trans. Amer. Math. Soc. **281** (1986), 129 – 146.

[9]   ———, *General correspondence for continued fractions*, J. Comp. Appl. Math. **19** (1987), 171 – 177.

[10] L. Jacobsen and D.R. Masson, *On the convergence of limit periodic continued fractions $k(a_n/1)$, where $a_n \to -\frac{1}{4}$. Part III*, Constr. Approx. **6** (1990), 363 – 374.

[11] L. Jacobsen and W.J. Thron, *Limiting structures for sequences of linear fractional transformations*, Proc. Amer. Math. Soc. **99** (1987), 141 – 146.

[12] W.B. Jones and W.J. Thron, *On the convergence of Padé approximants*, SIAM J. Math. Anal. **6** (1975), 9 – 16.

[13] _____ , *Sequences of meromorphic functions corresponding to formal laurent series*, SIAM J. Math. Anal. **10** (1979), 1 – 17.

[14] _____ , *Continued fractions. Analytic theory and applications. Encyclopedia of mathematics and its application, Vol. 11*, Addison-Wesley, Reading, Mass., 1980, now distributed by Cambridge University Press, New York.

[15] L. Lorentzen, *Computation of limit periodic continued fractions. A survey*, Numer. Algorithms **10** (1995), 69 – 110.

[16] _____ , *Ideas from continued fraction theory extended to Padé approximation and generalized iteration*, Proc. International Conference on Rational Approximation, vol. 61, 2000, pp. 185–206.

[17] _____ , *General convergence in quasi-normal families*, Proc. Edinburgh Math. Soc., to appear.

[18] L. Lorentzen and H. Waadeland, *Continued fractions with applications. Studies in computational mathematics, Vol. 3*, North - Holland, 1992.

[19] D.S. Lubinsky, *Padé tables of entire functions of very slow and smooth growth*, Constr. Approx. **1** (1985), 349 – 358.

[20] _____ , *Padé tables of entire functions of very slow and smooth growth, II*, Constr. Approx. **4** (1988), 321 – 339.

[21] _____ , *Power series equivalent to rational functions: A shifting - origin Kronecker type theorem, and normality of Padé tables*, Numer. Math. **54** (1988), 33 – 39.

[22] _____ , *Spurious poles in diagonal rational approximation*, Progress in Approximation Theory, Springer-Verlag, 1992, pp. 191 – 213.

[23] J. Nuttall, *The convergence of Padé approximants of meromorphic functions*, J. Math. Anal. Appl. **31** (1970), 147 – 153.

[24] J.F. Paydon and H.S. Wall, *The continued fraction as a sequence of linear transformations*, Duke Math. J. **9** (1942), 360 – 372.

[25] O. Perron, *Die Lehre von den Kettenbrüchen, Band II*, Teubner, Stuttgart, 1957.

[26] C. Pommerenke, *Padé approximants and convergence in capacity*, J. Math. Anal. Appl. **41** (1973), 775 – 780.

[27] J.L. Schiff, *Normal families*, Springer-Verlag, 1993.

[28] T.J. Stieltjes, *Reserches sur les fractions continues*, Ann. Fac. Sci. Toulouse **8J** (1894), 1 – 47, also in: Oeuvres, Vol. 2, pp. 402-566.

[29] W.J. Thron, *On parabolic convergence regions for continued fractions*, Math. Zeitschr. **69** (1958), 173 – 182.

[30] W.J. Thron and H. Waadeland, *Accelerating convergence of limit periodic continued fractions* $k(a_n/1)$, Numer. Math. **34** (1980), 72 – 90.

[31] ———, *Truncation error bounds for limit periodic continued fractions*, Math. Comp. **40** (1983), 589 – 597.

[32] H. Wallin, *The convergence of Padé approximants and the size of power series coefficients*, Appl. Anal. **4** (1974), 235 – 251.

Reducible Arithmetic Functions, Asymptotic Mean Behavior, and Polylogarithms

Lutz G. Lucht

1 Reducible Arithmetic Functions

Many results on arithmetic functions deal with the asymptotic behavior of their summatory functions under suitable conditions. Most of these results require that the functions be sufficiently nice multiplicative or additive functions with exponential weights. We present here a unified treatment which contains the additive and multiplicative functions as special cases.

Let $\mathcal{F} = \{f : \mathbb{N} \to \mathbb{C}\}$ denote the class of arithmetic functions; $\mathcal{M}$ the class of multiplicative functions, i.e., functions $f \in \mathcal{F}$ satisfying $f(1) = 1$ and $f(qn) = f(q)f(n)$ for all coprime $q, n \in \mathbb{N}$; $\mathcal{A}$ the class of additive functions, i.e., functions $f \in \mathcal{F}$ satisfying $f(qn) = f(q) + f(n)$ for all coprime $q, n \in \mathbb{N}$; and $\mathcal{R}$ the class of reducible functions in the sense of the following definition.

Definition 1. We call $f \in \mathcal{F}$ reducible if there exist $\alpha \in \mathcal{M}$ and $\beta \in \mathcal{F}$ such that

$$f(qn) = \alpha(q)f(n) + \beta(q) \quad \text{for all coprime } q, n \in \mathbb{N}. \tag{1.1}$$

We call $f \in \mathcal{F}$ almost constant if there exists some $q \in \mathbb{N}$ such that f is constant on the set $\mathbb{N}_q := \{n \in \mathbb{N} : (n, q) = 1\}$.

We see that $\mathcal{M} \subset \mathcal{R}$, by taking $\alpha = f$ and $\beta = 0$, and $\mathcal{A} \subset \mathcal{R}$, by taking $\alpha = 1$ and $\beta = f$.

Notice that formula (1.1) must hold for any decomposition of a number m as qn with $\gcd(q, n) = 1$. Now assume that f is not almost constant. Then f and 1 are linearly independent functions on $\mathbb{N}_q$ for all q. It follows that α and β are uniquely determined. By applying (1.1) to $q = q_1 q_2$ with coprime q_1, q_2 and $n \in \mathbb{N}_q$ we obtain $f(q_1 q_2 n) = \alpha(q_1 q_2)f(n) + \beta(q_1 q_2)$ and $f(q_1 q_2 n) = \alpha(q_1)\alpha(q_2)f(n) + \alpha(q_1)\beta(q_2) + \beta(q_1)$. This gives

$$\big(\alpha(q_1 q_2) - \alpha(q_1)\alpha(q_2)\big) f + \big(\beta(q_1 q_2) - \alpha(q_1)\beta(q_2) - \beta(q_1)\big) 1 = 0 \tag{1.2}$$

on $\mathbb{N}_q$, which implies that the coefficients vanish. In particular, we have $\alpha \in \mathcal{M}$. Therefore the condition $\alpha \in \mathcal{M}$ is not a severe restriction; it merely guarantees that even for almost constant reducible functions f the functions α and β are unique.

Let $\mathbb{P}$ denote the set of primes and let $\mathbb{P}^*$ be the set of prime powers p^k with $p \in \mathbb{P}$ and $k \in \mathbb{N}$. From equation (1.2) the following theorem is easily derived.

Theorem 1. *Every function $f \in \mathcal{R}$ is associated with a pair of uniquely determined functions $\alpha \in \mathcal{M}$, $\beta \in \mathcal{F}$ satisfying*

$$\alpha(qq') = \alpha(q)\,\alpha(q')\,, \quad \beta(qq') = \alpha(q)\,\beta(q') + \beta(q) \tag{1.3}$$

for all coprime $q, q' \in \mathbb{N}$. In particular, $\beta \in \mathcal{R}$. Conversely, every pair of functions α, $\beta : \mathbb{P}^* \to \mathbb{C}$ satisfying

$$\alpha(1) = 1\,, \quad \beta(1) = 0\,, \quad (1 - \alpha(q))\,\beta(q') = (1 - \alpha(q'))\,\beta(q)$$

for all coprime $q, q' \in \mathbb{P}^*$ determines unique extensions $\alpha \in \mathcal{M}$, $\beta \in \mathcal{R}$ with (1.3), such that $f = f(1)\,\alpha + \beta \in \mathcal{R}$.

Under suitable conditions on the distribution of values of $f \in \mathcal{R}$, we determine the asymptotic behavior of the summatory function

$$M(g, x) = \sum_{n \le x} g(n) \qquad (x \to \infty)$$

with $g = f e_{a/q}$ for coprime $a, q \in \mathbb{N}$, where $e_y(n) = e^{2\pi i y n}$.

For $q \in \mathbb{N}$, let $\langle q \rangle$ denote the multiplicative semigroup generated by the prime divisors of q, $G(q)$ the multiplicative group mod q, and $\widehat{G}(q)$ the Dirichlet character group mod q. For $f \in \mathcal{F}$, $p \in \mathbb{P}$, $k \in \mathbb{N}_0 := \mathbb{N} \cup \{0\}$ and $s \in \mathbb{C}$ we write

$$\widetilde{f}_{p,k}(s) = \sum_{\kappa=k}^{\infty} f(p^\kappa)\,p^{-\kappa s}\,,$$

if this Dirichlet series converges absolutely.

Definition 2. The class $\mathcal{K}$ consists of all functions $f \in \mathcal{R}$ having the following properties:

(A) There exists a constant $s \in \mathbb{C}$ with $\sigma := \operatorname{Re} s \ge 0$ and a slowly oscillating function $\ell : \mathbb{R}_+ \to \mathbb{C}$ such that for every $q \in \mathbb{N}$ and every $\chi \in \widehat{G}(q)$ the asymptotic estimate

$$M(f\chi, x) = \begin{cases} c_q\,x^s\,\ell(x) + o\!\left(x^\sigma\,|\ell(x)|\right) & \text{if } \chi = \chi_0 \\ o\!\left(x^\sigma\,|\ell(x)|\right) & \text{if } \chi \ne \chi_0 \end{cases} \qquad (x \to \infty)$$

holds.

(B) $c_q \neq 0$ for some $q \in \mathbb{N}$.

(C) The limit $\lim_{x \to \infty} x^s \ell(x)$ does not exist.

(D) For every $p \in \mathbb{P}$ there exists $\varepsilon > 0$ such that the series $\tilde{\alpha}_{p,0}(\sigma - \varepsilon)$ and $\widetilde{\beta}_{p,0}(\sigma - \varepsilon)$ converge absolutely, where $\alpha, \beta \in \mathcal{F}$ are the functions which give rise to f.

From the general theory of slowly oscillating functions (see, for instance, Seneta [5]), it follows that the number $s = s_f$ is uniquely determined by $f \in \mathcal{K}$. The class $\mathcal{K}$ is very large. For example, it contains all multiplicative functions f of Wirsing or Halász or Elliott type that have a non-vanishing main term in the asymptotic expansion of their summatory functions (cf. [3]).

2 Asymptotic Formulae

In the sequel we determine the asymptotic behavior of $M(fe_{a/q}, x)$ for $f \in \mathcal{K}$ and coprime $a, q \in \mathbb{N}$. For $s \in \mathbb{C}$ we set

$$\widehat{f}_q(s) = \sum_{t|q} \frac{\mu(q/t)}{\varphi(q/t)} \sum_{\substack{m \in \langle t \rangle \\ (m, q/t) = 1}} \frac{f(tm)}{(tm)^s}, \qquad (2.1)$$

where μ is Möbius' function and φ is Euler's function.

Theorem 2. *Let $f \in \mathcal{K}$ be generated by $\alpha \in \mathcal{M}$, $\beta \in \mathcal{R}$ as in Theorem 1, and let $s = s_f$, $\sigma = \operatorname{Re} s$. Then for coprime $a, q \in \mathbb{N}$,*

$$M(fe_{a/q}, x) = \begin{cases} \left(c_q \, \widehat{\alpha}_q(s) + o(1)\right) x^s \, \ell(x) & \text{if } \sigma > 1, \\[2mm] \left(c_q \, \widehat{\alpha}_q(s) + o(1)\right) x^s \, \ell(x) + \frac{\varphi(q)}{q} \, \widehat{\beta}_q(1) \, x & \text{if } \sigma \leq 1. \end{cases}$$

Proof. We decompose

$$M(fe_{a/q}, x) = \sum_{1 \leq b \leq q} e_{b/q}(1) \sum_{\substack{n \leq x \\ an \equiv b \bmod q}} f(n)$$

$$= \sum_{t|q} \sum_{\substack{1 \leq b \leq q \\ (b,q)=t}} e_{b/q}(1) \frac{1}{\varphi(q/t)} \sum_{\chi \in \widehat{G}(q/t)} \overline{\chi}\left(\frac{ab}{t}\right) \sum_{tn \leq x} f(tn) \chi(n)$$

and evaluate the innermost sum. With the principal character $\chi_0' \in \widehat{G}(t)$, it follows that

$$\sum_{n \leq y} f(tn)\,\chi(n) = \sum_{\substack{m \leq y \\ m \in \langle t \rangle}} \alpha(tm)\,\chi(m) \sum_{n \leq y/m} f(n)\,\chi_0'(n)\,\chi(n)$$

$$+ \sum_{\substack{m \leq y \\ m \in \langle t \rangle}} \beta(tm)\,\chi(m) \sum_{n \leq y/m} \chi_0'(n)\,\chi(n).$$

For $\chi \neq \chi_0 \in \widehat{G}(q/t)$, properties (A) and (C) lead to

$$\sum_{n \leq y} f(tn)\chi(n) = \sum_{\substack{m \leq y \\ m \in \langle t \rangle}} |\alpha(tm)|\, o\big((y/m)^\sigma |\ell(y/m)|\big) + \sum_{\substack{m \leq y \\ m \in \langle t \rangle}} |\beta(tm)|\, \mathcal{O}(1)$$

$$= o\big(y^\sigma |\ell(y)|\big),$$

whereas for $\chi = \chi_0 \in \widehat{G}(q/t)$

$$\sum_{n \leq y} f(tn)\chi_0(n) = c_q\, y^s \ell(y) \sum_{\substack{m \leq y \\ m \in \langle t \rangle}} \frac{\alpha(tm)}{m^s}\, \chi_0(m)$$

$$+ \frac{\varphi(q)}{q}\, y \sum_{\substack{m \leq y \\ m \in \langle t \rangle}} \frac{\beta(tm)}{m}\, \chi_0(m) + o\big(y^\sigma |\ell(y)|\big).$$

If $\sigma \leq 1$ then the series

$$\mathrm{A}_t(s) = \sum_{\substack{m \in \langle t \rangle \\ (m,q/t)=1}} \frac{\alpha(tm)}{(tm)^s}, \quad \mathrm{B}_t(1) = \sum_{\substack{m \in \langle t \rangle \\ (m,q/t)=1}} \frac{\beta(tm)}{tm}$$

are absolutely convergent, and we obtain

$$\sum_{n \leq y} f(tn)\chi_0(n) = c_q\, \mathrm{A}_t(s)\, t^s\, y^s \ell(y) + \frac{\varphi(q)}{q}\, \mathrm{B}_t(1)\, t\, y + o\big(y^\sigma |\ell(y)|\big). \quad (2.2)$$

If $\sigma > 1$ then at least $\mathrm{A}_t(s)$ converges absolutely, and the middle term on the right-hand side of equation (2.2) does not occur since the contribution of β is absorbed by the remainder term. In either case, by inserting this into the above representation of $M(fe_{a/q}, x)$ and observing that

$$\sum_{\substack{1 \leq b \leq q \\ (b,q)=t}} e_{b/q}(1) = \mu\Big(\frac{q}{t}\Big),$$

we obtain for $\sigma \leq 1$

$$M(fe_{a/q}, x) = c_q\, x^s \ell(x) \sum_{t|q} \frac{\mu(q/t)}{\varphi(q/t)}\, A_t(s) + \frac{\varphi(q)}{q}\, x \sum_{t|q} \frac{\mu(q/t)}{\varphi(q/t)}\, B_t(1)$$
$$+ o\big(x^\sigma |\ell(x)|\big)$$
$$= c_q\, \widehat{\alpha}_q(s)\, x^s \ell(x) + \frac{\varphi(q)}{q}\, \widehat{\beta}_q(1)\, x + o\big(x^\sigma |\ell(x)|\big).$$

For $\sigma > 1$ the middle term on the right side has to be omitted. $\qquad\square$

In particular, for multiplicative and additive functions Theorem 2 takes the following form.

Corollary 1. *Let $f \in \mathcal{K} \cap \mathcal{M}$, $s = s_f$, and let $a, q \in \mathbb{N}$ be coprime. Then $\alpha = f$, $\beta = 0$, and*

$$M(fe_{a/q}, x) = \big(c_q\, \widehat{f}_q(s) + o(1)\big)\, x^s \ell(x)\,.$$

Corollary 2. *Let $f \in \mathcal{K} \cap \mathcal{A}$, $s = s_f$, and let $a, q \in \mathbb{N}$ be coprime. Then $\alpha = 1$, $\beta = f$, and*

$$M(fe_{a/q}, x) = \begin{cases} \big(c_q\, \widehat{1}_q(s) + o(1)\big)\, x^s \ell(x) & \text{if } \sigma > 1, \\[2mm] \big(c_q\, \widehat{1}_q(s) + o(1)\big)\, x^s \ell(x) + \frac{\varphi(q)}{q}\, \widehat{f}_q(1)\, x & \text{if } \sigma \leq 1. \end{cases}$$

In particular, for $s = 1$ and $q > 1$

$$M(fe_{a/q}, x) = \frac{\varphi(q)}{q}\, \widehat{f}_q(1)\, x + o\big(x\,|\ell(x)|\big)\,.$$

3 Analysis of the Main Term

Observe that the main terms in the general asymptotic evaluation of the summatory function $M(fe_{a/q}, x)$ in Theorem 2 are independent of $a \in G(q)$. We investigate the case when the main terms vanish for almost all $q \in \mathbb{N}$. Let $\kappa(q)$ denote the square-free kernel of $q \in \mathbb{N}$, i.e., the product of the prime divisors of q. For $f \in \mathcal{K}$ we write, using the notation of Definition 2,

$$\Lambda_f = \lim_{x \to \infty} x^{s-1}\, \ell(x),$$

regardless of whether the limit exists. If $\Lambda_f \neq 0$ exists, then $s = 1$, and we may assume that $\ell(x) = 1$ when substituting c_q for $c_q\, \Lambda_f$. The following lemma relates the coefficients c_q and c_{pq}.

Lemma 1. *Let $f \in \mathcal{K}$, $s = s_f$ and $q \in \mathbb{N}$. Then $c_q = c_{\kappa(q)}$, and one of the following alternatives holds:*

(a) *If the limit Λ_f does not exist or vanishes, or if $\beta = 0$, then*

$$c_q = c_{pq}\,\tilde{\alpha}_{p,0}(s) \qquad (p \nmid q).$$

In this case, the set $T_\alpha := \{p \in \mathbb{P} : \tilde{\alpha}_{p,0}(s) = 0\}$ is finite, and $c_q \neq 0$ if and only if $q^* \mid q$, where $q^* = \prod_{p \in T_\alpha} p$.

(b) *If $\Lambda_f \neq 0$ exists then (with $s = 1$ and $\ell(x) = 1$)*

$$c_q = c_{pq}\,\tilde{\alpha}_{p,0}(1) + \frac{\varphi(pq)}{pq}\,\tilde{\beta}_{p,0}(1) \qquad (p \nmid q).$$

In this case, there are infinitely many square-free numbers q such that $c_q \neq 0$.

Proof. It is obvious that $c_q = c_{\kappa(q)}$ for all $q \in \mathbb{N}$. For primes $p \nmid q$ we conclude from (A) and (D) that

$$c_q\, x^s\, \ell(x) + o\big(x^\sigma\, |\ell(x)|\big) = \sum_{\substack{n \leq x \\ (q,n)=1}} f(n) = \sum_{k \geq 0} \sum_{\substack{m \leq xp^{-k} \\ (pq,m)=1}} f(p^k m)$$

$$= \sum_{k \geq 0} \alpha(p^k) \sum_{\substack{m \leq xp^{-k} \\ (pq,m)=1}} f(m) + \sum_{k \geq 0} \beta(p^k) \sum_{\substack{m \leq xp^{-k} \\ (pq,m)=1}} 1$$

$$= c_{pq}\,\tilde{\alpha}_{p,0}(s)\, x^s\, \ell(x) + \frac{\varphi(pq)}{pq}\, x \sum_{\substack{k \geq 0 \\ p^k \leq x}} \beta(p^k)\, p^{-k}$$

$$+ o\big(x^\sigma\, |\ell(x)|\big).$$

This gives

$$\big(c_q - c_{pq}\,\tilde{\alpha}_{p,0}(s) + o(1)\big)\, x^{s-1}\, \ell(x) = \begin{cases} 0 & \text{for } \sigma > 1 \\ \frac{\varphi(pq)}{pq}\,\tilde{\beta}_{p,0}(1) & \text{for } \sigma \leq 1, \end{cases}$$

from which the recursion formulae in (a) and (b) follow.

Next, in case (a), if $c_q \neq 0$ for some q, then $0 \neq c_q = c_{qp}\,\tilde{\alpha}_{p,0}(s)$ for all $p \nmid q$. Hence $\tilde{\alpha}_{p,0}(s) = 0$ at most for primes $p \mid q$, so that T_α is finite. Moreover, if $c_q \neq 0$ for some q then $p \mid q$ for all $p \in T_\alpha$, which shows that $q^* \mid q$ and

$$c_{q^*} = c_q \prod_{\substack{p \mid q \\ p \notin T_\alpha}} \tilde{\alpha}_{p,0}(s) \neq 0.$$

Conversely, it follows from $q^* \mid q$ that $c_q \neq 0$.

Finally, in case (b), assume that there are at most finitely many non-zero coefficients c_q, and let $q^* \in \mathbb{N}$ be the largest square-free number such that $c_{q^*} \neq 0$. Then, with distinct elements $p, p' \in \mathbb{P}$ that are relatively prime to q^*, the coefficients c_{pq^*}, $c_{p'q^*}$ and $c_{pp'q^*}$ vanish, and we have

$$c_{p'q^*} = c_{pp'q^*}\, \widetilde{\alpha}_{p,0}(1) + \frac{\varphi(pp'q^*)}{pp'q^*}\, \widetilde{\beta}_{p,0}(1) = \frac{\varphi(pp'q^*)}{pp'q^*}\, \widetilde{\beta}_{p,0}(1) = 0,$$

$$c_{q^*} = c_{pq^*}\, \widetilde{\alpha}_{p,0}(1) + \frac{\varphi(pq^*)}{pq^*}\, \widetilde{\beta}_{p,0}(1) = \frac{\varphi(pq^*)}{pq^*}\, \widetilde{\beta}_{p,0}(1).$$

Hence $\widetilde{\beta}_{p,0}(1) = 0$ and $c_{q^*} = 0$, which is a contradiction. $\qquad\square$

Let $f \in \mathcal{F}$ and $q \in \mathbb{N}$. The definition of the transform $f \mapsto \widehat{f}_q(s)$ in (2.1) shows, in particular, that

$$\widehat{f}_1(s) = f(1)\,, \quad \widehat{f}_{p^k}(s) = \widetilde{f}_{p,k}(s) - \frac{1}{\varphi(p)}\frac{f(p^{k-1})}{p^{(k-1)s}} \qquad (p^k \in \mathbb{P}^*)$$

if the series $\widetilde{f}_{p,0}(s)$ converges absolutely, say. The following lemma gives multiplicative properties of $\widehat{f}_q(s)$ for reducible functions f. The easy proof is based on (1.3).

Lemma 2. *Let $\alpha \in \mathcal{M}$ and $\beta \in \mathcal{F}$ be the generating functions of $f \in \mathcal{R}$ as given in Theorem 1. If the series $\widetilde{\alpha}_{p,0}(s)$ converges absolutely then $\widehat{\alpha}_q(s)$ is a multiplicative function of q. If both series $\widetilde{\alpha}_{p,0}(1)$ and $\widetilde{\beta}_{p,0}(1)$ converge absolutely then $\widehat{\beta}_q(1)$ satisfies*

$$\widehat{\beta}_1(1) = 0\,, \quad \widehat{\beta}_{qp^k}(1) = \widehat{\alpha}_q(1)\,\widehat{\beta}_{p^k}(1) \qquad (p^k \in \mathbb{P}^*).$$

It is convenient to introduce the class $\mathcal{H}$ of all arithmetic functions $h \in \mathcal{F}$ having finite support $\operatorname{supp} h = \{n \in \mathbb{N} : h(n) \neq 0\}$. Under the assumptions of Lemma 1 with $\sigma > 1$ the asymptotic main term in Theorem 2 vanishes if and only if $c_q\,\widehat{\alpha}_q(s) = 0$ for almost all q, which by Lemma 1 and Lemma 2 reduces to $\widehat{\alpha}_{p^k}(s) = 0$ for almost all $p^k \in \mathbb{P}^*$. For any fixed prime p, the corresponding system of these equations with p^k exceeding some constant can be recursively solved for the values of $\alpha(p^k)$. This gives

$$\alpha = g * I^{s-1}, \tag{3.1}$$

where $*$ denotes the Dirichlet convolution, with $I^{s-1} \in \mathcal{M}$ defined by $I^{s-1}(n) = n^{s-1}$ and some function $g \in \mathcal{H}$. In particular, $\alpha(p^k) \neq 1$ for some $p^k \in \mathbb{P}^*$. We infer from Theorem 1 that

$$\beta = c\,(1 - \alpha) \tag{3.2}$$

with some constant $c \in \mathbb{C}$, so that $f = c + h * I^{s-1}$ with some function $h \in \mathcal{H}$.

If $\sigma \leq 1$ and $\Lambda_f = 0$ then, under the assumptions of Lemma 1, the asymptotic main term in Theorem 2 vanishes if and only if $c_q \, \widehat{\alpha}_q(s) = 0$ and $\widehat{\beta}_q(1) = 0$ for almost all q. As before we obtain (3.1), and $\widehat{\beta}_{p^k}(1) = 0$ for almost all $p^k \in \mathbb{P}^*$ implies that $\beta(p^k) = 0$ for almost all $p^k \in \mathbb{P}^*$. If $\alpha \neq 1$ then (3.2) holds again and comparison shows that $\beta = 0$. If $\alpha = 1$ then $\beta \in \mathcal{A}$ and it follows that $\mu * \beta \in \mathcal{H}$. In any case we obtain $f = f(1)\,\alpha + \beta = h * I^{s-1}$ with some function $h \in \mathcal{H}$.

Finally, if $\beta = 0$ then $f = f(1)\,\alpha$ with $\alpha \in \mathcal{K} \cap \mathcal{M}$, and $c_q \, \widehat{\alpha}_q(s) = 0$ for almost all q gives $\alpha = g * I^{s-1}$ with some function $g \in \mathcal{H}$, so that $f = h * I^{s-1}$ with $h = f(1)\,g$. Combining the preceding expressions, we obtain the sufficiency part of the following result.

Theorem 3. *Let $f \in \mathcal{K}$ satisfy $s = s_f \neq 1$ or $\Lambda_f = 0$ or $\beta = 0$. Then*

$$M(fe_{a/q}, x) = o\big(x^\sigma \, |\ell(x)|\big) \qquad (x \to \infty) \qquad (3.3)$$

for almost all $q \in \mathbb{N}$ if and only if

$$f = c + h * I^{s-1} \qquad (3.4)$$

with some constant $c \in \mathbb{C}$ and some appropriate function $h \in \mathcal{H}$.

Proof. It remains to show that (3.4) implies (3.3). For x large enough, this follows from

$$M(fe_{a/q}, x) = c\, M(e_{a/q}, x) + \sum_{d=1}^{\infty} h(d)\, M\left(I^{s-1} e_{ad/q}, \frac{x}{d}\right)$$

combined with Corollary 1. Namely, we have $\ell(x) = 1$, and for $q > 1$ we find $\widehat{1}_q(1) = 0$ and, with $\alpha = I^{s-1}$, $\widehat{\alpha}_q(s) = 0$ as special cases of Lemma 2. Therefore $M(fe_{a/q}, x) = o(x^\sigma)$ for almost all $q \in \mathbb{N}$. $\qquad\Box$

The assumptions of Theorem 3 exclude $s_f = 1$ and either (i) $\Lambda_f = 1$, i.e., $\ell(x) = 1$, or (ii) Λ_f does not exist, i.e., $\ell(x)$ diverges as $x \to \infty$. In both cases it is impossible to determine the form of f precisely. This is seen by taking $f = c + \eta$ with a constant $c \neq 0$ and any function $\eta \in \mathcal{M}$ ($\alpha = \eta$, $\beta = c - \eta \neq 0$) having a smaller growth rate (e.g., $M(|\eta|, x) = o(x)$) in case (i), and $\eta \in \mathcal{A} \cap \mathcal{K}$ ($\alpha = 1$, $\beta = c + \eta \neq 0$) with $s_\eta = 1$ and non-existing Λ_η (e.g., $\eta = \omega$, the number of prime divisors function, or $\eta = \Omega$, the number of prime factors function) in case (ii). The latter case can be covered by refining condition (A) in the definition of the class $\mathcal{K}$ (see Lucht and Schmalmack [4]).

4   Polylogarithms

The polylogarithm function $L_s(z)$ is defined by

$$L_s(z) = \sum_{n=1}^{\infty} n^{-s} z^n \qquad (s \in \mathbb{C},\ |z| < 1)$$

as a holomorphic function of two complex variables. It is an entire function of s and analytically extends onto the cut plane $\mathbb{C} \setminus [1, \infty)$ with respect to z (cf. Jonquière [1]). This property carries over to the s-derivatives

$$L_{s,k}(z) := (-1)^k \frac{\partial^k}{\partial s^k} L_s(z) = \sum_{n=1}^{\infty} n^{-s} \log^k n\, z^n.$$

Consider the set

$$\mathsf{B} = \{z \mapsto L_{s,k}(z^d) : s \in \mathbb{C},\ k \in \mathbb{N}_0,\ d \in \mathbb{N}\}$$

and the complex vector space V generated by B.

Theorem 4. B *is a basis of* V.

Proof. See Lucht and Schmalmack [4], Theorem 1 .   □

Theorem 4   says that the polylogarithm functions $z \mapsto L_{s,k}(z^d)$ are linearly independent over $\mathbb{C}$. This is an interesting negative result in view of current attempts to explain existing linear relations between the values of polylogarithm functions at powers of certain algebraic numbers (compare Lewin [2]). In order to obtain an equivalent arithmetic reformulation of Theorem 4   we set

$$P(f,z) = \sum_{n=1}^{\infty} f(n)\, z^n.$$

Theorem 5. *The space* $\mathcal{G} = \{f \in \mathcal{F} : P(f,z) \in \mathsf{V}\}$ *is a unitary module over the algebra* $\mathcal{H} = \{h \in \mathcal{F} : |\mathrm{supp}\, h| < \infty\}$ *with the basis* $\mathcal{B} = \{I^s \log^k : s \in \mathbb{C},\ k \in \mathbb{N}_0\}$ *and the Dirichlet convolution as exterior multiplication.*

Proof. By definition, $P(f,z) \in \mathsf{V}$ if and only if

$$P(f,z) = \sum_{s \in S} \sum_{k \in K} \sum_{d=1}^{\infty} h_{s,k}(d)\, L_{-s,k}(z^d)$$

with finite sets $S \subset \mathbb{C}$, $K \subset \mathbb{N}_0$ and coefficient functions $h_{s,k} \in \mathcal{H}$ for all $s \in S$, $k \in K$. For $|z| < 1$ this means

$$P(f,z) = \sum_{s \in S} \sum_{k \in K} \sum_{d=1}^{\infty} \sum_{m=1}^{\infty} h_{s,k}(d)\, m^s \log^k m\, z^{dm}$$

$$= \sum_{n=1}^{\infty} \left(\sum_{s \in S} \sum_{k \in K} \left(h_{s,k} * I^s \log^k \right)(n) \right) z^n ,$$

from which the assertion follows by comparison of coefficients.   □

The following theorem determines all reducible functions $f \in \mathcal{G}$.

Theorem 6. *A function $f \in \mathcal{R}$ belongs to $\mathcal{G}$ if and only if there exist constants $s \in \mathbb{C}^{\times}$, $c \in \mathbb{C}$ and a function $h \in \mathcal{H}$ such that either $f = h * I^s + c$ or $f = h * 1 + c \log$.*

Proof. See Lucht and Schmalmack [4], Theorem 3.   □

Theorems 3 and 6 show that the polylogarithm functions are closely related to the functions $f \in \mathcal{K}$. They possess generating power series $P(f,z)$ having an analytic continuation beyond their circle of convergence.

Corollary 3. *Under the assumptions of Theorem 3, for $f \in \mathcal{K}$ the following assertions are equivalent:*

(a) $P(f,z)$ has a holomorphic continuation beyond $|z| = 1$,

(b) $f \in \mathcal{G}$,

(c) $f = c + h * I^{s-1}$ with some constant $s \in \mathbb{C}$ and some function $h \in \mathcal{H}$.

Proof. It follows from (a) that $P(f e_{a/q}, z)$ has a limit as $z \to 1-$, $z \in \mathbb{R}$, for a set of rational numbers $\frac{a}{q}$ that is dense in $[-\delta, \delta]$ for some $\delta > 0$. Since Abel summability implies Cesàro summability, we see that $M(f e_{a/q}, x)$ has a limit as $x \to \infty$ for the same set of rational numbers $\frac{a}{q}$. By Theorem 2, equation (3.1) holds for almost all $q \in \mathbb{N}$ and Theorem 3 yields (c). Obviously (c) implies (b), and (b) implies (a).   □

In particular, Corollary 3 covers all functions $f \in \mathcal{K} \cap \mathcal{M}$. Observe that the functions of the form $f = h * 1 + c \log$ with $c \neq 0$ and $h \in \mathcal{H}$ occurring in Theorem 6 belong to the subclass $\mathcal{K}'$ of exceptional functions $f \in \mathcal{K}$ with $\alpha = 1$, $s = s_f = 1$ and non-existing Λ_f, characterized in (ii). A closer investigation of $\mathcal{K}'$ leads, by refining condition (A), to a characterization similar to Corollary 3. For a detailed analysis, one must define the derivative of a slowly oscillating function near infinity; see Lucht and Schmalmack [4], Section 6.

References

[1] A. Jonquière, *Note sur le série* $\sum x^n/n^s$, Bull. Soc. Math. France **17** (1889), 142–152.

[2] L. Lewin, *Structural properties of polylogarithms*, Surveys and Monographs, vol. 37, American Math. Soc., Providence, 1991.

[3] G. Lucht, L. *Power series with multiplicative coefficients*, Math. Z. **177** (1981), 359–374.

[4] G. Lucht, L. and A. Schmalmack, *Polylogarithms and arithmetic function spaces*, Acta Arith. **95** (2001), 361–382.

[5] E. Seneta, *Regularly varying functions*, Lecture Notes in Math., vol. 508, Springer Verlag, New York, 1976.

On the Probability of Combinatorial Structures Without Some Components

E. Manstavičius[1]

1 Introduction and Statement of Results

We consider labeled combinatorial structures, so-called *assemblies*, which are defined as follows (see, for example, [1], [2]):

Let $\sigma = \{a_1, \ldots, a_N\}$ be an arbitrary set of N elements. Suppose σ is partitioned into subsets, and let $k_j = k_j(\sigma)$ denote the number of subsets of size j, $1 \leq j \leq N$, so that $1k_1 + \cdots + Nk_N = N$. Assume further that an additional structure is defined that to a subset of size j associates $m_j \geq 1$ different copies of the subset. We will call a subset having such a structure a *component* of σ. We assume that the number m_j depends only on the size j of the subset. A set σ with such a component structure is called an assembly. We define $\mathcal{S}_N$ as the set of assemblies of size N arising in the various decompositions of the set σ into subsets and using the same rule to define their structure, and we set $S_0 = \emptyset$.

The union of $\mathcal{S}_N$ over $N \geq 0$ is the class of assemblies, which (see [7]) is also called an *abelian partitional complex*. To determine $|\mathcal{S}_N| =: p(N)$, we observe that, for any fixed vector $\bar{k} = (k_1, \ldots, k_N)$ with nonnegative integer coordinates and satisfying $L(\bar{k}) := 1k_1 + \cdots + Nk_N = N$, there are

$$N! \prod_{j=1}^{N} \left(\frac{1}{j!}\right)^{k_j} \frac{1}{k_j!}$$

ways to partition a set of N elements into subsets so that, for each $j \leq N$, there are k_j subsets of size j. Thus, for each such $\bar{k}$ there are

$$Q_N(\bar{k}) := N! \prod_{j=1}^{N} \left(\frac{m_j}{j!}\right)^{k_j} \frac{1}{k_j!}$$

different assemblies. Hence the total number of assemblies of size N equals

$$p(N) = \sum_{L(\bar{k})=N} Q_N(\bar{k}).$$

[1]Supported by a Lithuanian State Stipend.

As can be verified directly, the exponential generating function of the sequence $p(N)$, $N \geq 0$, satisfies the formal relation

$$Z(z) := 1 + \sum_{N \geq 1} \frac{p(N)}{N!} z^N = \exp\left\{\sum_{j \geq 1} \frac{m_j}{j!} z^j\right\}.$$

Examples of classes of assemblies include the set of permutations, in which the components are cycles. In this case, we have $p(N) = N!$, $m_j = (j-1)!$, and $Z(z) = (1-z)^{-1}$.

Another interesting example is the class consisting of all mappings of a finite set into itself, with the components defined as the connected components of the corresponding functional digraphs. In this case, $p(N) = N^N$,

$$m_j = (j-1)! \sum_{s=0}^{j-1} \frac{j^s}{s!},$$

and $Z(z) = (1 - t(z))^{-1}$, where $t(z)$ is defined by the relation $t(z) = ze^{t(z)}$. For these and more sophisticated examples see [1] and [2].

Many assemblies satisfy the following condition:

Condition M. *There exist fixed parameters $q > 0$, $\theta > 0$ such that $\lambda_j := m_j q^j / j! \geq c/j$ with $c > 0$ and*

$$|r_j| := \left|\lambda_j - \frac{\theta}{j}\right| \leq \rho(j)$$

for $j \geq 1$. Here $\rho : \mathbf{R}^+ \to \mathbf{R}^+$ is a monotonically decreasing function satisfying

$$\rho(u) \leq Cu^{-1}, \quad \rho(u/2) \leq C_1\rho(u), \quad \sum_{j \geq 1} \rho(j) \leq C_2,$$

where C, C_1, and C_2 are positive constants.

The assemblies of permutations and of mappings of a finite set into itself satisfy this condition. In particular, in the second case we can take $q = e^{-1}$, $\theta = 1/2$, and $\rho(u) = 8u^{-3/2}$ (see [8]).

If Condition M is satisfied, the function

$$H(z) := \exp\left\{\sum_{j=1}^{\infty} r_j z^j\right\} = \sum_{n=0}^{\infty} \left(\sum_{L(\bar{k})=n} \prod_{j=1}^{n} \frac{r_j^{k_j}}{k_j!}\right) z^n =: \sum_{n=0}^{\infty} h_n z^n \qquad (1.1)$$

is analytic in $|z| < 1$ and continuous on $|z| = 1$. The last assertion follows from the inequalities

$$\sum_{n=0}^{\infty} |h_n| \leq \exp\left\{\sum_{j=1}^{\infty} |r_j|\right\} \ll 1. \tag{1.2}$$

Here, and in the sequel, all estimates and any constants c_i, C_k, $i \geq 1$, $k \geq 3$, depend at most on the parameters q and θ, and the constants c, C, C_1, and C_2 of Condition M. With the above notation we have

$$Z(qz) = \frac{H(z)}{(1-z)^\theta}, \quad z \in \{z \in \mathbf{C} : |z| \leq 1\} \setminus \{1\}.$$

Let ν_N denote the uniform probability measure on $\mathcal{S}_N$. The additive analogue of the main problem of classical probabilistic number theory (see [4]) is to find conditions under which the sequence of distribution functions

$$\nu_N \left(\sum_{j=1}^{N} h_{Nj}(k_j(\sigma)) - \alpha(N) < x\right)$$

converges weakly to some limit distribution, as $N \to \infty$. Here $h_{Nj}(k)$, $1 \leq j \leq N$, $N \in \mathbf{N}, k \in \mathbf{Z}^+$ is a three dimensional array of real numbers with $h_{Nj}(0) \equiv 0$, and $\alpha(N)$ is a centralizing sequence for this array. One might expect that the methods that have proved successful in the classical case will also lead to some progress in solving this problem. Thus, small sieve estimates as in [12] and [13], and in some other recent papers dealing with number-theoretical functions, should be useful here. In particular, in order to obtain necessary conditions, we can apply concentration function estimates to show that the sum of reciprocals of the sizes of components with large values $h_{Nj}(k_j(\sigma))$ is bounded, and we can eliminate these components by sieving techniques. Theorems 1 and 2 below show that the frequency of the remaining assemblies is large enough.

Papers applying sieve methods to combinatorial structures have appeared only recently. The sieve techniques used in these papers are analogs of the so-called fundamental lemma or Selberg's upper bound method (see [1], [2], [3], [14]).

Here we require estimates of a different kind. We consider the frequencies

$$\nu_N(J) := \nu_N \left(k_j(\sigma) = 0 \text{ for all } j \in J\right),$$

where J is an arbitrary subset of $\{1, \ldots, N\}$. The main difference to the sieve problems considered by previous authors lies in the fact that we do not restrict the position of the sieving subset J in advance. Moreover, our method is based on a probabilistic argument. Set

$$\mu_N(K) = \min_J \nu_N(J),$$

where minimum is taken over all subsets J satisfying the condition

$$\sum_{j \in J} \frac{1}{j} \le K, \quad K \ge 0. \tag{1.3}$$

One of the problems is to obtain a lower bound for

$$\liminf_{N \to \infty} \mu_N(K) =: \mu(K).$$

In [10] we proved:

Theorem A. *If S_N is the symmetric group, then*

$$\mu(K) \ge \exp\{-e^{7K}\}$$

for all $K \ge 0$.

We also showed that this lower bound cannot be improved beyond

$$\exp\{-(1 + o(1))Ke^K\}, \quad K \to \infty.$$

Here we will prove the following three theorems.

Theorem 1. *Assume that condition M holds. Let $0 < \delta < 1$ be given, and let $\mu_N(K; \delta)$ and $\mu(K; \delta)$ be defined in the same way as $\mu_N(K)$ and $\mu(K)$, but with the set J restricted by $J \subset [1, (1 - \delta)N)$. Then*

$$\mu(K; \delta) \ge c_1 \delta^\theta e^{-\theta K}$$

for all $K \ge 0$.

Set for brevity $\theta_1 = \min\{1, \theta\}$.

Theorem 2. *Assume that condition M holds. Let $\gamma_N(K)$ and $\gamma(K)$ be defined in the same way as $\mu_N(K)$ and $\mu(K)$ but with (1.3) replaced by*

$$\sum_{j \in J} \frac{1}{j^{\theta_1}} \le K, \quad K \ge 0. \tag{1.4}$$

Then

$$\gamma(K) \ge c_2 \exp\{-e^{C_3 K}\}$$

for all $K \ge 0$.

We expect that, as in the case of the symmetric group, an investigation of frequencies of assemblies without large components will give some information about the quality of the bounds for $\nu_N(J)$ given in Theorem 2. In the case when $|J|$ is small enough and Condition M holds, the analytic results obtained in the author's paper [11] yield asymptotic formulas for $\nu_N(J)$. We conjecture that the assertion of Theorem 2 does not hold with $\mu(K)$ in place of $\gamma(K)$, but the results of [11] do not give any information in this regard. To obtain an upper bound for $\nu_N(J)$ without any condition on the position of J, one can use moment estimates for additive functions defined on assemblies. This idea goes back to Erdős and Turán [6], who for the case of the symmetric group have exploited the second moment. Using, in addition, the author's estimates for higher moments, would yield the bound

$$\nu_N(J) \ll_s K_1^{-s}$$

for any $s \geq 1$, where K_1 is any number satisfying

$$\sum_{j \in J} \frac{1}{j} \geq K_1. \tag{1.5}$$

However, there is an alternative approach that yields the following better estimate.

Theorem 3. *Assume that Condition M and* (1.5) *hold. Then*

$$\nu_N(J) \ll e^{-\theta K_1}.$$

This shows that the bound of Theorem 1 is of the right order of magnitude with respect to K, provided that the constant δ is fixed.

The proof of Theorems 1 and 2 uses the "small sieve" of Erdős and Ruzsa [5].

2 Lemmas

Let ξ_j, $j \leq N$, be independent Poisson random variables with parameters λ_j, and let $X = L(\bar{\xi})$, where $\bar{\xi} = (\xi_1, \ldots, \xi_N)$. For a subset J of indices $j \leq N$, we set $\bar{J} = \{1, \ldots, N\} \setminus J$,

$$X_J = \sum_{j \in J} j\xi_j, \quad \bar{X}_J = X - X_J.$$

Lemma 1. *We have*

$$\nu_N(J) = \frac{P(\bar{X}_J = N)}{P(X = N)} \exp\left\{ -\sum_{j \in J} \lambda_j \right\}.$$

Proof. Direct calculation yields

$$\nu_N(J) = \frac{N!}{p(N)} \sum_{\substack{L(\bar{k})=N \\ k_j=0 \text{ for all } j \in J}} \prod_{j=1}^{N} \left(\frac{m_j}{j!}\right)^{k_j} \frac{1}{k_j!} \tag{2.1}$$

$$= P(X_J = 0 \,|\, X = N) = \frac{P(X_J = 0)P(\bar{X}_J = N)}{P(X = N)}.$$

Since

$$P(X_J = 0) = P(\xi_J = 0 \text{ for all } j \in J) = \exp\left\{-\sum_{j \in J} \lambda_j\right\},$$

the desired relation follows. $\qquad\square$

Lemma 2. *Let h_n be defined by* (1.1), *and suppose Condition M holds. Then*

$$h_n \ll n^{-1}, \quad n \geq 1.$$

Proof. We adapt the argument given in the proof of Lemma 4 of [9]. Consider the n-th Taylor coefficient of the function

$$W^l(z) := \left(\sum_{j=1}^{\infty} r_j z^j\right)^l =: \sum_{n=1}^{\infty} w_{ln} z^n.$$

Using the hypotheses of the lemma, we obtain

$$|w_{2n}| = \left|\sum_{1 \leq k < n} r_k r_{n-k}\right| \leq \sum_{1 \leq k < n} \rho(k)\rho(n - k)$$

$$\leq 2\rho(n/2) \sum_{1 \leq k \leq n/2} \rho(k) \leq 2C_1 C_2 \rho(n)$$

$$=: C_4 \rho(n).$$

By induction it follows that

$$|w_{nl}| \leq C_4^{l-1} \rho(n) \tag{2.2}$$

for each $l \geq 1$. Since h_n is the n-th Taylor coefficient of the function

$$\sum_{l=1}^{n} \frac{1}{l!} W^l(z),$$

(2.2) gives the bound

$$|h_n| \leq \sum_{l=1}^{n} \frac{|w_{ln}|}{l!} \leq \rho(n) \sum_{l=1}^{n} \frac{C_4^{l-1}}{l!} \ll \frac{1}{n},$$

as desired. □

Lemma 3. *Let $p(n)$, h_k, and q be defined as above, and suppose Condition M holds. Then*

$$\frac{p(n)q^n}{n!} = \sum_{k=0}^{n} \binom{\theta + k - 1}{k} h_{n-k} = \frac{n^{\theta-1}}{\Gamma(\theta)}(H(1) + o(1)),$$

as $n \to \infty$, where Γ denotes the Euler function.

Proof. We focus on the more difficult case $\theta < 1$. By Stirling's formula we have

$$\binom{\theta + k - 1}{k} = \frac{\Gamma(\theta + k)}{\Gamma(\theta)\Gamma(k + 1)} = \frac{k^{\theta-1}}{\Gamma(\theta)}(1 + O(k^{-1})), \quad k \geq 1, \qquad (2.3)$$

Let $M \geq 1$ and $0 < \varepsilon < 1$ be arbitrary fixed constants with $M \leq \varepsilon n$. Using the hypotheses of the lemma, Lemma 2, (1.2), and (2.3) we obtain

$$\frac{p(n)q^n}{n!} = \left(\sum_{0 \leq k \leq M} + \sum_{M < k \leq \varepsilon n} + \sum_{\varepsilon n < k \leq n-M} + \sum_{n-M < k \leq n} \right) \binom{\theta + k - 1}{k} h_{n-k}$$

$$= O \left(\frac{M}{n} + \varepsilon^\theta n^{\theta-1} + \varepsilon^{\theta-1} n^{\theta-1} \sum_{k \geq M} |h_k| \right)$$

$$+ \frac{n^{\theta-1}}{\Gamma(\theta)} \left(1 + O \left(\frac{M}{n} \right) \right) \left(H(1) + O \left(\sum_{k \geq M} |h_k| \right) \right).$$

Letting, in turn, $n \to \infty$, $M \to \infty$, and $\varepsilon \to 0$, we obtain the asserted estimate. □

3   Proof of Theorem 1

We need to estimate the probabilities appearing in Lemma 1. Let $0 \leq n \leq N$ and $0 < r < 1$ be given. Then

$$
P(X = n) = \exp\left\{ -\sum_{j=1}^{N} \lambda_j \right\} \frac{1}{2\pi i} \int_{|z|=r} \exp\left\{ \sum_{j=1}^{n} \lambda_j z^j \right\} \frac{dz}{z^{n+1}}
$$

$$
= \exp\left\{ -\sum_{j=1}^{N} \lambda_j \right\} \frac{1}{2\pi i} \int_{|z|=r} \frac{H(z)\,dz}{z^{n+1}(1-z)^\theta}
$$

$$
= \exp\left\{ -\sum_{j=1}^{N} \lambda_j \right\} \frac{p(n)q^n}{n!}
$$

$$
= H(1)^{-1} e^{-\theta\gamma} N^{-\theta} (1 + o(1)) \frac{p(n)q^n}{n!}, \tag{3.1}
$$

where γ denotes the Euler constant. Applying Lemma 3, we obtain

$$
P(X = n) \asymp n^{\theta-1} N^{-\theta} \tag{3.2}
$$

for $1 \leq n \leq N$. Summing these probabilities over $n \in [1, \delta N]$, we get

$$
P(X \leq \delta N) \gg \delta^\theta. \tag{3.3}
$$

Without loss of generality, we may assume that $\delta < 1/2$. For $(1-\delta)N \leq j \leq N$, we set $J_j := J \cup \{j\}$ and consider the events

$$
\{\omega : \xi_j = 1, \bar{X}_J = N\}.
$$

These events are pairwise disjoint (some may be empty). Since

$$
\{\omega : \bar{X}_J = N - j\} = \{\omega : \bar{X}_{J_j} = N - j, \ \xi_j = 0\},
$$

we have

$$
P(\bar{X}_J = N - j) = P(\bar{X}_{J_j} = N - j)e^{-\lambda_j}.
$$

Using the independence of summands, the conditions of the theorem, and (3.3) we obtain

$$
P(\bar{X}_J = N) \geq \sum_{(1-\delta)N \leq j \leq N} P(\xi_j = 1, \ \bar{X}_J = N)
$$

$$
= \sum_{(1-\delta)N \leq j \leq N} P(\xi_j = 1) P(\bar{X}_{J_j} = N - j)
$$

$$
= \sum_{(1-\delta)N \leq j \leq N} \lambda_j P(\bar{X}_J = N - j) \geq cN^{-1} P(\bar{X}_J \leq \delta N)
$$

$$
\geq cN^{-1} P(X \leq \delta N) \gg \delta^\theta N^{-1}. \tag{3.4}
$$

Since, by (3.2), $P(X = N) \asymp N^{-1}$, the desired result follows from Lemma 1 and (3.4).

This completes the proof of Theorem 1. $\qquad\qquad\qquad\qquad\qquad\qquad$ □

4   Proof of Theorem 2

For a proof of Theorem 2, we need the following averaging inequality.

Lemma 4. *Let* $I := \bar{J} \cap (N/k, N/2]$, $k \geq 2$, *and set*

$$S_i := (J \cap [1, N - i]) \cup \{i\}.$$

Then

$$\nu_n(J) \geq \frac{1 + o(1)}{k} \sum_{i \in I} \lambda_i \left(1 - \frac{i}{N}\right)^{\theta - 1} \nu_{N-i}(S_i)$$

as $N \to \infty$.

Proof. We start with the inequality

$$\nu_N(J) \geq \nu_N \left(k_j(\sigma) = 0 \; \forall j \in J \; \& \; \exists i \in I : \; k_i(\sigma) = 1\right). \qquad (4.1)$$

The indicator of the event in parentheses is the following function on $\mathcal{S}_N$:

$$\prod_{j \in J} \mathbf{1}(k_j(\sigma) = 0) \max_{i \in I} \mathbf{1}(k_i(\sigma) = 1) =: \max_{i \in I} \chi(i, \sigma).$$

Since $1k_1(\sigma) + \cdots + i k_i(\sigma) + \cdots + N k_N(\sigma) = N$ and $i \geq N/k$, for any given σ the sequence $\chi(i, \sigma)$, $i \in I$, contains at most k nonzero numbers. Hence

$$\max_{i \in I} \chi(i, \sigma) \geq \frac{1}{k} \sum_{i \in I} \chi(i, \sigma).$$

This inequality, (4.1), and (2.2) imply

$$\nu_N(J) \geq \frac{1}{k} \sum_{i \in I} \nu_N \left(k_j(\sigma) = 0 \;\text{ for all } j \in J, \; k_i(\sigma) = 1\right)$$

$$= \frac{1}{k} \sum_{i \in I} \frac{P(\xi_i = 1) P(X_J = 0) P\left(X_{\bar{J}_i} = N - i\right)}{P(X = N)}. \qquad (4.2)$$

We will evaluate the probabilities in the sum by applying (2.1) to the frequency $\nu_{N-i}(S_i)$.

Set

$$X = \left(\sum_{j \leq N-i} + \sum_{N-i<j\leq N} \right) j\xi_j =: Y + Z.$$

As in the proof of Theorem 1, we obtain, using (3.1) and Lemma 3,

$$P(X = N) = P(X = N - i)\frac{p(N)q^N}{N!}\frac{(N-i)!}{p(N-i)q^{N-i}}$$

$$= P(Y = N - i)P(Z = 0)\left(1 - \frac{i}{N}\right)^{\theta-1}(1 + o(1)), \quad (4.3)$$

uniformly in $i \in I$ as $N \to \infty$. In what follows we let Y_S (resp. Z_T) denote the sum of the random variables $j\xi_j$ over $j \in S \subset \{1,\ldots,N-i\}$ (resp. $j \in T \subset \{N-i+1,\ldots,N\}$), and set $\bar{Y}_S = Y - Y_S$ and $\bar{Z}_T = Z - Z_T$. Taking $S = J \cap [1, N-i]$ and $T = J \cap (N-i, N]$, we have

$$P(X_J = 0) = P(Y_S = 0)P(Z_T = 0)$$

$$= P(Y_{S_i} = 0)P(\xi_i = 0)^{-1}P(Z_T = 0). \quad (4.4)$$

If

$$\bar{S}_i := \{1,\ldots,N-i\} \setminus S_i = \bar{J}_i \cap [1, N-i]$$

and

$$\bar{T} = \{N-i+1,\ldots,N\} \setminus T = \bar{J} \cap (N-i+1, N],$$

then $\bar{X}_{J_i} = \bar{Y}_{S_i} + \bar{Z}_T$ and

$$P\left(\bar{X}_{J_i} = N - i\right) \geq P\left(\bar{Y}_{S_i} = N - i\right) P\left(\bar{Z}_T = 0\right). \quad (4.5)$$

Inserting (4.3), (4.4), and (4.5) into (4.2) we obtain

$$\nu_N(J) \geq \frac{1 + o(1)}{k} \sum_{i \in I} \lambda_i \frac{P(Y_{S_i} = 0)\, P\left(Y_{\bar{S}_i} = N - i\right)}{P(Y = N - i)}.$$

By (2.1) this is the desired estimate.   □

Lemma 5. *For each* $1 \leq j \leq N$*, we have*

$$\nu_N(k_j(\sigma) = 0) = \frac{N!}{p(N)q^N} \sum_{0 \leq k \leq N/j} \frac{(-\lambda_j)^k}{k!}\frac{p(N-kj)q^{N-kj}}{(N-kj)!}. \quad (4.6)$$

Proof. The frequency

$$\nu_N(k_j(\sigma) = 0) = \frac{N!}{p(N)} \sum_{\substack{L(\bar{k})=N \\ k_j=0}} \prod_{i=1}^{N} \left(\frac{m_i}{i!}\right)^{k_i} \frac{1}{k_i!}$$

is just the N-th Taylor coefficient of the function

$$\frac{N!}{p(N)} \exp\left\{ \sum_{i \geq 1, i \neq j} \frac{m_i z^i}{i!} \right\} = \frac{N!}{p(N)} \exp\left\{ -m_j z^j/j! \right\} Z(z).$$

Expanding the right-hand side and extracting the N-th coefficient gives the desired formula. □

It is interesting to note that, if Condition M is satisfied, then by Lemma 3 we have $Np(N-1)/(qp(N)) = 1 + o(1)$. Thus by Lemma 1 of [1] the finite dimensional distributions of the process $(k_1(\sigma), \ldots, k_N(\sigma), 0, \ldots)$ converge to those of the process $(\xi_1, \ldots, \xi_N, \xi_{N+1}, \ldots)$.

Proof of Theorem 2. If the set $T^* := J \cap ((1 - 1/k)N, N]$, $k > 2$, is sparse in the sense that the quantity

$$A := \sum_{j \in T^*} \frac{1}{j^{\theta_1}}$$

is small, we can apply Theorem 1. Setting $J^* = J \cap [1, (1 - 1/k)N]$, we have

$$\nu_N(J) \geq \nu_N(J^*) - \sum_{j \in T^*} (1 - \nu_N(k_j(\sigma) = 0)).$$

The frequencies $\nu_N(k_j(\sigma) = 0)$ can be evaluated by Lemma 5. Since $(1 - 1/k)N \leq j \leq N$, there are only two terms in (4.6), and thus

$$\nu_N(k_j(\sigma) = 0) = 1 - \lambda_j \frac{p(N-j)q^{N-j}}{(N-j)!} \frac{N!}{p(N)q^N}.$$

Applying Lemma 3 for $n = N - j \geq B$ with some sufficiently large B, depending on the parameters and constants in Condition M only, we obtain

$$\sum_{j \in T^*} (1 - \nu_N(k_j(\sigma) = 0)) \ll \sum_{\substack{j \in T^* \\ j \leq N-B}} \frac{1}{j} \left(1 - \frac{j}{N}\right)^{\theta-1}$$

$$+ N^{1-\theta} \sum_{\substack{j \in T^* \\ N-B < j \leq N}} \frac{1}{j} \leq C_5 A.$$

Applying Theorem 1 we obtain

$$\nu_N(J) \geq 2^{-1} c_1 k^{-\theta} e^{-\theta K} - C_5 A$$

for sufficiently large N. Setting $k := e^{K+\alpha}$ with some constant $\alpha > 0$ (to be specified later) and

$$A \leq \frac{c_1}{4C_5} k^{-\theta} e^{-\theta K} =: C_6 e^{-2\theta K - \theta \alpha}, \tag{4.7}$$

we obtain

$$\nu_N(J) \geq c(\alpha) e^{-2\theta K} \geq c(\alpha) \exp\left\{ -e^{2\theta K} \right\}$$

for some constant $c(\alpha) > 0$. This implies the assertion of Theorem 2 with any $c_2 \leq c(\alpha)$ and $C_3 \geq 2\theta$.

It remains to consider the case when (4.7) is not satisfied. In this case, we use a "real induction" type argument with respect to the parameter $K \geq 0$. We will recursively define a function $N_0 : \mathbf{R}^+ \to \mathbf{N}$ such that

$$\inf_{N \geq N_0(K)} \gamma_N(K) \geq c_2 \exp\left\{ -e^{C_3 K} \right\} \tag{4.8}$$

for all $K \geq 0$.

Let $K_0 > 0$ be defined by $K_0 = C_6 \exp\{-2\theta K_0 - \theta \alpha\}$. If $K \leq K_0$, then we are in the case considered above.

Suppose now that $K \geq K_0$ and that we have proved (4.8) with L in the place of K, for $K_0 \leq L \leq K - \Delta$, with some positive numbers Δ and $N_0(L)$. Our goal is to prove (4.8) for $L = K$ and to extend the function N_0 to the interval $(K - \Delta, K]$.

We begin by applying Lemma 4 with $k = e^{K+\alpha}$, where α is as above. If J satisfies condition (1.4) and $i \in I$, then $N/k < i \leq N/2$ and

$$K_i := \sum_{j \in S_i} \frac{1}{j^{\theta_1}} \leq \left(\frac{k}{N}\right)^{\theta_1} + K - \sum_{j \in J, j > N-i} \frac{1}{j^{\theta_1}} \leq \left(\frac{k}{N}\right)^{\theta_1} + K - A$$
$$\leq K - C_6 2^{-1} e^{-2\theta K - \theta \alpha} =: K - \Delta,$$

provided that

$$N \geq (2C_6^{-1} e^{2\theta K + \theta \alpha})^{1/\theta_1} e^{K+\alpha} =: N'(K).$$

Applying the induction hypothesis with $L = K - \Delta$, we obtain

$$\nu_{N-i}(S_i) \geq \gamma_{N-i}(K - \Delta) \geq c_2 \exp\left\{ -e^{C_3(K-\Delta)} \right\} \tag{4.9}$$

provided that

$$N \geq \max\{N'(K), 2N_0(K - \Delta)\} =: N''(K).$$

Combinatorial Structures Without Some Components

Now, for such N,

$$\sum_{i \in I} \lambda_i \left(1 - \frac{i}{N}\right)^{\theta - 1} \geq c \min\{1, 2^{1-\theta}\} \sum_{i \in I} \frac{1}{i}$$

$$\geq c \min\{1, 2^{1-\theta}\} \left(\log \frac{k}{2} - K - \frac{k}{N}\right)$$

$$\geq c \min\{1, 2^{1-\theta}\} \left(\alpha - \log 2 - (C_6 2^{-1} e^{-2\theta K - \theta \alpha})^{1/\theta_1}\right).$$

Choosing α independent of K and large enough, the last expression becomes ≥ 2. Inserting the resulting bound and (4.9) into the inequality of Lemma 4, we obtain

$$\nu_N(J) \geq 2c_2(1 + o(1)) \exp\left\{-e^{C_3(K-\Delta)}\right\} e^{-K - \alpha} \geq c_2 \exp\left\{-e^{C_3 K}\right\}.$$

provided that $N \geq N'''(K)$ and C_3 is large enough. For the last estimate, it suffices to ensure that

$$(K + \alpha) e^{-xK} + e^{-x\Delta} \leq 1 \tag{4.10}$$

if $K \geq K_0$ and x is large enough. If $x\Delta \leq 1$, inequality (4.10) follows from

$$(K + \alpha) e^{-xK} \leq \frac{x\Delta}{2} = \frac{C_6 x}{4} e^{-2\theta K - \theta \alpha}.$$

Thus the choice $x = 3\theta$ ensures (4.10) for $K \geq C_7$. For all $K_0 \leq K \leq C_7$, inequality (4.10) holds for some sufficiently large $x = x_0$. So the lower bound for $\nu_N(J)$ holds with $C_3 = \max\{3\theta, x_0\}$. Setting $N_0(L) = \max\{N''(K), N'''(K)\}$ for $K - \Delta < L \leq K$, we obtain (4.8).

This completes the proof of Theorem 2. $\qquad\square$

5 Proof of Theorem 3

The indicator function of the event in question is

$$\chi_1(J, \sigma) := \prod_{j=1}^{N} d_j^{k_j(\sigma)},$$

where $d_j = 0$ if $j \in J$ and $d_j = 1$ if $j \notin J$. Thus

$$\nu_N(J) = \frac{1}{p(N)} \sum_{\sigma \in \mathcal{S}_N} \chi_1(J, \sigma) = \frac{1}{p(N)} \sum_{L(\bar{k}) = N} \prod_{j=1}^{N} \left(\frac{d_j m_j}{j!}\right)^{k_j} \frac{1}{k_j!}.$$

Let $M_N := q^N p(N) \nu_N(J)/N!$ for $N \geq 1$ and set $M_0 = 1$. Then

$$\sum_{n=0}^{\infty} M_n z^n = \exp\left\{\sum_{j=1}^{\infty} d_j \lambda_j z^j\right\}. \tag{5.1}$$

Hence by Condition M,

$$\sum_{n=0}^{N-1} M_n \leq \exp\left\{\sum_{j=1}^{N} d_j \lambda_j\right\}$$

$$= \exp\left\{\sum_{j \leq N, j \notin J} \left(\frac{\theta}{j} + r_j\right)\right\} \ll e^{-\theta K_1} N^\theta. \tag{5.2}$$

On the other hand, differentiating both sides of (5.1) and comparing the coefficients in the resulting equality yields

$$M_N = \frac{1}{N} \sum_{n=0}^{N-1} d_{N-n} \lambda_{N-n} (N-n) M_n.$$

By Condition M and (5.2), this implies

$$M_N \ll e^{-\theta K_1} N^{\theta-1}.$$

In view of Lemma 3, this gives the desired estimate.

This completes the proof of Theorem 3.   □

Acknowledgments. The author thanks the organizers of the Millennial Conference on Number Theory for enabling him to attend this conference and for their hospitality during the conference. Also, the author is deeply indebted to Professor A. Hildebrand for improving the exposition of this paper.

References

[1] R. Arratia, D. Stark, and S. Tavaré, *Total variation asymptotics for Poisson process approximations of logarithmic combinatorial assemblies*, Ann. Probab. **23** (1995), 1347–1388.

[2] R. Arratia and S. Tavaré, *Independent process approximations for random combinatorial structures*, Adv. Math. **104** (1994), 90–154.

[3] T. Y. Chow, *The combinatorics behind number-theoretic sieves*, Adv. Math. **138** (1998), 293–305.

[4] P. D. T. A. Elliott, *Probabilistic number theory. I, II*, Springer-Verlag, Berlin, 1979/1980.

[5] P. Erdős and I. Z. Ruzsa, *On the small sieve. I. Sifting by primes*, J. Number Theory **12** (1980), 385–394.

[6] P. Erdős and P. Turán, *On some problems of a statistical group-theory. II*, Acta Math. Acad. Sci. Hungar. **18** (1967), 151–163.

[7] D. Foata, *La série génératrice exponentielle dans les problèmes d'énumération*, Les Presses de l'Université de Montréal, Montreal, Que., 1974, Avec un chapitre sur les identités probabilistes dérivées de la formule exponentielle, par B. Kittel, Séminaire de Mathématiques Supérieures, No. 54 (Été, 1971).

[8] J. C. Hansen, *A functional central limit theorem for random mappings*, Ann. Probab. **17** (1989), 317–332.

[9] H.-K. Hwang, *Asymptotics of Poisson approximation to random discrete distributions: an analytic approach*, Adv. in Appl. Probab. **31** (1999), 448–491.

[10] E. Manstavičius, *On random permutations without cycles of some lengths*, Period. Math. Hungar. **42** (2001), 37–44.

[11] ———, *Mappings on decomposable combinatorial structures: Analytic approach*, Combinatorics, Probability, and Computing **11** (2002), 61–78.

[12] I. Z. Ruzsa, *The law of large numbers for additive functions*, Studia Sci. Math. Hungar. **14** (1979), 247–253 (1982).

[13] Ĭ. Shyaulis, *Compactness of distributions of a sequence of additive functions*, Lith. Math. J. **27** (1987), 168–178.

[14] W.-B. Zhang, *Probabilistic number theory in additive arithmetic semigroups. I*, Analytic number theory, Vol. 2 (Allerton Park, IL, 1995), Birkhäuser Boston, Boston, MA, 1996, pp. 839–885.

Asymmetries in the Shanks–Rényi Prime Number Race

Greg Martin

1 Background

Let $\pi(x; q, a)$ denote the number of primes not exceeding x that are congruent to a modulo q. We know from the prime number theorem in arithmetic progressions that the two counting functions $\pi(x; q, a)$ and $\pi(x; q, b)$ are asymptotically equal as x tends to infinity (as long as a and b are both coprime to q). However, more complicated behavior emerges when we compare these counting functions for finite values of x. Imagine $\pi(x; q, a)$ and $\pi(x; q, b)$ as representing the two contestants in a race; as the primes are listed in order, the contestant $\pi(x; q, a)$ takes a step each time a prime is congruent to a mod q, and similarly for $\pi(x; q, b)$. How often is the first contestant ahead of the second? This "race game" is easily extended to include several contestants.

As a prime example (!), consider the two contestants $\pi(x; 4, 1)$ and $\pi(x; 4, 3)$. (We won't pay any attention to the other contestant $\pi(x; 4, 2)$, who, while quick out of the starting blocks, is rather lacking in endurance.) Chebyshev was the first to note that there are "many more" primes congruent to 3 (mod 4) than to 1 (mod 4). Indeed, the first value of x for which $\pi(x; 4, 1) > \pi(x; 4, 3)$ is $x = 26{,}861$ (Leech [7]). Even this victory for $\pi(x; 4, 1)$ is short-lived, as 26,861 is the first of a pair of twin primes, and so $\pi(x; 4, 3)$ catches right back up and does not relinquish the lead again until $x = 616{,}481$.

Similar biases are observed in race games to other moduli, especially to small moduli. For example, $\pi(x; 3, 1)$ does not exceed $\pi(x; 3, 2)$ for the first time until $x = 608{,}981{,}813{,}029$ (Bays and Hudson [1]). We can also compare the four counting functions $\pi(x; 8, a)$ for $a \in \{1, 3, 5, 7\}$. By the time x equals 271, each of $\pi(x; 8, 3)$, $\pi(x; 8, 5)$, and $\pi(x; 8, 7)$ has been in first place; but $\pi(x; 8, 1)$ does not even obtain undisputed possession of third place in this four-way race until $x = 588{,}067{,}889$ (Bays and Hudson [2]).

All of these biases just mentioned are instances of a universal tendency for contestants $\pi(x; q, a)$ where a is not a square modulo q to run ahead of contestants $\pi(x; q, b)$ where b is a square modulo q. We briefly indicate why

403

this is the case through an analytic argument (though see Hudson [5] for a different approach), at the same time establishing some of the notation to be used throughout this paper. We will always assume that the modulus q is fixed, and therefore we will not care about the dependence of implicit O-constants on q.

Let $\psi(x; q, a)$ have its usual meaning,

$$\psi(x; q, a) = \sum_{\substack{n \le x \\ n \equiv a \pmod{q}}} \Lambda(n) = \sum_{\substack{p^r \le x \\ p^r \equiv a \pmod{q}}} \log p,$$

and set $\psi(x) = \psi(x; 1, 1)$. Under the Generalized Riemann Hypothesis (GRH), the explicit formula from the proof of the prime number theorem for arithmetic progressions (see [3]) gives

$$\psi(x; q, a) = \frac{\psi(x)}{\phi(q)} - \frac{1}{\phi(q)} \sum_{\substack{\chi \pmod{q} \\ \chi \ne \chi_0}} \bar{\chi}(a) \sum_\gamma \frac{x^{1/2 + i\gamma}}{1/2 + i\gamma} + O(\log^2 x) \quad (1.1)$$

as long as a is coprime to q. Here the inner sum is indexed by the imaginary parts γ of the nontrivial zeros of the Dirichlet L-function corresponding to the character χ, and should be interpreted as

$$\lim_{T \to \infty} \sum_{|\gamma| < T} \frac{x^{1/2 + i\gamma}}{1/2 + i\gamma} \quad (1.2)$$

so that it will converge.

We isolate the contribution to $\psi(x; q, a)$ from the primes themselves, defining

$$\theta(x; q, a) = \sum_{\substack{p \le x \\ p \equiv a \pmod{q}}} \log p$$

so that

$$\theta(x; q, a) = \psi(x; q, a) - \sum_{b^2 \equiv a \pmod{q}} \theta(x^{1/2}; q, b)$$

$$- \sum_{c^3 \equiv a \pmod{q}} \theta(x^{1/3}; q, c) - \dots \quad (1.3)$$

Notice that the number of terms in the first sum on the right-hand side of this equation equals the number of square roots modulo q possessed by a (in particular, the sum is empty if a is a non-square modulo q). Let us define

$$c(q, a) = -1 + \#\{1 \le b \le q : b^2 \equiv a \bmod q\}, \quad (1.4)$$

so that $c(q, a)$ is an extension of the Legendre symbol $\left(\frac{a}{q}\right)$ to all moduli q (though not an extension with nice multiplicativity properties); then the number of terms in the first sum on the right-hand side of equation (1.3) is exactly $c(q, a) + 1$.

Invoking the prime number theorem for arithmetic progressions, we have $\theta(y; q, a) = y/\phi(q) + O(\sqrt{y} \log^2 y)$ for a fixed modulus q (assuming GRH). Using this fact together with the explicit formula (1.1), equation (1.3) becomes

$$\theta(x; q, a) = \frac{\psi(x)}{\phi(q)} - \frac{1}{\phi(q)} \sum_{\substack{\chi \ (\mathrm{mod}\ q) \\ \chi \neq \chi_0}} \bar{\chi}(a) \sum_{\gamma} \frac{x^{1/2 + i\gamma}}{1/2 + i\gamma}$$

$$- (c(q, a) + 1) \frac{\sqrt{x}}{\phi(q)} + O(x^{1/3}). \quad (1.5)$$

In particular, we have $\theta(x) = \theta(x; 1, 1) = \psi(x) - \sqrt{x} + O(x^{1/3})$.

Converting equation (1.5) to a formula for $\pi(x; q, a)$ involves only a straightforward partial summation argument. We phrase the final result in terms of a normalized error term for $\pi(x; q, a)$, namely

$$E(x; q, a) = \frac{\log x}{\sqrt{x}} \left(\phi(q)\pi(x; q, a) - \pi(x)\right). \quad (1.6)$$

From equation (1.5) applied to both $\theta(x; q, a)$ and $\theta(x)$, one can derive [8, Lemma 2.1]

$$E(x; q, a) = -c(q, a) - \sum_{\substack{\chi \ (\mathrm{mod}\ q) \\ \chi \neq \chi_0}} \bar{\chi}(a) E(x, \chi) + O\left(\frac{1}{\log x}\right), \quad (1.7)$$

where we have defined

$$E(x, \chi) = \sum_{\gamma} \frac{x^{i\gamma}}{1/2 + i\gamma} \quad (1.8)$$

(interpreted similarly to (1.2) to ensure its conditional convergence). The behavior of $E(x; q, a)$ is therefore that of a function oscillating in a roughly bounded fashion about the mean value $-c(q, a)$, which is positive if a is a non-square (mod q) and negative if a is a square (mod q). These two different possible mean values are the source of the bias towards nonsquare contestants.

Rubinstein and Sarnak [8] have quantified these biases. Define $\delta_{q;a_1,\ldots,a_r}$ to be the logarithmic density of the set of real numbers x such that the inequalities

$$\pi(x; q, a_1) > \pi(x; q, a_2) > \cdots > \pi(x; q, a_r)$$

hold, where the logarithmic density of a set S is

$$\lim_{x \to \infty} \frac{1}{\log x} \int_{[2,x] \cap S} \frac{dt}{t} \tag{1.9}$$

assuming the limit exists. Let us assume not only GRH, but also that the nonnegative imaginary parts of the nontrivial zeros of all Dirichlet L-functions are linearly independent over the rationals, a hypothesis we shall abbreviate LI. Under these assumptions, Rubinstein and Sarnak proved that $\delta_{q;a_1,\ldots,a_r}$ always exists and is strictly positive. They also proved that $\delta_{q;a,b} > \frac{1}{2}$ if and only if a is a nonsquare (mod q) and b is a square (mod q), and calculated several of these densities; for example, $\delta_{4;3,1} = 0.9959\ldots$ and $\delta_{3;2,1} = 0.9990\ldots$.

In joint work with Feuerverger [4], we calculated many other densities (under the hypotheses GRH and LI). One of our discoveries was that the numerical values of the densities can vary even when the modulus q is fixed. For example, modulo 8 the only square is 1 while the three nonsquares are 3, 5, and 7, and modulo 12 the only square is 1 while the three nonsquares are 5, 7, and 11; we calculated that

$$
\begin{array}{llll}
\delta_{8;3,1} = 0.999569 & & \delta_{12;11,1} = 0.999977 & \\
\delta_{8;7,1} = 0.998938 & \text{and} & \delta_{12;5,1} = 0.999206 & (1.10) \\
\delta_{8;5,1} = 0.997395 & & \delta_{12;7,1} = 0.998606. &
\end{array}
$$

We also found that for race games involving more than two contestants, certain orderings of the contestants are more likely than others even if the residue classes involved are all squares or all nonsquares (mod q), a situation that was foreshadowed in [8]. For example, we calculated that

$$
\begin{array}{llll}
\delta_{8;3,5,7} = \delta_{8;7,5,3} = 0.192801 & & \delta_{12;5,7,11} = \delta_{12;11,7,5} = 0.198452 & \\
\delta_{8;3,7,5} = \delta_{8;5,7,3} = 0.166426 & \text{and} & \delta_{12;7,5,11} = \delta_{12;11,5,7} = 0.179985 & \\
\delta_{8;5,3,7} = \delta_{8;7,3,5} = 0.140772 & & \delta_{12;5,11,7} = \delta_{12;7,11,5} = 0.121563. &
\end{array}
$$
$$(1.11)$$

The goal of this paper is to begin to understand these recently discovered types of asymmetries. We will focus on the densities listed in equations (1.10) and (1.11), explaining how we could have predicted that $\delta_{8;5,1}$ would be smaller than both $\delta_{8;3,1}$ and $\delta_{8;7,1}$, for example. The hope is that a very concrete inspection of these special cases will function as a starting point for a future analysis of the general case.

2   Error Terms and Random Variables

The great utility of the hypothesis LI, concerning the linear independence of the nonnegative imaginary parts of the nontrivial zeros of Dirichlet L-functions, is in facilitating calculations involving those zeros that are based on harmonic analysis. In this paper we will phrase these calculations in terms of random variables, focusing on underscoring the ideas involved rather than belaboring the analytic details.

Notice that for real characters χ we can write

$$E(x, \chi) = \sum_{\gamma > 0} \left(\frac{x^{i\gamma}}{1/2 + i\gamma} + \frac{x^{-i\gamma}}{1/2 - i\gamma} \right) = 2 \sum_{\gamma > 0} \frac{\sin(\gamma \log x + \alpha_\gamma)}{\sqrt{1/4 + \gamma^2}}$$

for certain real numbers α_γ independent of x. The hypothesis LI implies that $1/2$ can never be a zero of $L(s, \chi)$, which is why we do not have to consider $\gamma = 0$. Also, under LI, any vector of the form

$$\{\sin(\gamma_1 \log x + \alpha_{\gamma_1}), \dots, \sin(\gamma_k \log x + \alpha_{\gamma_k})\}$$

becomes uniformly distributed over the k-dimensional torus as x tends to infinity. (It is the presence of $\log x$ in this statement that requires us to define the $\delta_{q;a_1,\dots,a_r}$ as logarithmic densities rather than natural densities.) It can be shown that $E(x, \chi)$ has a limiting (logarithmic) distribution as x tends to infinity; moreover, this distribution can also be described in terms of random variables. We now give this description.

For any positive number β, let Z_β be a random variable that is uniformly distributed on the unit circle in the complex plane; we make the convention that $Z_{-\beta} = \overline{Z_\beta}$. We stipulate that any collection $\{Z_{\beta_i}\}$ with all of the β_i distinct and positive is an independent collection of random variables. For any Dirichlet character χ (mod q), define the random variable

$$X(\chi) = \sum_\gamma \frac{Z_\gamma}{\sqrt{1/4 + \gamma^2}},$$

where again the sum is indexed by the imaginary parts of the nontrivial zeros of $L(s, \chi)$. We can also write

$$X(\chi) = 2 \sum_{\gamma > 0} \frac{X_\gamma}{\sqrt{1/4 + \gamma^2}} \tag{2.1}$$

(since $L(\frac{1}{2}, \chi) \neq 0$), where the $X_\gamma = \operatorname{Re} Z_\gamma$ are independent random variables each distributed on $[-1, 1]$ with the sine distribution. One can then show (see [8]), assuming GRH and LI, that *the limiting distribution* of $E(x, \chi)$ *is identical to the distribution of the random variable* $X(\chi)$. Similarly, it follows from equation (1.7) that the limiting distribution of $E(x; q, a)$ is the same as the distribution of the random variable

$$-c(q, a) + \sum_{\substack{\chi \pmod q \\ \chi \neq \chi_0}} X(\chi),$$

where the definition of $X(\chi)$ must be modified slightly if χ is not real-valued. (One might expect the summand to be something like $\bar{\chi}(a)X(\chi)$ rather than simply $X(\chi)$, but the coefficient $\bar{\chi}(a)$ disappears early in the argument because $\bar{\chi}(a)Z_\gamma$ is the same random variable as Z_γ itself.) We remark that the hypothesis LI implies that the various $X(\chi)$ are mutually independent random variables.

Let us examine these normalized error terms and random variables more concretely for the moduli $q = 8$ and $q = 12$. For a fundamental discriminant D, let $\chi_D(n) = (\frac{D}{n})$ using Kronecker's extension of the Legendre symbol. Then the three nonprincipal characters (mod 8) are χ_{-8}, χ_{-4}, and χ_8, while the three nonprincipal characters (mod 12) are χ_{-4}, χ_{-3}, and χ_{12}. (Table 1 explicitly lists the values taken by these characters. We shall abuse notation a bit and also denote by χ_D a character modulo 8 or 12 that is induced by the primitive character χ_D, whose conductor is $|D|$.)

Now, if we want to consider how often $\pi(x; 8, 3)$ exceeds $\pi(x; 8, 1)$, for example, we can look at the limiting distribution of the normalized difference $E(x; 8, 3) - E(x; 8, 1)$ and ask what proportion of that distribution lies above 0. From the explicit formula (1.7), we have

$$E(x; 8, 3) - E(x; 8, 1) = 4 + \sum_{\substack{\chi \pmod q \\ \chi \neq \chi_0}} (1 - \bar{\chi}(3))E(x, \chi) + O\left(\frac{1}{\log x}\right)$$

$$= 4 + 2E(\chi_{-4}) + 2E(\chi_8) + O\left(\frac{1}{\log x}\right).$$

Thus $E(x; 8, 3) - E(x; 8, 1)$ has the same limiting distribution as the random

χ	$\chi(3)$	$\chi(5)$	$\chi(7)$
χ_{-8}	1	-1	-1
χ_{-4}	-1	1	-1
χ_8	-1	-1	1

χ	$\chi(5)$	$\chi(7)$	$\chi(11)$
χ_{-4}	1	-1	-1
χ_{-3}	-1	1	-1
χ_{12}	-1	-1	1

Table 1. Values of the nonprincipal characters (mod 8) and (mod 12)

variable $4 + 2X(\chi_{-4}) + 2X(\chi_8)$, where $X(\chi)$ is as in equation (2.1). In particular, the density $\delta_{8;3,1}$ equals the mass given to the interval $(0,\infty)$ by this limiting distribution, or in other words simply $\Pr(4 + 2X(\chi_{-4}) + 2X(\chi_8) > 0)$. In fact, if we define

$$X_{8;3,1} = 4 + 2X(\chi_{-4}) + 2X(\chi_8)$$
$$X_{8;5,1} = 4 + 2X(\chi_{-8}) + 2X(\chi_8) \tag{2.2}$$
$$X_{8;7,1} = 4 + 2X(\chi_{-8}) + 2X(\chi_{-4})$$

and

$$X_{12;5,1} = 4 + 2X(\chi_{-3}) + 2X(\chi_{12})$$
$$X_{12;7,1} = 4 + 2X(\chi_{-4}) + 2X(\chi_{12}) \tag{2.3}$$
$$X_{12;11,1} = 4 + 2X(\chi_{-4}) + 2X(\chi_{-3}),$$

then in each case, the distribution of the random variable $X_{q;a,1}$ is the same as the limiting distribution of the difference $E(x;q,a) - E(x;q,1)$, and $\delta_{q;a,1} = \Pr(X_{q;a,1} > 0)$.

If we have several random variables, each with mean 4 and symmetric about that mean, which ones will take positive values most often? If the random variables have roughly the same shape, then we expect the ones with the smallest variance to stay above 0 the most. So let's compute the variances of the random variables $X_{q;a,1}$.

For any $c > 0$, the variance of the random variable cX_γ is simply $\mathrm{Var}(cX_\gamma) = \frac{1}{2}c^2$, and the various X_γ are independent; so if we define $V(\chi) = \mathrm{Var}(X(\chi))$, we see from the definition (2.1) of $X(\chi)$ that

$$V(\chi) = \sum_{\gamma>0} \frac{2}{1/4 + \gamma^2}. \tag{2.4}$$

We know that the larger the conductor of a character is, the more numerous and low-lying (close to the real axis) the zeros of the corresponding $L(s,\chi)$ will be. In fact, the order of magnitude of the sum in equation (2.4) is known to be the logarithm of the conductor of χ, at least on GRH; one can see this from the formula (see Davenport [3, p. 83])

$$V(\chi) = \log\frac{q}{\pi} - \gamma_0 - (1 + \chi(-1))\log 2 + 2\,\mathrm{Re}\,\frac{L'(1,\chi)}{L(1,\chi)} \tag{2.5}$$

when χ is a primitive character (mod q), where γ_0 is Euler's constant. Therefore, $V(\chi)$ will be larger when the conductor of χ is large. In particular, we should expect

$$V(\chi_{12}) > V(\chi_{-8}) > V(\chi_8) > V(\chi_{-4}) > V(\chi_{-3}), \tag{2.6}$$

χ	$V(\chi)$
χ_{-3}	0.11323
χ_{-4}	0.15557
χ_8	0.23543
χ_{-8}	0.31607
χ_{12}	0.33017

Table 2. Values of $V(\chi) = \sum_{\gamma>0} \frac{2}{1/4+\gamma^2}$

and the numerical computation of the variances verifies these expectations (see Table 2).

Why do we say that we expect $V(\chi_{-8}) > V(\chi_8)$, when the two characters have the same conductor? There is a secondary phenomenon, namely that the zeros of L-functions corresponding to even characters tend to be not as low-lying as those of L-functions corresponding to odd characters (the trivial zero at $s = 0$ of an L-functions associated to an even character seems to have a repelling effect on the nontrivial zeros). Indeed, the term $-(1 + \chi(-1))\log 2$ in the formula (2.5) for $V(\chi)$, which vanishes for odd characters χ, slightly lowers the value of $V(\chi)$ for even characters χ.

Of course this observation would be spurious if the behavior of the real part of $L'(1,\chi)/L(1,\chi)$ were much different for odd and even characters. While there is no reason to suspect that this should be the case, it seems hard to say anything substantial about the distribution of these values (this is a subject that warrants further investigation). Nevertheless, a look at lists of the first several zeros of Dirichlet L-functions with small conductor does confirm that the zeros of $L(s,\chi)$ are lower-lying when χ is odd than when χ is even.

Returning to equations (2.2) and (2.3), we can easily compute the variance of the random variables $X_{q;a,1}$ (again since the various $X(\chi)$ are independent by LI). For example.

$$\mathrm{Var}(X_{12;5,1}) = 4V(\chi_{-3}) + 4V(\chi_{12}) = 4W_{12} - 4V(\chi_{-4}),$$

where we define

$$W_q = \sum_{\substack{\chi \ (\mathrm{mod}\ q) \\ \chi \neq \chi_0}} V(\chi).$$

In general, we obtain

$$\text{Var}(X_{8;3,1}) = 4W_8 - 4V(\chi_{-8})$$
$$\text{Var}(X_{8;5,1}) = 4W_8 - 4V(\chi_{-4})$$
$$\text{Var}(X_{8;7,1}) = 4W_8 - 4V(\chi_8)$$

and

$$\text{Var}(X_{12;5,1}) = 4W_{12} - 4V(\chi_{-4})$$
$$\text{Var}(X_{12;7,1}) = 4W_{12} - 4V(\chi_{-3})$$
$$\text{Var}(X_{12;11,1}) = 4W_{12} - 4V(\chi_{12}).$$

Given the relative sizes of the $V(\chi)$ as listed in equation (2.6), we see that

$$\text{Var}(X_{8;5,1}) > \text{Var}(X_{8;7,1}) > \text{Var}(X_{8;3,1})$$
$$\text{Var}(X_{12;7,1}) > \text{Var}(X_{12;5,1}) > \text{Var}(X_{12;11,1}).$$

This in turn suggests that

$$\Pr(X_{8;3,1} > 0) > \Pr(X_{8;7,1} > 0) > \Pr(X_{8;5,1} > 0)$$
$$\Pr(X_{12;11,1} > 0) > \Pr(X_{12;5,1} > 0) > \Pr(X_{12;7,1} > 0),$$

or equivalently

$$\delta_{8;3,1} > \delta_{8;7,1} > \delta_{8;5,1} \qquad \text{and} \qquad \delta_{12;11,1} > \delta_{12;5,1} > \delta_{12;7,1}.$$

This is exactly what is observed in equation (1.10).

We emphasize that although the justification ventured into the analytic realm, in the end these predictions of the relative sizes of the $\delta_{q;a,1}$ depended upon only algebraic properties of the various residue classes a modulo q. To each residue class a was associated a particular character based on the values of the characters at a, and the conductor of this character is what correlated with the size of $\delta_{q;a,1}$. This is in the same spirit as Chebyshev's bias: the sign of $\delta_{q;a,b} - 1/2$ was shown by Rubinstein and Sarnak [8] to be determined by whether the residues a and b are squares in the multiplicative group modulo q.

A similar sort of analysis can also explain the relative sizes of the densities listed in equation (1.11), for which it is convenient to define a slightly differently normalized error term for $\pi(x; q, a)$. When $q = 8$ or $q = 12$ and a is one of the three nonsquare residue classes (mod q), we define $\tilde{E}(x; q, a) = E(x; q, a) + E(x; q, 1)$; again, investigating the relative sizes of the various $\pi(x; q, a)$ is the same as investigating the relative sizes of the

$\tilde{E}(x; q, a)$. For example,

$$\tilde{E}(x; 12, 5) = E(x; 12, 5) + E(x; 12, 1)$$

$$= 2 - \sum_{\substack{\chi \pmod{12} \\ \chi \neq \chi_0}} (\bar{\chi}(5) + 1)E(x, \chi) + O\left(\frac{1}{\log x}\right)$$

$$= 2 - 2E(x, \chi_{-4}) + O\left(\frac{1}{\log x}\right),$$

which has the same limiting distribution as the random variable $2 + 2X(\chi_{-4})$. In fact, if we define the random variables

$$\tilde{X}_{8;3} = 2 + 2X(\chi_{-8}) \qquad\qquad \tilde{X}_{12;5} = 2 + 2X(\chi_{-4})$$
$$\tilde{X}_{8;5} = 2 + 2X(\chi_{-4}) \quad \text{and} \quad \tilde{X}_{12;7} = 2 + 2X(\chi_{-3})$$
$$\tilde{X}_{8;7} = 2 + 2X(\chi_8) \qquad\qquad \tilde{X}_{12;11} = 2 + 2X(\chi_{12}),$$

then in each case the distribution of $\tilde{E}(x; q, a)$ is the same as that of the random variable $\tilde{X}_{q;a}$. Note also that the three random variables $\tilde{X}_{8;3}$, $\tilde{X}_{8;5}$, and $\tilde{X}_{8;7}$ are mutually independent due to the hypothesis LI, and the same is true of $\tilde{X}_{12;5}$, $\tilde{X}_{12;7}$, and $\tilde{X}_{12;11}$.

If we have three independent random variables each with the same mean, which one would we expect to take values between those of the other two most frequently? Our intuition tells us that the random variable with smallest variance will prefer to stay in the middle, while the one with largest variance will more frequently be in first or last place. We easily see that the variances for these random variables are

$$\mathrm{Var}(\tilde{X}_{8;3}) = 4V(\chi_{-8}) \qquad\qquad \mathrm{Var}(\tilde{X}_{12;5}) = 4V(\chi_{-4})$$
$$\mathrm{Var}(\tilde{X}_{8;5}) = 4V(\chi_{-4}) \quad \text{and} \quad \mathrm{Var}(\tilde{X}_{12;7}) = 4V(\chi_{-3}) \qquad (2.7)$$
$$\mathrm{Var}(\tilde{X}_{8;7}) = 4V(\chi_8) \qquad\qquad \mathrm{Var}(\tilde{X}_{12;11}) = 4V(\chi_{12}).$$

Once again, our knowledge (2.6) of the relative sizes of the quantities $V(\chi)$ tells us that

$$\mathrm{Var}(\tilde{X}_{8;3}) > \mathrm{Var}(\tilde{X}_{8;7}) > \mathrm{Var}(\tilde{X}_{8;5})$$
$$\mathrm{Var}(\tilde{X}_{12;11}) > \mathrm{Var}(\tilde{X}_{12;5}) > \mathrm{Var}(\tilde{X}_{12;7}).$$

Therefore, we expect that of the three prime counting functions $\pi(x; 8, a)$ with $a \in \{3, 5, 7\}$, the function $\pi(x; 8, 3)$ spends more time in first and last place than the other two while the function $\pi(x; 8, 5)$ spends the most time in second place; similarly, of the prime counting functions $\pi(x; 12, a)$ with

$a \in \{5, 7, 11\}$, the function $\pi(x; 12, 11)$ spends more time in first and last place than the other two while the function $\pi(x; 12, 7)$ spends the most time in second place. All of these predictions match the observed densities in equation (1.11).

We emphasize how important it was that the trios of random variables $\{\tilde{X}_{8;3}, \tilde{X}_{8;5}, \tilde{X}_{8;7}\}$ and $\{\tilde{X}_{12;5}, \tilde{X}_{12;7}, \tilde{X}_{12;11}\}$ were independent, so that we could draw conclusions about their relative positions in the three-way race based solely on their individual variances. We could certainly have normalized the error terms in an artificial way so that one of the resulting random variables in a trio equaled zero, for example! But then the other two random variables would not have been independent, and the correlation between them would have ruined any chance at such a straightforward analysis.

We plan to generalize these observations and arguments, as much as possible, to general moduli q in a future paper. The situation regarding densities of the form $\delta_{q;a,1}$ for nonsquares a (mod q) will be complicated by the greater complexity of the multiplicative groups to higher moduli, but we believe that the analysis for the relative sizes of these two-way densities can be successfully generalized. At the moment, however, the analysis of the three-way races above relied on the fact that for every character χ (mod q), at least two of the three values $\chi(a_i)$ were equal; this is a property that only special triples $\{a_1, a_2, a_3\}$ (mod q) can possess. While these special cases of three-way races to higher moduli can be treated as above, a new idea will be needed to generalize further.

3 Densities and Equalities

Since this is a conference proceedings, it seems appropriate to record here some comments made at the Millennial Conference regarding the subject of this paper. First, Gérald Tenenbaum mentioned that the density

$$\delta_{q;a_1,\ldots,a_r} = \lim_{x \to \infty} \frac{1}{\log x} \int_{\substack{2 \leq t \leq x \\ \pi(t;q,a_1) > \cdots > \pi(t;q,a_r)}} \frac{dt}{t}, \qquad (3.1)$$

as defined in equation (1.9) and the preceding lines, is not the only possible quantity to study when measuring the biases of the various orderings of the prime counting functions $\pi(x; q, a_i)$. Indeed, he noted for example that for any real number $k > -1$, the related density

$$\delta_{q;a_1,\ldots,a_r}^{(k)} = \lim_{x \to \infty} \frac{k+1}{(\log x)^{k+1}} \int_{\substack{2 \leq t \leq x \\ \pi(t;q,a_1) > \cdots > \pi(t;q,a_r)}} \frac{(\log t)^k \, dt}{t}$$

will also exist in this context.

Surprisingly, it turns out that these densities $\delta_{q;a_1,\dots,a_r}^{(k)}$ are independent of the parameter $k > -1$. One can prove this by hand, using the fact that for any function of the form $f(x) = \alpha x^\beta$ with α and β positive, the (natural) density of those positive real numbers x for which the fractional part of $f(x)$ lies in an interval $[\gamma, \eta] \subset [0, 1]$ is exactly $\eta - \gamma$. In fact, the lack of dependence on the parameter k is a consequence of a more general result of Lau [6] regarding distributions of error terms of number-theoretic functions. So although the particular definition (3.1) is not canonical, the density values themselves seems to be natural quantities to consider.

On another topic, Rubinstein and Sarnak [8] showed that under the assumptions GRH and LI, the density of the set of positive real numbers x such that $\pi(x; q, a) = \pi(x; q, b)$ equals zero (in fact they prove something rather stronger). Carl Pomerance asked whether one could prove this particular statement unconditionally. This is an excellent question, and while it certainly might be possible to establish unconditionally that $\pi(x; q, a)$ and $\pi(x; q, b)$ are "almost never" equal, this author does not know how to do so.

Since we know (conditionally) that the equality $\pi(x; q, a) = \pi(x; q, b)$ has arbitrarily large solutions, one can ask whether a system of equalities of the form $\pi(x; q, a_1) = \cdots = \pi(x; q, a_r)$ also has arbitrarily large solutions. A conjecture of the author (see [4]), resulting from an analogy to random walks on lattices, is that the answer is yes when $r = 3$ but no when $r \geq 4$. One can refine this conjecture in the following way: if $\{a_1, \dots, a_r\}$ are mutually incongruent reduced residues (mod q), then we believe that

$$\liminf_{x \to \infty} \left(\max_{1 \leq i < j \leq r} |\pi(x; q, a_i) - \pi(x; q, a_j)| \right) = \begin{cases} 0, & \text{if } r \leq 3, \\ \infty, & \text{if } r \geq 4. \end{cases} \quad (3.2)$$

Of course this raises the issue as to what function $f(x)$ should be chosen so that for $r \geq 4$, the quantity

$$\liminf_{x \to \infty} \frac{\max_{1 \leq i < j \leq r} |\pi(x; q, a_i) - \pi(x; q, a_j)|}{f(x)} \quad (3.3)$$

would be finite and nonzero (and whether the order of magnitude of this function $f(x)$ depends on r). It follows directly from the fact that the difference $E(x; q, a) - E(x; q, b)$ possesses a limiting distribution that for any integers $q, r \geq 2$, there exists some constant $C = C(q, r) > 0$ such that the density of those positive real numbers x with

$$\frac{|E(x; q, a) - E(x; q, b)|}{\phi(q)} = \frac{|\pi(x; q, a) - \pi(x; q, b)|}{\sqrt{x}/\log x} > C$$

is less than $1/r^2$ for any pair a, b of distinct reduced residues modulo q. Thus more than half of the time we must have

$$\frac{\max_{1 \le i < j \le r} |\pi(x; q, a_i) - \pi(x; q, a_j)|}{\sqrt{x}/\log x} \le C,$$

since there are only $r(r-1)/2$ terms in the maximum.

This argument shows that the expression in equation (3.3) is finite when $f(x) = \sqrt{x}/\log x$. However, nothing immediately ensures that the expression is nonzero, in which case the proper choice of $f(x)$ would be somewhat smaller than $\sqrt{x}/\log x$. In any case, we should begin by trying to establish equation (3.2) in the first place, perhaps even in an extreme case such as $r = \phi(q)$.

References

[1] C. Bays and R. H. Hudson, *Details of the first region of integers x with* $\pi_{3,2}(x) < \pi_{3,1}(x)$, Math. Comp. **32** (1978), 571–576.

[2] ———, *Numerical and graphical description of all axis crossing regions for the moduli 4 and 8 which occur before 10^{12}*, Internat. J. Math. Math. Sci. **2** (1979), 111–119.

[3] H. Davenport, *Multiplicative number theory*, Springer, Berlin, 1980.

[4] A. Feuerverger and G. Martin, *Biases in the Shanks-Rényi prime number race*, Experiment. Math. **9** (2000), 535–570.

[5] R. H. Hudson, *A common combinatorial principle underlies Riemann's formula, the Chebyshev phenomenon, and other subtle effects in comparative prime number theory. I*, J. Reine Angew. Math. **313** (1980), 133–150.

[6] Y.-K. Lau, *On the existence of limiting distributions of some number-theoretic error terms*, preprint.

[7] J. Leech, *Note on the distribution of prime numbers*, J. London Math. Soc. **32** (1957), 56–58.

[8] M. Rubinstein and P. Sarnak, *Chebyshev's bias*, Experiment. Math. **3** (1994), 173–195.

On the Analytic Continuation of Various Multiple Zeta-Functions

Kohji Matsumoto

0 Introduction

In this article we give a survey of the history of the problem of analytic continuation of multiple zeta-functions, and we prove some new results in this connection. We begin in Section 1 by describing the work of E. W. Barnes and H. Mellin at the turn of the 20th century. In Sections 2 and 3 we discuss the Euler sum and its multi-variable generalization, which recently have become again the subject of active research. In Section 4 we describe a new method of M. Katsurada which uses the classical Mellin-Barnes integral formula to establish the analytic continuation of the Euler sum. In the final two sections we present new results of the author, obtained by applying the Mellin-Barnes formula to more general multiple zeta-functions.

1 Barnes Multiple Zeta-Functions

The problem of analytic continuation of multiple zeta-functions was first considered by Barnes [7][8] and Mellin [48][49]. Barnes [7] introduced the double zeta-function of the form

$$\zeta_2(s; \alpha, (w_1, w_2)) = \sum_{m_1=0}^{\infty} \sum_{m_2=0}^{\infty} (\alpha + m_1 w_1 + m_2 w_2)^{-s}, \qquad (1.1)$$

where α, w_1, w_2 are complex numbers with positive real parts, and s is the complex variable. The series (1.1) is absolutely convergent in the half-plane $\Re s > 2$. Actually Barnes first defined his function as the contour integral

$$\zeta_2(s; \alpha, (w_1, w_2)) = -\frac{\Gamma(1-s)}{2\pi i} \int_{\mathcal{C}} \frac{e^{-\alpha z}(-z)^{s-1}}{(1 - e^{-w_1 z})(1 - e^{-w_2 z})} dz, \qquad (1.2)$$

where $\mathcal{C}$ is the contour which consists of the half-line on the positive real axis from infinity to a small positive constant δ, a circle of radius δ oriented counterclockwise around the origin, and the other half-line on the positive real axis from δ to infinity. It is easy to see that (1.2) coincides with (1.1) when $\Re s > 2$. The expression (1.2) gives the meromorphic continuation of $\zeta_2(s; \alpha, (w_1, w_2))$ to the whole s-plane. Moreover, Barnes [7] studied very carefully how to extend the definition of $\zeta_2(s; \alpha, (w_1, w_2))$ to the case when the real parts of α, w_1, w_2 are not necessarily positive.

Barnes introduced his double zeta-function for the purpose of constructing the theory of double gamma-functions. As for the theory of double gamma-functions, there were several predecessors such as Kinkelin, Hölder, Méray, Pincherle, and Alexeiewsky, but Barnes developed the theory most systematically. Barnes [8] then proceeded to develop a theory of more general multiple gamma-functions, and he introduced the multiple zeta-function defined by

$$\zeta_r(s; \alpha, (w_1, \ldots, w_r)) = \sum_{m_1=0}^{\infty} \cdots \sum_{m_r=0}^{\infty} (\alpha + m_1 w_1 + \cdots + m_r w_r)^{-s}, \quad (1.3)$$

where r is a positive integer and α, $w_1, \ldots, w_r$ are complex numbers. Barnes assumed the following condition to ensure the convergence of the series. Let ℓ be any line in the complex s-plane crossing the origin. Then ℓ divides the plane into two half-planes. Let $H(\ell)$ be one of those half-planes, not including ℓ itself. The assumption of Barnes is that

$$w_j \in H(\ell) \qquad (1 \leq j \leq r). \tag{1.4}$$

Then, excluding the finitely many tuples $(m_1, \ldots, m_r)$ satisfying $m_1 w_1 + \cdots + m_r w_r = -\alpha$ from the sum, we see easily that (1.3) is absolutely convergent when $\Re s > r$. Barnes [8] proved an integral expression similar to (1.2) for $\zeta_r(s; \alpha, (w_1, \ldots, w_r))$, which yields the meromorphic continuation.

On the other hand, Mellin [48][49] studied the meromorphic continuation of the multiple series

$$\sum_{m_1=1}^{\infty} \cdots \sum_{m_k=1}^{\infty} P(m_1, \ldots, m_k)^{-s}, \tag{1.5}$$

where $P(X_1, \ldots, X_k)$ is a polynomial of k indeterminates and of complex coefficients with positive real parts. Mellin's papers include a prototype of the method in the present paper, though he treated the one variable case only. For example, the formula (4.1) appears on p. 21 of [48]. Following Mellin's work, many authors have investigated the series (1.5) and its generalizations; the main contributors include K. Mahler, P. Cassou-Noguès, P. Sargos, B. Lichtin, M. Eie and M. Peter. Most of these authors concentrated on the one variable case, and we do not discuss the details of this work. However, Lichtin's series of papers ([36], [37], [38], [39], and [40]) should be mentioned here. In [36] Lichtin proposed the problem of studying the analytic continuation of the Dirichlet series in several variables

$$\sum_{m_1=1}^{\infty} \cdots \sum_{m_k=1}^{\infty} P_0(m_1, \ldots, m_k)$$
$$\times P_1(m_1, \ldots, m_k)^{-s_1} \cdots P_r(m_1, \ldots, m_k)^{-s_r}, \qquad (1.6)$$

where $P_0, P_1, \ldots, P_r$ are polynomials of k indeterminates, and he carried out this investigation in [37], [38], [39], and [40]. In particular, Lichtin proved that the series (1.6) can be continued meromorphically to the whole space when the associated polynomials are hypoelliptic (and also satisfy some other conditions).

2   The Euler Sum

The two-variable double sum

$$\zeta_2(s_1, s_2) = \sum_{m=1}^{\infty} \sum_{n=1}^{\infty} m^{-s_1}(m+n)^{-s_2} \qquad (2.1)$$

is absolutely convergent if $\Re s_2 > 1$ and $\Re(s_1 + s_2) > 2$. The investigation of this sum goes back to Euler, who was interested in the values of (2.1) when s_1 and s_2 are positive integers. Various properties of the values of (2.1) at positive integers were given in Nielsen's book [54]. Ramanujan also had an interest in such problems, and some of their formulas were later rediscovered; see the comments on pp. 252–253 of Berndt [9]. In recent years, the Euler sum has again become an object of active research; see, for instance, [10] and [17].

As far as the author knows, the first investigation of the analytic continuation of $\zeta_2(s_1, s_2)$ was made by Atkinson [6], in his work on the mean square of the Riemann zeta-function $\zeta(s)$. When $\Re s_1 > 1$ and $\Re s_2 > 1$, one has

$$\zeta(s_1)\zeta(s_2) = \zeta(s_1 + s_2) + \zeta_2(s_1, s_2) + \zeta_2(s_2, s_1). \qquad (2.2)$$

Atkinson's aim was to integrate the left-hand side with respect to t, when $s_1 = \frac{1}{2} + it$ and $s_2 = \frac{1}{2} - it$. Hence he was forced to show the analytic continuation of the right-hand side. Atkinson [6] used the Poisson summation formula to deduce a certain integral representation which enabled him to obtain the analytic continuation.

On the other hand, Matsuoka [47] obtained the analytic continuation of the series

$$\sum_{m=2}^{\infty} m^{-s} \sum_{n<m} n^{-1},$$

which is equal to $\zeta_2(1,s)$. Independently, Apostol and Vu [4] proved that $\zeta_2(s_1, s_2)$ can be continued meromorphically with respect to s_1 for each fixed s_2, and also with respect to s_2 for each fixed s_1. The proofs of Matsuoka and Apostol-Vu are both based on the Euler-Maclaurin summation formula. The main aim of those papers was the investigation of special values of $\zeta_2(s_1, s_2)$ at (not necessarily positive) integer points, and the papers give various formulas for such values.

Note that Apostol and Vu [4] also considered the series

$$T(s_1, s_2) = \sum_{m=1}^{\infty} \sum_{n<m} \frac{1}{m^{s_1} n^{s_2} (m+n)} \tag{2.3}$$

and discussed its analytic continuation.

Let q be a positive integer (≥ 2), $\varphi(q)$ the Euler function, χ a Dirichlet character mod q, and $L(s, \chi)$ the corresponding Dirichlet L-function. Inspired by Atkinson's work [6], Meurman [50] and Motohashi [53] independently considered the sum

$$Q((s_1, s_2); q) = \varphi(q)^{-1} \sum_{\chi \bmod q} L(s_1, \chi) L(s_2, \bar{\chi}).$$

Corresponding to (2.2), the decomposition

$$Q((s_1, s_2); q) = L(s_1 + s_2, \chi_0) + f((s_1, s_2); q) + f((s_2, s_1); q)$$

holds, where χ_0 is the principal character mod q and

$$f((s_1, s_2); q) = \sum_{\substack{1 \leq a \leq q \\ (a,q)=1}} \sum_{m=0}^{\infty} \sum_{n=1}^{\infty} (qm+a)^{-s_1} (q(m+n)+a)^{-s_2}. \tag{2.4}$$

This is a generalization of the Euler sum (2.1). Meurman [50] proved the analytic continuation of (2.4) by generalizing the argument of Atkinson [6]. On the other hand, Motohashi derived a double contour integral representation for (2.4), which yields the analytic continuation. By refining Motohashi's argument, Katsurada and the author [32][33] obtained asymptotic expansions for

$$\sum_{\chi \bmod q} |L(s, \chi)|^2 \quad (s \neq 1) \qquad \text{and} \qquad \sum_{\substack{\chi \bmod q \\ \chi \neq \chi_0}} |L(1, \chi)|^2 \tag{2.5}$$

with respect to q. See also Katsurada [28], where a somewhat different argument using confluent hypergeometric functions is given.

Let $\zeta(s,\alpha)$ be the Hurwitz zeta-function defined by the analytic continuation of the series $\sum_{n=0}^{\infty}(\alpha+n)^{-s}$, where $\alpha > 0$. Katsurada and the author [34] gave an asymptotic expansion of the mean value

$$\int_0^1 |\zeta(s,\alpha) - \alpha^{-s}|^2 d\alpha \qquad (2.6)$$

with respect to $\Im s$. The starting point of the argument in [34] is the following generalization of (2.2):

$$\zeta(s_1,\alpha)\zeta(s_2,\alpha) = \zeta(s_1+s_2,\alpha) + \zeta_2((s_1,s_2);\alpha) + \zeta_2((s_2,s_1);\alpha), \qquad (2.7)$$

where

$$\zeta_2((s_1,s_2);\alpha) = \sum_{m=0}^{\infty}\sum_{n=1}^{\infty}(\alpha+m)^{-s_1}(\alpha+m+n)^{-s_2}. \qquad (2.8)$$

This is again a generalization of (2.1). In [34], the meromorphic continuation of $\zeta_2((s_1,s_2);\alpha)$ was shown using the formula

$$\zeta_2((s_1,s_2);\alpha) = \frac{\Gamma(s_1+s_2-1)\Gamma(1-s_1)}{\Gamma(s_2)}\zeta(s_1+s_2-1)$$

$$+\frac{1}{\Gamma(s_1)\Gamma(s_2)(e^{2\pi i s_1}-1)(e^{2\pi i s_2}-1)}\int_C \frac{y^{s_2}-1}{e^y-1}$$

$$\times \int_C h(x+y;\alpha)x^{s_1-1}dxdy, \qquad (2.9)$$

where

$$h(z;\alpha) = \frac{e^{(1-\alpha)z}}{e^z-1} - \frac{1}{z}.$$

This formula is an analogue of Motohashi's integral expression for (2.4).

The author [42] considered the more general series

$$\zeta_2((s_1,s_2);\alpha,w) = \sum_{m_1=0}^{\infty}\sum_{m_2=0}^{\infty}(\alpha+m_1)^{-s_1}(\alpha+m_1+m_2 w)^{-s_2}, \qquad (2.10)$$

where $w > 0$, and proved its analytic continuation using a similar method, which also gives the asymptotic expansion of $\zeta_2((s_1,s_2);\alpha,w)$ with respect to w when $w \to +\infty$. In particular, this yields an asymptotic expansion of the Barnes double zeta-function $\zeta_2(s;\alpha,(1,w))$ with respect to w, because this function is just $\zeta_2((0,s);\alpha,w)$. These results and also the asymptotic expansion of the double gamma-function are proved in [42]. Note that some claims in [42] on the uniformity of the error terms are not true; they are corrected in [44] (see also [43]).

3   Multi-variable Euler-Zagier Sums

The r-variable generalization of the Euler sum (2.1), defined by

$$\zeta_r(s_1, \ldots, s_r) \tag{3.1}$$

$$= \sum_{m_1=1}^{\infty} \sum_{m_2=1}^{\infty} \cdots \sum_{m_r=1}^{\infty} m_1^{-s_1} (m_1 + m_2)^{-s_2} \cdots (m_1 + \cdots + m_r)^{-s_r},$$

is absolutely convergent in the region

$$\mathcal{A}_r = \{(s_1, \ldots, s_r) \in \mathbf{C}^r \mid \Re(s_{r-k+1} + \cdots + s_r) > k \quad (1 \leq k \leq r)\}, \tag{3.2}$$

as will be shown in Theorem 3 below (in Section 6). (The condition of absolute convergence given by Proposition 1 of Zhao [66] is not sufficient.) In connection with knot theory, quantum groups and mathematical physics, the properties of (3.1) have been recently investigated by Zagier [64][65], Goncharov [19] and others, and the series is called the Euler-Zagier sum or the multiple harmonic series. The case $r = 3$ of (3.1) had, in fact, already been studied by Sitaramachandrarao and Subbarao [58]. The Euler-Zagier sum also appears in work of Butzer, Markett and Schmidt ([15], [16], [41]).

There are various interesting relations among values of (3.1) at positive integers. Some of these (for small r) can be found in earlier references, but systematic studies were undertaken by Hoffman [22] (see also [23]), who proved a number of relations, including some previous results and conjectures, and stated the sum conjecture and the duality conjecture. The sum conjecture, originally due to M. Schmidt (see Markett [41]) and also to C. Moen, was proved by Granville [21] and Zagier (unpublished). On the other hand, the duality conjecture has turned out to be an immediate consequence of iterated integral representations of Drinfel'd and Kontsevich (cf. Zagier [65]). Further generalizations were given by Ohno [55] and Hoffman-Ohno [24]. Other families of relations, coming from the theory of knot invariants, were discovered by Le-Murakami [35] and Takamuki [60]. Various relations were also discussed by Borwein et al. [11][12], Flajolet-Salvy [18] and Minh-Petitot [51]. For instance, a conjecture mentioned in Zagier [65] was proved in [12][13]. For the latest developments we refer to [14] and [56]. (The recent developments in this field are enormous, and it is impossible to mention all of them here.)

The papers mentioned above were mainly devoted to the study of the values of $\zeta_r(s_1, \ldots, s_r)$ at positive integers. On the other hand, except for the case $r = 2$ mentioned in the preceding section, the study of the analytic continuation of $\zeta_r(s_1, \ldots, s_r)$ has begun only very recently. First, Arakawa and Kaneko [5] proved that if $s_1, \ldots, s_{r-1}$ are fixed, then (3.1) can be continued meromorphically with respect to s_r to the whole complex plane. The analytic continuation of (3.1) to the whole $\mathbf{C}^r$-space as an r-variable function was established by Zhao [66], and independently by Akiyama, Egami and Tanigawa [1]. Zhao's proof is based on properties of generalized functions in the sense of Gel'fand and Shilov. The method in [1] is more elementary and is based on an application of the Euler-Maclaurin summation formula. Akiyama, Egami and Tanigawa [1] further studied the values of $\zeta_r(s_1, \ldots, s_r)$ at non-positive integers (see also Akiyama and Tanigawa [3]). Note that the statements about the trivial zeros of ζ_2 in Zhao [66] are incorrect. T. Arakawa pointed out that the method of Arakawa and Kaneko [5] can also be refined to give an alternative proof of the analytic continuation of $\zeta_r(s_1, \ldots, s_r)$ as an r-variable function.

Akiyama and Ishikawa [2] considered the multiple L-function

$$
L_r((s_1, \ldots, s_r); (\chi_1, \ldots, \chi_r)) \tag{3.3}
$$
$$
= \sum_{m_1=1}^{\infty} \sum_{m_2=1}^{\infty} \cdots \sum_{m_r=1}^{\infty} \frac{\chi_1(m_1)}{m_1^{s_1}} \frac{\chi_2(m_1 + m_2)}{(m_1 + m_2)^{s_2}} \cdots \frac{\chi_r(m_1 + \cdots + m_r)}{(m_1 + \cdots + m_r)^{s_r}},
$$

where $\chi_1, \ldots, \chi_r$ are Dirichlet characters. This series had been introduced earlier by Goncharov [20], but the main goal of Akiyama and Ishikawa [2] was to prove the analytic continuation of (3.3). For this purpose, they first wrote (3.3) as a linear combination of functions of the form

$$
\zeta_r((s_1, \ldots, s_r); (\alpha_1, \ldots, \alpha_r)) = \sum_{m_1=1}^{\infty} \sum_{m_2=1}^{\infty} \cdots \sum_{m_r=1}^{\infty} (\alpha_1 + m_1)^{-s_1}
$$
$$
\times (\alpha_2 + m_1 + m_2)^{-s_2} \cdots (\alpha_r + m_1 + \cdots + m_r)^{-s_r}, \tag{3.4}
$$

where $\alpha_1, \ldots, \alpha_r$ are positive, and considered the analytic continuation of the latter functions. They established this continuation by generalizing the argument in Akiyama, Egami and Tanigawa [1]. Ishikawa [26] derived additional properties in the special case $s_1 = \cdots = s_r$ of (3.3) and applied those results to the study of certain multiple character sums (Ishikawa [27]).

4 Katsurada's Idea

In Section 2 we mentioned the work of Katsurada and the author on asymptotic expansions of (2.5) and (2.6). The key ingredient in this work is the treatment of the functions (2.4) and (2.8); in [32] and [34] these functions are expressed by certain double contour integrals.

Katsurada [29][30] reconsidered this problem, and discovered a simple elegant alternative way of proving the expansions of (2.5) and (2.6). The key tool of Katsurada's method is the Mellin-Barnes integral formula

$$\Gamma(s)(1+\lambda)^{-s} = \frac{1}{2\pi i} \int_{(c)} \Gamma(s+z)\Gamma(-z)\lambda^z dz, \qquad (4.1)$$

where s and λ are complex with $\Re s > 0$, $|\arg \lambda| < \pi$, $\lambda \neq 0$, and c is real with $-\Re s < c < 0$. The path of integration is the vertical line from $c - i\infty$ to $c + i\infty$. This formula is known (see, e.g., Whittaker and Watson [62], Section 14.51, p. 289, Corollary), or can easily be proved as follows. First assume $|\lambda| < 1$, and shift the path to the right. The relevant poles of the integrand are located at $z = n$ ($n = 0, 1, 2, \ldots$) with the residue $(-1)^{n+1}\Gamma(s+n)\lambda^n/n!$. Hence the right-hand side of (4.1) is equal to

$$\Gamma(s) \sum_{n=0}^{\infty} \binom{-s}{n} \lambda^n = \Gamma(s)(1+\lambda)^{-s},$$

which is the left-hand side. The extension to $|\lambda| \geq 1$ now follows by analytic continuation.

Katsurada [30] used (4.1) to give a simple proof of the analytic continuation and the asymptotic expansion of the function (2.4). Subsequently, Katsurada [29] (this article was published earlier, but written later than [30]) showed that the same idea can be applied to the function (2.8) to obtain its analytic continuation. In [29], this idea is combined with some properties of hypergeometric functions, and hence the technical details are not so simple. Therefore, to illustrate the essence of Katsurada's idea, we give here a simple proof of the analytic continuation of the Euler sum (2.1) by Katsurada's method.

Assume $\Re s_2 > 1$ and $\Re(s_1 + s_2) > 2$. Putting $s = s_2$ and $\lambda = n/m$ in (4.1), and dividing the both sides by $\Gamma(s_2)m^{s_1+s_2}$, we obtain

$$m^{-s_1}(m+n)^{-s_2} = \frac{1}{2\pi i} \int_{(c)} \frac{\Gamma(s_2+z)\Gamma(-z)}{\Gamma(s_2)} m^{-s_1-s_2-z} n^z dz. \qquad (4.2)$$

We may assume $\max\{-\Re s_2, 1 - \Re(s_1 + s_2)\} < c < -1$. Then we can sum both sides of (4.2) with respect to m and n to obtain

$$\zeta_2(s_1, s_2) = \frac{1}{2\pi i} \int_{(c)} \frac{\Gamma(s_2+z)\Gamma(-z)}{\Gamma(s_2)} \zeta(s_1 + s_2 + z)\zeta(-z)dz. \qquad (4.3)$$

We now shift the path to $\Re z = M - \varepsilon$, where M is a positive integer and ε is a small positive number. This shifting is easily justified using Stirling's formula. The relevant poles of the integrand are at $z = -1, 0, 1, 2, \ldots, M -$

1. Counting the residues of those poles, we get

$$\zeta_2(s_1, s_2) = \frac{1}{s_2 - 1}\zeta(s_1 + s_2 - 1) + \sum_{k=0}^{M-1}\binom{-s_2}{k}\zeta(s_1 + s_2 + k)\zeta(-k)$$

$$+\frac{1}{2\pi i}\int_{(M-\varepsilon)}\frac{\Gamma(s_2 + z)\Gamma(-z)}{\Gamma(s_2)}\zeta(s_1 + s_2 + z)\zeta(-z)dz. \quad (4.4)$$

The last integral can be continued holomorphically to the region

$$\left\{(s_1, s_2) \in \mathbf{C}^2 \mid \Re s_2 > -M + \varepsilon, \Re(s_1 + s_2) > 1 - M + \varepsilon\right\},$$

because in this region the poles of the integrand are not on the path of integration. Hence (4.4) gives the analytic continuation of $\zeta_2(s_1, s_2)$ to this region. Since M is arbitrary, the proof of the continuation to the whole $\mathbf{C}^2$-space is complete. Moreover, from (4.4) we can see that the singularities of $\zeta_2(s_1, s_2)$ are located only on the subsets of $\mathbf{C}^2$ defined by one of the equations

$$s_2 = 1, \quad s_1 + s_2 = 2 - \ell \quad (\ell \in \mathbf{N}_0), \quad (4.5)$$

where $\mathbf{N}_0$ denotes the set of non-negative integers.

Katsurada applied (4.1) to various other types of problems. Here we mention his short note [31], in which he introduced (inspired by [42]) the double zeta-function

$$\sum_{m=0}^{\infty}\sum_{n=0}^{\infty}e^{2\pi i(ms_1 + ns_2)}(\alpha + m)^{-s_1}(\alpha + \beta + m + n)^{-s_2},$$

expressed it as an integral similar to (4.3), and obtained some asymptotic results in the domain of absolute convergence.

5   The Mordell-Tornheim Zeta-Function and the Apostol-Vu Zeta-Function

Let $\Re s_j > 1$ $(j = 1, 2, 3)$ and define

$$\zeta_{MT}(s_1, s_2, s_3) = \sum_{m=1}^{\infty}\sum_{n=1}^{\infty}m^{-s_1}n^{-s_2}(m + n)^{-s_3}. \quad (5.1)$$

This series was first considered by Tornheim [61], and the special case $s_1 = s_2 = s_3$ was studied independently by Mordell [52]. We call (5.1) the Mordell-Tornheim zeta-function. Tornheim himself called it the harmonic double series. Zagier [65] quoted Witten's paper [63] and studied (5.1) calling it the Witten zeta-function.

The analytic continuation of $\zeta_{MT}(s_1, s_2, s_3)$ was established by S. Akiyama and also by S. Egami in 1999. Akiyama's method is based on the Euler-Maclaurin summation formula, while Egami's proof is a modification of the method of Arakawa and Kaneko [5]. Both of these proofs are still unpublished.

Here, by using the method explained in the preceding section, we give a simple proof of the following result:

Theorem 1. *The function $\zeta_{MT}(s_1, s_2, s_3)$ can be meromorphically continued to the whole $\mathbf{C}^3$-space, and all of its singularities are located on the subsets of $\mathbf{C}^3$ defined by one of the equations $s_1 + s_3 = 1 - \ell$, $s_2 + s_3 = 1 - \ell$ ($\ell \in \mathbf{N}_0$) and $s_1 + s_2 + s_3 = 2$.*

Proof. Assume $\Re s_1 > 1$, $\Re s_2 > 0$ and $\Re s_3 > 1$. Then the series (5.1) is absolutely convergent. Putting $s = s_3$ and $\lambda = n/m$ in (4.1), and dividing the both sides by $\Gamma(s_3)m^{s_1+s_3}n^{s_2}$, we obtain

$$m^{-s_1-s_3}n^{-s_2}\left(1 + \frac{n}{m}\right)^{-s_3} = \frac{1}{2\pi i}\int_{(c)}\frac{\Gamma(s_3+z)\Gamma(-z)}{\Gamma(s_3)}m^{-s_1-s_3-z}n^{-s_2+z}dz.$$

We may assume $-\Re s_3 < c < \min\{\Re s_2 - 1, 0\}$. Summing with respect to m and n we get

$$\zeta_{MT}(s_1, s_2, s_3) = \frac{1}{2\pi i}\int_{(c)}\frac{\Gamma(s_3+z)\Gamma(-z)}{\Gamma(s_3)}\zeta(s_1+s_3+z)\zeta(s_2-z)dz. \quad (5.2)$$

Let M be a positive integer greater than $\Re s_2 - 1 + \varepsilon$, and shift the path to $\Re z = M - \varepsilon$. First assume that s_2 is not a positive integer. Then all the relevant poles are simple, and we obtain

$$\zeta_{MT}(s_1, s_2, s_3) = \frac{\Gamma(s_2+s_3-1)\Gamma(1-s_2)}{\Gamma(s_3)}\zeta(s_1+s_2+s_3-1)$$

$$+ \sum_{k=0}^{M-1}\binom{-s_3}{k}\zeta(s_1+s_3+k)\zeta(s_2-k)$$

$$+ \frac{1}{2\pi i}\int_{(M-\varepsilon)}\frac{\Gamma(s_3+z)\Gamma(-z)}{\Gamma(s_3)}\zeta(s_1+s_3+z)\zeta(s_2-z)dz. \quad (5.3)$$

When $s_2 = 1 + h$ ($h \in \mathbf{N}_0$, $h \leq M - 1$), the right-hand side of (5.3) contains two singular factors, but they cancel each other. In fact, we obtain

$$\zeta_{MT}(s_1, 1 + h, s_3)$$

$$= \binom{-s_3}{h} \left\{ \left(1 + \frac{1}{2} + \cdots + \frac{1}{h} - \psi(s_3 + h) \right) \zeta(s_1 + s_3 + h) \right.$$

$$\left. - \zeta'(s_1 + s_3 + h) \right\}$$

$$+ \sum_{\substack{k=0 \\ k \neq h}}^{M-1} \binom{-s_3}{k} \zeta(s_1 + s_3 + k)\zeta(1 + h - k)$$

$$+ \frac{1}{2\pi i} \int_{(M-\varepsilon)} \frac{\Gamma(s_3 + z)\Gamma(-z)}{\Gamma(s_3)} \zeta(s_1 + s_3 + z)\zeta(1 + h - z)dz,$$

$$(5.4)$$

where $\psi = \Gamma'/\Gamma$ and an empty sum is to be interpreted as zero. The desired assertions of Theorem 1 now follow from (5.3) and (5.4), as in the argument described in the preceding section. □

After the papers of Tornheim [61] and Mordell [52], the values of $\zeta_{MT}(s_1, s_2, s_3)$ at positive integers have been investigated by many authors (Subbarao and Sitaramachandrarao [59], Huard, Williams and Zhang [25], and Zagier [65]). In view of Theorem 1, it is an interesting problem to study the properties of the values of $\zeta_{MT}(s_1, s_2, s_3)$ at non-positive integers.

Next, recall the series (2.3) considered by Apostol and Vu [4] who, inspired by the work of Sitaramachandrarao and Sivaramasarma [57], obtained various formulas on the special values of (2.3).

Here we introduce the following three-variable Apostol-Vu zeta-function:

$$\zeta_{AV}(s_1, s_2, s_3) = \sum_{m=1}^{\infty} \sum_{n<m} m^{-s_1} n^{-s_2}(m + n)^{-s_3} \qquad (\Re s_j > 1). \quad (5.5)$$

Note that there is the following simple relation between ζ_{AV} and ζ_{MT}:

$$\zeta_{MT}(s_1, s_2, s_3) = 2^{-s_3}\zeta(s_1 + s_2 + s_3)$$

$$+ \zeta_{AV}(s_1, s_2, s_3) + \zeta_{AV}(s_2, s_1, s_3). \quad (5.6)$$

Also, there is a simple relation between $\zeta_{AV}(s_1, s_2, 1)$ and $\zeta_2(s_1, s_2)$ (see (17) of Apostol and Vu [4]).

We now prove the analytic continuation of the (three-variable) Apostol-Vu zeta-function $\zeta_{AV}(s_1, s_2, s_3)$. The proof is based on the same principle as that of Theorem 1, but the details are somewhat more complicated.

Theorem 2. *The function* $\zeta_{AV}(s_1, s_2, s_3)$ *can be continued meromorphically to the whole* $\mathbf{C}^3$-*space, and all of its singularities are located on the subsets of* $\mathbf{C}^3$ *defined by one of the equations* $s_1 + s_3 = 1 - \ell$, *and* $s_1 + s_2 + s_3 = 2 - \ell$ $(\ell \in \mathbf{N}_0)$.

Proof. Assume $\Re s_j > 1$ $(j = 1, 2, 3)$. Similar to (5.2), we obtain

$$\zeta_{AV}(s_1, s_2, s_3) = \frac{1}{2\pi i} \int_{(c)} \frac{\Gamma(s_3 + z)\Gamma(-z)}{\Gamma(s_3)} \sum_{m=1}^{\infty} \sum_{n < m} m^{-s_1 - s_3 - z} n^{-s_2 + z} dz$$

$$= \frac{1}{2\pi i} \int_{(c)} \frac{\Gamma(s_3 + z)\Gamma(-z)}{\Gamma(s_3)} \zeta_2(s_2 - z, s_1 + s_3 + z) dz, \quad (5.7)$$

where $-\Re s_3 < c < 0$. We now shift the path of integration to $\Re z = M - \varepsilon$. It is not difficult to show using (4.4) that $\zeta_2(s_1, s_2)$ is of polynomial order with respect to $\Im s_1$ and $\Im s_2$. Hence this shifting is possible. From (4.5) we see that the only pole of $\zeta_2(s_2 - z, s_1 + s_3 + z)$ (as a function in z), under the assumption $\Re s_j > 1$ $(j = 1, 2, 3)$, is $z = 1 - s_1 - s_3$. This pole is located to the left of the line $\Re z = c$, and hence irrelevant now. Counting the residues of the poles at $z = 0, 1, \ldots, M - 1$, we get

$$\zeta_{AV}(s_1, s_2, s_3) = \sum_{k=0}^{M-1} \binom{-s_3}{k} \zeta_2(s_2 - k, s_1 + s_3 + k)$$

$$+ \frac{1}{2\pi i} \int_{(M-\varepsilon)} \frac{\Gamma(s_3 + z)\Gamma(-z)}{\Gamma(s_3)} \zeta_2(s_2 - z, s_1 + s_3 + z) dz. \quad (5.8)$$

This formula already implies the meromorphic continuation except in the case when $s_1 + s_2 + s_3 = 2 - \ell$ $(\ell \in \mathbf{N}_0)$, where $\zeta_2(s_2 - z, s_1 + s_3 + z)$ is singular. To investigate the behaviour of the above integral on this polar set, we substitute the formula (4.4) into the integrand on the right-hand side of (5.8). We obtain

$$\zeta_{AV}(s_1, s_2, s_3) = \sum_{k=0}^{M-1} \binom{-s_3}{k} \zeta_2(s_2 - k, s_1 + s_3 + k)$$

$$+ \zeta(s_1 + s_2 + s_3 - 1) P(s_1, s_3)$$

$$+ \sum_{j=0}^{M-1} \zeta(s_1 + s_2 + s_3 + j)\zeta(-j) Q_j(s_1, s_3) + R(s_1, s_2, s_3), \quad (5.9)$$

where

$$P(s_1, s_3) = \frac{1}{2\pi i} \int_{(M-\varepsilon)} \frac{\Gamma(s_3 + z)\Gamma(-z)}{\Gamma(s_3)} \frac{dz}{s_1 + s_3 + z - 1},$$

$$Q_j(s_1, s_3) = \frac{1}{2\pi i} \int_{(M-\varepsilon)} \frac{\Gamma(s_3 + z)\Gamma(-z)}{\Gamma(s_3)} \begin{pmatrix} -s_1 - s_3 - z \\ j \end{pmatrix} dz,$$

and

$$R(s_1, s_2, s_3) = \frac{1}{(2\pi i)^2} \int_{(M-\varepsilon)} \frac{\Gamma(s_3 + z)\Gamma(-z)}{\Gamma(s_3)}$$

$$\times \int_{(M-\varepsilon)} \frac{\Gamma(s_1 + s_3 + z + z')\Gamma(-z')}{\Gamma(s_1 + s_3 + z)} \zeta(s_1 + s_2 + s_3 + z')\zeta(-z')dz'dz.$$

It is easy to see that

(i) $P(s_1, s_3)$ is holomorphic if $\Re s_3 > -M + \varepsilon$ and $\Re(s_1 + s_3) > 1 - M + \varepsilon$,

and

(ii) $Q_j(s_1, s_3)$ is holomorphic for $0 \leq j \leq M - 1$ if $\Re s_3 > -M + \varepsilon$.

Also, since the inner integral of $R(s_1, s_2, s_3)$ is holomorphic if $\Re(s_1 + s_3 + z) > -M + \varepsilon$ and $\Re(s_1 + s_2 + s_3) > 1 - M + \varepsilon$ as a function of the four variables (s_1, s_2, s_3, z), we see that

(iii) $R(s_1, s_2, s_3)$ is holomorphic if $\Re s_3 > -M + \varepsilon$, $\Re(s_1 + s_3) > -2M + 2\varepsilon$ and $\Re(s_1 + s_2 + s_3) > 1 - M + \varepsilon$.

From (i), (ii), (iii) and (5.9), we find that $\zeta_{AV}(s_1, s_2, s_3)$ can be continued meromorphically to the region

$$\{(s_1, s_2, s_3) \in \mathbf{C}^3 \mid \Re s_3 > -M + \varepsilon, \Re(s_1 + s_3) > 1 - M + \varepsilon,$$

$$\Re(s_1 + s_2 + s_3) > 1 - M + \varepsilon\}.$$

Since M is arbitrary, we obtain the analytic continuation of $\zeta_{AV}(s_1, s_2, s_3)$ to the whole $\mathbf{C}^3$-space. The information on singularities can be deduced from the representation (5.9). The proof of Theorem 2 is complete. $\square$

6 Generalized Multiple Zeta-Functions

Let $s_1, \ldots, s_r$ be complex variables. Let $\alpha_1, \ldots, \alpha_r, w_1, \ldots, w_r$ be complex parameters, and define the multiple series

$$\zeta_r((s_1, \ldots, s_r); (\alpha_1, \ldots, \alpha_r), (w_1, \ldots, w_r))$$

$$= \sum_{m_1=0}^{\infty} \cdots \sum_{m_r=0}^{\infty} (\alpha_1 + m_1 w_1)^{-s_1} (\alpha_2 + m_1 w_1 + m_2 w_2)^{-s_2}$$

$$\times \cdots \times (\alpha_r + m_1 w_1 + \cdots + m_r w_r)^{-s_r}. \tag{6.1}$$

We will explain later (in the proof of Theorem 3) how to choose the branch of the logarithms.

When $s_1 = \cdots = s_{r-1} = 0$, then the above series (6.1) reduces to the Barnes multiple zeta-function (1.3). The Euler-Zagier sum (3.1) and its generalization (3.4) are also special cases of (6.1). The multiple series of the form (6.1) was first introduced in the author's article [45], and the meromorphic continuation in the special case $0 < \alpha_1 < \alpha_2 < \cdots < \alpha_r$ and $w_j = 1$ $(1 \leq j \leq r)$ of (6.1) to the whole $\mathbf{C}^r$-space was proved in [45].

To ensure the convergence of (6.1), we impose condition (1.4) on the w_j's, which was first introduced by Barnes for his multiple series (1.3). However, we do not require any condition on the α_j's. If $\alpha_j \notin H(\ell)$ for some j, then there might exist finitely many tuples $(m_1, \ldots, m_j)$ for which

$$\alpha_j + m_1 w_1 + \cdots + m_j w_j = 0 \tag{6.2}$$

holds. We adopt the convention that the terms corresponding to such tuples are removed from the sum (6.1). Under this convention, we now prove:

Theorem 3. *If the condition* (1.4) *holds, then the series* (6.1) *is absolutely convergent in the region* $\mathcal{A}_r$, *defined by* (3.2), *uniformly in any compact subset of* $\mathcal{A}_r$.

Proof. We prove the theorem by induction. When $r = 1$, the assertion is obvious. Assume that the theorem is true for ζ_{r-1}. In what follows, an empty sum is to be interpreted as zero.

Let $\theta \in (-\pi, \pi]$ be the argument of the vector contained in $H(\ell)$ and orthogonal to ℓ. Then the line ℓ consists of the points whose arguments are $\theta \pm \pi/2$ (and the origin), and

$$H(\ell) = \left\{ w \in \mathbf{C} \setminus \{0\} \,\middle|\, \theta - \frac{\pi}{2} < \arg w < \theta + \frac{\pi}{2} \right\}.$$

We can write $w_j = w_j^{(1)} + w_j^{(2)}$, with $\arg w_j^{(1)} = \theta - \pi/2$ or $\theta + \pi/2$ (or $w_j^{(1)} = 0$) and $\arg w_j^{(2)} = \theta$. Similarly we write $\alpha_j = \alpha_j^{(1)} + \alpha_j^{(2)}$ with $\arg \alpha_j^{(1)} = \theta - \pi/2$ or $\theta + \pi/2$ (or $\alpha_j^{(1)} = 0$) and $\arg \alpha_j^{(2)} = \theta$ or $-\theta$ (or $\alpha_j^{(2)} = 0$). If the set

$$\mathcal{E} = \left\{ \alpha_j^{(2)} \,\middle|\, \arg \alpha_j^{(2)} = -\theta \quad \text{or} \quad \alpha_j^{(2)} = 0 \right\}$$

is not empty, we denote by $\tilde{\alpha}$ (one of) the element(s) of this set whose absolute value is largest. Let μ be the smallest positive integer such that

$\tilde{\alpha} + m_1 w_1^{(2)} \in H(\ell)$ for any $m_1 \geq \mu$, and split the sum (6.1) into

$$\zeta_r((s_1, \ldots, s_r); (\alpha_1, \ldots, \alpha_r), (w_1, \ldots, w_r))$$

$$= \sum_{m_1=0}^{\mu-1} \sum_{m_2=0}^{\infty} \cdots \sum_{m_r=0}^{\infty} + \sum_{m_1=\mu}^{\infty} \sum_{m_2=0}^{\infty} \cdots \sum_{m_r=0}^{\infty} = T_1 + T_2, \quad (6.3)$$

say. (If $\mathcal{E} = \emptyset$, we put $\mu = 0$.) For any $m_1 \leq \mu - 1$, we put $\alpha_j'(m_1) = \alpha_j + m_1 w_1$. Then

$$T_1 = \sum_{m_1=0}^{\mu-1} \alpha_1'(m_1)^{-s_1} \sum_{m_2=0}^{\infty} \cdots \sum_{m_r=0}^{\infty} (\alpha_2'(m_1) + m_2 w_2)^{-s_2}$$

$$\times \cdots \times (\alpha_r'(m_1) + m_2 w_2 + \cdots + m_r w_r)^{-s_r}$$

$$= \sum_{m_1=0}^{\mu-1} \alpha_1'(m_1)^{-s_1}$$

$$\times \zeta_{r-1}((s_2, \ldots, s_r); (\alpha_2'(m_1), \ldots, \alpha_r'(m_1)), (w_2, \ldots, w_r)). \quad (6.4)$$

To evaluate T_2, we put $\alpha_j'(\mu) = \alpha_j + \mu w_1$ and $m_1' = m_1 - \mu$. Then

$$T_2 = \sum_{m_1'=0}^{\infty} \sum_{m_2=0}^{\infty} \cdots \sum_{m_r=0}^{\infty} (\alpha_1'(\mu) + m_1' w_1)^{-s_1} (\alpha_2'(\mu) + m_1' w_1 + m_2 w_2)^{-s_2}$$

$$\times \cdots \times (\alpha_r'(\mu) + m_1' w_1 + m_2 w_2 + \cdots + m_r w_r)^{-s_r}. \quad (6.5)$$

Since $\alpha_j'(\mu) = (\alpha_j^{(1)} + \mu w_1^{(1)}) + (\alpha_j^{(2)} + \mu w_1^{(2)})$, the definitions of $\tilde{\alpha}$ and μ imply that $\alpha_j'(\mu) \in H(\ell)$. The right-hand side of (6.4) is absolutely convergent by induction assumption. Hence we only need to show the absolute convergence of (6.5). In other words, our remaining task is to prove the absolute convergence of (6.1) under the additional assumption that $\alpha_j \in H(\ell)$ ($1 \leq j \leq r$). Then all terms $\alpha_j + m_1 w_1 + \cdots + m_r w_r$ are in $H(\ell)$. Each factor on the right-hand side of (6.1) is to be understood as

$$(\alpha_j + m_1 w_1 + \cdots + m_j w_j)^{-s_j} = \exp(-s_j \log(\alpha_j + m_1 w_1 + \cdots + m_j w_j)),$$

where the branch of the logarithm is that defined by the condition

$$\theta - \frac{\pi}{2} < \arg(\alpha_j + m_1 w_1 + \cdots + m_j w_j) < \theta + \frac{\pi}{2}.$$

Let $\sigma_j = \Re s_j$, $t_j = \Im s_j$, and define $J_+ = \{j \mid \sigma_j \geq 0\}$ and $J_- =$

$\{j \mid \sigma_j < 0\}$. Since

$$
\begin{aligned}
|\alpha_j + m_1 w_1 + \cdots + m_j w_j| \\
\geq |\alpha_j^{(2)} + m_1 w_1^{(2)} + \cdots + m_j w_j^{(2)}| \\
= |\, |\alpha_j^{(2)}| e^{i\theta} + m_1 |w_1^{(2)}| e^{i\theta} + \cdots + m_j |w_j^{(2)}| e^{i\theta}| \\
= |\alpha_j^{(2)}| + m_1 |w_1^{(2)}| + \cdots + m_j |w_j^{(2)}|,
\end{aligned}
$$

we have

$$
|\alpha_j + m_1 w_1 + \cdots + m_j w_j|^{-\sigma_j} \leq (|\alpha_j^{(2)}| + m_1 |w_1^{(2)}| + \cdots + m_j |w_j^{(2)}|)^{-\sigma_j}
$$

for $j \in J_+$. On the other hand, it is clear that

$$
|\alpha_j + m_1 w_1 + \cdots + m_j w_j|^{-\sigma_j} \leq (|\alpha_j| + m_1 |w_1| + \cdots + m_j |w_j|)^{-\sigma_j}
$$

for $j \in J_-$. Therefore, setting

$$
\alpha_j^* = \begin{cases} |\alpha_j^{(2)}| & \text{if } j \in J_+, \\ |\alpha_j| & \text{if } j \in J_-, \end{cases}
$$

and

$$
w_j^* = \begin{cases} |w_j^{(2)}| & \text{if } j \in J_+, \\ |w_j| & \text{if } j \in J_{-,,} \end{cases}
$$

we find that $\alpha_j^* > 0$, $w_j^* > 0$ for all j and that

$$
\begin{aligned}
|(\alpha_j + m_1 w_1 + \cdots + m_j w_j)^{-s_j}| \\
= |\alpha_j + m_1 w_1 + \cdots + m_j w_j|^{-\sigma_j} \exp(t_j \arg(\alpha_j + m_1 w_1 + \cdots + m_j w_j)) \\
\leq (\alpha_j^* + m_1 w_1^* + \cdots + m_j w_j^*)^{-\sigma_j} \exp(2\pi |t_j|).
\end{aligned}
$$

Hence

$$
\begin{aligned}
|\zeta_r((s_1, \ldots, s_r); (\alpha_1, \ldots, \alpha_r), (w_1, \ldots, w_r))| \\
\leq \exp(2\pi(|t_1| + \cdots + |t_r|)) \\
\times \sum_{m_1=0}^{\infty} \cdots \sum_{m_r=0}^{\infty} (\alpha_1^* + m_1 w_1^*)^{-\sigma_1} (\alpha_2^* + m_1 w_1^* + m_2 w_2^*)^{-\sigma_2} \\
\times \cdots \times (\alpha_r^* + m_1 w_1^* + \cdots + m_r w_r^*)^{-\sigma_r}. \qquad (6.6)
\end{aligned}
$$

We claim that for any positive integers $k \le r$, the series

$$S(k) = \sum_{m_{r-k+1}=0}^{\infty} \sum_{m_{r-k+2}=0}^{\infty} \cdots \sum_{m_r=0}^{\infty}$$

$$(\alpha_{r-k+1}^* + m_1 w_1^* + \cdots + m_{r-k+1} w_{r-k+1}^*)^{-\sigma_{r-k+1}}$$

$$\times (\alpha_{r-k+2}^* + m_1 w_1^* + \cdots + m_{r-k+2} w_{r-k+2}^*)^{-\sigma_{r-k+2}}$$

$$\times \cdots \times (\alpha_r^* + m_1 w_1^* + \cdots + m_r w_r^*)^{-\sigma_r}$$

is convergent in the region $\sigma_r > 1, \sigma_{r-1} + \sigma_r > 2, ..., \sigma_{r-k+1} + \cdots + \sigma_r > k$, and the estimate

$$S(k) \ll (\beta_1(k) + m_1 w_1^* + \cdots + m_{r-k} w_{r-k}^*)$$

$$\times (\beta_2(k) + m_1 w_1^* + \cdots + m_{r-k} w_{r-k}^*)^{c(k)} \qquad (6.7)$$

holds, where $\beta_1(k) > \beta_2(k) > 0$,

$$c(k) = k - 1 - (\sigma_{r-k+1} + \cdots + \sigma_r), \qquad (6.8)$$

and the implied constant depends on σ_j, α_j^* and w_j^* $(r - k + 1 \le j \le r)$. Note that $c(k) < -1$.

We prove this claim by induction. For any positive real numbers a, b and $\sigma > 1$, we have

$$\sum_{m=0}^{\infty} (a + bm)^{-\sigma} \le a^{-\sigma} + \int_0^{\infty} (a + bx)^{-\sigma} dx \ll \left(1 + \frac{a}{b}\right) a^{-\sigma}, \qquad (6.9)$$

where the implied constant depends only on σ. Using (6.9) with $m = m_r$, $\sigma = \sigma_r$, $a = \alpha_r^* + m_1 w_1^* + \cdots + m_{r-1} w_{r-1}^*$ and $b = w_r^*$, we easily obtain the case $k = 1$ of the claim with $\beta_1(1) = \alpha_r^* + w_r^*$ and $\beta_2(1) = \alpha_r^*$. Now assume that the claim is true for $S(k - 1)$. Then we have

$$S(k) \ll \sum_{m_{r-k+1}=0}^{\infty} (\alpha_{r-k+1}^* + m_1 w_1^* + \cdots + m_{r-k+1} w_{r-k+1}^*)^{-\sigma_{r-k+1}}$$

$$\times (\beta_1(k-1) + m_1 w_1^* + \cdots + m_{r-k+1} w_{r-k+1}^*)$$

$$\times (\beta_2(k-1) + m_1 w_1^* + \cdots + m_{r-k+1} w_{r-k+1}^*)^{c(k-1)}.$$

If $-\sigma_{r-k+1} \ge 0$, we replace α_{r-k+1}^* and $\beta_1(k-1)$ by $\max\{\alpha_{r-k+1}^*, \beta_1(k-1)\}$. If $-\sigma_{r-k+1} < 0$, we replace α_{r-k+1}^* and $\beta_2(k-1)$ by $\min\{\alpha_{r-k+1}^*, \beta_2(k-1)\}$. In either case, we get an estimate of the form

$$S(k) \ll \sum_{m_{r-k+1}=0}^{\infty} (B_1 + m_1 w_1^* + \cdots + m_{r-k+1} w_{r-k+1}^*)^{C_1}$$

$$\times (B_2 + m_1 w_1^* + \cdots + m_{r-k+1} w_{r-k+1}^*)^{C_2}, \qquad (6.10)$$

where $B_1 > B_2 > 0$, $C_1 \geq 0$, $C_2 < 0$, and

$$C_1 + C_2 = -\sigma_{r-k+1} + 1 + c(k-1) = c(k). \qquad (6.11)$$

Since

$$(B_1 + m_1 w_1^* + \cdots + m_{r-k+1} w_{r-k+1}^*)^{C_1}$$
$$= (B_2 + m_1 w_1^* + \cdots + m_{r-k+1} w_{r-k+1}^*)^{C_1}$$
$$\times \left(1 + \frac{B_1 - B_2}{B_2 + m_1 w_1^* + \cdots + m_{r-k+1} w_{r-k+1}^*} \right)^{C_1}$$
$$\leq \left(1 + \frac{B_1 - B_2}{B_2} \right)^{C_1} (B_2 + m_1 w_1^* + \cdots + m_{r-k+1} w_{r-k+1}^*)^{C_1},$$

(6.10) and (6.11) imply that

$$S(k) \ll \sum_{m_{r-k+1}=0}^{\infty} (B_2 + m_1 w_1^* + \cdots + m_{r-k+1} w_{r-k+1}^*)^{c(k)}. \qquad (6.12)$$

The claim for $S(k)$ now follows by applying (6.9) to the right-hand side of (6.12), with $\beta_1(k) = B_2 + w_{r-k+1}^*$ and $\beta_2(k) = B_2$. Hence by induction we find that the claim is true for $1 \leq k \leq r$, and the claim for $k = r$ implies the absolute convergence of the right-hand side of (6.6). This completes the proof of Theorem 3. $\qquad\qquad\qquad\qquad\qquad\qquad\qquad\qquad\square$

We now apply the method explained in Sections 4 and 5 to the generalized multiple zeta-function (6.1). In addition to (1.4), we assume

$$\alpha_j \in H(\ell) \ (1 \leq j \leq r) \quad \text{and} \quad \alpha_{j+1} - \alpha_j \in H(\ell) \ (1 \leq j \leq r-1). \quad (6.13)$$

We use (4.1) with $s = s_r$ and

$$\lambda = \frac{\alpha_r - \alpha_{r-1} + m_r w_r}{\alpha_{r-1} + m_1 w_1 + \cdots + m_{r-1} w_{r-1}}.$$

Under the assumption (6.13) both the numerator and the denominator of λ are the elements of $H(\ell)$, and hence $|\arg \lambda| < \pi$. Similar to (4.3), (5.2) or (5.7), we obtain

$$\zeta_r((s_1, \ldots, s_r); (\alpha_1, \ldots, \alpha_r), (w_1, \ldots, w_r))$$
$$= \frac{1}{2\pi i} \int_{(c)} \frac{\Gamma(s_r + z)\Gamma(-z)}{\Gamma(s_r)} \zeta_{r-1}((s_1, \ldots, s_{r-2}, s_{r-1} + s_r + z);$$

$$(\alpha_1, \ldots, \alpha_{r-1}), (w_1, \ldots, w_{r-1}))\zeta\left(-z, \frac{\alpha_r - \alpha_{r-1}}{w_r}\right) w_r^z dz. \quad (6.14)$$

Hence, shifting the path of integration, we can prove:

Theorem 4. *Under the conditions* (1.4) *and* (6.13), *the multiple zeta-function* (6.1) *can be continued meromorphically to the whole* $\mathbf{C}^r$*-space.*

In the present article we confined ourselves to the above very brief outline of the method. The details of the proof, which is by induction on r, will be given in [46].

Finally we mention the analytic continuation of Mordell multiple zeta-functions. In Section 5 we quoted Mordell's paper [52], in which he studied the special case $s_1 = s_2 = s_3$ of (5.1). In the same paper, Mordell also considered the multiple series

$$\sum_{m_1=1}^{\infty} \cdots \sum_{m_r=1}^{\infty} \frac{1}{m_1 m_2 \cdots m_r (m_1 + m_2 + \cdots + m_r + a)}, \quad (6.15)$$

where $a > -r$. By using Mordell's result on (6.15), Hoffman [22] evaluated the sum

$$\sum_{m_1=1}^{\infty} \cdots \sum_{m_r=1}^{\infty} \frac{1}{m_1 m_2 \cdots m_r (m_1 + m_2 + \cdots + m_r)^s} \quad (6.16)$$

when s is a positive integer.

Here we introduce the following multi-variable version of (6.16), which is at the same time a generalization of the Mordell-Tornheim zeta-function (5.1):

$$\zeta_{MOR,r}(s_1, \ldots, s_r, s_{r+1})$$
$$= \sum_{m_1=1}^{\infty} \cdots \sum_{m_r=1}^{\infty} m_1^{-s_1} \cdots m_r^{-s_r} (m_1 + \cdots + m_r)^{-s_{r+1}}. \quad (6.17)$$

Theorem 5. *The series* (6.17) *can be continued meromorphically to the whole* $\mathbf{C}^{r+1}$*-space.*

This and related results will be discussed in a forthcoming paper.

Acknowledgements. The author expresses his sincere gratitude to Professor M. Kaneko, Professor Y. Ohno and the referee for valuable comments and information on recent results concerning multiple zeta values.

References

[1] S. Akiyama, S. Egami, and Y. Tanigawa, *Analytic continuation of multiple zeta functions and their values at non-positive integers*, Acta Arith. **98** (2001), 107–116.

[2] S. Akiyama and H. Ishikawa, *On analytic continuation of multiple L-functions and related zeta-functions*, Analytic Number Theory (C. Jia and K. Matsumoto, eds.), Kluwer, to appear.

[3] S. Akiyama and Y. Tanigawa, *Multiple zeta values at non-positive integers*, Ramanujan J., to appear.

[4] T. M. Apostol and T. H. Vu, *Dirichlet series related to the Riemann zeta function*, J. Number Theory **19** (1984), 85–102.

[5] T. Arakawa and M. Kaneko, *Multiple zeta values, poly-Bernoulli numbers, and related zeta functions*, Nagoya Math. J. **153** (1999), 189–209.

[6] F. V. Atkinson, *The mean-value of the Riemann zeta function*, Acta Math. **81** (1949), 353–376.

[7] E. W. Barnes, *The theory of the double gamma function*, Philos. Trans. Roy. Soc. (A) **196** (1901), 265–387.

[8] _____, *On the theory of multiple gamma function*, Trans. Cambridge Phil. Soc. **19** (1904), 374–425.

[9] B. C. Berndt, *Ramanujan's Notebooks, Part I*, Springer, New York, 1985.

[10] D. Borwein, J. M. Borwein, and R. Girgensohn, *Explicit evaluation of Euler sums*, Proc. Edinburgh Math. Soc. **38** (1994), 277–294.

[11] J. M. Borwein, D. M. Bradley, and D. J. Broadhurst, *Evaluations of k-fold Euler/Zagier sums: a compendium of results for arbitrary k*, Electron. J. Comb. **4** (1997), 21 pp.

[12] J. M. Borwein, D. M. Bradley, D. J. Broadhurst, and P. Lisoněk, *Combinatorial aspects of multiple zeta values*, Electron. J. Comb. **5** (1998), 12 pp.

[13] _____, *Special values of multiple polylogarithms*, Trans. Amer. Math. Soc. **353** (2001), 907–941.

[14] D. Bowman and D. M. Bradley, *Resolution of some open problems concerning multiple zeta evaluations of arbitrary depth*, Compositio Math., to appear.

[15] P. L. Butzer, C. Markett, and M. Schmidt, *Stirling numbers, central factorial numbers, and representations of the Riemann zeta function*, Results in Math. **19** (1991), 257–274.

[16] P. L. Butzer and M. Schmidt, *Central factorial numbers and their role in finite difference calculus and approximation*, Approximation Theory (J. Szabados and K. Tandori, eds.), North-Holland, Amsterdam, 1991, pp. 127–150.

[17] R. E. Crandall and J. P. Buhler, *On the evaluation of Euler sums*, Experimental Math. **3** (1995), 275–285.

[18] P. Flajolet and B. Salvy, *Euler sums and contour integral representations*, Experimental Math. **7** (1998), 15–35.

[19] A. B. Goncharov, *Polylogarithms in arithmetic and geometry*, Proc. Intern. Cong. Math. (Zurich, 1994), Vol. 1, Birkhäuser, Basel, 1995, pp. 374–387.

[20] _____, *Multiple polylogarithms, cyclotomy and modular complexes*, Math. Res. Letters **5** (1998), 497–516.

[21] A. Granville, *A decomposition of Riemann's zeta-function*, Analytic Number Theory (Y. Motohashi, ed.), Cambridge Univ. Press, Cambridge, 1997, pp. 95–101.

[22] M. E. Hoffman, *Multiple harmonic series*, Pacific J. Math. **152** (1992), 275–290.

[23] _____, *The algebra of multiple harmonic series*, J. Algebra **194** (1997), 477–495.

[24] M. E. Hoffman and Y. Ohno, *Relations of multiple zeta values and their algebraic expression*, preprint.

[25] J. G. Huard, K. S. Williams, and Nan-Yue Zhang, *On Tornheim's double series*, Acta Arith. **75** (1996), 105–117.

[26] H. Ishikawa, *On analytic properties of multiple L-functions*, Analytic extension formulas and their applications (S. Saitoh, ed.), Kluwer Acad. Publ., Dordrecht, 2001, pp. 105–122.

[27] _____, *A multiple character sum and a multiple L-function*, Arch. Math., to appear.

[28] M. Katsurada, *Asymptotic expansions of the mean values of Dirichlet L-functions II*, Analytic Number Theory and Related Topics (K. Nagasaka, ed.), World Sci. Publishing, River Edge, NJ, 1993, pp. 61–71.

[29] _____, *An application of Mellin-Barnes' type integrals to the mean squares of Lerch zeta-functions*, Collect. Math. **48** (1997), 137–153.

[30] _____, *An application of Mellin-Barnes type of integrals to the mean square of L-functions*, Liet. Mat. Rink. (Lithuanian Math. J.) **38** (1998), 98–112.

[31] _____, *Power series and asymptotic series associated with the Lerch zeta-function*, Proc. Japan Acad. **74A** (1998), 167–170.

[32] M. Katsurada and K. Matsumoto, *Asymptotic expansions of the mean values of Dirichlet L-functions*, Math. Z. **208** (1991), 23–39.

[33] _____, *The mean values of Dirichlet L-functions at integer points and class numbers of cyclotomic fields*, Nagoya Math. J. **134** (1994), 151–172.

[34] _____, *Explicit formulas and asymptotic expansions for certain mean square of Hurwitz zeta-functions I*, Math. Scand. **78** (1996), 161–177.

[35] T. Q. T. Le and J. Murakami, *Kontsevich's integral for the Homfly polynomial and relations between values of multiple zeta functions*, Topology and its Appl. **62** (1995), 193–206.

[36] B. Lichtin, *Poles of Dirichlet series and D-modules*, Théorie des Nombres/Number Theory (J.-M. DeKoninck, ed.), De Gruyter, Berlin, 1989, pp. 579–594.

[37] _____, *The asymptotics of a lattice point problem associated to a finite number of polynomials I*, Duke Math. J. **63** (1991), 139–192.

[38] _____, *Volumes and lattice points - proof of a conjecture of L. Ehrenpreis*, Singularities, Lille 1991 (J.-P. Brasselet, ed.), Cambridge Univ. Press, Cambridge, 1994, pp. 211–250.

[39] _____, *The asymptotics of a lattice point problem associated to a finite number of polynomials II*, Duke Math. J. **77** (1995), 699–751.

[40] _____, *Asymptotics determined by pairs of additive polynomials*, Compositio Math. **107** (1997), 233–267.

[41] C. Markett, *Triple sums and the Riemann zeta function*, J. Number Theory **48** (1994), 113–132.

[42] K. Matsumoto, *Asymptotic series for double zeta, double gamma, and Hecke L-functions*, Math. Proc. Cambridge Phil. Soc. **123** (1998), 385–405.

[43] _____, *Asymptotic expansions of double gamma-functions and related remarks*, Analytic Number Theory (C. Jia and K. Matsumoto, eds.), Kluwer, to appear.

[44] _____, *Corrigendum and addendum to "Asymptotic series for double zeta, double gamma, and Hecke L-functions"*, Math. Proc. Cambridge Phil. Soc., to appear.

[45] _____, *Asymptotic expansions of double zeta-functions of Barnes, of Shintani, and Eisenstein series*, preprint.

[46] _____, *The analytic continuation and the asymptotic behaviour of multiple zeta-functions I*, preprint.

[47] Y. Matsuoka, *On the values of a certain Dirichlet series at rational integers*, Tokyo J. Math. **5** (1982), 399–403.

[48] H. Mellin, *Eine Formel für den Logarithmus transcendenter Funktionen von endlichem Geschlecht*, Acta Soc. Sci. Fenn. **29** (1900), no. 4.

[49] _____, *Die Dirichlet'schen Reihen, die zahlentheoretischen Funktionen und die unendlichen Produkte von endlichem Geschlecht*, Acta Math. **28** (1904), 37–64.

[50] T. Meurman, *A generalization of Atkinson's formula to L-functions*, Acta Arith. **47** (1986), 351–370.

[51] H. N. Minh and M. Petitot, *Lyndon words, polylogarithms and the Riemann ζ function*, Discrete Math. **217** (2000), 273–292.

[52] L. J. Mordell, *On the evaluation of some multiple series*, J. London Math. Soc. **33** (1958), 368–371.

[53] Y. Motohashi, *A note on the mean value of the zeta and L-functions I*, Proc. Japan Acad. Ser. A Math. Sci. **61** (1985), 222–224.

[54] N. Nielsen, *Handbuch der Theorie der Gammafunktion*, Teubner, 1906.

[55] Y. Ohno, *A generalization of the duality and sum formulas on the multiple zeta values*, J. Number Theory **74** (1999), 39–43.

[56] Y. Ohno and D. Zagier, *Multiple zeta values of fixed weight, depth, and height*, Indag. Math., to appear.

[57] R. Sitaramachandrarao and A. Sivaramasarma, *Some identities involving the Riemann zeta function*, Indian J. Pure Appl. Math. **10** (1979), 602–607.

[58] R. Sitaramachandrarao and M. V. Subbarao, *Transformation formulae for multiple series*, Pacific J. Math. **113** (1984), 471–479.

[59] M. V. Subbarao and R. Sitaramachandrarao, *On some infinite series of L. J. Mordell and their analogues*, Pacific J. Math. **119** (1985), 245–255.

[60] T. Takamuki, *The Kontsevich invariant and relations of multiple zeta values*, Kobe J. Math. **16** (1999), 27–43.

[61] L. Tornheim, *Harmonic double series*, Amer. J. Math. **72** (1950), 303–314.

[62] E. T. Whittaker and G. N. Watson, *A Course of Modern Analysis*, 4th ed., Cambridge Univ. Press, 1927.

[63] E. Witten, *On quantum gauge theories in two dimensions*, Commun. Math. Phys. **141** (1991), 153–209.

[64] D. Zagier, *Periods of modular forms, traces of Hecke operators, and multiple zeta values*, Studies on automorphic forms and *L*-functions, Sūrikaiseki Kenkyūsho Kōkyūroku, vol. 843, RIMS, Kyoto Univ., 1993, pp. 162–170.

[65] ———, *Values of zeta functions and their applications*, First European Congress of Mathematics, Vol. II (Paris, 1992), Progr. Math., vol. 120, Birkhäuser, Basel, 1994, pp. 497–512.

[66] J. Zhao, *Analytic continuation of multiple zeta functions*, Proc. Amer. Math. Soc. **128** (2000), 1275–1283.

Three-Term Relations for Some Analogues of the Dedekind Sum

Jeffrey L. Meyer

1 Introduction

R. Dedekind introduced what is now called the Dedekind sum in [6] as part of the multiplier system in the transformation of the eta-function. The Dedekind sum is defined by

$$s(d,c) = \sum_{j=1}^{c} \left(\left(\frac{j}{c}\right)\right) \left(\left(\frac{dj}{c}\right)\right),$$

with

$$((x)) = \begin{cases} 0, & \text{if } x \in \mathbb{Z}, \\ x - [x] - \frac{1}{2}, & \text{otherwise.} \end{cases}$$

Dedekind proved, among other results, a reciprocity theorem for the sum. The theorem is widely applied whenever results about Dedekind sums are proved.

Theorem 1 (Reciprocity Law, Dedekind). *If $(c,d) = 1$ and $c, d > 0$, then*

$$s(c,d) + s(d,c) = -\frac{1}{4} + \frac{1}{12}\left(\frac{d}{c} + \frac{c}{d} + \frac{1}{cd}\right). \tag{1.1}$$

Rademacher [15] proved the following three-term relation for $s(d,c)$.

Theorem 2 (Three-term relation, Rademacher). *If $(a,b) = (b,c) = (c,a) = 1$ and $aa' \equiv 1 \pmod{b}$, $bb' \equiv 1 \pmod{c}$, and $cc' \equiv 1 \pmod{a}$, then*

$$s(bc',a) + s(ca',b) + s(ab',c) = -\frac{1}{4} + \frac{1}{12}\left(\frac{a}{bc} + \frac{b}{ac} + \frac{c}{ab}\right). \tag{1.2}$$

Other authors (see [7], [1], [10] and [5], for example) have generalized Rademacher's relation. For general reference on the subject the reader should consult the monograph by Rademacher and Grosswald [16]. J. E. Pommersheim [14] extended his version by induction to give an explicit n-term relation.

Theorem 3 (Pommersheim). *Let* $p, q, u, v \in \mathbb{N}$ *with* $(p, q) = (u, v) = 1$. *Set* $t = pv + qu$ *and let* u', v' *be any integers with* $uu' + vv' = 1$. *Then*

$$s(p,q) + s(u,v) = s(pu' - qv', t) - \frac{1}{4} + \frac{1}{12}\left(\frac{q}{vt} + \frac{v}{tq} + \frac{t}{qv}\right). \qquad (1.3)$$

Rademacher showed that (1.1) follows from Theorem 2. Pommersheim derives (1.2) from Theorem 3. In a recent paper, K. Girstmair [8] proved that (1.3) is in fact equivalent to Dedekind's reciprocity theorem. Since there are several elementary proofs of (1.1), Girstmair notes that the results on Dedekind sums from [14] are not as deep as previously thought.

We use analytic methods to give the same type of relations for certain analogues of the Dedekind sum.

2   A Theta-Function Analogue of the Dedekind Sum

B. C. Berndt gave transformation formulas for the logarithms of the classical theta-functions in [3].

For $\text{Im}(z) > 0$, define $\theta(z) = \sum_{n=-\infty}^{\infty} \exp(\pi i n^2 z)$. Berndt proved the following formula.

Theorem 4 (Berndt). *If* $\text{Im}(z) > 0$ *and* $c > 0$, *then*

$$\log \theta(Vz) = \log \theta(z) + \frac{1}{2}\log(cz + d) - \frac{1}{4}\pi i + \frac{1}{4}\pi i S(d, c), \qquad (2.1)$$

where

$$S(d,c) = \sum_{j=1}^{c-1}(-1)^{j+1+[dj/c]}.$$

For additional arithmetic and analytic properties of $S(d,c)$ and the other sums arising in the formulas proved by Berndt in [3] see [9], [4], [12] and [13]. We need the following property of $S(d,c)$, proved in [9] and [12], in the proofs of Theorem 5 and Theorem 6.

Lemma 1. *Let* $(d,c) = 1$ *with* $c > 0$. *Let* $dd' \equiv 1 \pmod{c}$. *We have the following:*

    (i) *If* c *is odd and both* d *and* d' *are even, then* $S(d',c) = S(d,c)$.
    (ii) *If* c *is even and* $dd' \equiv 1 \pmod{2c}$, *then* $S(d',c) = S(d,c)$.
    (iii) *If* c *is even and* $dd' \equiv c + 1 \pmod{2c}$, *then* $S(d' + c, c) = S(d,c)$.

The next theorem is the analogue of Pommersheim's three-term relation.

Theorem 5. *Let $p, q, u, v \in \mathbb{N}$ with $(p, q) = (u, v) = 1, q + p$ odd, u odd and v even. Set $t = pv + qu$ and let u', v' be any integers with $uu' + vv' = 1$ and v' even. Then*

$$S(p, q) + S(u, v) = S(pu' - qv', t) + 1.$$

Proof. Since $(p, q) = 1$ we can choose a and b such that $ap - bq = 1$. Set

$$V = \begin{bmatrix} a & b \\ q & p \end{bmatrix} \quad \text{and} \quad W = \begin{bmatrix} u & -v' \\ v & u' \end{bmatrix}.$$

Then

$$VW = \begin{bmatrix} au + bv & -av' + bu' \\ t & pu' - qv' \end{bmatrix}.$$

First we write Berndt's formula (2.1) for V:

$$\log \theta(Vz) = \log \theta(z) + \frac{1}{2} \log(qz + p) - \frac{1}{4}\pi i + \frac{1}{4}\pi i S(p, q). \qquad (2.2)$$

Next we apply (2.1) with V replaced by VW to see that

$$\log \theta(VWz) = \log \theta(z) + \frac{1}{2} \log(tz + pu' - qv')$$
$$- \frac{1}{4}\pi i + \frac{1}{4}\pi i S(pu' - qv', t). \qquad (2.3)$$

Then we use (2.2) with z replaced by Wz to deduce that

$$\log \theta(V(Wz)) = \log \theta(Wz) + \frac{1}{2} \log(qWz + p)$$
$$- \frac{1}{4}\pi i + \frac{1}{4}\pi i S(p, q). \qquad (2.4)$$

Finally we replace V by W in (2.1) to see that

$$\log \theta(Wz) = \log \theta(z) + \frac{1}{2} \log(vz + u') - \frac{1}{4}\pi i + \frac{1}{4}\pi i S(u', v). \qquad (2.5)$$

We replace $\log \theta(Wz)$ in (2.4) with (2.5) and then combine the result with (2.3) to conclude that

$$\frac{1}{4}\pi i S(p, q) + \frac{1}{4}\pi i S(u', v) = \frac{1}{4}\pi i S(pu' - qv', t) + \frac{\pi i}{4}. \qquad (2.6)$$

We have used the following lemma [11] to conclude that there are no branch changes with the logarithms, so the complete cancellation is justified.

Lemma 2. *Let $A, B < C$, and D be real with A and B not both zero and $C > 0$. Then for $\mathrm{Im}(z) > 0$,*

$$\arg((Az + B)/(Cz + D)) = \arg(Az + B) - \arg(Cz + D) + 2\pi k,$$

where k is independent of z and

$$k = \begin{cases} 1, & \text{if } A \leq 0 \text{ and } AD - BC > 0, \\ 0, & \text{otherwise.} \end{cases}$$

From Lemma 1 and the fact that $S(-d, c) = -S(d, c)$ in (2.6), we deduce the desired result. $\qquad\square$

From Theorem 5, we derive the analogue of Rademacher's three-term relation discovered by L. A. Goldberg [9].

Theorem 6 (Goldberg). *Let $a, b, c \in \mathbb{N}$ be pairwise coprime with $aa' \equiv 1 \pmod{b}$, $bb' \equiv 1 \pmod{c}$, $cc' \equiv 1 \pmod{2a}$ and a, a' even. Then*

$$S(bc', a) + S(ca', b) + S(ab', c) = 1.$$

Proof. Set $v = a$, $q = b$. Then find $m, n \in \mathbb{Z}$ such that $c = mb + na$. Next set $p = n$ and $u = m$. Finally find m^* and a^* with a^* even and $mm^* + aa^* = 1$. Then we apply Theorem 5 to obtain

$$S(n, b) + S(m, a) = S(nm^* - ba^*, c) + 1.$$

Since $S(-d, c) = -S(d, c)$, this becomes

$$S(n, b) + S(m, a) + S(ba^* - nm^*, c) = 1. \qquad (2.7)$$

It is clear from the definition that $S(d + 2c, c) = S(d, c)$. Since $n \equiv ca' \pmod{2b}$, we have that

$$S(n, b) = S(ca', b). \qquad (2.8)$$

A straight-forward calculation shows that $m^* \equiv bc' \pmod{2a}$. Since $mm^* \equiv 1 \pmod{2a}$ we have, from Lemma 1, that

$$S(m, a) = S(bc', a). \qquad (2.9)$$

Finally, note that $2^{-1}(b'a - n'm)$ is the inverse of $ba^* - nm^* \pmod{2c}$, where 2^{-1} is the inverse of 2 $\pmod{c}$ Thus, from Lemma 1 again, we deduce that

$$S(ba^* - nm^*, c) = S(ab', c). \qquad (2.10)$$

Substituting (2.8), (2.9) and (2.10) into (2.7) we have proved the theorem.
$\qquad\square$

Berndt gives a reciprocity theorem in [3] for the sum $S(d,c)$.

Theorem 7. *If* $(c,d) = 1$, $c + d$ *is odd and* $c, d > 0$, *then*

$$S(d,c) + S(c,d) = 1. \tag{2.11}$$

Theorem 7 can be derived from Theorem 6. However, because of the parity restrictions, it is unlikely that there exists the same equivalence between the reciprocity theorem and the three-term relation as Girstmair showed for the Dedekind sum.

3 A Character Analogue of the Dedekind Sum

Berndt introduced a character analogue of the Dedekind sum in [2]. This sum arises in the transformation formula for a character analogue of the eta-function $A(z; \chi)$ defined by

$$A(z; \chi) = \sum_{m=1}^{\infty} \sum_{n=1}^{\infty} \frac{\chi(m)\chi(n)e\,(nmz/k)}{n}.$$

First, the generalized Bernoulli functions introduced in [2], $\overline{B}_n(x, \chi)$, $n \geq 1$, $-\infty < x < \infty$, can be alternatively defined by

$$\overline{B}_n(x, \chi) = k^{n-1} \sum_{h=1}^{k-1} \overline{\chi}(h)\overline{B}_n\left(\frac{x+h}{k}\right),$$

where $\overline{B}_n(x)$ is the standard periodic Bernoulli function. We now define the Dedekind character sum by

$$s(d, c; \chi) = \sum_{n \bmod ck} \chi(n)\overline{B}_1\left(\frac{dn}{c}, \chi\right)\overline{B}_1\left(\frac{n}{ck}\right).$$

We state Berndt's transformation formula.

Theorem 8 (Berndt). *Let* $\mathrm{Im}(z) > 0$ *and let* χ *be a real primitive character mod* k. *If* $a \equiv d \equiv 0 \pmod{k}$, *then*

$$G(\overline{\chi})A(Vz; \chi) = \overline{\chi}(b)\chi(c)\{G(\chi)A(z; \overline{\chi}) + \pi i\chi(-1)s(d, c; \overline{\chi})\};$$

if $b \equiv c \equiv 0 \pmod{k}$, *then*

$$G(\overline{\chi})A(Vz; \chi) = \overline{\chi}(a)\chi(d)\{G(\overline{\chi})A(z; \chi) + \pi i\chi(-1)s(d, c; \chi)\}.$$

Here $G(\chi)$ *is the ordinary Gauss sum*

$$G(1, \chi) = \sum_{h=1}^{k-1} \chi(h)e(hz/k).$$

The following lemma is analogous to Lemma 1.

Lemma 3. *If χ is a real primitive character mod k, $dd' \equiv 1 \pmod{ck}$ and $c \equiv 0 \pmod{k}$, then*

$$s(d, c; \chi) = s(d', c; \chi).$$

Proof. Let $V = \dfrac{d'z + b}{cz + d}$ where b is chosen so that $dd' - bc = 1$. Next we use $V^{-1} = \dfrac{-dz + b}{cz - d'}$. Then, from Theorem 8, we see that

$$G(\chi)A(z, \chi) = G(\chi)A(V(V^{-1}z), \chi)$$

$$= \chi(d')\chi(d)\{G(\chi)A(V^{-1}z, \chi) + \pi i \chi(-1)s(d, c; \chi)\}$$

and

$$G(\chi)A(V^{-1}z, \chi) = \chi(-d)\chi(-d')\{G(\chi)A(z, \chi) + \pi i \chi(-1)s(-d', c; \chi)\}.$$

Since $s(-d, c; \chi) = s(d, c; \chi)$, the lemma follows. $\qquad\square$

Theorem 9 (Three-term relation). *Let χ be a real primitive character with modulus k. Let $p, q, u, v \in \mathbb{N}$ with $p \equiv v \equiv 0 \pmod{k}$. Let u' and v' be any integers with $v' \equiv 0 \pmod{k}$ and $uu' + vv' = 1$. Then*

$$s(p, q; \chi) + s(u, v; \chi) = s(pu' - qv', qu + pv; \chi).$$

Proof. We apply Berndt's transformation formulas as in the method of Theorem 5 except that the congruences modulo k need to be noted. $\qquad\square$

References

[1] T. Asai, *The reciprocity of Dedekind sums and the factor set for the universal covering group of $SL(2, \mathbb{R})$*, Nagoya. Math. J. **37** (1970), 67–80.

[2] B. C. Berndt, *Character transformation formulae similar to those for the Dedekind eta-function*, Analytic Number Theory, Proc. Sympos. Pure Math., vol. XXIV, American Math. Soc., St. Louis., 1973, pp. 9–30.

[3] _____, *Analytic Eisenstein series, theta-functions, and series relations in the spirit of Ramanujan*, J. Reine Angew. Math. **303/304** (1978), 332–365.

[4] B. C. Berndt and L. A. Goldberg, *Analytic properties of arithmetic sums arising in the theory of the classical theta-functions*, SIAM J. Math. Anal. **15** (1984), 143–150.

[5] R. Bruggeman, *On the distribution of Dedekind sums*, Contemp. Math. **166** (1994), 197–210.

[6] R. Dedekind, *Erläuterungen zu zwei Fragmenten von Riemann — Riemann's Gesammelte Math. Werke*, 2nd ed., Dover, New York, 1892.

[7] U. Dieter, *Beziehungen zwischen Dedekindschen Summen*, Abh. Math. Sem., vol. 21, 1957, pp. 109–125.

[8] K. Girstmair, *Some remarks on Rademacher's three-term relation*, Arch. Math. **73** (1999), 205–207.

[9] L. A. Goldberg, *Transformations of Theta-functions and Analogues of Dedekind Sums*, Ph.D. thesis, University of Illinois, Urbana, 1981.

[10] R. R. Hall and M. N. Huxley, *Dedekind sums and continued fractions*, Acta Arith. **63** (1993), 79–90.

[11] J. Lewittes, *Analytic continuation of Eisenstein series*, Trans. Amer. Math. Soc. **171** (1972), 469–490.

[12] J. L. Meyer, *Analogues of Dedekind Sums*, Ph.D. thesis, University of Illinois, Urbana, 1997.

[13] _____, *Properties of certain integer-valued analogues of Dedekind sums*, Acta Arith. **82** (1997), 229–242.

[14] J. E. Pommersheim, *Toric varieties, lattice points and Dedekind sums*, Math. Ann. **295** (1993), 25–49.

[15] H. Rademacher, *Generalization of the reciprocity formula for Dedekind sums*, Duke Math. J. **21** (1954), 391–397.

[16] H. Rademacher and E. Grosswald, *Dedekind sums*, Carus Math. Monogr., vol. 16, Mathematical Association of America, Washington, D.C., 1972.